Reef Fisheries

CHAPMAN & HALL FISH AND FISHERIES SERIES

Amongst the fishes, a remarkably wide range of fascinating biological adaptations to diverse habitats has evolved. Moreover, fisheries are of considerable importance in providing human food and economic benefits. Rational exploitation and management of our global stocks of fishes must rely upon a detailed and precise insight of the interaction of fish biology with human activities.

The *Chapman & Hall Fish and Fisheries Series* aims to present authoritative and timely reviews which focus on important and specific aspects of the biology, ecology, taxonomy, physiology, behaviour, management and conservation of fish and fisheries. Each volume will cover a wide but unified field with themes in both pure and applied fish biology. Although volumes will outline and put in perspective current research frontiers, the intention is to provide a synthesis accessible and useful to both experts and non-specialists alike. Consequently, most volumes will be of interest to a broad spectrum of research workers in biology, zoology, ecology and physiology, with an additional aim of the books encompassing themes accessible to non-specialist readers, ranging from undergraduates and postgraduates to those with an interest in industrial and commercial aspects of fish and fisheries.

Applied topics will embrace synopses of fishery issues which will appeal to a wide audience of fishery scientists, aquaculturists, economists, geographers and managers in the fishing industry. The series will also contain practical guides to fishery and analysis methods and global reviews of particular types of fisheries.

Books already published and forthcoming are listed below. The Publisher and Series Editor would be glad to discuss ideas for new volumes in the series.

Available titles

1. **Ecology of Teleost Fishes**
   Robert J. Wootton
2. **Cichlid Fishes**
   Behaviour, ecology and evolution
   Edited by Miles A. Keenlyside
3. **Cyprinid Fishes**
   Systematics, biology and exploitation
   Edited by Ian J. Winfield and Joseph S. Nelson
4. **Early Life History of Fish**
   An energetics approach
   Ewa Kamler
5. **Fisheries Acoustics**
   David N. MacLennan and E. John Simmonds
6. **Fish Chemoreception**
   Edited by Toshiaki J. Hara
7. **Behaviour of Teleost Fishes**
   Second edition
   Edited by Tony J. Pitcher

8. **Genetics and Fish Breeding**
Colin R. Purdom
9. **Fish Ecophysiology**
J. Cliff Rankin and Frank B. Jensen
10. **Fish Swimming**
John J. Videler
11. **On the Dynamics of Exploited Fish Populations**
Raymond J.H. Beverton and Sidney J. Holt
(Facsimile reprint)
12. **Sea Bass**
Graham D. Pickett and Michael G. Pawson
13. **Fish Bioenergetics**
Malcolm Jobling
14. **On the Sex of Fish and Gender of Scientists**
Daniel Pauly
15. **Hake**
Fisheries, products and markets
Edited by Jürgen Alheit and Tony J. Pitcher
16. **Environmental Biology of Fishes**
Malcolm Jobling
17. **Electric Fishes**
History and behavior
Peter Moller
18. **The Impact of Species Changes in African Lakes**
Edited by Tony J. Pitcher and Paul B. Hart
19. **Pike**
Biology and exploitation
Edited by John F. Craig
20. **Reef Fisheries**
Edited by Nicholas V.C. Polunin and Callum M. Roberts

Forthcoming titles

Fisheries Ecology
Second edition
Edited by T.J. Pitcher and P.J. Hart

Early Life History and Recruitment in Fish Populations
Edited by R.C. Chambers and E.A.Trippel

Deep Demersal Fish and Fisheries
N.R. Merrett and R. Haedrich

Reef Fisheries

Edited by

Nicholas V.C. Polunin

Department of Marine Sciences and Coastal Management
University of Newcastle
UK

and

Callum M. Roberts

Department of Environmental Economics and Environmental Management
University of York
UK

Springer-Science+Business Media, B.V.

First edition 1996

Originally published by Chapman & Hall in 1996.
Softcover reprint of the hardcover 1st edition 1996

Typeset in 10/12 Phontina by Acorn Bookwork, Salisbury, Wiltshire

ISBN 978-94-015-8781-5 ISBN 978-94-015-8779-2 (eBook)
DOI 10.1007/978-94-015-8779-2

A catalogue record for this book is available from the British Library

Library of Congress Catalog Card Number: 95–83399

Printed on permanent acid-free text paper, manufactured in accordance with ANSI/NISO Z39.48-1992 and ANSI/NISO Z39.48-1984 (Permanence of Paper).

Contents

Contributors

Timothy J.H. Adams
Coastal Fisheries Programme, South Pacific Commission, B.P. D5, Nouméa cedex, New Caledonia

Richard S. Appeldoorn
Department of Marine Sciences, University of Puerto Rico, Mayagüez, Puerto Rico 00681–5000, USA

George W. Boehlert
NOAA/NMFS, Southwest Fisheries Science Center, Pacific Environmental Group, 1352 Lighthouse Ave., Pacific Grove, CA 93950-2097, USA

James A. Bohnsack
NOAA NMFS, Southeast Fisheries Center, Miami Laboratory, Coastal Resources Division, 75 Virginia Beach Drive, Miami, Florida FL 33149, USA

Paul Dalzell
Coastal Fisheries Programme, South Pacific Commission, B.P. D5, Nouméa cedex, New Caledonia

Simon Jennings
Department of Marine Sciences & Coastal Management, University of Newcastle, Newcastle upon Tyne NE1 7RU, United Kingdom

John M. Lock
Ministry of Agriculture Fisheries and Food, Nobel House, 7 Smith Square, London SW1P 3JR, United Kingdom

John W. McManus
ReefBase Project, Coastal & Coral Reef Resource Systems Program, ICLARM, MC PO Box 2631, Makati, Metro Manila 0718, Philippines

John L. Munro
ICLARM, European Liaison Office, Laboratoire de Biologie Marine et Malacologie, Université de Perpignon, Perpignon 66860 cedex, France

Daniel Pauly
ICLARM, MC PO Box 2631, Makati, Metro Manila 0718, Philippines, and Fisheries Centre, University of British Columbia, 2204 Main Mall, Vancouver, BC, Canada V6T 1Z4

Nicholas V.C. Polunin
Department of Marine Science and Coastal Management, University of Newcastle, Newcastle upon Tyne NE1 7RU, United Kingdom

Callum M. Roberts
Eastern Caribbean Center, University of the Virgin Islands, St Thomas, US Virgin Islands 00802, USA
Present address: Department of Environmental Economics and Environmental Management, University of York, Heslington, York YO1 5DD, United Kingdom

Kenneth Ruddle
School of Policy Studies, Kwansei Gakuin University, 2-1 Gakuen, Sanda, Hyogo-ken 669-13, Japan

Yvonne J. Sadovy
Department of Zoology, University of Hong Kong, Hui Oi Chow Science Building, Pokfulam Road, Hong Kong

Series foreword

Among the fishes, a remarkably wide range of biological adaptations to diverse habitats has evolved. As well as living in the conventional habitats of lakes, ponds, rivers, rock pools and the open sea, fish have solved the problems of life in deserts, in the deep sea, in the cold Antarctic, and in warm waters of high alkalinity or of low oxygen. Along with these adaptations, we find the most impressive specializations of morphology, physiology and behaviour. For example we can marvel at the high-speed swimming of the marlins, sailfish and warm-blooded tunas, air-breathing in catfish and lungfish, parental care in the mouth-brooding cichlids and viviparity in many sharks and toothcarps.

Moreover, fish are of considerable importance to the survival of the human species in the form of nutritious, delicious and diverse food. Rational exploitation and management of our global stocks of fishes must rely upon a detailed and precise insight of their biology.

The *Chapman and Hall Fish and Fisheries Series* aims to present timely volumes reviewing important aspects of fish biology and fisheries. Most volumes will be of interest to research workers in biology, zoology, ecology, physiology and fisheries, but an additional aim is for the books to be accessible to a wide spectrum of non-specialist readers ranging from undergraduates and postgraduates to those with an interest in industrial and commercial aspects of fish and fisheries.

Reef Fisheries comprises volume number 20 in the *Chapman & Hall Fish and Fisheries Series.* Despite a great deal of published research on the ecology of reefs, including some excellent multi-author volumes, such as Peter Sale's *The Ecology of Fishes on Coral Reefs*, this is the first book to focus exclusively on the problems of understanding and managing the human exploitation of reef resources. In editing this book, Nick Polunin and Callum Roberts have assembled a team of 14 authors from 8 countries, covering the topic in 14 chapters. As an assurance of quality and relevance, like all multi-author books in the *Chapman & Hall Fish and Fisheries Series*, the chapters are fully peer-reviewed.

There are two main reasons why reef fisheries are difficult to understand and consequently deserve special attention. The first is our ignorance of the production system of reefs. To be sustainable, human fisheries must take only that portion of ecological production that does not compromise either future recruitment or ecosystem stability. But in comparison to lakes, upwellings or ocean shelf systems, the production ecology of reefs is not well understood. For example, corals entail a subtle symbiosis between zooids and algae that is hard to quantify. A coral reef has been described as a 'wall of mouths' designed to intercept food transported on ocean currents in the midst of low productivity areas, but we do not clearly understand how to predict growth and production in these systems. These areas of ecological ignorance mean that the fundamental basis for assessment, evaluation and management of reef fisheries suffers from serious deficiencies that impede insight into the effects of harvest.

The second problem is the artisanal nature of reef fisheries, which typically entail a multi-species catch and employ many different types of gear, all of which have different implications for habitat structure and fish community composition. Data about catches and effort is often hard to come by, or lack essential spatial information. Fish harvests usually involve dispersed operations by a large number of poor small-scale fishers. Artisanal fisheries are therefore difficult to assess and even more difficult to manage. Understanding of human as well as biological dimensions are crucial for fisheries management, which is why this book includes chapters that describe work in the areas of fisheries anthropology and sociology.

Furthermore, many reefs world-wide are currently in a parlous state, impacted by explosives, other destructive methods of fishing, long term overexploitation, and adverse effects of unrestricted tourism and other uses. Sadly, we are forced to question if there is enough time left to preserve them and what international institutions may be needed to achieve this.

In common with other books in the *Chapman & Hall Fish and Fisheries Series*, our aim in publishing *Reef Fisheries* is to provide a synopsis and evaluation of current work that may be useful for reference, together with pointers to areas where further research could be profitably directed. As Series Editor I am confident that the quality and scope of *Reef Fisheries* will enable it to meet this aim, and I hope that the book will find its way into the library of institutes and individuals concerned with the biology of fish and fisheries.

Reference

Sale, P.F. (Ed.) *The Ecology of Fishes on Coral Reefs*. Academic Press, London, UK, 754pp.

Professor Tony J. Pitcher
Editor, Chapman and Hall Fish and Fisheries Series
Director, Fisheries Centre, University of British Columbia, Vancouver, Canada

Preface

Reef habitats reach their greatest extent in tropical waters and the fisheries of tropical reefs are considered especially complex. This complexity applies to biodiversity, habitat variety, ecosystem intricacy, a little-known environment and a multiplicity of fishing gears. Not only, therefore, is stock assessment difficult, but a high proportion of target species change sex, and there is a high level of biological interactions, such as through predation. The 'precautionary principle' suggests that we should not tamper with what we do not understand, but fisheries have already defiled the reef wilderness. Indeed, fishing is probably the single most widespread source of human disturbance to tropical reefs. It is important to appreciate the consequences of fishing and to improve management where impacts are undesirable.

Tropical reefs have frequently been compared with tropical rain forest. Both are manifestly productive, are rich in living things and are showing signs of sensitivity to human interference. But scientific understanding and environmental awareness of reefs lag that of rain forest. There are some constructive contrasts between reef and rain forest when it comes to exploitation. Human exploitation in rain forest is principally of the dominant plants for wood, with drastic consequences for the biota, habitat and ecosystem. Tropical reef fisheries, however, exploit primarily the animal populations. This use does not have the global habitat effects of indiscriminate logging in rain forest, nor therefore should it have similar drastic consequences for the biota. But ecosystem consequences of reef fishing have long been suspected all the same bolstered no doubt by known instances of habitat damage.

As a selective range of disturbances affecting a poorly known system, fishing represents a valuable opportunity to better understand the extent to which tropical reefs are more than the sum of their parts. This task is facilitated by the visibility of many of the organisms and accessibility is further facilitated by warmth of water. In fact the conditions for direct observation are scarcely equalled by those in any other extensive aquatic system. There are increasingly frequent claims, even from within the scientific community, that reefs everywhere are in decline. Given the comparative novelty of reef investigation, this assertion will not be easily substantiated. There has been contention also that reef fishery stocks are especially prone to collapse, but the manner and extent of this supposed sensitivity has also scarcely been critically examined.

Many tropical reef fisheries are set on coasts which are already overburdened with people. Questions arise of how many more mouths can be fed, of gross impacts and of sustainability in general. This book arose out of our recognition that tropical reefs are productive and that their fisheries are potentially of world-scale importance. Yet the literature on them is scattered and much of it is inaccessible. A lot has been written about the population and community ecology of reef fishes, but as yet there has been no synthesis on the fisheries, their biological basis, conduct and management. Ecologists and fishery scientists have taken very different approaches and contributed to somewhat different parts of this literature. We hope our attempt to assemble a broad cross-section of investigations and their output will encourage a *rapprochement* between these fields. The social science of management is also much neglected and another aim of this book is to draw attention to the many aspects of reef fishery investigations and management outside the natural sciences.

With such a broad remit, we sought contributors capable of presenting geographically-broad views as well as up-to-date summaries of subject areas in which they are active. The result is a book which we hope will be useful to a wide range of people, from managers and policy makers to consultants, researchers and students.

Aside from the authors, many others have contributed directly to the production of this book. Manuscripts were reviewed by fellow authors and by specialist referees who were not contributing to the book. We are especially grateful to the following external reviewers: Tim Bayliss-Smith, Jim Beets, John Caddy, Julian Caley, James Carrier, Gary Davis, Ed DeMartini, Al Edwards, Richard Grigg, Tony Hawkins, Mark Hixon, Bob Johannes, Geoff Kirkwood, Dave Klumpp, Michel Kulbicki, Jeff Leis, Tim McClanahan, Paul Medley, Chris Mees, Jim Parrish, Jeff Polovina, Bob Rowley, Saul Saila, Doug Shapiro, John Tarbit, Rob van Ginkel and Drew Wright. We thank Martin Tribe and Chuck Hollingworth for their unstinting help with production and editing. We thank our respective wives, Cally and Julie, for their patience when we have been otherwise engaged. Of course our thanks go too to the reef fishers, without whom this book would never have arisen!

Nick Polunin
Callum Roberts

Chapter one

The scope of tropical reef fisheries and their management

John L. Munro

SUMMARY

Coral reef resource systems extend throughout the tropics and are exploited primarily by subsistence fishers, supplying food for millions of people. The magnitude of harvests per unit area taken from coralline shelves approximates those taken by trawlers from temperate shelves. In view of this, the current estimated potential global annual harvest from tropical reef fisheries of 6 million metric tonnes (t) is probably conservative. The relative composition of reef fishery catches changes in response to increasing effort, largely due to the different vulnerability of predatory and herbivorous species to fishing gears. In extreme cases, this change can result in dramatically reduced value of the total catch. Marine protected areas, either transitory or permanent, appear to offer the best prospects for management of reef fisheries, particularly if they are allied to community-based systems.

1.1 REEF RESOURCE SYSTEMS

Tropical reef fisheries provide employment and sustenance for millions of coastal dwellers (Salvat, 1992). Their distribution is discontinuous; only minor reef systems are found in the eastern sides of the Atlantic and

Reef Fisheries. Edited by Nicholas V.C. Polunin and Callum M. Roberts.
Published in 1996 by Chapman & Hall, London. ISBN 0 412 60110 9.

Pacific Oceans, while they reach their greatest development in the clear coastal waters of the Indo–West Pacific (Figs 1.1–1.3).

For the purpose of this book a reef fishery is defined as one that is conducted in an area in which the presence of reef building (hermatypic) corals largely precludes the commercial-scale use of mobile fishing gears such as trawls and seine nets. The lower bound of such fisheries is set by the depth at which hermatypic corals can grow and, depending upon water clarity, can extend beyond 80 m in oceanic areas and around islands (Goreau and Wells, 1967). Non-hermatypic corals and sponges often extend this boundary. On continental shelves and in relatively turbid waters the lower limits often lie at 30–60 m (Clausade *et al.*, 1971; Done, 1982).

Along continental margins, the seabed normally slopes away to depths of around 200 m before descending more rapidly towards the ocean depths. However, there are often carbonate shelves which have developed for reasons relating to change in relative sea level in recent geological time and to growth of corals. An abrupt break in the shelf usually occurs at depths of 50 m or less, sometimes even close to the sea surface, beyond which the shelf descends precipitously to the bathyal zone. Windward or upcurrent portions of such shelves often support prolific coral growth, leeward or downcurrent areas usually have less coral, and large areas are covered by calcareous sediments. Characteristically, a sill reef is present at the windward margins of an insular shelf and vast areas of luxuriant coral growth can be found 20–50 m below the surface.

The growth of corals towards the surface to form an emergent reef and reef flat is a process that culminates in the formation of an offshore ribbon or bank reef, a barrier reef surrounding a lagoon, or in the development of a fringing reef close to shore. The characteristics of either a lagoon or a reef flat depend upon exposure, tidal regime and turbidity. Ecological conditions may develop in lagoons such that they no longer support coral growth and the lagoon floor is covered by soft sediments, although sometimes interspersed with emergent coral heads.

Thus it is that between the conspicuous reef crest and reef flats which emerge at the lowest spring tides and the lower limits to coral growth there are, in many tropical seas, large areas which can be described either as coralline shelves or as lagoons (Fig. 1.4).

Fig. 1.1 (pg. 3) Diagrammatic map of the central Atlantic Ocean and far-eastern Pacific Ocean, showing geographic areas with major reef systems.
Fig. 1.2 (pg. 4) Diagrammatic map of the Indian Ocean, including western South East Asia and the far-western Pacific Ocean, showing geographic areas with major reef systems.
Fig. 1.3 (pg. 5) Diagrammatic map of the central Pacific Ocean, showing geographic areas with major reef systems.

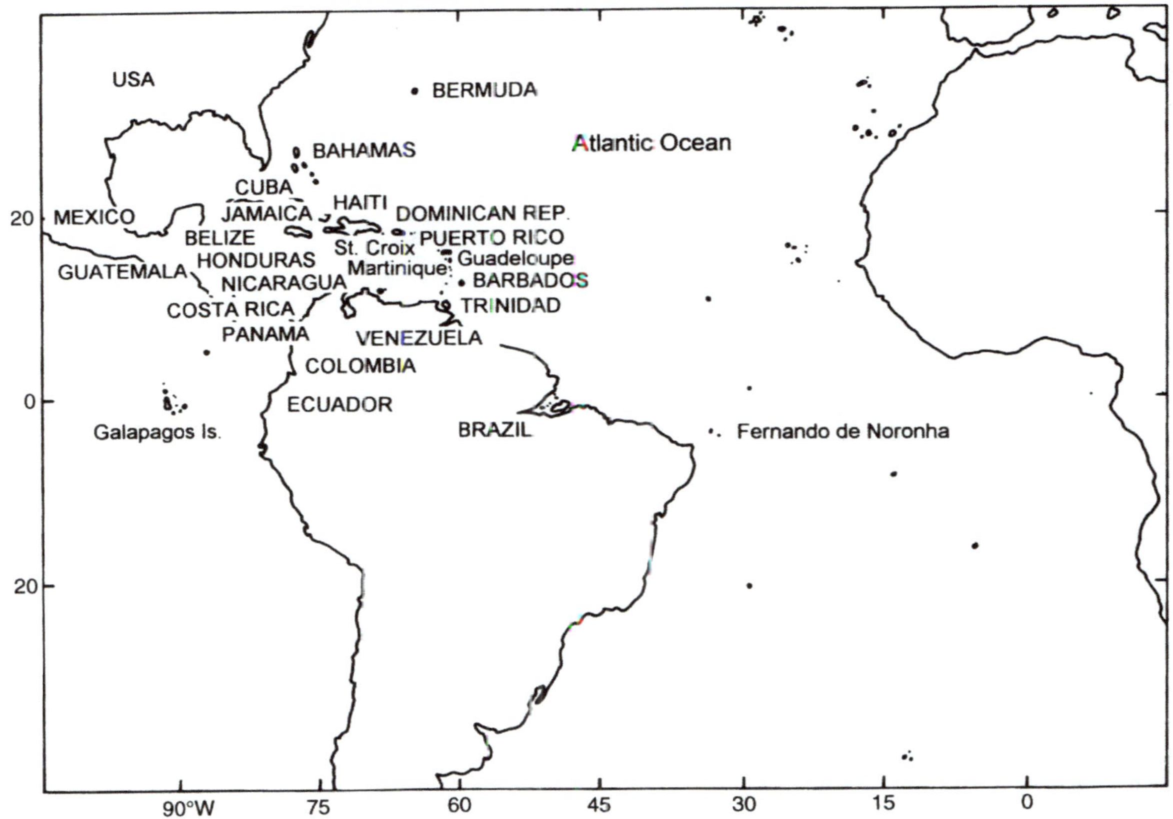
USA
BERMUDA
Atlantic Ocean
BAHAMAS
CUBA
HAITI
MEXICO
JAMAICA
DOMINICAN REP.
BELIZE
PUERTO RICO
St. Croix
HONDURAS
Guadeloupe
GUATEMALA
Martinique
BARBADOS
NICARAGUA
TRINIDAD
COSTA RICA
PANAMA
VENEZUELA
COLOMBIA
ECUADOR
Galapagos Is.
BRAZIL
Fernando de Noronha
20
0
20
90°W
75
60
45
30
15
0

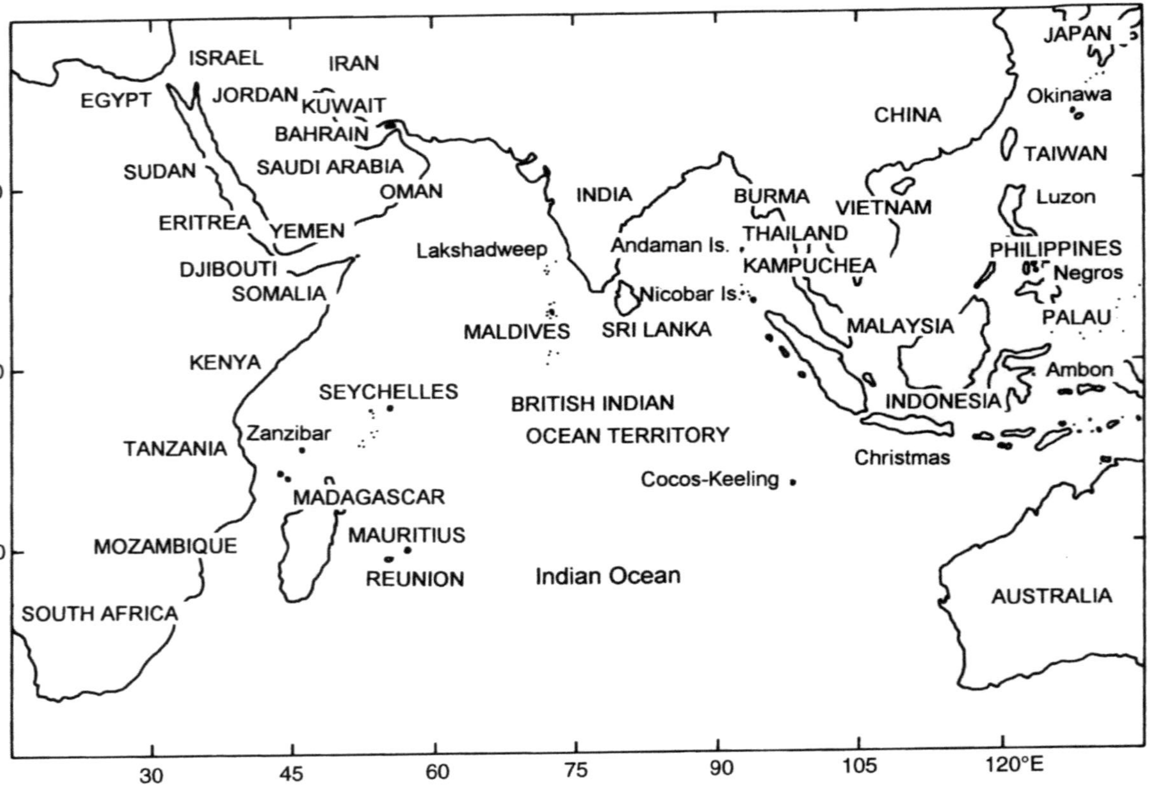

ISRAEL
IRAN
JAPAN
EGYPT
JORDAN
KUWAIT
Okinawa
CHINA
BAHRAIN
TAIWAN
SAUDI ARABIA
SUDAN
OMAN
INDIA
BURMA
VIETNAM
Luzon
ERITREA
YEMEN
THAILAND
Andaman Is.
PHILIPPINES
Lakshadweep
DJIBOUTI
KAMPUCHEA
Negros
SOMALIA
Nicobar Is.
PALAU
MALDIVES
SRI LANKA
MALAYSIA
KENYA
Ambon
SEYCHELLES
BRITISH INDIAN
OCEAN TERRITORY
INDONESIA
Zanzibar
TANZANIA
Christmas
Cocos-Keeling
MADAGASCAR
MAURITIUS
MOZAMBIQUE
REUNION
Indian Ocean
AUSTRALIA
SOUTH AFRICA
20
0
20
30
45
60
75
90
105
120°E

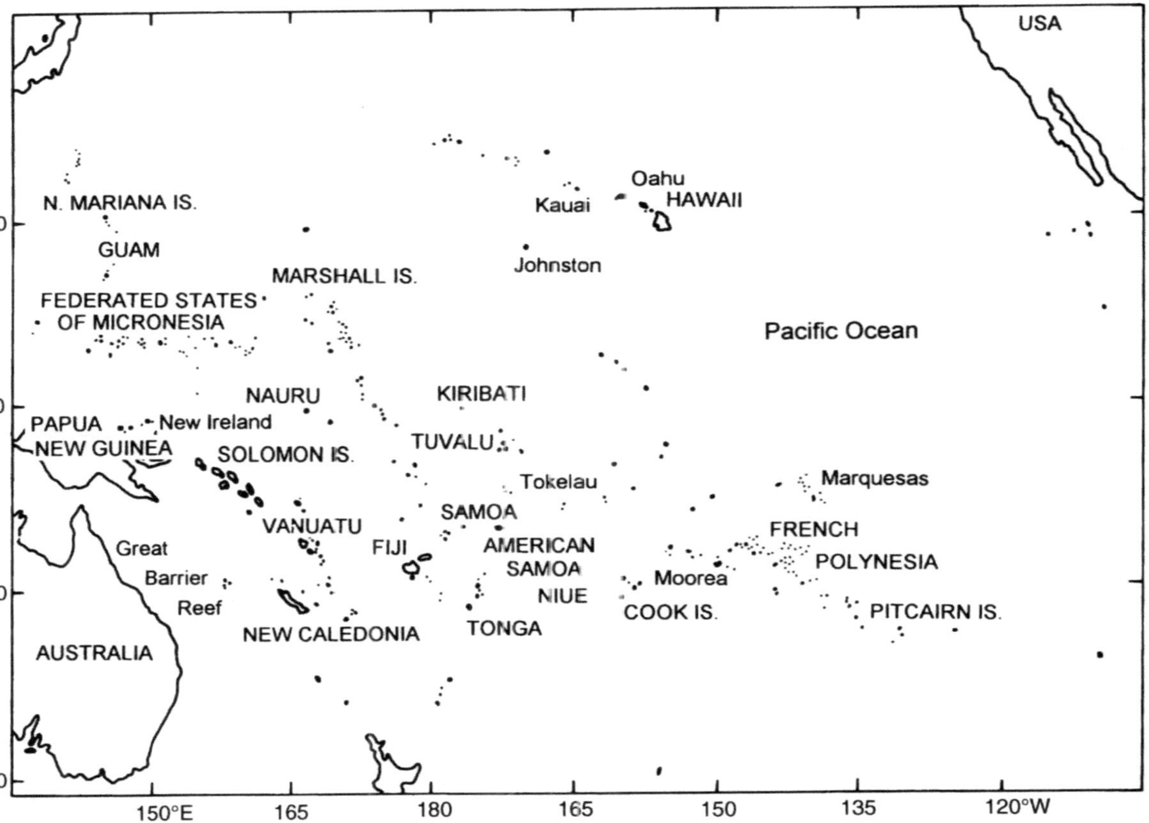
USA
Oahu
HAWAII
Kauai
N. MARIANA IS.
GUAM
Johnston
MARSHALL IS.
FEDERATED STATES
OF MICRONESIA
Pacific Ocean
NAURU
KIRIBATI
PAPUA
NEW GUINEA
New Ireland
SOLOMON IS.
TUVALU
Tokelau
Marquesas
SAMOA
VANUATU
FRENCH
POLYNESIA
FIJI
AMERICAN
SAMOA
Great
Barrier
Reef
Moorea
NIUE
COOK IS.
PITCAIRN IS.
TONGA
NEW CALEDONIA
AUSTRALIA
20
0
20
40
150°E
165
180
165
150
135
120°W

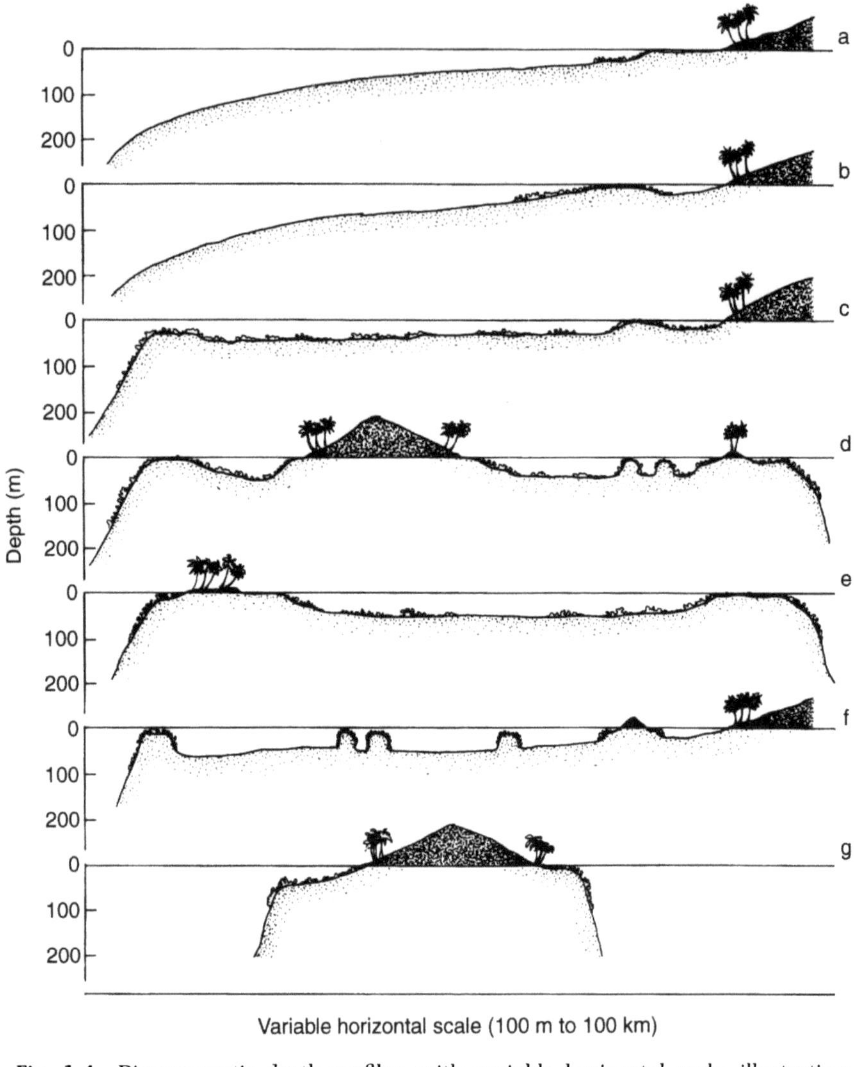

Fig. 1.4 Diagrammatic depth profiles, with variable horizontal scale, illustrating the range of forms that reefs may take, including (a) limited reef growth in the shallows of a gently sloping shelf, (b) more extensive shallow fringing reef development close to shore, (c) extensive reef flat, (d) atoll-like reef surrounding a high island, (e) sea-level oceanic atoll, (f) lagoonal reef system and (g) fringing reef of oceanic high islands.

Many valuable species of fishes occur in areas of hard bottom between the limit of growth of hermatypic corals and the upper edge of the bathyal zone (300 m). Some of these species are reef dwellers in their early life stages, but most are not obligate reef inhabitants and are merely inhabitants of a characteristic zone of hard seabed. Because of the steepness of the zone the actual area covered by 'deep reefs' is very limited, being of moderate extent only along continental margins or areas where geological upheavals have produced aberrant shallow areas.

A resource system can be defined as that set of entities, factors and parameters which control the productivity of a resource in human terms. Tropical reef fisheries are such a system. The primary factor is the area covered by that system. In the case of coral reefs, there is little agreement about what constitutes a coral community; for tropical reefs as a whole, the limits of the system are still more vague.

Smith (1978) estimated the global area covered by tropical reefs to be around 617 000 km^2, and this value has been widely cited. The figure was derived by taking the proportion of the world's continental and island shelves (to 200 m) which are in areas where reefs could grow, assuming that shelves sloped uniformly to 200 m and setting 30 m as the lower limit to coral growth. Thus only 15% of the coralline shelves were assumed to support reef systems. However, as described above, very large areas of tropical shelves slope fairly gently to 30–50 m and then drop precipitously towards the bathyal zone. This description would cover most of the Caribbean and Bahamas, all of the island shelves and banks of the Indian and Pacific Oceans and most of the Red Sea. Additionally, Munro (1977) estimated that the shelf areas of the Caribbean and Bahamas totalled 660 000 km^2, almost all of which is non-trawlable coralline shelf. Thus coralline shelves of the western Atlantic alone appear to exceed Smith's (1978) estimate. On the other hand, a substantial proportion of the tropical shelves of East Africa and South and South East Asia consist of various muddy or silty substrata where hermatypic corals could not grow or which are too deep for reef growth. On balance, it still seems likely that Smith (1978) underestimated the global area of tropical reefs.

Of the total area estimated by Smith, 30% is in the Asiatic Mediterranean, 25% in the tropical Pacific, 30% in the Indian Ocean (including the Red Sea and Gulf) and 15% in the tropical Atlantic.

The lack of detailed data on the bathymetry of coralline regions makes extrapolations of total harvests infeasible and this is a priority area for research now being addressed by ICLARM's ReefBase project (Froese, 1994). This is important, because even the most elementary calculations show that the potential productivity is large.

1.2 IMPORTANCE OF TROPICAL REEF FISHERIES

The area of the globe that can be described as coralline shelf is certainly a very small fraction of the earth's surface, but the relative importance of tropical reef fisheries on a human scale is very great indeed.

Because industrial fishing gears are largely precluded, reef fisheries are the domain of small-scale fishers. These fishers often use gears that are precisely adapted to season or moon phase and often target particular species. However, there are so many species of harvestable organisms on reefs that they can invariably be characterized as multispecies, multigear, fisheries. Production is governed primarily by two features: distance from centres of human population, and population densities and hence demand for seafoods. Many remote Pacific atolls have perhaps never been fished by anything more than a passing yacht. Some have been pulse-fished by distant-water fishing vessels for particularly valuable species such as spiny lobsters, giant clams, groupers or snappers. By contrast, a reef complex in the Philippines consisting of 26 km^2 of reef flat and 42 km^2 of reef slope provides employment for 17 000 people (McManus *et al.*, 1992). In the latter case, virtually every component of the reef community has value as a commodity, can be consumed or used as bait and, hence, could be traded.

Many of the reef fisheries of Asia and the Caribbean are now heavily exploited by large populations. Frequently the fishers, usually landless peasants, cannot afford to consume their harvests. All except the most inferior portion of the catch must be sold to purchase rice or other staples.

As human pressures on reef resource systems increase, they induce changes in the composition of exploited communities and changes in the reefs themselves. This is particularly so when overcrowded conditions and poverty lead to deforestation and poor land use, and consequent siltation and degradation of the reefs. Extreme conditions, in which fishers resort to the use of poisons or explosives, further exacerbate the problem.

A number of authors have provided estimates of harvests of reef fishes and invertebrates per unit area of reef (Stevenson and Marshall, 1974; Munro, 1977; Marshall, 1980; Marten and Polovina, 1982; Alcala and Gomez, 1985; Munro and Williams, 1985; Russ, 1991; Jennings and Polunin, in press a). Confusion or disagreement over the outer limits of the area to be included, the habitats subsumed in coralline shelf and the taxa to be incorporated, make comparisons difficult. However, it is clear from these studies that limited areas of relatively shallow water with dense cover of live corals can produce extremely high yields (Medley *et al.*, 1993). The greatest yield reported so far is 44.0 $t\,km^{-2}\,year^{-1}$ from American Samoa (Wass, 1982).

If the definition of a coralline shelf is spread wider to encompass the entire 'super-ecosystem' (Marshall, 1980), which extends from the inner margins of mangroves, across seagrass beds, sand and coral, to the outer limit of the deep reef at 200 m (or the nearly equivalent untrawlable coralline shelf; Munro, 1977), the harvests per unit area are very much less. They are still of the same order of magnitude as those once taken in trawl fisheries on northern Atlantic and Pacific shelves before mismanagement reduced the bulk of the catches to worthless animals low in the food chains.

Current harvests of reef fishes and invertebrates are unknown for most countries, but are probably well below their potential in many areas. This is because of the remoteness of many reef systems from significant markets in some cases and the degradation and overexploitation of reef systems in others.

The size of harvests per unit area is clearly inversely related to the average depth of the area exploited. Shallow areas of actively growing coral are capable of producing sustained harvests of 20–30 $t\,km^{-2}\,year^{-1}$ (see Munro, 1984a; Marshall, 1985; Russ, 1991; Medley *et al.*, 1993). This is particularly so if the intertidal zone is intensively gleaned for invertebrates and if planktivorous fishes, particularly fusiliers (Caesionidae), form a significant part of the catch. These points can partly explain the lower productivity of Caribbean reef systems in which planktivorous fishes are not very important in harvests. The Caribbean also has little tidal amplitude and consequently few substantial reef flats.

Munro (1977) found that harvests of demersal and neritic pelagic species from the Caribbean probably amounted to less than 0.4 $t\,km^{-2}\,year^{-1}$ but that harvests exceeding 1.2 $t\,km^{-2}\,year^{-1}$ were readily obtained in many areas. The greatest harvests were from the intensively fished northern coast of Jamaica, amounting to 3.7 $t\,km^{-2}\,year^{-1}$.

McManus *et al.* (1992) found that the harvests of fish and invertebrates from the intensely exploited 42 km^2 reef slope (2–60 m) and the 26 km^2 reef flat at Bolinao in the northern Philippines amounted to 2.7 $t\,km^{-2}\,year^{-1}$ and 12 $t\,km^{-2}\,year^{-1}$, respectively. Acosta and Recksiek (1989), in an earlier study of part of the same system, estimated hook-and-line catches of fish from the reef slope (2–18 m) to be 1.1 $t\,km^{-2}\,year^{-1}$ and that trap and spear-fishing on the reef flat yielded 6.4 $t\,km^{-2}\,year^{-1}$. The fish yield from the entire 11 km^2 study area was 4.4 $t\,km^{-2}\,year^{-1}$.

There is thus a high degree of convergence in the available information. Yields in excess of 20 $t\,km^{-2}\,year^{-1}$ are reported from areas of coral, reef flat and seagrass beds which are intensively fished for a wide range of reef fishes and invertebrates. Shallow reef flats yield more than 10 $t\,km^{-2}\,year^{-1}$, even under serious overfishing, and if shelf areas below the reef crests are considered separately, harvests of 1–3 $t\,km^{-2}\,year^{-1}$ are common, but diminish with increasing depth.

Munro (1977) concluded that potential annual harvests from the coralline shelves (within the 200 m isobath) in the tropical Atlantic were around 750 000 t (1.1 $t\,km^{-2}\,year^{-1}$) if all fisheries were moderately exploited. S.V. Smith (1978) extrapolated from that value to conclude that global harvests of reef-related fishes and invertebrates might be raised to around 6×10^6 t year^{-1}. The production of the demersal and neritic pelagic fishes and invertebrates from the coralline shelves of the Caribbean and Bahamas appeared to be around 280 000 t in the early 1970s, or about 0.4 t km^{-2} year^{-1}. If the area of the world's reefs is 617 000 km^2 and they were fished on roughly the same scale as those in the Caribbean, the total world catch would have been about 2×10^6 t, potentially worth about $4 billion. However, we know that at that time many reef areas were yielding harvests vastly in excess of 0.4 t km^{-2} year^{-1} and can conclude that 6×10^6 t year^{-1} is probably an underestimate, particularly if the total reef area is greater than supposed. Unfortunately, as mentioned previously, we lack sufficiently detailed knowledge of the bathymetry of coralline shelves to extrapolate reasonably from these figures.

1.3 COMPLEXITY OF REEF FISHERIES

Reef fisheries differ in complexity according to the relative wealth of the communities that exploit them. Fisheries in Australia and the USA target groupers (Serranidae), snappers (Lutjanidae) and related fish groups and spiny lobsters, as do the export-orientated fisheries of Belize. In contrast, unless a fish is actually poisonous, there is apparently no such thing as a trash fish in many impoverished developing countries.

The harvest from an intensively exploited reef system in the Indo–Pacific will typically include 200–300 species of fishes, and somewhat fewer (around 100) in the Caribbean. To this can be added a huge array of comestible, decorative or otherwise useful invertebrates and seaweeds. However, as a general rule, fewer than 20 species will make up around 75% of the weight of the catch.

It has been shown that progressive changes in the reef fish community composition will result from sustained fishing with a variety of relatively unselective fishing gears (Munro, 1980; Munro and Smith, 1984; Munro and Williams, 1985; Koslow *et al.*, 1988; Russ and Alcala, 1989; Russ, 1991). This change in composition arises because the larger predatory species usually tend to be vulnerable to a wider range of fishing gears than herbivores or planktivores and are generally considered more desirable and therefore are specifically targeted. For example, groupers are highly vulnerable to virtually every sort of artisanal fishing gear whereas parrotfishes (Scaridae) can only be caught by traps, nets or spears and are generally less catchable. The fraction of the stock killed by a given amount of fishing is greater in the case of groupers than for parrotfishes. Additional factors that militate against survival of the larger species of predatory fishes include their relatively large size at first maturity combined with a small size at recruitment to the fishery. In the case of the groupers, protogynous hermaphroditism (Shapiro *et al.*, 1993b) could lead to the relative numbers of males decreasing to negligible levels and consequent collapse of the spawning stock (Bannerot *et al.*, 1987).

Changes in the relative abundances of competing species as a result of one species being more catchable than the other should result in substantial changes in community composition. The end result of these pressures will be an exploited community composed largely of the least catchable or desirable species of fishes and in which the most prized are extinct or nearly so. For example, Koslow *et al.* (1988) found that fish communities in the vicinity of the Pedro Cays, near Jamaica, had been reduced over a period of 15 years to a community dominated by boxfishes (Ostraciidae), pufferfishes (Tetraodontidae) and squirrelfishes (Holocentridae).

Successful development of the ECOPATH models (Polovina, 1984; Christensen and Pauly, 1992b) has provided a tool for examining the basis, and possibly predicting the consequences for the community, of different rates and types of exploitation. For many reef organisms, however, we still lack sufficient information on food consumption, electivity and other parameters governing trophic webs.

1.4 MANAGEMENT OPTIONS

Reef fisheries have a dismal management record. In the most heavily populated regions, overwhelming pressures of poverty and intensive competition have depleted resources to extraordinarily low levels. In many of these fisheries, there is almost no economic constraint on effort. The fishing gears are simple and inexpensive and the opportunity cost of a fisher's labour is close to zero. The possibility of a decent catch, however remote, is sufficient incentive for unremitting exploitation of the resources. These are consequently overfished in terms of fish growth rates, reproductive capacity and of a decent return on human effort.

Even where fishing pressure is less intense, there are still singularly few examples of effective management. Bermuda provides an example of a reef fishery managed in fine detail, but the political will to implement necessary management policies was only brought about by its virtual collapse as a result of unrestrained exploitation (Burnett-Herkes *et al.*, 1989).

Management options are the same as for any other fishery. They include limits on total catch, on the fishing effort, on fishing areas or on seasons. All of these may reduce fishing mortality, restrict size at first capture, and thus prevent or limit growth or recruitment overfishing (Munro and Williams, 1985). The suitability of various management options will be determined by local social, economic and political considerations, but as a general case it can be concluded that well-established practices such as catch quotas and minimum size limits are impracticable in multispecies fisheries. Periodic closures or permanent reserves have greater potential, particularly if they are allied with traditional or community-based management systems (Ruddle *et al.*, 1992).

The relative merits of marine protected areas (MPAs) have been discussed by Bohnsack (Plan Development Team, 1990) and there is much support for the concept given current concerns about loss of biodiversity. However, DeMartini (1993) has shown that benefits will depend on several factors. These include the types of fishes targeted, the relative size of the reserve and the position of the reserve relative to other unprotected reef systems to which larvae may be taken by ocean currents. Benefits in terms of increased catches will be zero or negative if net emigration from a permanent MPA does not occur, or if the size of the reserve becomes so large that it severely restricts the available fishing area. The prime alternative would be long-term rotational closures of large tracts of reefs and coralline shelves.

The scope for establishing marine protected areas is very large. The affinity of tourists and nature lovers for coral reefs results in powerful lobby groups and consequent political support for their establishment.

A gathering force is the development of 'co-management' in which fisheries management is devolved or delegated to local communities, with the government playing only an advisory role. There are few examples of successful co-management, but enabling legislation has been implemented in places as disparate as the Eastern Caribbean (through the Organization of Eastern Caribbean States) and the Philippines. In much of the South Pacific, community-based management of a sort is firmly rooted in customary practice and only lacks the active support of Governments to make it more effective.

1.5 REEF FISHERIES RESEARCH

Research on reef fisheries has a short history. Prior to 1955, little thought was given to their management and most works were concerned with taxonomy, natural history (e.g. Longley and Hildebrand, 1941) or development. In most countries, populations were not sufficiently large to have greatly affected harvests and few urgent concerns were felt for the future.

Additionally, there were very negative perceptions of the potential productivity of coral reefs, which were perceived to be closed ecosystems surviving in a barren sea by parsimoniously recycling their meagre nutrient resources (Odum and Odum, 1955). The role of nitrifying bluegreen algae and of planktivorous organisms in entrapping energy was only recognized later (Johannes, 1974; Johannes and Gerber, 1975; Webb *et al.*, 1975; Wiebe *et al.*, 1975). Nevertheless, important ecological papers appeared which laid the foundations for much of reef fish ecology (Odum and Odum, 1955; Hiatt and Strasburg, 1960; Randall, 1967).

Thus, it was only in the last 40 years that serious attempts have been made to manage fisheries, notably in Bermuda (Bardach, 1958, 1959; Bardach *et al.*, 1958), Cuba (Buesa Mas, 1960, 1961), the Virgin Islands (Randall, 1962; Swingle *et al.*, 1970; Sylvester and Dammann, 1972) and Jamaica (Munro, 1983g). Fisheries in the Indo–Pacific received even less attention and seminal works appeared much later, pertaining to New Caledonia (Loubens, 1978, 1980a,b), the Philippines (Carpenter, 1977; Carpenter and Alcala, 1977) and French Polynesia (Galzin, 1987). By the mid 1960s it was becoming apparent that reef fishery resources were under threat and resource systems seriously degraded in many countries (Munro, 1969). This realization coincided with the widespread availability of scuba equipment, the development of underwater visual census techniques (Brock, 1982; Bell *et al.*, 1985b; Harmelin-Vivien *et al.*, 1985; Bohnsack and Bannerot, 1986; Bellwood and Alcala, 1988; St John *et al.*, 1990) and the discovery of daily rings in otoliths of numerous species of fishes (Pannella, 1971, 1974; Brothers, 1980; Brothers *et al.*, 1983). This led to a flowering of reef fishery science and funding of greatly increased numbers of projects.

Concurrently, reef fish biology flourished, with numerous academic studies of the smaller elements of reef fish communities, notably those of Sale, Russell, Talbot and their co-workers. Substantial advances were made in the understanding of community dynamics, although issues tended to be somewhat clouded by questions of scale. Fishery scientists dealt with scales of tens of kilometres and the fish biologists were concerned with several orders of magnitude less (Roberts, Chapter 4).

Major advances in our understanding of reef fish biology, in the estimation of population parameters (Pauly, 1980a, 1984a) and development of multispecies models have kept up the momentum (Christensen and Pauly, 1992b). In particular, recognition that global ecosystems are under threat and that biodiversity is diminishing appears to have sparked a massive upsurge of interest in the conservation or, at least, the sustainable exploitation of tropical reefs.

Marine protected areas in particular are now seen as a prime solution, satisfying the needs of fishing communities, tourists and conservation interests (Russ, 1985; Alcala, 1988; Alcala and Russ, 1990; Plan Development Team, 1990; Roberts and Polunin, 1991, 1993). However, there are still insufficient data on which to base calculations of optimal size of MPAs, nor do we have sufficient knowledge of the dispersal or retention of larval or postlarval marine organisms (Boehlert, Chapter 3).

The flowering of research on reef fish biology also led to a vigorous debate on the stability of reef fish communities, centred on the very large differences in species composition and abundance of fishes which can be observed on seemingly similar reefs or in successive years on the same reef (Sale, 1978, 1980, 1985; Doherty and Williams, 1988; Roberts, 1991; Fowler *et al.*, 1992). After some false starts, it seems now to be recognized that populations of reef fishes are often settlement limited. Differences and changes in community composition are the result of complex processes, in which chance plays a major role. These changes govern the numbers of postlarval reef fishes that actually manage to find a reef onto which they can settle (Doherty and Williams, 1988; Roberts, 1991; Doherty and Fowler, 1994a).

Settlement limitation has an important bearing on the selection of conservation and fishing areas. Areas that regularly receive abundant settlement might be reserved as fishing grounds. Areas that are major sources of potential recruits should be conserved as marine protected areas. The problem, however, is that our knowledge of the degree of connectivity of reef systems is distinctly limited and far more research is needed before we will be able to understand larval dispersal and settlement processes (Boehlert, Chapter 3; Roberts, Chapter 4). The connectivity is in any event species-specific, being dependent upon spawning strategies, duration of the planktonic larval stage and the ability or inability to delay settlement in the postlarval stage.

Finally, the technical basis for reef fisheries enhancement or ranching is rapidly falling into place as a result of advances in hatchery technology and understanding of larval and juvenile nutrition, all leading to reduced costs per fingerling (Polovina, 1991a; Tucker *et al.*, 1991; Sorgeloos and Sweetman, 1993; Bohnsack, Chapter 11).

The essential feature of stocking systems for tropical reefs is that if the species stocked is reef-bound or if the habitat stocked is bounded by deep water or other non-reef habitats, the reef can function as an unfenced fish farm. The evidence that reef fisheries are often settlement limited and the wide variation in standing stocks on unexploited reef systems all suggest that, within limits, trophic resources are not limiting, particularly at the lower end of the food chain (Munro and Williams, 1985).

The following chapters of this book will show two things. Substantial advances in understanding of tropical reef fish and fisheries have been achieved in recent years. We are, however, still a long way from knowing how to optimize and sustainably manage the most complex fisheries which lie within the seas.

Chapter two

Reproduction of reef fishery species

Yvonne J. Sadovy

SUMMARY

Although reproductive patterns in tropical fishery species are characterized by diversity, some interesting trends are clearly emerging, despite the few families studied in detail. Reproductive output is highly variable, both within and among species, years and individuals. Especially in larger species, the differential between the fecundities of different-sized conspecifics may be orders of magnitude. Environmental and biological factors influence when and where reproduction takes place, with annual spawning seasons ranging from as little as a week or two, to much of the year; the notion that tropical marine species are all characterized by protracted spawning is unfounded. Most species produce pelagic eggs and, characteristic of reef fishes in general, all produce pelagic larvae. Females may spawn a few times to many times annually. Spawning often takes place towards dusk, in some cases at specific times in the lunar or tidal cycle. Evidence for lunar cyclicity in most larger species, however, is not strong except among a number of species which aggregate to spawn. Spawning occurs in areas of residence or at well-defined aggregation sites, metres or many kilometres away from home sites. Males and females mate either pairwise or in small groups which characteristically comprise one female and several males; the size of the testis relative to the body (gonadosomatic index) of ripe males apparently accords with mating pattern – larger for group and smaller for pair spawners. Larger, more mobile, species within

Reef Fisheries. Edited by Nicholas V.C. Polunin and Callum M. Roberts.
Published in 1996 by Chapman & Hall, London. ISBN 0 412 60110 9.

a family tend to migrate to reproduce in aggregations which are frequently, but by no means exclusively, located offshore and close to deep waters. Such sites may be quite distinct but there is no evidence to indicate that they are unique. It is not clear whether aggregation spawning and the preference for specific spawning locations evolved largely for the benefit of larval dispersal or survival, to enable males and females that live somewhat dispersed to come together or for some other reason.

Morphological factors such as body form and size are also important determinants of reproductive output, with more compressed forms exhibiting lower relative fecundity at length than more rounded body forms. These factors influencing egg output per spawn, combined with maturation size and spawning frequency and duration, are all pertinent to our understanding of reproductive patterns and strategies and inter- and intra-annual variation in reproductive output. By elucidating the variables that determine egg production we might better understand relationships between numbers and activities of spawning adults and the resulting temporal, spatial and abundance patterns of larval and juvenile appearance on reefs.

From a fishery perspective, there is an urgent need to better estimate and maintain reproductive output in exploited populations. For example, spawning aggregations, because they are often consistent in time and space, may be heavily targeted by fishers. Of particular concern is evidence that aggregations can be decimated by heavy fishing with a possible concomitant loss of larger individuals or disruption of reproductive activity. Because the effects of intense aggregation-fishing, in both the short and long term, are largely unknown but may seriously compromise reproductive output, they should be accorded particular attention for research and management. Similarly, quite a few commercially exploited reef families are hermaphroditic. The impact of fishing or conventional management practices on hermaphroditic species is not understood and will be difficult to assess until we come to understand, among other things, the factors that induce sexual changeover. Research and monitoring programmes designed specifically to address these issues provide a rich and exciting challenge to the curious ecologist and the innovative fishery biologist.

2.1 INTRODUCTION

Fish numbers vary in time and space. It is of interest to both coral reef ecologists and fishery biologists to understand how and why they do so and the relationships between patterns of egg production and subsequent population size. Traditionally, population abundance was thought to be limited by the carrying capacity of the environment (Smith and Tyler, 1972; Sale, 1977; Doherty and Williams, 1988). More recent studies, however, have indicated that populations may not reach carrying capacity, but are limited by the number of individuals that settle out of the plankton (Doherty, 1982; Victor, 1983; Doherty and Fowler, 1994a). Temporal and spatial patterns of spawning are linked with patterns of subsequent settlement in some damselfishes (Doherty, 1991; Meekan *et al.*, 1993; Robertson *et al.*, 1993), and egg production with subsequent numbers of settling larvae, at least at local spatial scales in a species with a relatively short larval life (Meekan *et al.*, 1993).

In fishery biology, 'recruitment' refers to the point at which organisms first become vulnerable to capture. The relationship between population size and subsequent recruitment has long been of concern (Beverton and Holt, 1957; Ricker, 1975). Fishery biologists have been perplexed by the apparent lack of relationships between spawners and recruits (Rothschild, 1986). None the less, some link clearly exists; reduction of adult populations by heavy fishing pressure is followed by declines in subsequent recruitment if spawning biomass (usually assessed as adult biomass) falls below certain critical levels (Rothschild, 1986; Goodyear, 1989). To better understand the relationships between numbers of spawners and numbers of recruits, we must evaluate the roles of egg production and subsequent egg, larval and juvenile mortalities. In this chapter I focus on the reproductive biology of tropical fishery species by examining patterns of egg output, temporal and spatial patterns of spawning and mating patterns.

The reproductive biology of reef fishery species has received remarkably little serious attention compared with other aspects of their natural history. Information on reproduction, with the exception of behaviour, in larger tropical reef species, is fragmented and confined to a few notable surveys and syntheses (Breder and Rosen, 1966; J.L. Munro *et al.*, 1973; Erdman, 1976; Johannes, 1978a; Thresher, 1984; Grimes, 1987; Shapiro, 1987a; A.D. Munro *et al.*, 1990).

I have limited this review to the principal taxa of commercially important reef fishes, which are exploited largely as a food resource and are closely associated with reefs for their post-settlement life. These not only include the more obvious and economically important species, but increasingly encompass species such as squirrelfishes (Holocentridae), parrotfishes (Scaridae), surgeonfishes (Acanthuridae) and even some of the larger angelfishes (Pomacanthidae) and wrasses (Labridae), as preferred species decline. Fisheries do not only target larger individuals. Juveniles of a number of species, including porgies (Sparidae) and groupers (Serranidae), are increasingly taken in some regions to supply mariculture operations. The marine aquarium and curio trades have also grown rapidly since the mid 1970s, further adding pressure to depleted communities. The reproductive biology of many of these smaller species, however, has been partially addressed elsewhere and is beyond the scope of this chapter (e.g. Thresher, 1984). Examples from deeper-water, non-tropical or non-commercial, taxa are occasionally included for comparative purposes, to illustrate specific points or to identify general patterns. Taxonomic nomenclature is according to Heemstra and Randall (1993) for groupers, Carpenter and Allen (1989) for emperors (Lethrinidae) and Allen (1985) for snappers (Lutjanidae). I use the term recruitment in its fishery sense.

2.2 REPRODUCTIVE OUTPUT

'Lifetime reproductive output' refers to the number of eggs produced by a female throughout her lifetime. It depends on a number of factors, including her longevity, the size or age of sexual maturity, duration of annual spawning season, spawning frequency during that season and fecundity. A reef ecologist might be concerned with correlates of these reproductive attributes and mating systems or behaviour (e.g. Warner, 1984, 1991; Turner, 1993), whilst a fishery biologist would be interested in relationships between numbers of spawners and the resulting numbers of recruits. Stock assessment models such as 'spawning stock biomass per recruit' (SSBR) also require data on reproductive output. The potential annual reproductive output of an average recruit is an important correlate of population growth potential. As 'recruitment overfishing' (reduction of population size causing reduced egg production and increased chance of recruitment failure) is detected in a growing number of fished populations, the importance of SSBR analyses for management decisions to ensure the maintenance of sufficient reproductive adults in exploited populations increases (Goodyear, 1989). In this section, I summarize information on sexual maturation, annual frequency and duration of spawning activity and fecundity.

Maturation

Sexual maturation is an important transition point because it represents the period during which individuals enter the reproductive population and acquire the potential to contribute to future generations. Fishery managers are interested in the size of maturation relative to capture size to reduce capture of pre-reproductive individuals or to determine how quickly individuals enter the reproductive population.

Sexual maturation is expressed in several ways. The minimum size of sexual maturity is the smallest length (L_{min}) at which maturation is

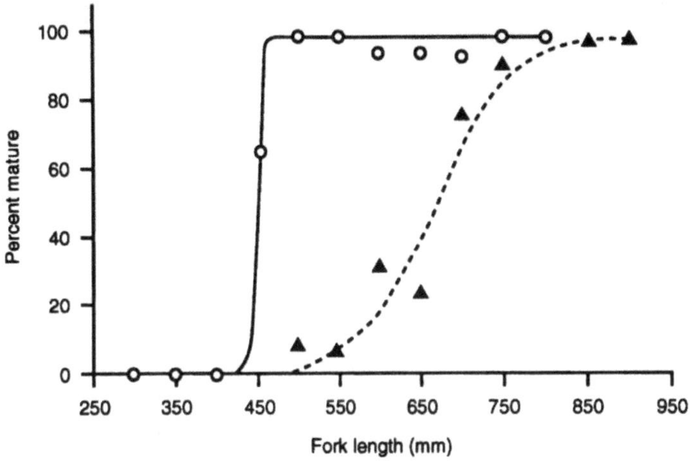

Fig. 2.1 Maturation curves for two Hawaiian snappers, *Aprion virescens* (circles) and the deep-water *Etelis coruscans* (filled triangles). Proportion of sexually mature females within each size class is plotted using a logistic function to predict the proportion of mature females for each species at size. Redrawn from Everson *et al.* (1989).

observed in a population. The size at which 50% of individuals have attained sexual maturity (L_{50}) best represents a mean population value. The smallest size at which all fish are mature is also useful because it enables determination, in concert with minimum maturation size, of the shape of the maturity curve. The form of the curve facilitates estimation of L_{50} and the probabilities of maturity at different sizes. Amongst teleost reef fishes, the size range over which sexual maturation occurs may be narrow or broad, producing similar estimates for L_{min} and L_{50} in the former, and dissimilar estimates in the latter case (Fig. 2.1). In general, larger, slower-growing species tend to mature over a broader size range than faster-growing and shorter-lived species, with age of sexual maturation among different species varying widely, from 1–2 years up to 6–7 years (Munro, 1983h; Bullock *et al.*, 1992).

For a given population, the minimum size at which fish become sexually mature is often a predictable proportion of the maximum mean size (L_{max}) attained in the population. This proportion L_{min}/L_{max} usually ranges between 0.4 and 0.9 and is constant within families comprising fish of similar dimension (e.g. Beverton and Holt, 1959; Blaxter and Hunter, 1982; Longhurst and Pauly, 1987). Pauly (1984b) proposed that such consistency is partly due to relationships between oxygen supply and demand, as mediated by species-specific gill surface area, such that the same basic mechanism that determines maximum size also determines that of sexual maturation.

Most data on size at maturation are available for groupers and snappers. Grimes (1987) found in snappers that L_{min}/L_{max} ranged from 0.23 to 0.84 (males and females combined), that the relationship between L_{min} and L_{max} was linear, and that, in some cases, males mature at a slightly smaller size than females. The data indicated that populations associated with islands mature at a somewhat larger size relative to those associated with continents (ratios of 0.51 and 0.41, respectively), although there are exceptions (e.g. McPherson *et al.*, 1992). For data on female snappers provided by Grimes (1987) and supplemented by data on snappers in Table 2.1, the relationship between L_{min} and L_{max} ($r^2 = 0.61$; $n = 48$; $P < 0.01$) is:

$$L_{min} = 11.63 + 0.50\, L_{max}. \tag{2.1}$$

Data from 27 grouper populations indicate that L_{min}/L_{max} (again, based on females only) varied from 0.33 to 0.74 (mean 0.51) and that, as for snappers, there was a significant correlation ($r^2 = 0.87$; $n = 27$; $P < 0.01$) between L_{min} and L_{max} (for females):

$$L_{min} = -9.00 + 0.51\, L_{max} \tag{2.2}$$

The regressions (Fig. 2.2) for female snappers and groupers did not differ significantly ($F_{1,69} = 0.06$; $P > 0.05$) with the pooled equation ($r^2 = 0.76$; $n = 75$; $P < 0.01$) being:

$$L_{min} = 3.61 + 0.51\, L_{max}. \tag{2.3}$$

For grouper males, because many species are characterized by protogyny (where adult females change sex to become adult males; see below), the size of sexual transition essentially represents that of male maturation and loss of reproductive females. Shapiro (1987a) analysed data on groupers by producing a ratio that expressed the size at which sex change occurs as a proportion of maximum fish size. He determined that the size range over which males appear can be quite broad, spanning between 33% and 100% of the size range of males and females combined in a given population. These data indicate that sex change takes place over a wide range of sizes within a population and suggest that, in some cases, males may develop directly from the juvenile phase rather than derive from adult females via sex change. In both the Nassau grouper, *Epinephelus striatus*, and jewfish, *E. itajara*, male and female conspecifics mature at about the same size, and the sexual pattern of the Nassau appears to be essentially gonochoristic (most individuals reproduce as only male or female in their lifetime), while that of the jewfish has yet to be confirmed (Bullock *et al.*, 1992; Sadovy and Colin, 1995).

Table 2.1 Length at first sexual maturity (L_{min}) and maximum length (L_{max}) of females of 22 species of grouper and of nine species of snapper, as a supplement to Grimes (1987). Lengths were determined from visual staging of gonads and from maximum female lengths sampled in each study. All lengths are in mm and expressed as SL (standard length), FL (fork length) or TL (total length)

Species	Location	L_{min}	L_{max}		Source
Groupers (Serranidae)					
Cephalopholis (*Epinephelus*) *cruentata*	Netherlands Antilles	145	305	TL	Nagelkerken (1979)
C. fulva	Puerto Rico	145	275	FL	Sadovy, unpubl.
Epinephelus aeneus	Tunisia	400	880	SL	Bouain (1980)
E. costae (= *alexandrinus?*)	Tunisia	270	530	SL	Bouain (1980)
E. areolatus	New Caledonia	128	325	SL	Loubens (1980b)
E. chlorostigma	Red Sea	220	550	TL	Ghorab *et al.* (1986)
	Seychelles	280	560	TL	de Moussac (1986a)
E. diacanthus	Taiwan	125	210	SL	Chen *et al.* (1980)
E. fasciatus	New Caledonia	116	228	SL	Loubens (1980b)
E. guaza (= *marginatus?*)	Tunisia	360	800	SL	Bouain (1980)
E. guttatus	Bermuda	191	405	SL	Burnett-Herkes (1975)
	Puerto Rico	190	490	FL	Sadovy *et al.* (1994a)
E. hoedti (= *cyanopodus?*)	New Caledonia	460	625	SL	Loubens (1980b)
E. itajara	Gulf of Mexico	1250	2065	TL	Bullock *et al.* (1992)
E. maculatus	New Caledonia	246	495	SL	Loubens (1980b)
E. microdon (= *polyphekadion?*)	New Caledonia	316	455	SL	Loubens (1980b)
E. morio	Gulf of Mexico	260	790	SL	Moe (1969)
E. rhyncholepis (= *rivulatus?*)	New Caledonia	187	287	SL	Loubens (1980b)
E. striatus	Bahamas	425	730	SL	Sadovy and Colin (1995)
E. tauvina	Singapore	475	850	SL	Tan and Tan (1974)
	Kuwait	450	945	SL	Hussain and Abdullah (1977)
Mycteroperca microlepis	Gulf of Mexico	400	1150	SL	Koenig *et al.* (in press)
	Gulf of Mexico	450	1128	TL	Hood and Schleider (1992)
	East USA	600	1050	TL	Collins *et al.* (1987)

Table 2.1 continued

Species	Location	L_{min}	L_{max}		Source
Plectropomus leopardus	East Australia	240	480	SL	Goeden (1978)
P. maculatus	East Australia	300	550	SL	Ferreira (1993)
Variola louti	New Caledonia	287	390	SL	Loubens (1980b)
Snappers (Lutjanidae)					
Aprion virescens	Hawaii	450	1050	FL	Everson *et al.* (1989)
Etelis coruscans	Hawaii	500	950	FL	Everson *et al.* (1989)
Lutjanus bohar	New Guinea	370	650	FL	Wright *et al.* (1986)
L. coccineus	Arabian Gulf	400	750	TL	Lee and Al-Baz (1989)
L. erythropterus	East Australia	500	650	FL	McPherson *et al.* (1992)
L. malabaricus	East Australia	540	800	FL	McPherson *et al.* (1992)
L. sebae	East Australia	485	850	FL	McPherson *et al.* (1992)
L. synagris	Trinidad	195	455	TL	Manickchand-Dass (1987)
L. vitta	NW Australia	140	350	FL	Davis and West (1993)

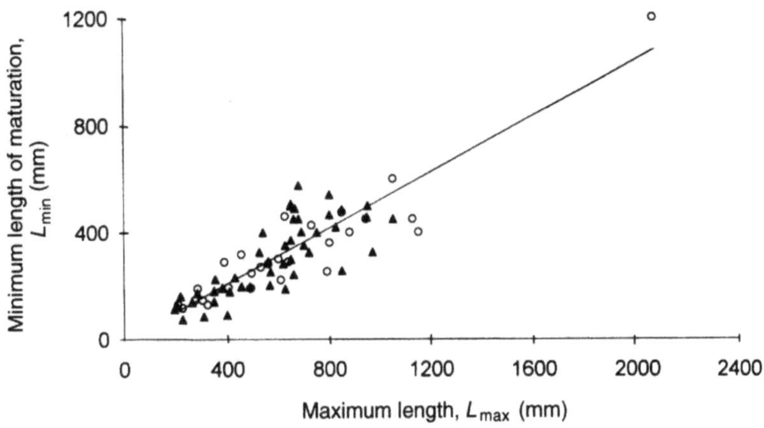

Fig. 2.2 Relationship between minimum length of maturation (L_{min}) and maximum length (L_{max}), both in mm, attained by females in study populations for groupers (circles) and snappers (filled triangles). The relationship is still significant after exclusion of the largest animal, the jewfish, *Epinephelus itajara*. From literature cited in Table 2.1 and Grimes (1987).

Given the developmental plasticity typical of fishes, it is not surprising that the size and age of sexual maturation may vary markedly within a species, both within a population over time, and among neighbouring populations. For example, reduced abundance can produce an increase in growth rate leading to earlier sexual maturation through a combination of genotypic and environmental (including fishing-induced) effects (Stearns and Crandall, 1984; Craig, 1985; Rothschild, 1986; Nelson and Soulé, 1987; Grimes *et al.*, 1988; Jennings and Lock, Chapter 8). High population density, conversely, can reduce growth rates, resulting in later maturation, decreased fecundity and increased longevity (Craig, 1985; Rothschild, 1986). Growth and sexual development in small, non-breeding, anemonefishes (Pomacentridae) are inhibited by cohabiting conspecific breeding adults (Allen, 1972). Adult sex change is socially mediated in a number of tropical species (Shapiro, 1989), in response to the numbers and sizes of adjacent reproductive males and females (page 56) (Shapiro and Lubbock, 1980; Ross, 1990).

Patterns of egg production

Tropical fishery species are typified by the production of pelagic larvae and, almost invariably, pelagic eggs. Some species spawn a few times each year; others spawn repeatedly over many months. A few families produce and tend demersal eggs (e.g. triggerfishes, Balistidae). Rabbitfishes (Siganidae) are a special case in that they scatter adhesive demersal eggs over the substratum; some eggs may enter the plankton and no parental care is involved (Thresher, 1984). Clearly, frequency and duration of spawning are key factors in estimating reproductive output (e.g. for SSBR analysis) and I here examine the methods of assessing, and available data on, patterns of egg production.

Spawning frequency

Spawning frequency, or the number of times a female spawns in a spawning season, is not well quantified in tropical species. Although information on spawning frequency is generally sparse, available data suggest that many, if not all, species are batch spawners, i.e. individual females spawn multiple batches of eggs during each reproductive season. Evidence for batch or multiple spawning derives from direct observations of repeated spawning of individually identified females (Brown *et al.*, 1994; Sadovy *et al.*, 1994b), or through indirect methods such as macroscopic examination of egg (oocyte) sizes and stages of development. Direct observations are relatively rare, however. For example, of the larger grouper species known to aggregate to spawn, only four have been observed to spawn under natural conditions (Colin *et al.*, 1987; Colin, 1992; Samoilys and Squire, 1994; Sadovy *et al.*, 1994b). Results from tagging indicated that identified individuals aggregate, and possibly spawn, more than once a year (Brown *et al.*, 1994). Spawning has apparently not been observed in emperors or grunts (Haemulidae) and has rarely been observed in snappers. In species of parrotfishes, wrasses, angelfishes, butterflyfishes (Chaetodontidae) and lizardfishes (Synodontidae), females are reproductively active over extended periods and are likely to spawn many times (Bauer and Bauer, 1981; Thresher, 1984; Colin and Clavijo, 1988; Hourigan, 1989; Donaldson, 1990; Colin and Bell, 1991).

Given the difficulties of direct observation, indirect methods are usually necessary to assess spawning frequency (Grimes, 1987; Hoffman and Grau, 1989; West, 1990). Oocytes increase in size through accumulation of yolk as they mature (i.e. the eggs become 'vitellogenic' or yolked) and take up water (hydrate), causing further expansion shortly prior to spawning. Size–frequency distributions of mature egg diameters are thus used to indicate the number of batches of similar-staged eggs developing simultaneously in the ovary. However, because simple summation of graphical modes of mature egg diameters can produce both over- and underestimates of the number of times a female is likely to spawn, even to infer multiple batch production from multiple modes is risky and other techniques must be applied (Foucher and Beamish, 1980; Kartas and Quignard,

1984; West, 1990; Hunter *et al.*, 1992; Davis and West, 1993). Assessment of spawning frequency is further complicated, at least in some longer-lived species and at the population level of interest to fishery biologists, because individual adults may not spawn every year (e.g. the grouper *Mycteroperca microlepis* and porgy *Acanthopagrus australis*), may reduce reproductive activity with increasing age or size (e.g. *A. australis* and snapper *Aprion virescens*), or may restrict spawning to certain times of the year, despite a protracted spawning season in the population, as in certain parrotfishes and groupers (Robertson and Warner, 1978; Pollock, 1982, 1984a; Everson *et al.*, 1989; Koenig *et al.*, in press).

Two techniques are well suited to the analysis of spawning frequency, but are rarely applied to tropical species. One method is to 'age' post-ovulatory follicles, which encase developing eggs, degenerate quickly and in a characteristic way following egg release, and hence may be used to back-calculate the time of spawning. The other method is to determine the monthly frequency of occurrence of females with hydrated oocytes in the ovaries (Hunter and Goldberg, 1980; Hunter and Macewicz, 1985; Hunter *et al.*, 1992; Davis and West, 1993). Where such techniques have been applied, the results are quite striking and have important implications for estimating annual reproductive output. Individuals of the snapper *Lutjanus vitta*, previously estimated to spawn 1–3 times in a season, may in fact spawn as many as 22 times a month for up to several lunar months (Davis and West, 1993). Collins *et al.* (1993) estimated that individual female *L. campechanus* may spawn 30–40 times a year. Both species are therefore indeterminate spawners, in which immature (pre-vitellogenic or unyolked) eggs continue to develop into yolked eggs during the spawning season. In such cases, annual fecundity is potentially extremely high because the number of eggs that could be spawned each season is far greater than the number of mature eggs present at the start of the season, or at any one time in the ovary. In indeterminate spawners, the assessment of potential annual egg output is fraught with difficulties because it requires knowledge both of the number of eggs produced each time a female spawns, and of the number of times she spawns each year.

Determinate spawners, by contrast, are those in which all eggs likely to be spawned in the annual season are mature at the start of the season, so pre-spawning counts of mature oocytes provide a good indication of potential annual egg output (Hunter *et al.*, 1992). Determinate spawning is most likely in those species or populations that spawn a few times over restricted periods each year. Histological analyses of ovaries of the red hind grouper, *Epinephelus guttatus*, for example, indicated that although yolked eggs were present over 4 months, the period of annual spawning was limited to approximately 10 consecutive days, and that spawning was determinate. In such cases, potential annual fecundity can be determined by counts of mature oocytes taken shortly prior to initiation of spawning (Sadovy *et al.*, 1994a). Whether other species that spawn over short periods annually are determinate or indeterminate spawners is unknown.

Duration of annual spawning

The duration of annual spawning is usually summarized monthly and refers to the length of time each year over which a population spawns. It is assessed in a number of ways. Histological or macroscopic analyses of oocytes or ovaries and gonad indices are most commonly applied (West, 1990). Histological descriptions provide the most complete information but are time consuming and costly. Macroscopic analyses of gonads may be subjective, with resolution of different stages somewhat limited. With both micro- and macroscopic approaches, female developmental classes may be divided into as few as two categories of maturation (ripe and ripening) or up to six (immature, mature inactive, mature active, ripe, ripe-running or spawning, spent), or more, stages. Ovaries best reflect the duration of spawning activity because testes often mature in advance of ovaries, yielding estimates of longer reproductive seasons.

Gonad indices (sometimes termed maturity factors), such as the gonado-somatic (gonosomatic) index (GSI), provide a measure of gonad weight (GW) per unit body weight (BW) (usually expressed as $GW/BW \times 100$). The GSI indicates the state of reproductive readiness because the gonad increases in weight, relative to body weight, during the spawning season. These indices are easy to calculate and are widely used. They not only provide a general indication of peaks in spawning activity but may also reflect aspects of mating patterns (page 54) (Yamamoto and Yoshioka 1964; Billard, 1987; West, 1990; McPherson *et al.*, 1992; Render and Wilson, 1992). Gonad indices may, however, be biased when samples of fish of different body sizes are compared. The GSI is dependent not only on stage of gonadal maturation, but also on body size (Nagelkerken, 1979; DeVlaming *et al.*, 1982; West, 1990; Wootton, 1990; Shapiro *et al.*, 1993a) (Fig. 2.3). Both body and gonad weight may change during spawning, for example when gonad weights and fat reserves exhibit inverse cycles (Reshetnikov and Claro, 1976; Claro, 1983). Progressively larger fish may also show a propensity for decreasing mean GSI (Everson *et al.*, 1989). Alternative gonad indices are available to compensate for some of these problems (DeVlaming *et al.*, 1982; Erickson *et al.*, 1985).

I have compiled data on spawning season duration for over 90 species in more than 100 populations based on studies of annual reproductive cycles in females which involved either macroscopic or histological (i.e. visual) evaluation of stages of ovarian maturation, or GSI (Table 2.2). Because, as we saw in the red hind, the simple summation of all months in which

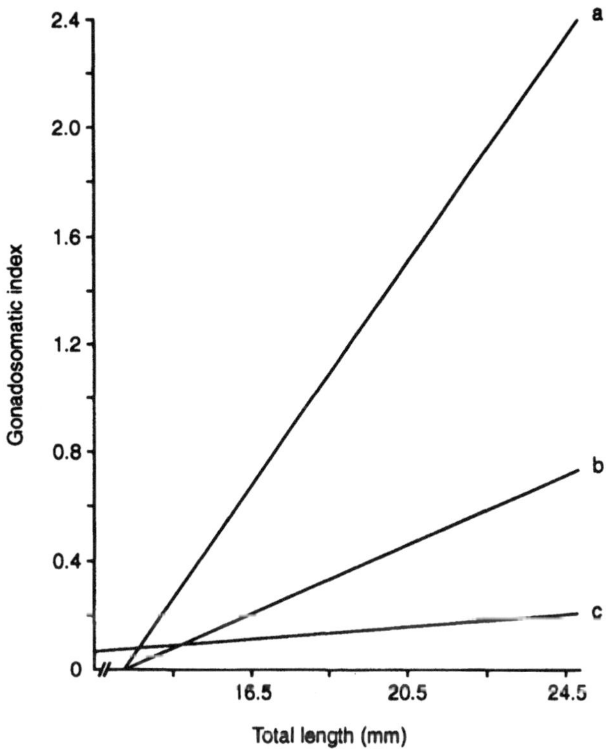

Fig. 2.3 Relationships between gonadosomatic index and total length for females of the graysby grouper, (*Cephanopholis* (*Epinephelus*) *cruentata*, at three different stages of ovarian maturation: (a) July–August, mainly mature, active; (b) May–June and September–October, mainly developing or early post-spawning; (c) January–April and November–December, mainly inactive. The relationships are significant and show that the GSI is dependent not only on the stage of ovarian maturation, but also on fish size. Modified from Nagelkerken (1979).

mature, yolked eggs occur may overestimate the length of the spawning season, peak spawning months were scored for visually evaluated samples, in addition to the total months in which mature ovaries were noted. 'Peak' spawning was assigned to months in which 50% or more of the females sampled contained yolked eggs, or eggs classified as ripe, and was taken to best represent the principal spawning season. In the cases of spawning seasonality evaluated by GSI data, peak spawning activity was assigned to those months in which the mean GSI attained was 50% or more of the maximum mean female GSI recorded in the study.

Because evaluation of ovarian stages included both visual inspections and those using GSI, it was necessary to establish to what extent GSI and

Table 2.2 Spawning seasons assessed for reef species on the basis of either visual staging of ovaries or ovarian gonadosomatic index (GSI). Only studies that included at least 8 months of data and 70 or more individuals were used. Months in which 50% or more of the females (only adult females were included where data distinguished adult from juvenile) were designated as 'ripe' were considered 'peak' spawning months and are denoted by an 'M' to indicate histologically or macroscopically examined ovaries. For months in which ripe ovaries were noted in fewer than 50%, and more than a few per cent, of females, an 'm' is used. For GSI data, peak spawning months were assigned when ovary GSI attained 50% or more of their maximum mean value and denoted by 'G'. For studies providing both GSI and visual data, only the latter were included in the table. A dash denotes lack of data for the month indicated

Species	Location	Month J	F	M	A	M	J	J	A	S	O	N	D	Source
Emperors (Lethrinidae)														
Lethrinus miniatus	New Caledonia						–	–	–	–	M	M	M	Loubens (1980b)
(= *chrysostomus*)	Great Barrier Reef									G	G	G		Brown *et al.* (1994)
L. choerorhynchus	East Africa	m		m					m			m		Nzioka (1979)
L. elongatus	Red Sea				m	m	M	M						Wassef and Bawazeer (1992)
L. lentjan	East Africa	M		m	m				m	m	m		M	Nzioka (1979)
	India	m	m			m	m	m				m	m	Toor (1964)
L. mahsena	Indian Ocean	m			m						M	m	m	Bertrand (1986)
L. nebulosus	New Caledonia	m						m	M	M	m			Loubens (1980b)
	Australia								G	G	G	G	G	Kuo (1988)
	Japan	M	M	M	M	M	M	m			m	m	m	Ebisawa (1990)
	East Africa	M	M					M	m		M		M	Nzioka (1979)
L. genivittatus (= *nematacanthus*)	New Caledonia	–				m	m	M	M	M	M	M	M	Loubens (1980b)
L. variegatus	New Caledonia	M	M	m	m				m	m	M	M	M	Loubens (1980b)
Gymnocranius euanus (= *japonicus*)	New Caledonia	–				–		–		–		M	M	Loubens (1980b)

Grunts (Haemulidae)														
Haemulon album	Cuba		–	–			m	m	m	m	m			García-Cagide and Claro (1983)
H. plumieri	Jamaica	m	M	M	M	m	m	m	m			m	m	Munro *et al.* (1973)
	Puerto Rico	M	M	M	M	M	M	m	M	M	m	m	M	Román (1991)
	Cuba	M	M	M	M	M		m			M	m	m	García-Cagide (1987)
H. sciurus	Jamaica	M	M	M	M	m	m	m	m	m	m	m	m	Munro *et al.* (1973)
	Cuba	m		m	m						m	M	m	García-Cagide (1987)
H. flavolineatum	Jamaica	–	m	m	M	m		m	m	m			–	Munro *et al.* (1973)
Pomadasys argenteus	Kuwait	M	M	M	M	m					M	m	m	Abu-Hakima (1984)
	Kuwait			G		G								Hussain and Abdullah (1977)
P. hasta	India	m	m	M	M	M	M	M	M	M	m	m	m	Deshmukh (1973)
Snappers (Lutjanidae)														
Aprion virescens	Hawaii		–	–	–	m	m	m		m				Everson *et al.* (1989)
Apsilus dentatus	Jamaica	M	M	m	M	m	–	m	m	M	M	M	–	Munro *et al.* (1973)
Etelis coruscans	Hawaii			–		–		m		m	m		–	Everson *et al.* (1989)
Lutjanus amabilis	New Caledonia	M	M						m	M	M	M	M	Loubens (1980b)
L. analis	Cuba				G	G	G	G						Claro (1983)
L. apodus	Jamaica	–	–	–	m	m	m			–				Munro *et al.* (1973)
L. bohar	Papua New Guinea	–	–		–	m					M	M	M	Wright *et al.* (1986)
L. buccanella	Jamaica		m	m	M	m	–		m	m	–	m	–	Munro *et al.* (1973)
L. erythropterus	Australia	G	G	–				–		G	G	G	G	McPherson *et al.* (1992)
L. griseus	Florida						G	G	G					Domeier *et al.* (in press)
	Venezuela					–		m	M		m		–	Campos and Bashirullah (1975)
L. kasmira	Andaman Sea	m	m				m		M	m	m		m	Rangarajan (1971)
L. malabaricus	Australia										G	G	G	McPherson *et al.* (1992)
L. quinquelineatus	New Caledonia	M	M		m						m	M	M	Loubens (1980b)
L. sebae	Australia	G	G						G	G	G	G	G	McPherson *et al.* (1992)

Table 2.2 continued

Species	Location	J	F	M	A	M	J	J	A	S	O	N	D	Source
							Month							
L. synagris	Venezuela			M	M	M	M	M						Mendez Rebolledo (1989)
	Trinidad		m	M	m	m	m	m	m	m	m	m	m	Manickchand-Dass (1987)
	Cuba				M	M	m	m	m					Reshetnikov and Claro (1976)
L. vitta (= *vittus*)	New Caledonia	M	M		M			m	m	M	M	M	M	Loubens (1980b)
Ocyurus chrysurus	Jamaica	–	M		M	m	–	–	M	M		m		Munro *et al.* (1973)
	Cuba			m	m	M	m	m	m	m	m			Piedra (1969)
Pristipomoides filamentosus	Seychelles	m	m	M	m	m		m		m	m	m	m	Mees (1993)
Rhomboplites aurorubens	Eastern USA					G	G	G	G	G				Grimes and Huntsman (1979)
Porgies (Sparidae)														
Acanthopagrus australis	Australia					G	G	G	G					Pollock (1984b)
A. latus	Kuwait		G	G										Hussain and Abdullah (1977)
	Kuwait	G	G	G										Abu-Hakima (1984)
A. cuvieri	Kuwait	G	G											Hussain and Abdullah (1977)
Archosargus probatocephalus	Gulf of Mexico			G	G									Render and Wilson (1992)
Calamus leucosteus	SE USA				m	m	m	m	m					Waltz *et al.* (1982)
Dentex tumifrons	Taiwan			–	G	G	G	–	G	G	G	G		Liu and Su (1971)
	South China Sea	G	G	G	G								G	Liu and Su (1971)
Diplodus sargus	Kuwait	m	m	m								M	M	Abou-Seedo *et al.* (1990)
Pagrus pagrus	SE USA	G	G	G										Manooch (1976)

Groupers (Serranidae)														
Cephalopholis	Jamaica		m	m	m	m	m						–	Munro *et al.* (1973)
(*Epinephelus*) *cruentata*	Curaçao				m	M	M	M	M	m				Nagelkerken (1979)
C. (*Epinephelus*) *fulva*	Jamaica	m	M	m	m		–	m				–		Munro *et al.* (1973)
	Puerto Rico	M	M										M	Sadovy, unpubl. data
C. taeniops	Senegal							G	G	G				Siau (1994)
Epinephelus aeneus	Mediterranean						M	M						Bouain and Siau (1983)
	Mediterranean						m	m	m					Vadiya (1984)
	Mediterranean							G						Bouain (1980
E. alexandrinus (?)	Mediterranean					m	M	M	M					Vadiya (1984)
	Mediterranean								M	M				Bouain and Siau (1983)
	Mediterranean							G	G					Bouain (1980)
E. areolatus	New Caledonia	–	M	M	m	–				m	M	M	M	Loubens (1980b)
E. chlorostigma	Red Sea					m	m	m						Ghorab *et al.* (1986)
E. diacanthus	Taiwan				G	G	G			–				Chen *et al.* (1980)
E. tauvina (= *coioides*)	Kuwait				G	G	G							Abu-Hakima (1987)
	Kuwait				G	G								Hussain and Abdullah (1977)
E. fasciatus	New Caledonia	–	m				–	–		m	M	M		Loubens (1980b)
E. guaza	Mediterranean						M	M						Bouain and Siau (1983)
	Mediterranean							G						Bouain (1980)
E. guttatus	Jamaica	m	M		M									Munro *et al.* (1973)
	Puerto Rico	M	M	m									M	Sadovy *et al.* (1994a)
	Venezuela	M	M										m	Villarroel (1982)
	Bermuda					m	M	m						Burnett-Herkes (1975)
E. itajara	Florida							m	m	m				Bullock *et al.* (1992)
E. maculatus	New Caledonia		–								m	M	M	Loubens (1980b)
E. morio	Gulf of Mexico	m	m	m	M	M	m				m		m	Moe (1969)
	Mexico	G	–	G				–		–		–		Brule *et al.* (in press)
E. striatus	Jamaica		M		M	M	–		–	–				Munro *et al.* (1973)
	Belize	M	m										m	Carter *et al.* (1994)

Table 2.2 continued

Species	Location	Month J	F	M	A	M	J	J	A	S	O	N	D	Source
Mycteroperca microlepis	Gulf of Mexico	m	m	M	m	m	m	m				m		Hood and Schlieder (1992)
	SE USA	M	M	M	M									Collins *et al.* (1987)
	Gulf of Mexico		G	G										Koenig *et al.* (in press)
Plectropomus leopardus	Australia	–				–		–			–	G	G	Goeden (1978)
Wrasses (Labridae)														
Bodianus perditio	New Caledonia				m	–	M	–	–		m			Loubens (1980b)
Parrotfish (Scaridae)														
Scarus croicensis	Jamaica		m	m	m	m	m							Munro *et al.* (1973)
S. guttatus	East Africa	m		M	M						m		m	Nzioka (1979)
Sparisoma aurofrenatum	Jamaica	m	–	m	m	m	m	m	m		m	m	m	Munro *et al.* (1973)
S. chrysopterum	Jamaica		m	m	m	m	m							Munro *et al.* (1973)
S. viride	Jamaica		m	m	m	m								Munro *et al.* (1973)
Angelfish (Pomacanthidae)														
Holacanthus ciliaris	Jamaica	m	m	–	m	m	m	m	m			–		Munro *et al.* (1973)
Pomacanthus arcuatus	Jamaica		m	M	m	m	m							Munro *et al.* (1973)
Fusiliers (Caesionidae)														
Pterocaesio pisang	Philippines	m	M	M	M	–	M	m	m	m	m	m	m	Cabanban (1984)
Threadfin bream (Nemipteridae)														
Nemipterus peroni	New Caledonia						m	m	m	m	M	M	M	Loubens (1980b)

N. japonicus	India						–	–		m	m	m		Krishnamoorthi (1972)
N. virgatus	South China Sea			G	G	G	G	G						Liu and Su (1992)
Scolopsis bimaculatus	Kenya	m		m	m	m				m	m	m		Nzioka (1985)
Goatfish (Mullidae)														
Parupeneus porphyreus	Hawaii	M	M	M	M	M	M	m					m	Walsh (1987)
Pseudupeneus maculatus	Jamaica		M	M	M	M	–	–	–		m	m		Munro *et al.* (1973)
Upeneus moluccensis	Hong Kong			G	G	G	G							Lee (1974)
Rabbitfish (Siganidae)														
Siganus canaliculatus	Arabian Gulf			M	m	m								Al-Ghais (1993)
	Caroline Islands		G	G	G									Hasse *et al.* (1977)
	Hong Kong			m	m	m	m							Tseng and Chan (1982)
S. rivulatus	Red Sea					m	M	M	m					Popper and Gundermann (1975)
	Mediterranean					M	M	M						Popper and Gundermann (1975)
S. sutor	East Africa	M	m	m	m	M	M	m						Ntiba and Jaccarini (1990)
S. vermiculatus	Fiji	G	G								G	G	G	Gundermann *et al* (1983)
Surgeonfish (Acanthuridae)														
Acanthurus bahianus	Jamaica		M	m	m									Munro *et al.* (1973)
A. chirurgus	Jamaica		m	m	m	m	m							Munro *et al.* (1973)
A. coeruleus	Jamaica		m	m	m	m	m							Munro *et al.* (1973)
Ctenochaetus striatus	French Polynesia		G										G	Montgomery and Galzin (1993)
Triggerfish (Balistidae)														
Balistes frenatus	New Caledonia	M	M	M			–	–	–	–			M	Loubens (1980b)
B. capriscus	Ghana											G	G	Ofori-Danson (1990)

Table 2.2 continued

Species	Location	Month J	F	M	A	M	J	J	A	S	O	N	D	Source
Trunkfish (Ostraciidae)														
Acanthostracion polygonius	Puerto Rico	M	M	M	M	m	m				m		m	Sadovy, unpubl. data
A. quadricornis	Jamaica	–	–	M	m	m	m	m	m	m	m	m		Munro *et al.* (1973)
Lactophrys triqueter	Jamaica	–	–	m	m	m	m	m						Munro *et al.* (1973)

visual data were comparable. I assessed 22 populations (21 species) for which both GSI and visual assessments had been applied simultaneously to determine reproductive activity. The relationship between peak maturity and peak GSI was significant, but weak ($r^2 = 0.29$; $n = 22$; $P < 0.01$), and took the form:

$$\text{peak GSI} = 1.65 + 0.35 \times \text{peak maturity} \qquad (2.4)$$

The relationship differed from unity because peak GSI tends to underestimate the length of seasons of longer duration, as assessed by peak maturity using visual techniques. This is probably because spawning activity is less clearly concentrated at any one time during more extended spawning seasons and changes in GSI are consequently less marked. The GSI of ripe ovaries varied widely among species with extremes of 2–30%, and the highest values were generally associated with shorter spawning seasons. This range is similar to that noted for temperate and freshwater fishes (Wootton, 1990).

Data from Table 2.2, partially summarized in Figs 2.4 and 2.5, indicate short to moderate seasons in groupers, emperors, porgies, surgeonfishes, rabbitfishes (e.g. Fig. 2.5(a)–(e)), goatfishes (Mullidae) and triggerfishes. Species such as many grunts and snappers (Fig. 2.5 (f) and (g)), fusiliers (Caesionidae) and threadfin breams (Nemipteridae), with moderate to extended spawning seasons, may exhibit intermittent peaks of activity which frequently include a lunar or semilunar component ('Lunar patterns', page 46). Many wrasses, parrotfishes and angelfishes also have extended spawning seasons, although the intensity of spawning activity, as determined by direct observation of spawning, varies seasonally (Table 2.2; Talbot, 1960; Eggleston, 1972; Burnett-Herkes, 1975; Colin, 1977; Dan, 1977; Johannes, 1978a, 1988a; Thresher, 1984; Bell and Colin, 1986; Walsh, 1987; Colin and Clavijo, 1988; Hunte von Herbing and Hunte, 1991; Samoilys and Squire, 1994).

One interesting pattern evident in some families is an inverse linkage between body size and length of spawning season. For example, smaller grouper species tend to spawn over more extended periods (e.g. *Cephalopholis* spp.) and do not aggregate to spawn. By contrast, larger species (e.g. *Epinephelus guttatus, E. striatus, Mycteroperca tigris* and *Plectropomus leopardus*) aggregate to spawn and exhibit short reproductive seasons (Thompson and Munro, 1978; Nagelkerken, 1979; Moore and Labisky, 1984; Colin, 1992; Samoilys and Squire, 1994; Sadovy *et al.*, 1994a,b). Among emperors, data suggest that smaller species, such as *Lethrinus genivittatus* (formerly *nematacanthus*) and *L. variegatus*, tend to have longer spawning seasons, while larger species, such as *L. miniatus* (formerly *chrysostomus) and L. nebulosus*, show a tendency towards shorter seasons (Loubens, 1980b; Brown *et al.*, 1994).

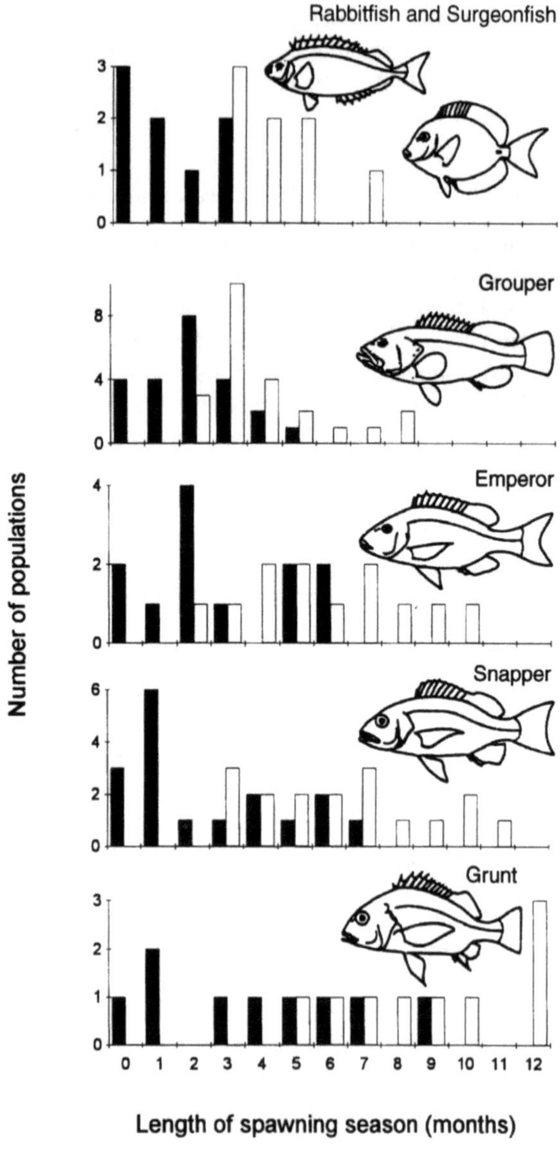

Fig. 2.4 Lengths of spawning seasons, in months, for populations in six families of tropical reef fishery species. Black bars indicate 'peak' spawning season in months, as determined by including only those monthly samples in which 50% or more of the females had vitellogenic (mature) oocytes. White bars indicate all months in which ovaries containing vitellogenic oocytes were found. Data sets indicating no peak spawning months received a zero score for 'peak' months. Only ovaries examined macroscopically or histologically are included. Redrawn from sources for each family cited in Table 2.2.

Length of spawning season in snappers also appears to be somewhat size related (Domeier *et al.*, in press). Smaller species of western Atlantic *Lutjanus* have protracted spawning seasons with no apparent spawning migrations, but larger species spawn over more limited periods and migrate to specific sites to spawn. However, the situation may be somewhat complex. Grimes (1987) reviewed spawning patterns in snappers and noted that the length of spawning season tends to depend on whether or not populations are predominantly insular or continental in distribution. In the former case, seasons are extended and last much of the year; in the latter they are more restricted in duration and concentrate over the summer months, although there are exceptions (e.g. Everson *et al.*, 1989). The picture is further complicated by the tendency for deeper-dwelling species to exhibit more extended spawning seasons, than shallow-water species (Grimes, 1987).

Comparisons between studies of spawning seasonality are difficult. Studies vary widely in their application of definitions of gonad maturation with terms such as 'mature', 'maturing' or 'ripe' often not clearly defined. In many cases, it is not evident whether monthly values of GSI or 'per cent mature females' are based only on mature females or included immature females. In some cases, monthly catches of juveniles are relatively substantial, which would lower mean monthly GSI values. Such shortcomings greatly reduce the value of much of the data. Care is needed when comparing studies based on gonad indices with those based on visual methods, although corrections such as alternative gonad indices may be applied.

Fecundity

Fecundity relates to the number of eggs produced by a female and can be defined in many ways (Bagenal, 1978; Kartas and Quignard, 1984). Fecundity is a critical component of reproductive output and is included in certain stock assessments. Despite the importance of estimating fecundity, little reliable information is available for tropical fishes. Yet studies on the reproductive biology of fishery species often include fecundity estimates. With few exceptions, sample sizes are small, size ranges limited and methodology inconsistent or not clearly specified. Variation in fecundity must also be an important factor in the evolution of mating systems, yet surprisingly, variability in fecundity among females has been given little attention. Indeed, females are typically considered to be equivalent in reproductive output in behavioural–ecological studies.

Fecundity is evaluated in a number of different ways but most commonly as total (standing stock of advanced yolked oocytes), batch (number of hydrated oocytes released per spawning) and relative (fecundity divided by female weight) fecundities (Hunter *et al.*, 1992). Batch

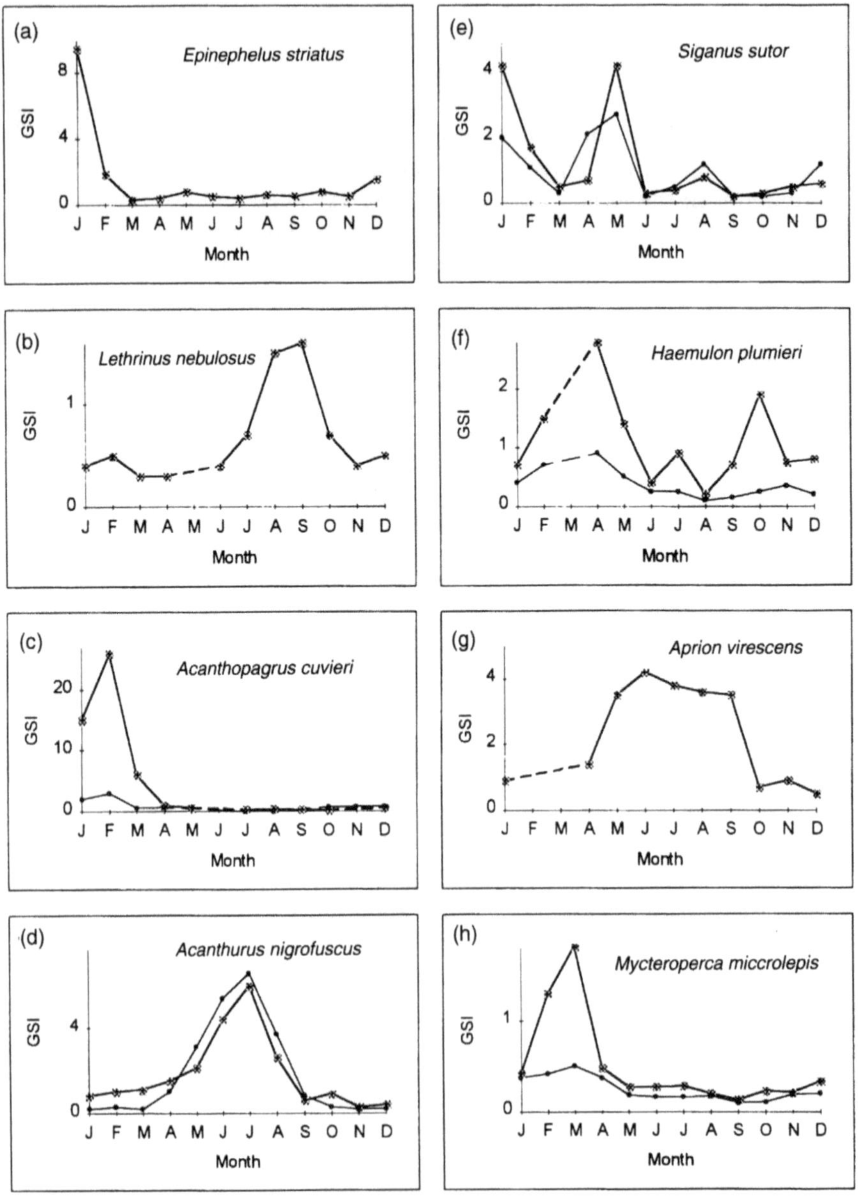
(a)
Epinephelus striatus
GSI
Month
(b)
Lethrinus nebulosus
(c)
Acanthopagrus cuvieri
(d)
Acanthurus nigrofuscus
(e)
Siganus sutor
(f)
Haemulon plumieri
(g)
Aprion virescens
(h)
Mycteroperca miccrolepis
J F M A M J J A S O N D

fecundity (*F*) is a function of body size, most typically expressed as body length (*L*), with the relationship between the two characteristically taking the form $F = aL^b$, where a and b are a constant and an exponent, respectively, derived from the data. Because batch fecundity relates to the volume of the body cavity available to accommodate the ripe ovaries, geometry indicates that the length exponent, b, should be around 3.0 (Bagenal, 1978; Wootton, 1990). Thresher (1984) determined the relationship between instantaneous (= total) fecundity and body size (standard length in cm) for 56 species of a wide size range of tropical pelagic and demersal spawners to be:

$$\log_{10} F = 1.44 + 2.37 \log_{10} L. \qquad (2.5)$$

This relationship did not differ according to spawning mode (i.e. pelagic versus demersal spawning). That the length exponent is somewhat lower than the 3.0 expected on allometric grounds indicates that, for some species, although larger individuals produce an absolutely larger total volume of eggs than smaller ones, fecundity per unit length is relatively less in larger than in smaller females (but see below). A length exponent of approximately unity is generally attributable to a limited sample size range or to a small L_{max}.

Wootton (1990) noted that the length exponent in fishes ranges from b = 1 to b = 5; the higher the exponent, the greater the relative increase in fecundity with increasing size. I found fecundity exponents of 15 species from nine families to vary from a low of 1.6 in a triggerfish (*Balistes*) (Ofori-Danson, 1990) to highs of 5.6 and 4.8 in croakers (Sciaenidae) (Haimovici and Cousin, 1988; Manickchand-Heileman and Kenny, 1990)and 5.1 in a parrotfish (Gonzalez *et al.*, 1993), with porgies at about 4.0 (Manooch, 1976; Waltz *et al.*, 1982; Chakroun and Kartas, 1987) and groupers, snappers, emperors, threadfin bream and trunkfishes (Ostraciidae) ranging from 2.0 to 4.0 (Rangarajan, 1971; Campos and Bashirullah, 1975; Dan, 1977; Chen *et al.*, 1980; Bertrand, 1986; Ghorab *et al.*, 1986; Kuo, 1988; Davis and West, 1993; Sadovy, unpubl. data). These should only be taken as general trends, however, because estimation

---

Fig. 2.5 Gonadosomatic indices (GSI) over the annual cycle of eight species from seven families of tropical reef fishery species: (a) grouper, Belize (Carter *et al.*, 1994); (b) emperor, New Caledonia (Loubens, 1980b); (c) porgy, Kuwait (Hussain and Abdullah, 1977); (d) surgeonfish, Gulf of Aqaba (Fishelson *et al.*, 1987); (e) rabbitfish, Kenya (Ntiba and Jaccarini, 1990); (f) grunt, Cuba (Garcia-Cagide, 1987); (g) snapper, Hawaii (Everson *et al.*, 1989); (h) grouper, Gulf of Mexico (Hood and Schleider, 1992). Redrawn from sources. Note differing vertical scales. Broken line, no data available over period indicated. In (c)–(f) and (h), dots denote males and crosses denote females.

of fecundity among studies was not based on a standard fecundity measure. For example, the length exponent was calculated on the basis of fecundity estimated in different ways, ranging from counts of all maturing eggs to total ripe eggs, mature eggs, eggs greater than a certain specified diameter, fully yolked and hydrated, as well as batch fecundity.

Groupers and snappers provide most of the estimates of fecundity. In groupers, species maxima ranged from 107 000 oocytes in a 285 mm (length type unspecified) *Epinephelus costae* (formerly *alexandrinus*) (Vadiya, 1984) to 260 million in 2115 mm TL *E. tauvina* (probably a misidentification of *E. coioides*) (Selvaraj and Rajagopalan, 1973). This latter fecundity approximates that of the ocean sunfish *Mola mola*, one of the most fecund of vertebrates (Moyle and Cech, 1988), and underscores the high fecundity of some of the largest reef fish. Data on groupers ranging from 200 mm to >2000 mm and gonad weights of between 10 g and 17 000 g show a significant relationship between total fecundity (F) (maximum values recorded per study) and length (L) (standard, total or unspecified) of the form ($r^2 = 0.96$; $n = 12$; $P < 0.01$):

$$\log_{10} F = -1.89 + 3.03 \log_{10} L \qquad (2.6)$$

(Moe, 1969; Selvaraj and Rajagopalan, 1973; Thompson and Munro, 1978; Chen *et al.*, 1980; Vadiya, 1984; Ghorab *et al.*, 1986; Abu-Hakima, 1987; Bullock and Smith, 1991; Carter *et al.*, 1994; Sadovy, unpubl. data).

Estimated fecundities in snappers were given by Grimes (1987). Maximum fecundities ranged from 203 000 at 445 mm (length type unspecified) for *Lutjanus buccanella* to 9.32 million ova at 605 mm FL (fork length) for *L. campechanus*. Grimes noted that there was apparently no consistent relationship between fecundity and body size among the snapper species but that data were inadequate to fully address this issue. None the less, it was clear that, given exponential increases in oocyte number with fish size, larger fish can produce by far the greatest absolute numbers of eggs. Also, because larger females may spawn more times and over a longer period than smaller females (Grimes, 1987; Plan Development Team, 1990), annual fecundity of the former is likely to be several orders of magnitude greater than that of the latter. A model of the red snapper, *L. campechanus*, life history calculated that one large red snapper female (601 mm FL) potentially produces the same number of eggs as 212 small (420 mm FL) females (Plan Development Team, 1990). These figures illustrate that the absolute numbers of eggs produced by conspecific females of different sizes can vary enormously, i.e. that females are clearly not equivalent in egg output.

Because fecundity depends partially on female size, comparisons of fecundities among species or among conspecific females should be standar-

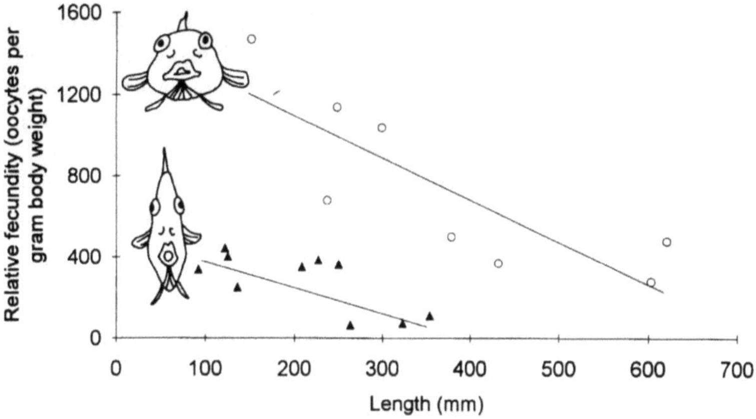

Fig. 2.6 Relative fecundity in fishes of different body form. Wide body form is represented by groupers (circles) ($Y = 1499 - 2.037X$; $n = 8$; $r^2 = 0.68$) and narrow body form by butterflyfishes, angelfishes, porgy, triggerfishes, grunts and a spadefish (Ephippidae) (filled triangles) ($Y = 520 - 1.528X$; $n = 10$; $r^2 = 0.52$). Each symbol represents the mean of relative fecundity and body size data provided for each species (Moe, 1969; Campos and Bashirullah, 1975; Mahadevan and Devadoss, 1975; Nagelkerken, 1979; Chen *et al.*, 1980; Bouain and Siau, 1983; Munro, 1983h; García-Cagide, 1986, 1987; Ghorab *et al.*, 1986; Chakroun and Kartas, 1987; Sadovy, unpubl. data).

dized for body size by employing a measure such as relative fecundity. Data suggest that relative fecundity is extremely variable among species, with higher values in the round groupers and lower values in more compressed species such as angelfish and grunts (Fig. 2.6). Smaller species of either body form evidently allocate relatively more energy per batch than larger ones. Relative fecundity in snappers tends to be highly variable and fits neither pattern. Unfortunately, few comparisons between body form and relative fecundity were possible because few workers supplied the necessary weight data. Indeed, considerable value was lost from much of the literature relating to fecundity because of missing information, not only of weight data, but also of body length type measured, and even the kind of fecundity (i.e. batch or total) being estimated.

Morphological factors that influence fecundity are body size and body form. Choat and Bellwood (1991) proposed that in teleosts of large adult size and with a body form that permits a relatively substantial ovarian mass, a few spawning episodes will produce large numbers of offspring. Conversely, in small teleosts, or those with little space in the visceral cavity, large numbers of eggs can only be produced by continual spawning and rapid production of eggs (Robertson, 1991a). The suggestion is that both body form and size constrain potential fecundity, with possible consequences for other aspects of life history such as reproductive strategies and abdominal anatomy. Among butterflyfishes and angelfishes, for example, larger species spawn less frequently but produce more eggs per spawn than smaller species (Moyer *et al.*, 1983; Hourigan, 1989). The post-anal placing of fat bodies in surgeonfishes (Fishelson *et al.*, 1985) and the post-anal location of the urogenital system in the moray eel (Muraenidae), *Rhinomuraena* spp. (Fishelson, 1990), have likewise been attributed to body cavity constraints.

Aside from body form and size, fecundity is also influenced by egg size. With few exceptions (rabbitfishes, triggerfishes, pufferfishes [Tetraodontidae]), larger reef species produce large numbers of small, spherical (in some parrotfishes, elongate), pelagic eggs. Hydrated egg diameters and volumes may be highly variable, even among closely related species, ranging from lows of 0.4–0.5 mm in rabbitfishes, wrasses and triggerfishes, to 1.5–2.0 mm in lizardfishes, pufferfishes and trunkfishes, and 2.0–4.0 mm in some moray eels. Egg diameters of a little less than 1.0 mm are characteristic of the majority of species (Breder and Rosen, 1966; Munro, 1983h; Johnson, 1984; Thresher, 1984, 1988a; Pauly and Pullin, 1988; Davis and West, 1993). Among angelfishes, smaller species tend to have smaller eggs than larger species (Thresher and Brothers, 1985). No such pattern was noted in wrasses and parrotfishes of Enewetak, with the smallest egg among the scarids found in the largest parrotfish species, *Bolbometopon muricatum* (Colin and Bell, 1991).

Fecundity is highly variable, both within species and between years (Bagenal, 1978). Its estimation requires wide size ranges of individuals and clear explanations of methods. Consideration is also necessary of factors that may curtail potential fecundity such as parasitism and oocyte atresia (resorption of eggs prior to spawning). Ovaries of grunts and groupers were frequently noted to contain trematodes in Jamaica and in one *Haemulon* grunt almost 30% of the ovary weight consisted of nematodes (Munro, 1983h). Eggleston (1972) noted that parasitism resulted in a 15% reduction in the number of eggs of threadfin breams. Small numbers of atretic oocytes have been noted in groupers and in the porgy, *Pagrus pagrus*, shortly after termination of spawning (Manooch, 1976; Bouain and Siau, 1983; Sadovy *et al.*, 1992; Ferreira, 1993). In the emperor, *Lethrinus nebulosus*, atresia of vitellogenic oocytes is common during both prespawning and spawning periods (Ebisawa, 1990). In the porgy, *Acanthopagrus australis*, a large proportion of adult females do not migrate to spawn each year despite the fact that their oocytes undergo vitellogenesis. In migrating females of this species, oocytes continue to hydrate, whereas in non-migrators, vitellogenic oocytes are resorbed (Pollock, 1984a). Fecundity and atresia are related to food supply and population density, although the relationships are not simple and large and little understood intraspecific variations among years have been noted for a number of temperate species (Bagenal, 1978).

2.3 SPAWNING PATTERNS

Tropical fishes exhibit a wide range of temporal and spatial patterns in reproduction, from spawning restricted to less than 1 month annually at specific sites, to spawning throughout the year at a diverse range of localities. Environmental factors that might be associated with when and where individuals spawn are relevant to our understanding of yearly and seasonal variations in spawning activity. Questions of interest include: to what extent do the timing, intensity, duration and location of spawning determine the profiles of larval settlement events on the reef? How critical are specific spawning sites and times for annual reproductive output, and what is the potential impact on the population of disturbance by fishing or other activities?

The timing and location of spawning activity in tropical reef fishes that produce pelagic eggs are widely considered to be a response to major problems facing larvae, i.e. finding food, maximizing dispersal and avoiding predation (Johannes, 1978a; Barlow, 1981; Doherty *et al.*, 1985). It has also been proposed that timing of reproduction is mediated by requirements of juvenile fishes at the beginning of their benthic existence (Walsh, 1987). Somewhat less attention has been given to the possibility that factors that influence the reproductive capacity of the adults directly or are critical for bringing widely dispersed adults together to spawn may also be important (Colin *et al.*, 1987; Shapiro *et al.*, 1988, 1993b; Robertson, 1991a).

In establishing causal agents for spawning patterns, we are greatly hampered by a lack of empirical data to evaluate the various hypotheses (Shapiro *et al.*, 1988; Robertson, 1991a; Appeldoorn *et al.*, 1994). For example, while preliminary data indicate that benefits derived from specific sites and times of spawning may apply in the short term, such as the first few hours following spawning (Appeldoorn *et al.*, 1994; Hensley *et al.*, 1994), longer-term studies do not support a dispersal advantage to a specific site and time of spawning (Colin, 1992). Robertson (1991a) concluded that to assess the relative importance of larval biology, juvenile biology, and adult biology factors as determinants of seasonal patterns of spawning, we need data on the most basic aspects of these fishes' life history as well as on relationships between environmental parameters and knowledge of times and places of spawning.

The general paucity of detailed data on reproductive periodicity precludes detailed analyses of temporal and spatial patterns of spawning for commercially important reef species. Incomplete sample series, or sampling less frequent than monthly, permit nothing more than a broad overview of the data. I confine myself to temperature and diurnal, lunar and spatial correlates of reproductive activity over broad geographic areas and across a range of species. Spawning patterns in specific regions, or for certain families, have been variously summarized (e.g. Munro *et al.*, 1973; Lam, 1974; Erdman, 1976; Johannes, 1978a; Sale, 1980; Thresher, 1984; Grimes, 1987; Shapiro, 1987a; Walsh, 1987; Colin and Clavijo, 1988; Colin and Bell, 1991; Robertson, 1991a).

Temporal patterns

Annual periodicities

Most island populations located in tropical and warm temperate areas are exposed to annual production cycles which are typically continuous and of low amplitude compared with colder climates (Cushing, 1975a). This has been suggested as an explanation for the less marked seasonality observed in many tropical and subtropical species. None the less, seasonality clearly exists and the factors that determine it are a complex of geography and biology (Robertson, 1991a). To examine annual trends in spawning activity I looked at a range of species in a variety of locations using the data from Table 2.2.

Assuming that regional water temperature profiles reflect those experienced by fishes in a particular area, spawning occurs between 19 and 29°C, within an overall annual range of 16–31°C (Table 2.3). There does seem to be a trend for decreasing strength of spawning seasonality with decreasing latitude (Munro, 1983h; Robertson, 1991a) but this is not obviously related to specific environmental variables (Robertson, 1991a). However, even when spawning activity extends over many months, there is often more intense activity in separate periods, with lesser activity in the intervening period (e.g. Munro *et al.*, 1973; Nzioka, 1979). In areas of extreme seasonal upheavals, the spawning season may be split, as in the East African area where two peaks of reproductive activity occur, each following a monsoon season.

Several patterns are indicated at the family level. Porgies, for example, tend to spawn during colder periods than other co-occurring reef fishes (e.g. Manooch, 1976; Render and Wilson, 1992), probably reflecting the predominantly subtropical/temperate distribution of this family. Among western Atlantic groupers the timing of spawning may be temperature related, because the latitudinal shifts in timing of spawning apparently do not correlate with patterns of change in other variables such as daylength and annual cycle of primary productivity (Robertson, 1991a). In the Car-

Table 2.3 Annual temperature range and approximate temperature range over which spawning occurs in seven regions ordered by latitude from 30°N to 20°S. Indicated are the months in which reproductive activity has been noted in a number of larger reef species. Porgies are not included (sources: Table 2.2 and Talbot, 1960; Erdman, 1976; Johannes, 1978, 1981; Walsh, 1987; Colin and Clavijo, 1988; Robertson, 1991a; Egretaud, 1992; Samoilys and Squire, 1994; Colin, pers. comm.)

Region	Spawning months	Annual temp. range (°C)	Approx. spawning temp. range (°C)	Latitude range
Gulf of Mexico, Florida and Bermuda	Feb.–Sept.	17–27	19–27	25–30°N
Kuwait and Red Sea	Feb.–July	16–31	20–27	20–30°N
Hawaii	Jan.–Aug.	22–28	24–26	20°N
Caribbean	Oct.–Apr.	25–30	25–28	10–20°N
Belau	Feb.–June*	28–30	28–29	0–10°N
East Africa and Indian Ocean	Feb.–May and Sept.–Nov.	25–29	26–29	0–10°S
New Caledonia, Fiji and northern Australia	Aug.–Feb.	20–30	21–27	10–20°S

*Months indicated are the principal spawning months although reproductive activity also occurs throughout all months of the year (see also Colin and Bell, 1991, for the Marshall Islands).

ibbean and Bahamas, *Epinephelus guttatus*, *E. striatus*, *E. morio* and *Mycteroperca tigris* spawn from December to April, when temperatures are minimal to rising, although these species do not show strict temporal overlap in spawning during this period. In Florida and more northerly Bermuda, these same species spawn at temperatures similar to those of lower latitudes, coinciding with the warmest annual temperatures later in the year at the higher latitudes (Oliver La Gorce, 1939; Moe, 1969; Munro *et al.*, 1973; Burnett-Herkes, 1975; Colin, 1992; Carter *et al.*, 1994; Sadovy *et al.*, 1994a,b; Brule *et al.*, in press). The Indo–Pacific coral trout, *Plectropomus leopardus*, spawns in temperatures of above 24°C and the timing of aggregation build-up appears to be dependent on attaining this temperature, or some factor correlated therewith (Brown *et al.*, 1994). A particularly interesting case is that of the Nassau grouper, *Epinephelus striatus*, which has a very restricted spawning period of a few weeks, over 1–2 months, each year. Irrespective of location and daylength, increasing or decreasing temperature, spawning occurs within a narrow band of 25–26°C (Colin, 1992; Tucker *et al.*, 1993). This species spawned spontaneously in an artificially lit Havana aquarium in April (i.e. out of season for the region) in water of 26°C (Guitart-Manday and Juarez-Fernandez,

1966). The relative importance of environmental and biological factors in determining timing and duration of spawning is unknown and may only be understood in a wider context of other selective pressures operating (e.g. larval versus juvenile or adult survival).

Sub-annual periodicities

Lunar patterns

Lunar-synchronized spawning occurs in four orders of fishes (Salmoniformes, Atheriniformes, Tetraodontiformes and Perciformes), in both intertidal and reef spawners (Taylor, 1984). Lunar cycles might be adaptive in various ways which, as for spawning seasonality, may be advantageous to either juveniles or adults (Johannes, 1978a; Thresher, 1984; Gladstone and Westoby, 1988; Robertson, 1991a). Tides are related to the lunar cycle and may flush larvae offshore and away from predators, or disperse them from the spawning site. Alternatively, larval dispersion may be minimized and larvae retained inshore. Food or light conditions associated with certain lunar phases may be important for larval survival. Reproductive activity of adults may be synchronized by certain moon phases. However, of the various hypotheses proposed to account for the seasonality of spawning, those invoked for lunar patterns have yet to be evaluated empirically and conclusions remain tentative until further detailed studies provide more data for comparative analyses.

While lunar cyclicity in spawning activity does occur in a number of reef fishes, one consistent pattern that emerges is that it is largely limited to pelagic spawners which migrate to spawn, and to a few demersal spawners. Many other families show no evidence of lunar cycling (Thresher, 1984). On the basis of direct observations of spawning behaviour, at an essentially atidal site, Colin and Clavijo (1988) noted a general lack of lunar cycling in the pelagic spawning activities of a number of species in the tropical western Atlantic. Colin and Clavijo (1988) found that lunar spawning is more characteristic of those species that migrate to spawning sites, while pelagic spawners that did not migrate (such as many wrasses and parrotfishes) showed far less indication of lunar synchronization at either Indo–Pacific or Atlantic sites (Johannes, 1978a; Thresher, 1984; Colin and Clavijo, 1988; Colin and Bell, 1991). Thresher (1984) noted that non-migratory surgeonfishes have no lunar cycle but that migratory species do. Rabbitfishes exhibit similar variability in associations between lunar patterns and migratory activity (Lam, 1974; Popper *et al.*, 1976, 1979; Johannes 1978a; Ntiba and Jaccarini, 1990; Al-Ghais, 1993). Grouper species that migrate to spawning aggregations may spawn at or around the full or new moon, but even within a species, moon phase at reproduction is variable (Brown *et al.*, 1994; Sadovy *et al.*, 1994a). Some of the smaller *Epinephelus* groupers, which do not aggregate to spawn, probably do not exhibit a lunar component to their reproductive activity (P. Colin, pers. comm.).

Lunar or semilunar cycles have been recorded for snappers, with increased spawning intensity at full or new moon, or both (Randall and Brock, 1960; Grimes, 1987; Davis and West, 1993; Domeier *et al.*, in press), in mullets (Mugilidae) and emperors (Johannes, 1978a), fusiliers (Bell and Colin, 1986), and possibly some triggerfishes (Johannes, 1978a; Fricke, 1980; Lobel and Johannes, 1980). Trunkfishes and angelfishes show no apparent lunar component (Moyer and Nakazono, 1978; Thresher, 1984; Colin and Clavijo, 1988). Little is known of possible lunar patterns, at least based on gonadal analyses, in grunts, porgies and goatfish. However, back-calculated spawning dates from otoliths of newly settled French grunts, *Haemulon flavolineatum*, in the US Virgin Islands showed a very marked lunar pattern (McFarland *et al.*, 1985; Roberts, Chapter 4, Fig. 4.2). This pattern was due either to spawning at the time of spring tides, or to strong temporal and periodic forcing of larval settlement or survival.

Diurnal patterns

There is, in general, a strong relationship between different modes of spawning and time of day. Pelagic spawners tend to reproduce in late afternoon/early night, while demersal spawning occurs at varying times during the day (Thresher, 1984). A pronounced peak at late afternoon was noted in groupers, goatfishes, angelfishes, trunkfishes, some porgies, a snapper, *Lutjanus synagris*, and a fusilier, *Caesio teres* (Wicklund, 1969; Colin and Clavijo, 1978, 1988; Johannes, 1981, 1988a; Thresher, 1984; Bell and Colin, 1986; Colin *et al.*, 1987; Robertson, 1991a; Colin, 1992; Sadovy *et al.*, 1994; Samoilys and Squire, 1994). Parrotfishes, surgeonfishes, some wrasses and a snapper, *Lutjanus vitta*, however, spawn at various times throughout the day, including dusk, often in response to the timing of the tides, although by no means all daytime spawners exhibit tidal spawning patterns even when tidal regimes are strong. Two mullets spawn between late evening and midnight and large parrotfishes may spawn in the early morning (Arnold and Thompson, 1958; Randall, 1961; Helfrich and Allen, 1975; Robertson, 1983; Thresher, 1984; Gladstone, 1986; Colin and Clavijo, 1988; Myrberg *et al.*, 1988; Colin and Bell, 1991; Davis and West, 1993). Little is known of the diurnal timing of spawning in emperors and grunts. The demersal spawners among commercially important reef fishes spawn before dawn (triggerfishes), or at various times at night, early morning or in the mid to late afternoon (rabbitfishes), probably depending on the tidal cycle (Lam, 1974; Johannes, 1981; Robertson, 1983; Thresher, 1984).

Several hypotheses have been proposed to account for diurnal patterns of spawning in reef fishes, especially the concentrated spawning at dusk noted in pelagic egg producers. These range from lowered levels of predation on eggs or spawners at dusk, compared with other times of the day (Robertson and Hoffman, 1977; Johannes, 1978a; Lobel, 1978; Thresher, 1984), to feeding constraints on adults. Robertson (1983), for example, proposed that the feeding biology of adults may be an important factor in diurnal timing of spawning in certain surgeonfish species, such as *Acanthurus lineatus*, which face stronger competition for food than do other surgeonfishes. Changing environmental conditions such as tides and light levels may simply act as cues for adults to synchronize spawning (Colin and Clavijo, 1988), or provide cues for the initiation of the final stages of egg maturation (Hoffman and Grau, 1989) to enable many adults to spawn at the same time on a given day.

Spatial patterns

Reef fishery species are characterized by pelagic spawning, either within areas of residence (e.g. small groupers, some wrasses and parrotfishes, trunkfishes and angelfishes) or after migrating to specific spawning sites (Thresher, 1984; Colin and Clavijo, 1988; Shapiro *et al.*, 1993b). As a rule, it is the larger species that migrate, with smaller species tending to spawn in areas of residence. Presumably, larger fishes are capable of moving greater distances, or live more dispersed in non-spawning seasons and must migrate to assemble in numbers. In some species, spawning may include certain individuals that remain resident, while others migrate to spawn. Identifying the locations at which fishes spawn, especially in species that aggregate at specific sites, is of interest for identifying sources of eggs and larvae and for understanding the significance of spawning, versus non-spawning, areas. It is also of importance when considering the siting of marine reserves and the potential impact on reproduction of disruption to spawning areas, or behaviours, by activities such as fishing, dredging or anchoring.

Some migrations involve short distances (metres) and occur daily, with spawning in pairs or small groups and a return to home sites thereafter. This happens in some wrasses, parrotfishes and surgeonfishes (Robertson and Hoffman, 1977; Colin and Clavijo, 1978; Robertson, 1983; Warner, 1988). Other migrations involve movements of up to a few kilometres on a daily or lunar basis, to form large aggregations of up to thousands of individuals, with subsequent return to home sites, often along consistent and well-defined migration routes, as in certain surgeonfishes, mullets, goatfishes, fusiliers and rabbitfishes (Helfrich and Allen, 1975; Johannes, 1978a; Robertson, 1983; Bell and Colin, 1986; Fishelson *et al.*, 1987; Myrberg *et al.*, 1988). Finally, larger groupers and snappers migrate much farther to form aggregations of hundreds or thousands of fish which may last for days at specific locations and times of the year, often with a lunar periodicity (Johannes, 1978a; Colin *et al.*, 1987; Carter, 1989; Sadovy, 1993a; Domeier *et al.*, in press).

Groupers provide the most extreme examples known among demersal reef species in terms of distance migrated to a spawning site, numbers aggregating and long-term persistence of aggregations at specific sites. Nassau grouper are known to assemble repeatedly at certain sites for decades (Sadovy, in press). Tagged Nassau grouper have been recorded to move at least 110 km to an aggregation site (Colin, 1992). The gag grouper, *Mycteroperca microlepis*, migrates greater distances although it is not clear whether a spawning migration is involved (Van Sant *et al.*, 1994). Aggregations number from lows of 20 or so, to as many as tens of thousands of fish. Upper estimates of aggregation sizes range from many hundreds in rabbitfishes, 2000 to >10 000 in *Acanthurus nigrofuscus*, 6000–7000 in *A. coeruleus*, 16 000 in *Thalassoma bifasciatum*, 20 000 in *Acanthurus bahianus* and from several hundreds up to an estimated 30 000, or more, in *Epinephelus striatus* (Smith, 1972; Hasse *et al.*, 1977; Johannes, 1978a; Warner and Robertson, 1978; Robertson, 1983, 1985; Fishelson *et al.*, 1987; Myrberg *et al.*, 1988; Colin, 1992; C. Roberts, pers. comm.). Aggregation densities have rarely been measured but are known to vary from 44 to 76 fish per 1000 m^2 for groupers dispersed close to the substrate, to many hundreds or even thousands over the same area but when assembled in a three-dimensional array (Smith, 1972; Colin, 1992; Shapiro *et al.*, 1993b; Samoilys and Squire, 1994).

Aggregations of ripe emperors, bigeye (Priacanthidae), grunts, and of the largest parrotfish, *Bolbometopon muricatum*, and wrasse, *Cheilinus undulatus*, have been noted, although spawning was not observed in any but the parrotfish (Moe, 1966; Johannes, 1981; Gladstone, 1986; Colin and Clavijo, 1988; L. Squire, pers. comm.). In the emperor, *Lethrinus nebulosus*, large numbers of ripe fish are caught in specific locations in March and April off Japan, suggesting aggregation spawning (Ebisawa, 1990).

For migrating species, spawning sites are often, but by no means always, located in deeper water than residential sites. In Palau, 19 out of 20 species of fishes reported by fishers to leave their normal habitat to spawn, moved to deeper water. The spawning sites ranged from inshore reef slopes, the interface of shallow and deep waters and channels through fringing or barrier reefs, to reef crests, outer reef slopes and blue water (Johannes, 1978a). Many aggregation sites are located on the edges of insular platforms or in channels leading to the open sea, at depths ranging from 5 to 100 m and often focusing on prominent aspects of reef structure or the shelf edge (Smith, 1972; Johannes, 1978a, 1988a; Olsen and LaPlace, 1979; Robertson, 1983; Bell and Colin, 1986; Colin and Clavijo, 1978, 1988; Colin, 1992; Gilmore and Jones, 1992; Sadovy *et al.*, 1994a; Samoilys and Squire, 1994). On the other hand, some species spawn in areas of habitat change at sandy/live reef boundaries and are unlikely to make long daily migrations to a distant shelf edge from inshore (Randall and Randall, 1963; Colin and Clavijo, 1978; Thresher, 1984; Myrberg *et al.*, 1988). Rabbitfish aggregate at traditional sites varying from mangrove stands and shallow tidal reef flats to deeper reefs (George, 1972; Lam, 1974; Johannes, 1978a). While aggregations of red hind grouper may be located at the edge of the insular shelf at a number of locations, aggregating individuals are somewhat dispersed and may extend for several kilometres inshore away from the drop-off to deeper water (Sadovy *et al.*, 1994a). Similarly, although offshore spawning has been noted in some species of mullet, individuals may migrate to specific spawning sites in estuaries or in fresh water. In the reef-associated mullet, *Crenimugil crenilabis*, aggregations of 500–1500 fish occur in shallow water (Helfrich and Allen, 1975; Thresher, 1984). Among porgies, *Lagodon rhomboides* apparently migrates offshore (Caldwell, 1957), and *Acanthopagrus australis* moves out of estuarine areas to surf bars at the junction of estuary and ocean, to spawn (Pollock, 1984a).

Some spawning sites show long-term persistence up to many decades and may be used by various species of groupers and snappers at different times. At certain sites in the Bahamas, off Honduras and in Puerto Rico, several different species of grouper aggregate, albeit not contemporaneously (Fine, 1990; Colin, 1992; Sadovy *et al.*, 1994b). The red hind aggregates in some cases in the same area as the rock hind, *Epinephelus adscensionis* (Colin *et al.*, 1987; Shapiro *et al.*, 1993b). Beets and Friedlander (1992) report areas in the Virgin Islands used by several species, including the red hind grouper, the Nassau grouper, the yellowfin grouper, *Mycteroperca venenosa*, and the lane snapper, *Lutjanus synagris*. Carter (1989) noted sites in Belize used by Nassau grouper, black grouper, *Mycteroperca bonaci*, and dog snapper, *Lutjanus jocu*. About 50 spawning sites are known in the Caribbean for the Nassau grouper, at one-third of which aggregations no longer form, probably due to heavy fishing pressure (Sadovy, 1993a, in press; Jennings and Lock, Chapter 8). In general, however, information on the locations of spawning areas is limited for tropical fishery species. Few spawning areas have apparently been identified in the South China Sea or in tropical waters of northern Australia, although the Great Barrier Reef, and the northern and northwestern areas of the Gulf of Thailand are noted as spawning areas for groupers, bigeye, jacks (Carangidae), threadfin bream and drum (Sciaenidae) (Chullasorn and Martosubroto, 1986; Kailola *et al.*, 1993; Brown *et al.*, 1994).

Little is known about spawning site selection. There are, in fact, no data supporting the idea of aggregation sites as 'superior', other than in the very short term (i.e. hours), and no support for the idea that such areas are unique (e.g. Colin, 1992; Appeldoorn *et al.*, 1994; Hensley *et al.*, 1994). The spawning sites may be selected for their physical location, inasmuch as water movements at the sites are proposed to be beneficial for egg survival or dispersal (Johannes, 1978a; Barlow, 1981; Lobel, 1989). Their location may be important for enhancing fertilization rates (Petersen *et al.*, 1992), or they may simply be convenient and distinctive physical focal points for assembling adults (Colin and Clavijo, 1988; Shapiro *et al.*, 1988). Alternatively, sites may not currently possess intrinsic value for spawning but constitute areas used traditionally, or determined by some past geological event. It has been suggested, for example, that migration to shelf-edge areas in the Caribbean for spawning may reflect an adaptation to glacial period reef conditions rather than to those of the present day (Colin and Clavijo, 1988). In such cases the location of such sites may be communicated socially, as in birds and mammals (Galef, 1976; Helfman and Schultz, 1984; Warner, 1988, 1990) or access thereto may entail the use of established routes, as in many surgeonfishes (Myrberg *et al.*, 1988).

It is also of interest to ask why many reef species aggregate to spawn. Several reasons have been proposed. Aggregations could function to maximize egg density such that egg-predators are swamped, thereby increasing the chance of any one egg to survive (Johannes, 1978a; Thresher, 1984). They may increase fertilization rates by concentrating spawners (Pennington, 1985; Levitan, 1991; Petersen *et al.*, 1992). Aggregations may be important in bringing together large numbers of individuals for mate selection, especially in cases where individuals live highly dispersed during non-reproductive periods, thereby synchronizing reproductive activities (Colin and Clavijo, 1988; Shapiro *et al.*, 1993a). Evaluation of the advantages of aggregation spawning and the sites selected as well as the role of selective pressures on adults and their progeny, combined with historic factors and their interrelationships, would benefit greatly from further comparative and empirical data.

2.4 MATING PATTERNS AND REPRODUCTIVE STRATEGIES

Because reviews and descriptions of reproductive behaviour based on field observations appear in a number of publications (Thresher, 1984; Colin and Clavijo, 1988; Colin, 1992; Sadovy *et al.*, 1994a; Domeier *et al.*, in press), I shall briefly summarize the mating patterns that characterize the larger reef species. In particular, I consider aspects that merit special consideration in the study of exploited fish populations and that might influence reproductive output.

Mating patterns

Mating patterns of reef fishery species entail pair spawning or group spawning, or both (Thresher, 1984; Warner, 1991). Pair spawning characteristically involves a male and female which engage in an upward spawning rush prior to egg and sperm release in the case of pelagic spawners, or pairing and egg deposition in demersal spawners (Thresher, 1984). Pair spawning largely occurs in species that live in (or form temporarily while aggregated) single-male/multiple-female reproductive units within which a male spawns consecutively with a number of different females (e.g. angelfishes, some groupers, parrotfishes and wrasses) (Fig. 2.7(a)). Such groups typically comprise between two and seven females (Thresher, 1984). Reported less frequently is pair spawning that involves a single male and female, either within mobile aggregations, as in the parrotfish *Bolbometopon muricatum*, in aggregations of a rabbitfish (Thresher, 1984) and in a jack (P. Colin, pers. comm.), or involving monogamous pairs outside aggregations, as in some species of butterflyfishes.

Group spawning, on the other hand, is defined as a number of males simultaneously fertilizing the eggs of one female (Warner and Robertson, 1978; Thresher, 1984; Colin, 1992) (Fig. 2.7(b)). It is possible, however, that more than one female engages in group spawning, although this has yet to be confirmed (Colin and Bell, 1991).

Many species in a number of families (e.g. parrotfishes, wrasses, surgeonfishes, goatfishes) exhibit mating patterns whereby both group and pair spawning occur. This mixed strategy was first described by Randall and Randall (1963). The predominance of one or other strategy may be a facultative response to population density and distribution of resources (e.g. Choat and Robertson, 1975; Colin and Clavijo, 1978; Thresher, 1984; Warner, 1991). Colin (1992) observed that courtship and colour

---

Fig. 2.7 Mating patterns characteristic of tropical fishery species. (a) Male grouper *Epinephelus guttatus*, probably defending a temporary territory within a spawning aggregation prior to pair spawning with a female. The semi-upright position of the male has only been observed in an aggregation and there is little sexual dichromatism (photo by Scott Bannerot). (b) Four stages in a group spawn of the grouper, *E. striatus*, in a spawning aggregation, from assembling close to the substrate (1) to briefly forming a spawning group probably led by a female in dark colour phase followed by males in a bicolour phase (2), to spawning (3) and return to substratum (4). From Colin (1992); figure drawn by Bonny Bower-Dennis.

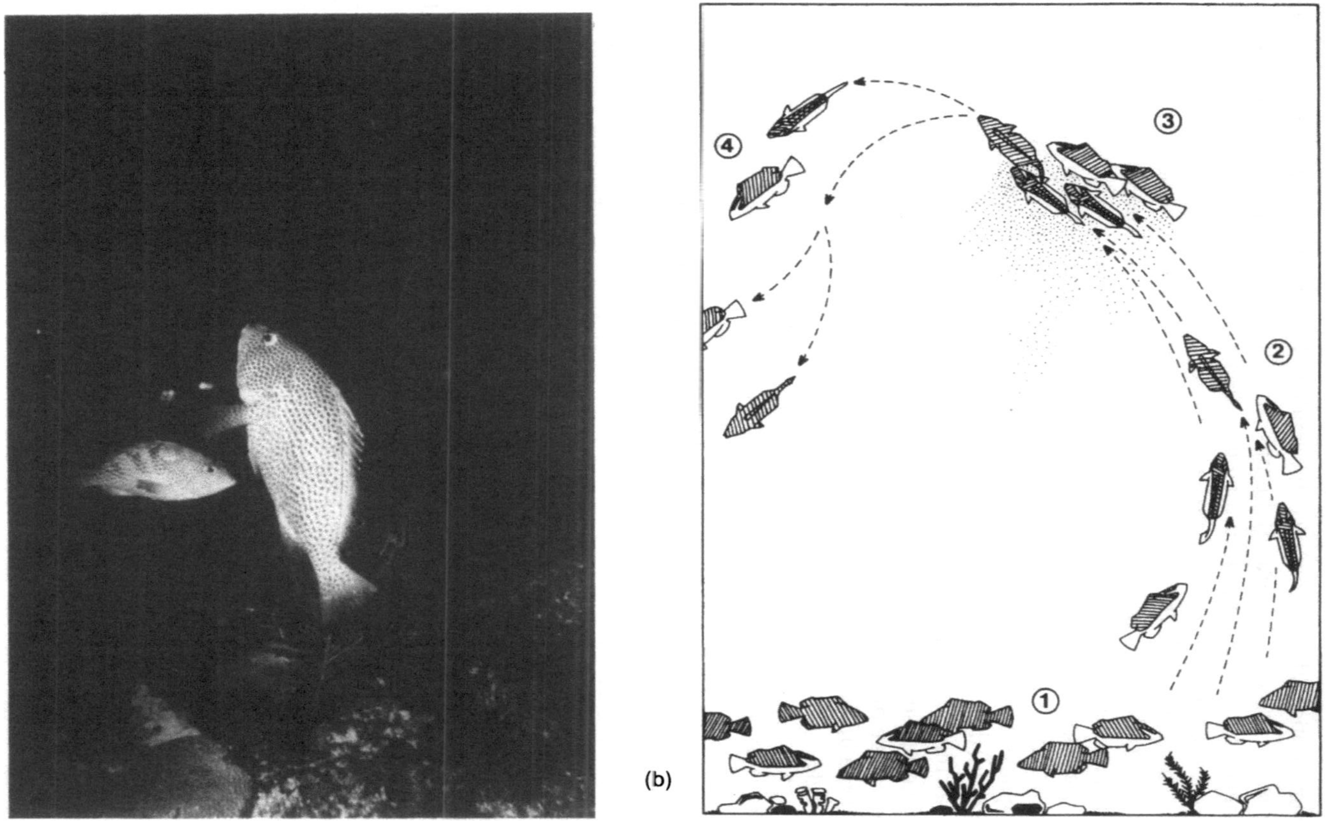

(a) (b)

were noticeably less intense in smaller, less dense, aggregations of the Nassau grouper in the Bahamas than in larger aggregations, which suggests that some minimum aggregation density, or fish abundance, may be necessary for successful group spawning to occur.

Choat and Robertson (1975) and Robertson and Warner (1978) noted for parrotfish species that spawned both in pairs and in small groups that the ripe testes of group-spawning males were relatively larger than those of pair spawners. They proposed that, because competition among group-spawning males to fertilize eggs was intense, these males developed testes capable of delivering more sperm than pair-spawners. Testis size was determined to be facultative because a change from group spawning to pair spawning was accompanied by a relative decrease in size. Similar patterns of group spawning with concomitant large testis size (Fig. 2.5(d)) and/or pair spawning associated with small testis size (Fig. 2.5(h); Sadovy *et al.*, 1994a,b), also occur in surgeonfishes and several groupers, temperate porgies, wrasses and some freshwater fishes (Warner and Robertson, 1978; Robertson *et al.*, 1979; Robertson, 1985; Billard, 1987; Buxton, 1990; Buxton and Clarke, 1992; Colin, 1992; Tucker *et al.*, 1993; Koenig *et al.*, in press; Sadovy and Colin, 1995). These patterns suggest that testis size, relative to body size (i.e. GSI) at the time of spawning, may provide a useful index of mating patterns in fishes.

Sexual patterns and sex ratios

Sexual patterns and adult sex ratios are far more diverse among tropical than temperate fishery species. Such attributes are especially significant to fisheries because of the potential disruption of reproduction brought about by fishing. Because stock assessment and management techniques, with the exception of marine fishery reserves (Plan Development Team, 1990), were developed with gonochoristic (separate sex) populations of equivalent numbers of males and females in mind, it is particularly important that populations that do not share these characteristics be carefully assessed. Effects on predictions of standard fisheries models, in such cases, need to be reviewed (Bannerot *et al.*, 1987).

I examined data on sex ratios in 13 families. Adult sex ratios were quite variable, with deviations from unity attributable to several likely causes. Among these were differential distributions, or movements, of males and females, variation between the sexes in mortality, sex change (see below) and size-selective trapping. Sex ratios of unity were noted for squirrelfishes (Holocentridae), jacks, snappers, most grunts, goatfishes, some parrotfishes, surgeonfishes and triggerfishes. Groupers, angelfishes, emperors, porgies, some wrasses and some parrotfishes tend to have female-biased sex ratios, largely due to a sexual pattern of protogynous hermaphroditism

(Truslé and Truslé, 1976; Loubens, 1980b; Young and Martin, 1982; Munro, 1983h; Sadovy and Shapiro, 1987; Shapiro, 1987a). However, deviations from unity were also noted in gonochoristic species. For example, in several species of snapper the proportion of females is greater among larger fish. In other snappers there is variation in the numbers of males and females throughout the course of the spawning season, with more females towards the end (e.g. Campos and Bashirullah, 1975; Reshetnikov and Claro, 1976; Grimes and Huntsman, 1979; Grimes, 1987; Everson *et al.*, 1989; Al-Ogaily *et al.*, 1992).

While it is important to evaluate sex ratios accurately, care is needed in interpreting much of the available data for four principal reasons. It is necessary to distinguish between adult and population (i.e. adult plus juvenile) sex ratios if we are interested in the ratio of reproductive males to reproductive females. Sex ratio can depend on location or time sampled. Aggregation sex ratios may differ from non-aggregation ratios, or according to time during the reproductive season (Hasse *et al.*, 1977; Sadovy and Shapiro, 1987; Al-Ghais, 1993; Shapiro *et al.*, 1994). Sex ratios in samples taken from aggregations may vary over the course of the aggregation if there is differential movement of males and females into the aggregation area (Johannes, 1988a; Bell *et al.*, 1992; Sadovy *et al.*, 1994a). Lastly, sample sizes should be large enough to minimize sampling errors.

While gonochorism characterizes temperate marine species, among reef fishery families almost 50% contain hermaphroditic species in which most individuals function as both sexes, either simultaneously or sequentially. Gonochorism typifies the surgeonfishes, rabbitfishes, grunts, snappers, goatfishes, jacks, mullets, trunkfishes and triggerfishes. Protogynous hermaphroditism has been noted in angelfishes, emperors, parrotfishes, wrasses, porgies and groupers. Protandrous hermaphroditism (male-to-female adult sex change) occurs in porgies, snooks (Centiopomidae) and flatheads (Platycephalidae) and both gonochorism and hermaphroditism occur in certain species of porgy, grouper, wrasse and parrotfish (Atz, 1964; Smith, 1975; Warner, 1978; Policansky, 1982; Nakazono and Kuwamura, 1987). Closely related species, however, do not necessarily exhibit the same sexual pattern; gonochores and hermaphrodites may occur within a single genus (Hoffman, 1983).

Particular problems may arise in the case of hermaphroditic populations that are heavily fished (Bannerot *et al.*, 1987; Jennings and Lock, Chapter 8). In protogynous species, for example, in which males tend to be larger than females on average, there are indications that size-selective fishing mortality may result in the differential loss of larger males. Buxton (1993) worked on two protogynous species of porgy, *Chrysoblephus laticeps* and *C. cristiceps*, that grow slowly and are long lived. He determined that, compared with a protected population in a reserve, *C. cristiceps* in exploited areas had increased growth rates and sex ratios became more female biased. In the gag grouper, *Mycteroperca microlepis*, the percentage of reproductive males taken from aggregations in the Gulf of Mexico dropped from 17% to 2% over about 10 years, suggesting the possibility that insufficient males may remain in the reproductive population to fertilize eggs from all females (Koenig *et al.*, in press). Other data are suggestive of a decrease in the proportion of males with increased exploitation (Bannerot *et al.*, 1987).

The factors that bring about sex change in larger reef species are little understood. One of the critical questions is to what extent change of sex is influenced by local demographic conditions, or depends on other factors such as absolute size or age (Bannerot *et al.*, 1987; Shapiro, 1989). An increase in female bias with increasing fishing effort is consistent with the idea that protogynous sex change occurs at a characteristic size. It is inconsistent with a mechanism based solely on behavioural induction of sex change because behavioural control might be expected to compensate for differential loss of males, thereby stabilizing the sex ratio, independent of any absolute size or age. For all species in which factors that induce sex change have been studied, however, either behaviour, sex ratio or relative size (or some combination) have been implicated, never absolute size or age (Ross, 1990). If these models also apply to commercial species, there are several possible explanations for the apparent inconsistency: removal of multiple males through fishing (i.e. a male mortality much greater than normal) could produce increasingly lengthy delays when multiple females must change sex in response (Shapiro, 1980); there may be a threshold sex ratio above which sex ratios fluctuate but sex change is not induced (Shapiro and Lubbock, 1980); sex change may take longer to initiate or complete in progressively smaller females (Lutnesky, 1994a), be density-dependent (Lutnesky, 1994b), or respond to both sex ratio and male size (Cowen, 1990). Clearly, a number of possible explanations may be applicable and controlled experiments are necessary to resolve which models apply to larger reef species.

2.5 REPRODUCTION IN FISHERIES SCIENCE

Importance and relevance

The reproductive biology of reef fishery species has received remarkably little serious attention compared with other aspects of their natural history. While reproduction in freshwater, estuarine, intertidal, temperate and smaller tropical marine fishes has been variously reviewed, information on the larger tropical reef species, with the exception of behaviour, is largely fragmented among a few notable surveys and syntheses (Breder and Rosen, 1966; Munro *et al.*, 1973; Erdman, 1976; Johannes, 1978a; Thresher, 1984; Grimes, 1987; Shapiro, 1987a; A.D. Munro *et al.*, 1990).

This lack of emphasis on the reproductive biology and ecology of larger tropical species has been further exacerbated by the difficulties of studying some of the larger fishes in terms of access to samples, restricted opportunities for direct observations, limited research funding and facilities in many tropical locations, as well as serious methodological difficulties of assessing spawning periodicity and fecundity in interdeterminate spawners. Moreover, fish ecologists and fishery biologists have much to learn from each other. For example, while changes in fecundity with body size are a critical part of assessing reproductive output for fisheries workers, ecologists have often regarded all female conspecifics as equivalent in this respect. Conversely, where ecologists are keenly aware that environmental and behavioural factors can strongly influence egg output, fisheries biologists have generally been little concerned with such details.

Exploited populations, if managed at all, have traditionally been managed to maximize growth, not reproductive potential. Increasingly recognized, however, is the importance of maintaining sufficient reproductive adults and, hence, egg production, for population maintenance. As fish populations become more depleted, understanding reproductive processes becomes ever more critical. Size-related parameters, such as fecundity or sexual maturation, are needed for determining appropriate capture sizes if juvenile mortality is to be minimized, or for estimating the reproductive potential of different-sized females for stock assessments. Different management approaches may become evident as our knowledge grows (Sadovy, 1994). Given the exponential increase in fecundity with body size in many species, it may, for example, be advisable to protect a small number of larger, rather than a large number of smaller, females (all else being equal). Knowledge of the time and place of spawning may be crucial for protecting spawners, to ensure that reproductive behaviour is undisturbed, or for decisions to do with the placing and size of marine reserves. A better understanding will also enable us to make more realistic inter- and intraspecific comparisons in annual or lifetime fecundity with which to develop hypotheses concerned with interrelationships between the temporal, spatial and quantitative aspects of egg production and of the resulting numbers of juveniles. We may even be able to improve estimates of the spawner side of the spawner-recruit equation.

We now appreciate that hermaphroditism and aggregation spawning characterize many exploited species and pose special problems for management. We need to understand whether hermaphroditic species are more or less vulnerable to heavy fishing pressure than gonochoristic ones. Given the prevalence of aggregation spawning, researchers must clarify the role of aggregations in annual reproductive output and the potential effects of fishing on large groups of fish congregating for the sole purpose of spawning.

Future research directions

A critical area that demands attention concerns better evaluation of reproductive output – the quantification of inter- and intraspecific, annual and individual, variations in egg production. This requires not only improvements in methods for assessing reproductive output, but also a greater appreciation of other factors that determine patterns of egg production. For example, how is egg output constrained by body form? To what extent are temporal and spatial patterns of spawning a response to selection for maximizing egg output under conditions believed to involve high levels of egg and larval mortalities? Could we predict that larger fish spawn less frequently, or over a shorter period each year, than smaller ones because they can produce more eggs each time they spawn? Given the extremely limited period of annual spawning in some species, might these be more vulnerable to heavy fishing pressure at critical times, or more subject to recruitment failures, than those with longer spawning seasons? The sooner we debunk the myth that protracted spawning characterizes tropical fishes, the faster we can use the wide interspecific variation in reproduction that we observe to develop and test hypotheses of reproductive patterns and strategies.

Many of the larger pelagic species aggregate to spawn or exhibit hermaphroditic sexual patterns. We must understand the effects of reduced numbers, or density, of adults on reproductive activity and the possible impact of the differential removal of one or other sex through fishing. We might predict, for example, that once males fall below some threshold number, either absolutely or relative to females, insufficient remain to fertilize females' eggs. If aggregation densities or fish numbers assembling fall too low, is there a reduction in reproductive activity or in the stimulus necessary for females to ovulate? How much of the total annual reproduction occurs in aggregations and does every mature female spawn each year? To find out, more effort must be directed towards non-aggregation areas during the reproductive season. For sequentially hermaphroditic species what, if any, is the role of the aggregation in the mediation of sex change? What are the responses of hermaphroditic populations to exploitation and is it valid to apply the standard stock assessment and management approaches to such species? The answers to such questions should enhance our ability to monitor and manage tropical populations and may ultimately benefit allied areas such as mariculture, which will draw heavily on a better appreciation of the factors that influence reproductive output.

Finally, care is clearly needed to address certain shortcomings in the reporting and methods of reproductive research. Many studies would have been far more valuable for the present review, for example, had they included simple size conversions (e.g. length to weight), or the size of the largest fish in the sample. Often not specified were the type of sexual maturation assessed (i.e. minimum or 50%), what was intended by the terms 'mature' or 'maturing' or what egg development stage (hydrated, vitellogenic, etc.) was used to assess fecundity, or even the kind of fecundity (batch, total, etc.) being evaluated. Sex ratio data often involved low sample sizes, or did not specify whether only adults, or both adults and juveniles, were included, making sex ratio comparisons difficult or impossible. Various methods such as corrections to the GSI or post-ovulatory follicle ageing would enhance our ability to characterize spawning frequency, fecundity, the timing of spawning and to distinguish indeterminate from determinate spawners.

ACKNOWLEDGEMENTS

I am most grateful to the following for material and assistance: R.S. Appeldoorn, F.C. Coleman, P.L. Colin, G. Garcia-Moliner, C.C. Koenig, G.R. Mitcheson, M.A. Samoilys, D.Y. Shapiro and the Caribbean Fishery Management Council.

Chapter three

Larval dispersal and survival in tropical reef fishes

George W. Boehlert

SUMMARY

Almost all reef fishery organisms have a pelagic larval dispersal phase, ranging from days to months depending on species. Dispersal is often argued to be an adaptation to avoid intense predation on the reef, but has also been claimed to promote persistence in unpredictable environments. Spatial and temporal patterns of spawning on reefs have long been interpreted as adaptations to maximize advection of eggs and larvae away from the reef but evidence for this has been criticized. Spawning site choice can be influenced by many other factors, including mate choice and predation risk for spawners. Nevertheless, spawning sites are often located on down-current sides of reefs, promoting transport off the reef, if not out of the reef system as a whole.

Studies of the pelagic phase have been hampered by the severe logistic difficulties of sampling, including very low larval densities and the need for multiple gear types to sample adequately. Recent progress has been made and several distinctive larval assemblages have been identified with increasing distance offshore: a lagoon and embayment assemblage, near-shore (<0.5 km), neritic (3–5 km), and oceanic (>5 km). Assemblages are dominated by taxa with demersal eggs close inshore, but include

Reef Fisheries. Edited by Nicholas V.C. Polunin and Callum M. Roberts.
Published in 1996 by Chapman & Hall, London. ISBN 0 412 60110 9.

increasing proportions of species with pelagic eggs moving offshore. Offshore assemblages also include oceanic species. Reef fishery species are typically most common in neritic and oceanic assemblages, but most are relatively rare in larval surveys. A broad range of oceanographic features may facilitate larval retention near reefs, often in concert with larval behaviour. The role of larval behaviour in distribution has been much debated but is evidently important in determining onshore–offshore distribution patterns. Mortality rates have been measured for the pelagic larvae of very few tropical species but values (around 25–40% mortality per day) are of the same order as for temperate species.

The degree to which populations on reefs are interconnected is dependent upon species-specific dispersal capabilities, hydrography, distances separating reefs, and availability of suitable habitat. Intriguingly, there does not seem to be a clear correspondence between pelagic larval duration and geographic distribution. Long-distance dispersal is important in preventing genetic differentiation of species but most replenishment of populations almost certainly comes from smaller-scale dispersal over distances up to hundreds of kilometres. Reefs may fall into categories of sources or sinks (areas of consistently high settlement) for larvae. The identification of larval sources and sinks is important for management strategies and marine fishery reserve placement, and constitutes a major research challenge.

3.1 INTRODUCTION

Reef fisheries exploit a wide range of tropical and subtropical species which vary in their life history patterns. Virtually all species have a pelagic larval or juvenile stage, and the larvae may be found at varying distances from the shore. For demersal reef species, these stages serve as the principal means of dispersal and are critical to the maintenance of local and regional populations. Fluctuations in populations of tropical reef animals may be controlled by variable mortality in the planktonic stage, resulting in variable, and often episodic, levels of return of settlement-stage animals back to the reef (Doherty and Williams, 1988; Robertson *et al.*, 1993; Doherty and Fowler, 1994a). Variable survival in post-settlement animals also contributes to fluctuations in adult numbers (Shulman and Ogden, 1987; Roberts, Chapter 4), but this process occurs at smaller local scales and these may have less impact on fisheries (Doherty, 1991).

Our knowledge of the ecology and dynamics of the larval stages of tropical reef fishery species is meagre relative to that in temperate and subpolar regions. Present understanding of the temporal dynamics from spawning (Sadovy, Chapter 2) to settlement (Roberts, Chapter 4) in reef fishes has been advanced by otolith research (Victor, 1991). The spatial dynamics, however, involving dispersal distances, source populations, and physical mechanisms, remains unclear. In this chapter, I describe processes affecting eggs, larvae, and juvenile stages prior to settlement to the reef habitat. I refer to this latter event as 'settlement', as opposed to 'recruitment', which in fisheries science denotes addition to a fished population.

Relatively little is known about dynamics of settlement in most commercially important species for several reasons. One factor is the difficulty of identifying eggs and larvae, although recent work (e.g. Leis and Rennis, 1983; Leis and Trnski, 1989) has improved the taxonomy of these stages. Other factors contributing to the lack of information are the relative scarcity of larvae of important species, and the dearth of adequate planktonic (and, for later stages, nektonic) sampling in tropical waters. I will thus draw examples from several non-reef species to illustrate the critical processes and then suggest research needed on this topic for species important in tropical reef fisheries.

3.2 EGG AND LARVAL PRODUCTION FROM REEFS

The fate of egg and larval production varies from local retention to long-distance dispersal, but there is uncertainty about which predominates. Species with pelagic eggs were often considered to be adapted for dispersal; this notion was supported in part by observed spawning of many species near reef passes or elsewhere where eggs might be dispersed offshore. By contrast, species with demersal eggs, or parental care, were thought to be adapted to retain local populations. This dichotomy is not entirely supported by distributional data (Leis, 1991a, 1994). A focus on long distance or 'teleplanic' larvae (Scheltema, 1986) helped form the idea that the role of pelagic eggs in reef species was for dispersal. Selection for such dispersal may be associated with minimizing the effect of local extinctions under uncertain environmental conditions (Barlow, 1981).

Johannes (1978a) suggested an alternative view, that the function of pelagic eggs and larvae was to minimize exposure of vulnerable, very small stages to high levels of predation on the reef. The premise was that advective mortality would be lower than that from predation if these stages had remained on the reef. He proposed that spawning during seasons with the lowest winds and currents, or when eddies (Sale, 1970) or other retention mechanisms were present, would serve to reduce advective loss. Since that time, retention mechanisms have been discussed in detail (Lobel and Robinson, 1986; Bakun, 1988; Boehlert *et al.*, 1992; Boehlert and Mundy, 1993; Cowen and Castro, 1994).

The short-term fate of eggs is influenced by spatial and temporal patterns of adult spawning (Hensley *et al.*, 1994). Demersal eggs are frequently larger and subject to different sources of predation from planktonic eggs. There are also adaptations for survival, including parental care (mouthbrooding in cardinalfishes, Apogonidae; nest care in damselfishes, Pomacentridae; brood pouches in pipefishes, Syngnathidae; Sadovy, Chapter 2), and tough or toxic eggs (Gladstone, 1987). The fate of larvae hatched from demersal eggs is partially dependent upon the size of the larva at hatching. Very large larvae may remain on or near the reef (Kingsford and Choat, 1989), whereas many other species are advected away (see below) and lead a pelagic existence prior to settlement. Robertson (1990) examined whether temporal spawning patterns of species with demersal eggs (eight damselfishes and one blenny, Blenniidae) are related to larval or adult biology, and suggested that temporal spawning patterns may not be related to periods of highest larval survival.

The fate of pelagic eggs is also dependent on the spatial and temporal distribution of spawning relative to the physical environment. Temporal spawning patterns may be adapted to tidal, diurnal, lunar or seasonal periodicities. The shorter time scales may be adaptations for offshore advection (tidal and lunar scale, Johannes, 1978a) and for reduction of predation on eggs (nocturnal spawning, Hobson and Chess, 1978). Seasonal variability in the production cycle of larval prey is less pronounced in tropical than temperate areas, but large-scale oceanographic features such as eddies or current patterns do show seasonal patterns. These features may either advect or facilitate retention of larvae, and thus have been proposed as selective agents in the evolution of spawning periodicity (Johannes, 1978a; Lobel, 1989).

The proposition that spatial patterns of spawning promote advection of eggs from the reef has recently been subjected to scrutiny. Shapiro *et al.* (1988) suggested that spawning location may not be an adaptation for egg dispersal. Specific studies with tracers and on egg distribution have shown that at small spatial scales, spawning location may be conducive to transport of eggs off patch reefs (Hensley *et al.*, 1994). On broader spatial and temporal scales, however, selection of spawning sites and times in wrasses does not appear to be adapted to transport off the broader reef platform (Appeldoorn *et al.*, 1994). The location of spawning aggregations on reefs suggests that they are not always closely related to hydrography at particular reef heads or patch reefs, but rather may represent cultural transmission of traditional spawning sites. Sites selected appear to facilitate mate choice while reducing predation hazard to spawners (Warner, 1988), or be better suited to enhance fertilization success (Petersen *et al.*, 1992).

Differences in egg sizes have implications for life in the pelagic environment, including incubation time, larval size at birth, and mortality (Ware, 1975). These traits are thought typically to be related to the nature of the environment in which the larvae must survive. Winemiller and Rose (1993) ran a simulation model with size-based rules for larvae and examined the outcome in terms of survival. In prey-poor environments, a small number of larger offspring was advantageous, whereas more small larvae were favoured in a prey-rich environment. They also suggested that large-scale patchiness of prey resources favours more small eggs.

Differences in egg and larval size vary on large, inter-oceanic scales within groups; Thresher and Brothers (1985) relate these differences to oceanic productivity as well as predator type and abundance. In the angelfishes (Pomacanthidae), species with larger maximum adult size tend to produce larger planktonic eggs, and egg size is inversely related to the pelagic larval duration (Thresher and Brothers, 1985). The maximum egg size, however, is less than 1 mm diameter; eggs of this size will hatch in less than a day given the high temperatures in the tropics (Pauly and Pullin, 1988). This contrasts with the larger eggs typical in higher latitudes which, in keeping with the lower temperatures, take from days to weeks before they hatch. Pauly and Pullin (1988) suggested that temperature has nearly a five-fold greater effect on incubation time than does egg size.

In the tropics, demersal eggs are typically larger than planktonic eggs and take longer to hatch (Thresher and Brothers, 1989). Although the association of recruitment variability with egg size and larval type is unknown in the tropics, Pepin and Myers (1991) found no correlation across 21 species in temperate waters. They did note a positive correlation between increase in larval length over the pelagic period and recruitment variability. A possible explanation for this is that longer larval durations result in more variable mortality.

Pelagic larval durations in tropical reef fishes have been studied in great detail since the early 1980s due to technical advances in understanding otolith daily increments. In an early survey of 43 taxa in 12 families, Brothers *et al.* (1983) provided estimates ranging from 2 to 12 weeks in the plankton. Subsequent studies provided more exhaustive estimates for 100 species in each of two families, the wrasses (Labridae; Victor, 1986b) and damselfishes (Wellington and Victor, 1989). The wrasses had longer larval durations (15–121 days) and higher variability within species as compared with the shorter (12–39 days) periods and low variability within species of damselfishes, for which size at settlement was correlated with larval duration (Wellington and Victor, 1989). For wrasses, delayed metamorphosis (Victor, 1986a) may contribute to greater variability in larval duration within species.

3.3 LARVAL DISTRIBUTION AND ABUNDANCE

Limitations of sampling

The low densities of reef fish larvae make the study of ocean dynamics of the pelagic stage problematical. Sampling problems alone are often such that some 10^5–10^6 m^3 of water must be filtered to collect numbers that are statistically adequate (Clarke, 1991). Differing larval habitats, behaviour, and changes in swimming ability during ontogeny (and thus net avoidance), also dictate that more than a single gear type be used (Leis, 1991a). Different gear types selectively capture different taxa and sizes within taxa (Clarke, 1983; Choat *et al.*, 1993). Additional problems are encountered with scale and technology of sampling. Only recently have tropical plankton studies graduated from ring net methodology. They now use samplers such as Tucker trawls (Boehlert *et al.*, 1992), MOCNESS nets (Boehlert and Mundy, 1994; Cowen and Castro, 1994; Fig. 3.1), and drop nets near the reef (Kobayashi, 1989) for small larvae, and light traps (Doherty, 1987; Smith *et al.*, 1987) for large near-settlement-stage larvae and juveniles. These new gears reduce (but cannot eliminate) net avoidance and sample older life stages, often with concurrent environmental information.

Distribution and larval assemblages

In the vicinity of reefs, larval distributional patterns differ among species or species groups. Distinct ichthyoplankton assemblages populate different scales of distance from an island or reef, and these have recently been reviewed by Boehlert and Mundy (1993) for seamounts and islands, and Leis (1993) for tropical reefs. The assemblage closest to the reef or island, and least subject to advective loss, is that in embayments and lagoons, where densities of shore fish larvae can be high (Leis, 1994). Although he captured 200 taxa of shore fish larvae in lagoons and nearshore waters, Leis (1994) found only 33 taxa from 15 families that complete their life cycle within the lagoon. Of these, 30 had demersal eggs (defined here to include brooded or otherwise non-pelagic eggs). In the offshore direction, 'nearshore assemblages' may extend from the reef to perhaps 0.5 km offshore and are dominated by larvae of small species, which often have demersal eggs, such as the triplefins (Tripterygiidae) and gobies (Gobiidae). Because of the proximity to difficult-to-sample reefs and near-bottom areas, this assemblage may be the most poorly characterized. At approximately 3–5 km offshore, 'neritic' assemblages include inshore species with both demersal and planktonic eggs, although larvae of oceanic species are more frequently encountered. Farther offshore 'oceanic' assemblages are not well defined but include several taxa of reef species, most frequently those derived from planktonic eggs (Leis, 1993).

Species important in tropical reef fisheries, such as snappers (Lutjanidae), groupers (Serranidae), grunts (Haemulidae), emperors (Lethrinidae), parrotfishes (Scaridae) and surgeonfishes (Acanthuridae), have pelagic eggs and are typically in greatest abundance in the neritic to oceanic assemblages (Boehlert and Mundy, 1993; Leis, 1993). A comparison of the relative abundance of shore fish larvae around Hawaii suggests that those important in fisheries are, with the exception of the jacks (Carangidae), relatively rare in larval surveys (Table 3.1). The same pattern holds for some of the conspicuous reef species such as butterflyfishes (Chaetodontidae) and angelfishes. Despite equal sampling effort with distance from shore (2–28 km, Boehlert and Mundy, unpublished data), small, cryptic species with demersal eggs, such as gobies, tend to dominate the collections (Table 3.1), often because of very high densities in nearshore waters. This dominance is likely to be related to the greater spatial area over which eggs and larvae of those taxa occurring offshore are dispersed.

Leis (1991a, 1993) has confirmed the general trend of declining percentages (of total shore fish larvae) of reef fish larvae from demersal eggs with increasing distance from shore. Results obtained from disparate sampling approaches indicate that larvae from pelagic eggs constitute a greater percentage of total shore fish larvae as distance offshore increases around Oahu, Hawaii (Fig. 3.2). That this may be a common phenomenon in fishes is suggested by the example of demersal fishes off Nova Scotia, where Suthers and Frank (1991) noted that an assemblage of larvae from pelagic eggs was more widely and uniformly distributed than that hatched from demersal eggs. The latter assemblage was more confined to shallow waters and more patchily distributed.

The differences may be related to the longer period of passive transport by pelagic eggs. Pelagic eggs and small yolk-sac larvae may be more subject to advection than the relatively well-developed larvae from demersal eggs. This is supported by the angelfishes, where species with smaller eggs have longer pelagic larval durations (Thresher and Brothers, 1985). Small larvae of yellowfin tuna are spawned in high numbers near islands but do not show behaviour that might promote retention (Boehlert and Mundy, 1994). Survival in this oceanic species, however, is not dependent upon remaining in, or returning to, the nearshore environment.

Is the distribution of reef fish larvae a function of spawning location, advection and dilution resulting in higher densities near shore? Leis (1982) identified four assemblages of larvae within 3 km of the coast of Oahu in the Hawaiian Islands, and suggested that maintenance of the three inshore assemblages indeed required behavioural orientation by the larvae.

Fig. 3.1 Examples of sampling gear used in ichthyoplankton studies. (a) MOCNESS net (multiple opening/closing net and environmental sensing system) (Wiebe *et al.*, 1985). This gear has multiple nets in the Tucker trawl design along with environmental sensors for depth, environmental data, and net performance. Real-time data collection allows the net to be fished more accurately than less sophisticated gear types. (b) Manta neuston net (Brown and Cheng, 1981), typically towed at night to capture animals in the upper tens of cm.

Behaviour is evidently important in maintaining onshore–offshore distributional patterns. Leis (1991a) suggested that four out of six factors important in determining larval distribution involved behaviour, and that larval behaviour, particularly that involving swimming, is dependent upon size and ontogenetic development.

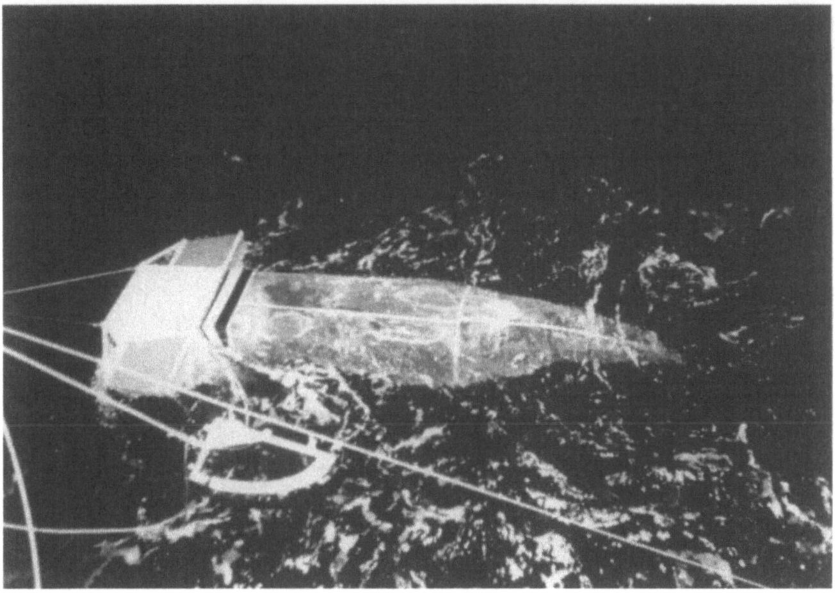

The role of larval behaviour in distribution can only be inferred. Vertical distribution has obvious consequences for advection, but has been poorly studied in the tropics with the exception of relatively shallow waters (Watson, 1974; Leis, 1986, 1991b). In the few studies farther offshore, reef fish larvae have been found in deeper water. Off Johnston Island, the greatest densities of reef fish larvae were between 50 and 100 m deep; densities in the 100–200 m stratum were approximately the same as in the 0–25 and 25–50 m strata (Boehlert *et al.*, 1992). The percentage of early (yolk-sac and preflexion) reef fish larvae tended to be higher in shallower strata. Around the Hawaiian island of Oahu, the vertical distribution of larval shore fishes in offshore waters is similar to that around Johnston Island (Fig. 3.3). In a station nearer shore, shore fish larvae from demersal eggs were in highest abundance at greater depths than those from pelagic eggs (Fig. 3.3(a)), whereas farther offshore they were not in markedly deeper water (Fig. 3.3(b)). These data combine several families, however, and an assessment of the role of behaviour in transport will require taxon-specific analysis.

Physical mechanisms of retention and modelling of larval distributions

Several physical mechanisms may serve to maintain planktonic stages in the vicinity of reefs for eventual settlement back to the local population (Boehlert and Mundy, 1993). The mechanisms include entrainment in

Table 3.1 Comparison of the importance of shore fishes in the fishery with larval abundance in Hawaii. Fishery data are from throughout Hawaii and represent commercial catch data for these families from 1980 to 1990 (Smith, 1993). Larval data are based upon over 90 000 shore fish larvae taken around the Hawaiian island of Oahu with samples taken on windward and leeward sides of the island at distances from 2 to 28 km offshore in four seasons of the year (Boehlert and Mundy, unpublished data)

	% in fishery	% of shore fish larvae
Fished species		
Carangidae	81.74	2.76
Lutjanidae	7.81	0.52
Mullidae	3.78	0.55
Holocentridae	2.45	0.51
Scaridae	2.98	0.37
Acanthuridae	1.25	0.54
Unfished species		
Gobiidae	–	73.69
Schindleriidae	–	6.27
Bleniidae	–	1.08
Apogonidae	–	1.00
Labridae	–	0.87
Callionymidae	–	0.66
Pomacentridae	–	0.49
Chaetodontidae	–	0.06
Pomacanthidae	–	0.05

boundary layers, small-scale frontal dynamics (Govoni, 1993), tidally-induced fronts and currents (Kingsford *et al.*, 1991), topographically produced eddies (Lobel and Robinson, 1986; Lee *et al.*, 1992), seasonally reduced or variable currents (Johannes, 1978a), regions of no or returning flow (Boehlert *et al.*, 1992; Graham *et al.*, 1992) and topographically steered flow (Cowen and Castro, 1994). Spawning periodicity, spawning location, and larval size, morphology, and related behaviour (e.g. vertical distribution, orientation to near-bottom features or boundary layers) may all influence the effect of these mechanisms, but most work to date has been speculative. Several examples of distributional data support different physical mechanisms of retention. Concentrations of shore fish larvae around certain areas of islands (Boehlert *et al.*, 1992; Cowen and Castro, 1994) are one such example, and the difference in relative abundance of larvae from demersal eggs on leeward versus windward sides of an island (Fig. 3.2) represents another.

Modelling studies have contributed to our understanding of advection and retention of larvae. Black and Moran (1991) modelled the drift of

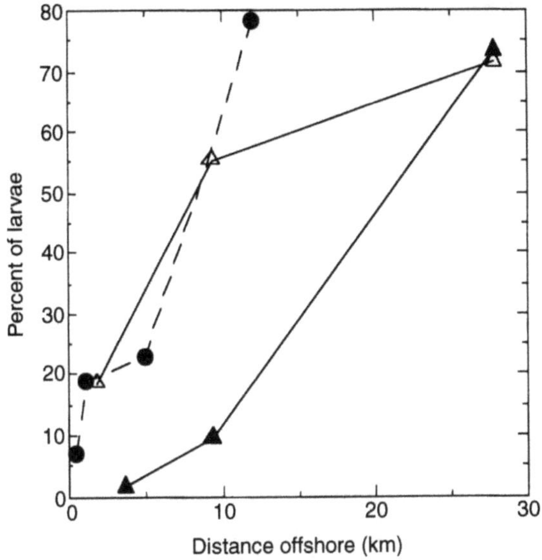

Fig. 3.2 Percentage of shore fish larvae from pelagic eggs as a function of distance offshore, Oahu, Hawaii. Data from Leis and Miller (1976; filled circles) are from various locations around the island. Data represented by triangles were taken at specific stations on windward (filled triangles) and leeward (open triangles) sides of the island in September 1985 (Boehlert and Mundy, unpublished). Larvae from demersal eggs are represented in the latter study by four of the most abundant shore fish families (Gobiidae, Schindleriidae, Pomacentridae, Apogonidae) and from pelagic eggs by nine of the most abundant families (Carangidae, Monacanthidae, Synodontidae, Labridae, Scaridae, Callionymidae, Acanthuridae, Bothidae, Tetraodontidae).

passive, neutrally buoyant particles in a complex circulation around reefs and showed specific areas where hydrodynamics alone may concentrate larvae and possibly allow self seeding (Black *et al.*, 1991). The general patterns observed in these modelling studies were similar to results of field studies on the short-lived larvae of corals (Sammarco and Andrews, 1989), where most settlement was local and occurred in areas with low flushing and high retention times. Application of models to fish larvae, with longer larval durations, presence in deeper, vertically structured water columns, and more complex larval behaviour, require greater complexity. Models with three-dimensional physics incorporating larval behaviour may demonstrate decreased advection from preferred habitats (Myers and Drinkwater, 1989; Werner *et al.*, 1993). More recent model applications have been used to track patches in the sea (Bartsch, 1993), or to examine advection in multi-year time series relative to recruitment (Berntsen *et al.*, 1994).

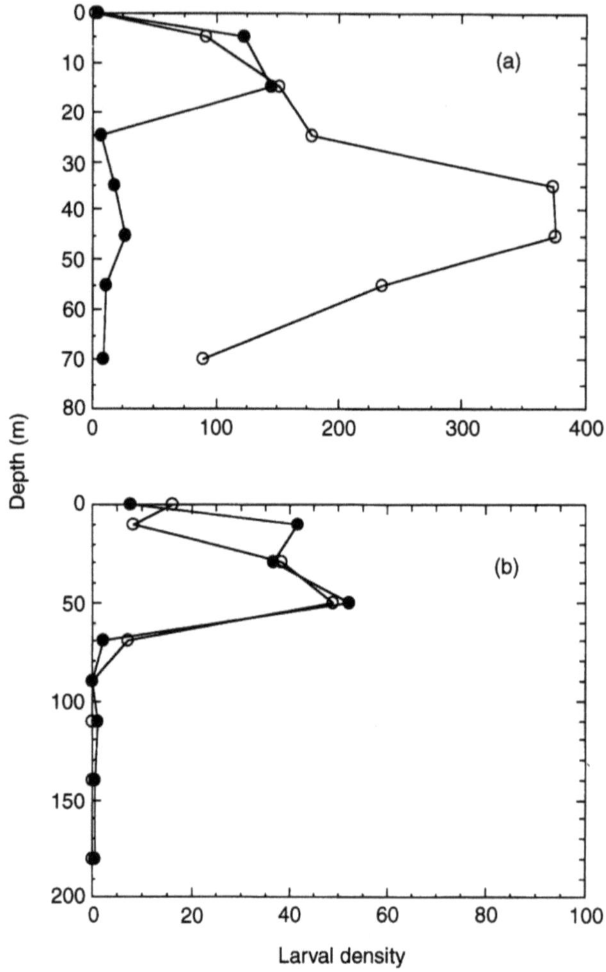

Fig. 3.3 Vertical distribution of shore fish larvae off leeward Oahu, Hawaii, sampled at distances of (a) 1.9 km and (b) 9.3 km offshore in September, 1985. Larvae from pelagic eggs (filled circles) and demersal eggs (open circles) are from families as defined in Fig. 3.2; larval density is larvae per thousand cubic metres. Data from Boehlert and Mundy (unpublished).

Undersampled environments

Our understanding of larval distributional features may be in error, partly due to the inability to sample environments of great importance to pre-settlement-stage larvae. Larval sampling in reef waters typically occurs either in very nearshore waters from small boats or in water sufficiently deep that larger vessels avoid going aground. There is in addition a real paucity of near-bottom sampling, in part due to bottom rugosity. Late larvae and juveniles have been observed in large numbers near sandy bottoms, coral heads and other features (Leis *et al.*, 1989), by drop-net sampling near patch reefs (Kobayashi, 1989), and by diver-steered nets (Brogan, 1994). In temperate habitats, near-bottom ichthyoplankton sampling has shown very high densities of selected taxa (Jahn and Lavenberg, 1986). Kobayashi (1989) observed spatial variations in the abundance of fish larvae near patch reefs within Kaneohe Bay, which were at spatial scales that most horizontal tows would not detect; moreover, the patterns were species specific.

Improved sampling in near-reef habitats will further our understanding of larval behaviours conducive to settlement. In an important study using new sampling tools (aggregation devices with plankton purse seines and light traps), McCormick and Milicich (1993) were able to provide new insights into the pelagic ecology of goatfishes (Mullidae). Late-stage animals occurred in pulses, leading the authors to propose the existence of schools with pelagic animals of several species covering broad size ranges. Based upon the sizes present, they proposed that an individual goatfish might spend as much as 70% of its pelagic period within these schools. The schools were proposed to be dynamic, with solitary pelagic individuals joining the school and larger animals departing as they settle onto the reef (McCormick and Milicich, 1993).

More evidence supporting high larval abundance in little-sampled habitats is provided by the episodic nature of settlement of some species to reefs. Sampling conducted in the surf zone of the reef face (Dufour and Galzin, 1993) and in reef passes (Shenker *et al.*, 1993) shows strong pulses of settlement, often with lunar periodicity. In some cases, settlement pulses may arise from the arrival of 'patches' of larvae advected near a reef (Victor, 1984; Williams and English, 1992) or perhaps from a school of the kind described by McCormick and Milicich (1993). Observations of large numbers of larvae that could support temporally consistent settlement events of single species are lacking; this may indicate that sampling has been insufficient. In fact, physical and/or behavioural mechanisms may exist to allow pooling of competent larvae in habitats intermediate between the pelagic and reef settlement habitat. Dufour (1991), for example, noticed cycles in the abundance of settlers that did not relate to larval supply sampled using plankton nets. In a shallower system, however, Milicich *et al.* (1992) found that abundance estimated by light trapping was a good predictor of settlement.

Even in later juvenile stages, the habitats of many reef species are poorly known. A good example is the Hawaiian pink snapper, *Pristipomoides filamentosus*, which until recently was not known near shore at sizes below 18 cm (F.A. Parrish, 1989). The deep-water, featureless bottom habitat occupied by young of this species is now recognized, but the dynamics of settlement and subsequent recruitment to the later-stage habitats are not understood. New types of sampling gear will be required for late larval stages and results may change our thinking about the spatial extent of dispersal and the temporal duration of the pelagic stage in reef populations.

3.4 OCEAN DYNAMICS OF THE PELAGIC STAGE

Low densities of reef fish larvae in offshore waters have generally precluded direct study of larval feeding, predation, and mortality rates. Most pertinent research in tropical waters has been conducted in estuarine areas or embayments where larval abundance is high (Houde and Lovdal, 1984; Leak and Houde, 1987). Consequently, understanding of larval ecology in offshore waters has largely been based upon inference from distributional analysis (Leis, 1991a) or from characteristics of animals surviving to settlement (Shapiro, 1987b). As will be described below, techniques have been developed to evaluate the characteristics of survivors. The most powerful tool for this approach is through the use of otoliths in larvae and settlement-stage animals (Methot, 1984; Robertson *et al.*, 1993). These techniques have improved rapidly over the past two decades.

Starvation and predation

Feeding by, and predation on, temperate and boreal fish larvae, and their effects on survival and fish year-class strength, have received a great deal of attention (Hunter, 1981; Bailey and Houde, 1989). The role of ocean dynamics in prey production and availability (through concentration) has been shown to be important in survival (Lasker, 1975). Few studies of feeding by reef fish larvae in tropical regions are available, except those in lagoon (Schmitt, 1986) or estuarine systems (Houde and Lovdal, 1984). Where examined, larvae typically fed selectively on appropriately sized zooplankton and taxonomic composition of prey was not remarkably different from that in temperate waters. Of note, however, is the high prey concentration. Using fine-mesh plankton nets, Houde and Lovdal (1984) noted mean prey (zooplankton smaller than 100 μm) densities of 273 per litre (105 per litre excluding tintinnids). These concentrations are much greater than those typical of coastal or open ocean waters (Hunter, 1981). Moreover, the prey abundance was relatively uniform (Houde and Lovdal, 1985). Such high prey levels may help explain the high densities of larvae seen in these systems, including atoll lagoons (Leis, 1994).

Although planktonic prey resources for larval fishes in waters away from reefs are lower than in waters above the reef or within lagoons (Johannes, 1978a), the high productivity of reefs (Polunin, Chapter 5) could potentially support high abundances immediately offshore. Primary productivity around islands is often higher than in the surrounding open ocean (the 'island mass effect' of Doty and Oguri, 1956), particularly around high islands with freshwater run-off (Dandonneau and Charpy, 1985). Waters near islands in tropical and subtropical areas may similarly be characterized by increased zooplankton abundance, with many species abundant only in insular waters (Jones, 1962; Hernandez-Leon, 1991). Theilacker (1986) noted lower percentages of starving larvae in collections taken near islands in temperate waters off California than in those taken further offshore. This is consistent with the localized spawning of tunas of the genus *Thunnus* around tropical and subtropical islands. High abundances of tuna larvae near islands may be associated with better feeding conditions (Boehlert and Mundy, 1994). The greater availability of prey in nearshore waters might likewise benefit the young stages of reef species, although it is worth noting that prey of larger tuna larvae and juveniles include late larval stages of shore fishes. The aggregation of reef fish larvae near shore (Leis, 1991a) could reflect higher prey abundance. Those species distributed farther offshore may be more dispersed; if in fact food abundance is lower in those waters, differing strategies for larval survival may characterize species in nearshore and offshore assemblages.

Predation on ichthyoplankton, like feeding and starvation, is difficult to study for relatively rare larval stages. Many papers have referred to the 'relatively high rates' of predation on the reef face, particularly the upcurrent side (Hamner *et al.*, 1988), or from demersal plankton at night (Alldredge and King, 1985). On the reef, however, diurnal planktivores are typically concentrated where currents are high, whereas nocturnal planktivores are most abundant in areas of slower currents where demersal plankton aggregate (Hobson and Chess, 1978). The habitat complexity of reefs may provide refuge from predation for late, well developed larvae, but not for eggs or early larvae which are unable to orientate to the complex habitat; this is analogous to the situation in seagrass beds (Olney and Boehlert, 1988).

Although data on offshore predator abundance are lacking (Bailey and Houde, 1989), it has been suggested that predation rates are significantly higher on reefs than offshore (Johannes, 1978a; Shapiro *et al.*, 1988). Predators on early larvae include chaetognaths, planktivorous fishes, medusae, euphausiids, amphipods and other large zooplankton. These have been studied in greater detail in temperate regions (Bailey and Houde, 1989). Offshore predators on early larvae are likely to be less abundant than in temperate waters but large, late-stage larvae may be more vulnerable to predation in the tropics. Stomach contents of tuna have often been found to consist of pre-settlement larvae of reef species (Dragovich and Potthoff, 1972). The abundance of late-stage larvae in predator diets and rarity in samples from open water indicates some inadequacy of net sampling. Predation on late larval stages, however, may be an important source of spatial and temporal variation in survival. Further predation on settling larvae by planktivorous fishes (Hamner *et al.*, 1988) may be minimized by settlement at night during the dark phase of the moon (Dufour and Galzin, 1993; Shenker *et al.*, 1993). The relationship of offshore larval stages of reef species with their zooplankton prey and predators is an important area for study. Such studies need to include better identification of the habitat of the larvae, improved sampling techniques, and further research on the distribution and abundance of the predators.

Pelagic survivorship in reef fish larvae

Estimates of mortality in tropical reef fishes are typically available only from the time of settlement (Roberts, Chapter 4). Their highly dispersed larvae make estimation for earlier stages impracticable. Doherty *et al.* (1985) conducted simulations of larval dispersal and survival under different conditions of prey availability, but no data were available to support the mortality rates used in modelling. The best candidates for accurate estimation of mortality are species that complete their life cycle within lagoons and are moderately abundant, such as cardinalfishes, blennies or damselfishes (Leis, 1994). The anchovy *Anchoa mitchilli* (Engraulidae) was subject to mean daily mortality rates of over 85% for eggs and 26–36% for larvae in the food-rich (and predator-rich?) environment of Biscayne Bay, Florida (Leak and Houde, 1987). These levels are not markedly different from those observed in temperate areas (Houde and Zastrow, 1993). Daily predation losses (18–28%) were greater than those for starvation (10–11%). It would be presumptuous to suggest rates for reef fish larvae in the offshore environment. Although mortality from advection away from suitable habitats must be considered, it is unknown whether it is lower than inshore predation mortality as proposed by Johannes (1978a).

An alternative approach to understanding processes important in mortality is to examine the characteristics of survivors. Is spawning timed to maximize larval survival of reef animals in the tropics? In temperate and subtropical upwelling systems, spawning seasonality is thought to be adapted to improve larval survival, in relation both to the production cycle and to physical factors. Coincidence of spawning and such 'optimal environmental windows' may promote higher survival (Cury and Roy, 1989). This is akin to the match–mismatch hypothesis proposed by Cushing (1975b) where a temporal match of prey and larval production are requisite to high survival. The optimal environmental window concept involves environmental variability which affects survival through mechanisms involving both the production and relative aggregation of food resources.

The temporal distribution of spawning and of birth dates of larvae surviving to the settlement stage have been compared. This can allow inferences to be made about the environmental factors responsible for variation in survival (Methot, 1984). Robertson *et al.* (1988, 1993) and Robertson (1990) used this approach on tropical reef species. The bulk of settlement for most species did not arise from spawning peaks, suggesting that some intervening variable(s) affected survival in the plankton. A problem with this approach on reefs, however, is the spatial domain of sampling for both spawners and newly settled fish. The extent of advection to other areas (or settlement by fish spawned elsewhere) may indicate that these two distributions are disjunct. Without a better understanding of the original source of settlers at the sampling location, this approach to determine those offshore factors that cause variability in survival will be difficult to validate. Addressing this question on appropriate scales (e.g. at an isolated reef or island supported principally by local recruitment) may be the best way to determine the linkage of survival with variation in offshore factors. It may also help to distinguish spawner versus offspring benefits in the temporal and spatial patterns of reproduction (Boehlert, 1988; Boehlert *et al.*, 1992; Cowen and Castro, 1994).

3.5 CONNECTIVITY AMONG REEFS

The degree to which populations on reefs are connected is dependent upon the dispersal capabilities of the species involved, the hydrographic regime, the distances separating reefs and the availability of suitable habitat. Oddly, however, there does not seem to be a clear correspondence between pelagic larval duration and geographic distribution of reef fishes (Victor, 1991). Knowledge of most long-range, or teleplanic, larvae (Scheltema, 1986) is frequently based on animals with pelagic eggs. For example, wrasses have been observed over 1000 km from the nearest source population (Leis, 1983). Based upon otolith estimates, mean advection speeds for these species may have been approximately 18 km per day (Victor, 1987). Leis (1983) suggested that growth cessation may occur in these animals far from appropriate habitat for survival, and delayed metamorphosis was later confirmed, also in a wrasse (Victor, 1986a). If otolith increment formation ceases with cessation of growth, pelagic larval durations may be minimal estimates. Long-distance dispersing larvae may occasionally contribute to recruitment pulses (Lutjeharms and Heydorn,

1981a), but even low levels of dispersal promote gene flow (Waples, 1987; Schultz and Cowen, 1994). Distributional patterns offshore and loss of animals from reef populations are of greater concern to fisheries, for at greater distances from shore the likelihood of advection of the water mass containing the larvae increases (Fiedler, 1986).

Animals dispersing far offshore face the task of detecting and reaching a habitat appropriate for settlement. In a modelling exercise with neutrally buoyant particles and specific rules for settlement, Black (1993) suggested that after 10 days, 8% of 'larvae' might have settled on the natal reef, whereas only 1% would have settled on downstream reefs. The low settlement downstream was largely due to advection around, rather than impingement on, those reefs. Although fish larvae and pre-settlement juveniles are probably not analogous to neutrally buoyant particles, there are some circumstances where drifting with currents may be adaptive. Aggregation behaviour, often under flotsam (Fig. 3.4), may be adaptive if advection through surface slicks or other mechanisms results in movement towards shore (Kingsford, 1990). Attraction to flotsam, 'marine snow', and physical features may provide greater structuring to the pelagic environment for fish larvae and pre-settlement juveniles than previously thought (Kingsford, 1993).

Larval behaviour plays an important role in detection of settlement habitat. Flow–topography interactions may result in physical perturbations, which may extend some distance from an island and either aggregate, or possibly serve as cues for, settlement-stage fishes (Kingsford, 1990; Kingsford *et al.*, 1991). Specific sensory abilities, such as antennal receptors in larval rock lobster (Phillips and MacMillan, 1987), may be used to detect habitat from a distance. This may allow directed swimming, which has been observed in larvae of American lobster (Cobb *et al.*, 1989). Enhanced swimming abilities in pre-settlement fishes (Stobutzki and Bellwood, 1994) may facilitate movement to preferred habitats once cues are detected.

The spatial component of larval dispersal and settlement represents one of the biggest gaps in knowledge of reef fish populations. Relatively little is known about the geographic source of settling animals at any location. Advances in otolith daily growth increment techniques have allowed important progress to be made. In particular, temporal distributions of spawning dates of newly settled fish and pelagic larval durations, which provide information on potential dispersal distances, have become better known (Victor, 1991). New molecular techniques such as DNA markers may allow improvements in the capability to assess the geographic origin of settling larvae (Powers *et al.*, 1990), but to date little progress has been made. Molecular techniques have been used to identify larvae to species, and a recent study combined this approach with otoliths. It showed that

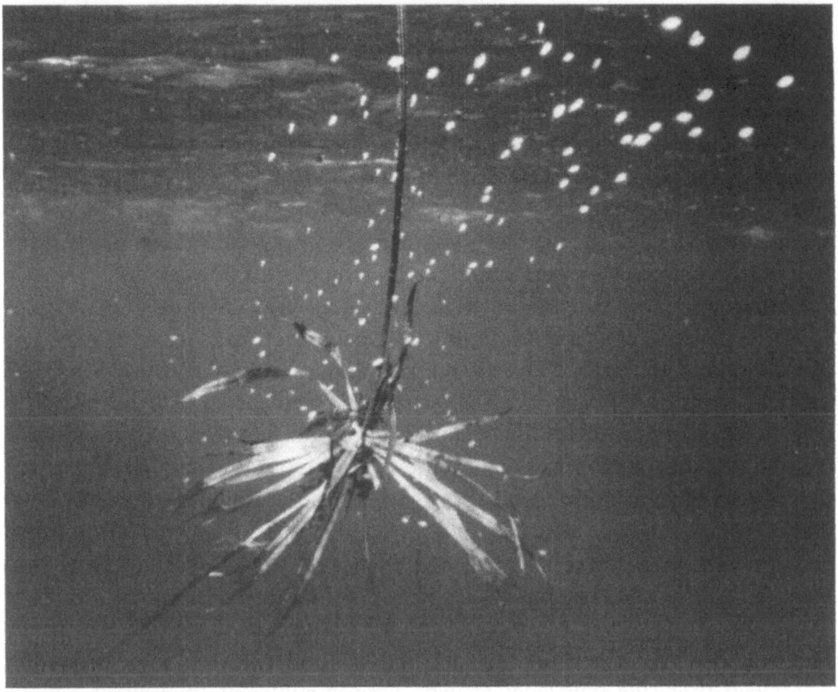

Fig. 3.4 Aggregation of pre-settlement fishes around a palm frond in surface waters offshore of Oahu, Hawaii. The silver fishes loosely aggregated above the frond are the damselfish, *Abudefduf abdominalis*. Below the frond, tightly aggregated fishes visible as dark spots are juveniles of the oceanic nomeid *Psenes* sp. These two taxa show different behaviour in the presence of a photographer. Attraction to such floating objects may facilitate advection towards shore in certain circumstances.

two larval morphotypes came from a single wrasse species, but were from two separate cohorts and temporally separated cross-shelf transport events (Hare *et al.*, 1994). The ability to identify the origins of settlers will allow the establishment of relationships between production and settlement and thereby greatly improve the ability to manage reef fisheries (Levin, 1990). As will be discussed further below, it also has significance in design and placement of reef fishery reserves.

Likely dispersal distances and potential contributions to fisheries

Although there is little question that the potential exists for long-distance dispersal in marine species, the consequences for population dynamics are unclear. I present two contrasting examples, one involving lobster in the South Atlantic, and the other reef fishes in the North Atlantic, to illustrate this uncertainty. Drifter tracks suggest that the lobster stock at Vema Seamount may be replenished by recruitment from the Tristan da Cunha Islands, some 2000 km upstream (Lutjeharms and Heydorn, 1981a,b). This study, however, was largely descriptive and did not seriously assess the potential role of local recruitment. The explanation also required that a large patch of larvae arrive at the island as a single, rare event. Because lobster abundance was derived from fisheries data, the possibility cannot be excluded of strong local recruitment from a moderately low population. This could be from localized source areas which are not exploited.

The second case applies a modelling approach. Schultz and Cowen (1994) examined the potential contribution of reef fish larvae advected to populations at Bermuda. The source population was more than 1000 km away at Cape Hatteras on the US east coast. Schultz and Cowen examined pelagic larval durations of likely species and modelled the current patterns, including initial entrainment into the Gulf Stream, subsequent transport by the Gulf Stream and mechanisms leading to Bermuda waters. Although their detailed analyses erred by design on the side of long-distance transport, Schultz and Cowen (1994) suggested that the likely numerical contribution from Cape Hatteras to reef fish populations in Bermuda would be minor. Bermuda populations could not be sustained if long-distance transport was the sole source of recruits. From a genetic standpoint, however, it is likely that this low level of gene flow is sufficient to prevent genetic differentiation; spiny lobsters, for example, lack genetic differentiation between the Caribbean and Bermuda (Hateley and Sleeter, 1993).

The contrasting conclusions from Lutjeharms and Heydorn (1981a,b) and Schultz and Cowen (1994) suggest that further research on different species, hydrographic systems and spatial scales would be useful. The rigorous approach applied by Schultz and Cowen (1994) serves as an excellent basis for such research. From the fisheries management point of view, however, it may be better to consider local recruitment (on the scale of hundreds of kilometres) to be the norm rather than the exception. In tropical reef systems, the utility of recruitment studies to fisheries management will require improved knowledge of the source of recruits, the oceanographic features important in larval dispersal and survival, and the interannual variability in all these factors. Some papers have begun to examine rare recruitment events (Choat *et al.*, 1988) and longer time series of recruitment on varied geographic scales (Fowler *et al.*, 1992; Doherty and Fowler, 1994a).

Populations and metapopulations

Larval dispersal and its directionality are questions central to the spatial dynamics of recruitment. They are important to our understanding of population dynamics and production of tropical reef fishes. They are also pertinent to fisheries issues because they affect the resilience to fishing on separate, localized populations, which combine to form the larger metapopulation. The metapopulation is defined here as a collection of smaller, relatively independent subpopulations occurring within a system of patches (Hanski, 1991), for example islands within an archipelago. In reef fish populations, settlement intensity at small scales may be independent of local population size or spawning activity (Doherty, 1991), implicating source populations from elsewhere.

From a metapopulation standpoint, dispersal and colonization of patches will not be a random process but rather a directed one under the influence of the advective oceanographic regime. Unusual environmental conditions may result in anomalous settlement events giving rise to a spatially expanded metapopulation; to a local observer, however, these may appear as patches of species otherwise uncommon in the region, and these may become extinct in the absence of further settlement events (Choat *et al.*, 1988). The more common feature is settlement from an upstream source (Lutjeharms and Heydorn, 1981a; Schultz and Cowan, 1994). Relating settlement variability (e.g. Fowler *et al.*, 1992) in the larger metapopulation to oceanography on similar temporal and spatial scales would be a fruitful approach.

Over greater geographic scales, and dependent upon dispersal capabilities, the geographic region and its isolation, and the oceanographic regime, genetic differentiation may occur. However, its relationships with life history patterns are not always clear. In the Hawaiian Archipelago, analysis of genetic variability in three species with different life history characteristics (damselfish, snapper and lobster) showed in each case a single, panmictic stock (Shaklee, 1984; Shaklee and Samollow, 1984; Seeb *et al.*, 1990). Over a broader spatial scale, Winans (1985) observed three distinct groupings of milkfish in the Pacific, including the broad Indo–Pacific region, the Philippine Archipelago and the Hawaiian Archipelago. Likewise Lacson (1994) noted a high degree of genetic structuring over the 19° of latitude between Okinawa and Palau for two damselfishes. Spatial distance evidently does not alone determine genetic heterogeneity. Lacson and Morizot (1991) observed a loss of genetic heterogeneity present in Florida Keys damselfish observed only 3 years previously at the same site; they proposed that population bottlenecks (e.g. drastic reductions in population size) such as hurricanes or other environmental perturbations were responsible for the low level of heterogeneity.

The concept that rate of gene flow might depend upon larval duration (Waples, 1987) is also not supported in all cases. Planes *et al.* (1993a) compared populations across various geographic scales in French Polynesia (within reef, within island, within archipelago, between archipelago).

Dascyllus aruanus, a damselfish species with a relatively short larval duration (26 days), showed genetic heterogeneity only at the between-archipelago scale. A later comparison of the surgeonfish *Acanthurus triostegus*, with a 60 day larval duration, however, showed significant between-island differences in allele frequency (Planes, 1993). Other forms of dispersal, such as rafting, may complement larval dispersal to confound this pattern (Jokiel, 1989).

3.6 IMPORTANCE OF PELAGIC LARVAL DISPERSAL AND MORTALITY IN FISHERIES MANAGEMENT

The establishment of local populations may be the result of specific, large settlement events (Choat *et al.*, 1988; Robertson, 1988). These may replenish selected habitats on relatively rare occasions, and the frequency of these events can define the success of local fisheries (Lutjeharms and Heydorn, 1981a; Cavarivière, 1982). Life history characteristics of the animal will influence the persistence of the population (Schultz and Cowen, 1994). The difficulty in identifying sources of settling juveniles has been discussed above, but settlement sinks can be established and are perhaps easier to document. Self-recruiting populations may exist on varying spatial scales but it is not unusual to see appropriate habitats seemingly depauperate of certain species (Ebert and Russell, 1988). Various physical mechanisms promoting retention (Bakun, 1988; Boehlert and Mundy, 1993) could make settling juveniles more or less available in some areas than others. For example, settlement may be low in areas of consistent upwelling near headlands (Ebert and Russell, 1988) or it may be enhanced in upwelling shadows (Graham *et al.*, 1992). In tropical systems, time-series data document temporal variability of settlement at given locations (Williams, 1983), but surveys over broad spatial scales have identified areas where settlement is consistently good (Fowler *et al.*, 1992). The spatial areas of enhanced settlement, or sinks, however, may not necessarily coincide spatially with the sources. The frequency of settlement events relative to the life cycle of the species considered is important, and periodic extinction of local populations may occur naturally (Choat *et al.*, 1988). When combined with fishing activity, populations could be rapidly depleted. A coincident source–sink could support fishing, whereas a source without consistent settlement would require a significantly different management strategy, such as establishment of a reserve.

On a regional basis, understanding of sources and sinks for recruitment, along with the underlying knowledge of oceanography and advective patterns, will be crucial to identifying potential reef reserves, currently under consideration as fishery management tools (Roberts and Polunin,

1993; Bohnsack, Chapter 11). Choosing the wrong area (e.g. a recruitment sink) could make the value of such a reserve far lower than that of an area clearly identified as a source of recruitment to other areas. Life history characteristics that differ among taxa may present differing population responses to reserves (DeMartini, 1993), and a single reserve may not protect all life stages of a species (Carr and Reed, 1993). The dynamics of the larval stage plays a critical role in answering these questions. Understanding these relationships will become a major challenge to reef fisheries management in the future.

3.7 WAYS FORWARD IN LARVAL RESEARCH

Understanding of reef fish ecology, and performance of reef fisheries management, will be much improved by better understanding of factors controlling population dynamics. Fisheries are operated at large spatial and temporal scales, and at these scales, recruitment variability is likely to regulate population numbers (Doherty and Williams, 1988; Doherty and Fowler, 1994a). From a management standpoint, assessment of juvenile abundance (as a recruitment index) may be sufficient to predict year-class strength for a fishery several years in advance (Phillips, 1986). Understanding the factors responsible for recruitment strength remains an important, if elusive, goal of fisheries science. Significant advances have been made in some temperate and subtropical systems, but typically on abundant, pelagic fishes (Peterman *et al.*, 1988; Cury and Roy, 1989). That both temperate and tropical reef species have lagged behind in this regard is understandable given their complexity.

In the coming decade, increased understanding of larval and pre-settlement pelagic stages should improve management. For early larval stages, characterization of the physical habitat for particular species (Boehlert and Mundy, 1994) will allow directed, stratified sampling to improve ecological information (distribution, inferred behaviour). These data will support more realistic numerical models incorporating three-dimensional ocean flow and larval behaviour, which can be run either in a retrospective manner to help understand past recruitment (Berntsen *et al.*, 1994), or so as to examine present, or predict future, patterns (Bartsch, 1993; Werner *et al.*, 1993). With improved ecological understanding, studies of larval prey and predators and their patchiness around reefs and assessment of the nutritional condition of larvae (Theilacker, 1986) would be a productive approach to assess spatial patterns of starvation and predation mortality. Subsequent analysis of the role of environmental variability in this component of mortality would allow environmental variables to be inserted into models.

Research on pelagic, pre-settlement larvae or juveniles will require improved sampling techniques that show correspondence between their abundance and subsequent settlement. This has been done for phototactic larvae by using light traps (Meekan *et al.*, 1993). Determining the extent to which larval pooling occurs will require more effort in habitats that have proven difficult to sample. Assessment of the geographic source of larvae using new molecular technologies could be applied to newly settling animals to provide information that will complement advective modelling studies.

ACKNOWLEDGEMENTS

For reviews and comments on an earlier version of the paper, I thank Tom Clarke, Ed DeMartini, Richard Parrish, Nick Polunin and Callum Roberts. Bruce Mundy kindly provided the photo for Fig. 3.4.

Chapter four

Settlement and beyond: population regulation and community structure of reef fishes

Callum M. Roberts

SUMMARY

This chapter examines processes acting at and following settlement from the pelagic phase to reefs. Settlement sets the starting conditions which post-settlement events modify. Recent studies have revealed quite precise habitat selection at settlement, facilitated in some cases by olfactory cues deriving from adults. Many species settle directly to reefs, but some undergo juvenile development in adjacent, nursery habitats. Seagrass beds and mangroves have been widely claimed to be important nurseries for many reef species. However, very few studies have quantified the proportion of reef populations to have passed through such nursery habitats. Available data suggest that the importance of such species is greater in the Caribbean than in the Indo–Pacific. Many species found in these habitats as juveniles undergo daily foraging migrations from the reef to the same habitats in later life; fishery species are important among them. Decline of nursery and foraging habitats close to reefs has been suggested as one cause of recent declines of fisheries in the Caribbean.

A series of models have been proposed to account for the structure of

Reef Fisheries. Edited by Nicholas V.C. Polunin and Callum M. Roberts.
Published in 1996 by Chapman & Hall, London. ISBN 0 412 60110 9.

fish communities on reefs, invoking varying degrees of control by post-settlement processes from none to great. Initial models focused on the role of competition for resources. Later models sought to account for small-scale unpredictability in assemblage composition and population size, invoking stochastic or limited settlement from the plankton. None of these models is sufficient alone. Predation following settlement has been shown to be a particularly potent force in structuring assemblages. Competition has been demonstrated to limit population sizes of a number of species, including fishery organisms. Present views converge more on multi-factorial explanations of population regulation based on a keener appreciation of the scale-dependence of processes. Mortality rates and patterns vary among species and within and among reef habitats. Many patterns of habitat use, including those of nursery habitats, can be explained in part as adaptations to reducing predation risk. A broad and robust generalization is that structurally complex habitats support abundant and diverse faunas resulting from protection from predation and greater availability of resources. Reef degradation reduces reef structural complexity and so has serious consequences for fishery productivity. Future research will benefit fisheries by focusing more effort on the effects of reef degradation and on processes that have so far been neglected, such as the importance of post-settlement movement.

4.1 INTRODUCTION

This chapter takes a selective look at processes acting on reef organisms at settlement and beyond. Because it would be impossible to cover the entire spectrum of processes, I focus on the ones that are most important to reef fishery science. Those seeking a more detailed coverage of topics relating to reef fish ecology would find no better starting point than chapters in Sale (1991).

Ecology and fishery biology are disciplines which have much in common but their practitioners only sometimes meet. Ecologists often feel that fisheries scientists are too remote from the natural history of species being exploited to make informed judgements about how to manage stocks. For example, despite a wealth of fisheries data, management of Georges Bank cod stocks was based on only a rudimentary understanding of cod ecology. This deficiency is only now being rectified following the collapse of stocks. By contrast, fisheries scientists may feel that ecologists are too bothered about small-scale details to properly comprehend the larger perspective which management necessitates. Both disciplines will gain much from closer integration. This chapter explores the insights that ecology can give into reef fishery management.

4.2 SETTLEMENT AND RECRUITMENT

The terms settlement and recruitment are often used interchangeably by ecologists. Settlement is the point at which an animal moves from the pelagic realm onto the reef. It is usually, but not always, the time of metamorphosis from larva to juvenile. To those ecologists who do distinguish settlement from recruitment, recruitment generally means a point shortly after metamorphosis when an animal becomes fully established on the reef. To fishery biologists, recruitment has a very specific and different meaning. It is the point at which an animal first becomes vulnerable to capture by fishing gear. This is the point at which fishers first become aware of its existence. Fishery recruitment thus depends not only on processes of larval production, dispersal and survival, but also on post-settlement processes such as growth, competition and predation. Throughout this chapter I use the term settlement to define the point of entry to reef populations, and recruitment refers to fisheries recruitment.

Larval supply to reefs may be highly variable both in space and time (Boehlert, Chapter 3). Once competent larvae arrive at a reef, there are a number of processes operating which may further modify the spatial distribution of settlers and their abundance.

Habitat selection at settlement

The majority of species show distinct zonation patterns across reefs. Some are entirely restricted to particular zones (for example, the spur-and-groove zone) while other, more widely distributed, species generally peak in abundance within specific zones. These distribution patterns are usually predictable and conserved no matter what variations there are in abundance of a species (Clarke, 1977; Waldner and Robertson, 1980; Russ, 1984a; Williams, 1991). Zonation patterns may be generated in several ways: habitat selection at settlement, post-settlement movement and selection or differential mortality among zones after settlement. In some species, juvenile distributions follow those of adults, while in others they may be distinct, with juveniles inhabiting a separate, nursery habitat or different reef zone (Sweatman, 1983; McFarland *et al.*, 1985; Harmelin-Vivien, 1989a; Planes *et al.*, 1993b).

In the past it has often been assumed that larval fishes will settle immediately on to a reef when competent to do so. However, studies of reef fishes suggest that habitat selection at settlement is very important in setting distribution patterns (Sweatman, 1983, 1985, 1988; Meekan, 1988; Carr, 1991; Kaufman *et al.*, 1992; Wellington, 1992; Planes *et al.*, 1993b). Larvae may delay settlement until they locate suitable sites. Kaufman *et al.* (1992) found that in at least 31 of 68 species observed on

Caribbean reefs, settlement was into specific habitats or microhabitats. Their observations suggested that a 'transition phase' may exist between pelagic and demersal stages which could be an important period during which larvae actively search for suitable habitat. Several species apparently settled for periods of minutes to hours before re-entering the pelagic phase.

Within habitats, precise microhabitat selection patterns have also been documented for some species. For example, fish larvae may settle only into particular types of coral, the algal mats within damselfish territories, or stands of macroalgae (Sale *et al.*, 1984; Carr, 1991; Green, 1993). Sweatman, in a series of elegant experiments, has shown that juvenile damselfishes may select sites based on the presence of conspecifics, presumably a good indicator that a site is capable of supporting the species (Sweatman, 1983, 1985). By contrast, Stimson (1990) found that settlement by the butterflyfish *Chaetodon miliaris* in Hawaii was density dependent, with apparent avoidance of adults by settling juveniles.

Some species may be able to detect settlement habitats by olfaction. Sweatman (1988) provided preliminary data suggesting that settlers of three species of damselfishes detected the presence of conspecifics or heterospecifics by olfactory cues. Boudreau *et al.* (1993) found that larvae of the temperate lobster *Homarus americanus* were attracted to odour plumes of adults and macroalgae (a settlement habitat), but avoided odour plumes from a predatory fish species. Likewise Murata *et al.* (1986) found that anemonefishes (Pomacentridae) can apparently smell out their hosts. Zimmer-Faust and Tamburri (1994) demonstrated that planktonic oyster larvae, *Crassostrea virginica*, were able to detect settlement sites based on olfaction of low-molecular-weight peptides originating from adults. There is no information available to assess the generality of olfaction as a mechanism of settlement site selection. However, cues almost certainly vary greatly among species (Sweatman, 1988; Leis, 1991a; Victor, 1991).

Nursery habitats

Juveniles of many species, such as lobster, *Panulirus* spp., and queen conch, *Strombus gigas*, inhabit separate habitats from adults, either different zones within habitats, or different habitats entirely. For example, adults of the surgeonfish *Acanthurus sohal* in the Red Sea defend territories at the seaward edge of reefs, whilst juveniles are restricted to shallow areas of the reef flat close to shore (C.M. Roberts, 1986). Juvenile queen conch in the Bahamas inhabit areas of seagrass with lower shoot densities than adults and migrate to deeper water as they grow (Stoner and Waite, 1990; Stoner and Sandt, 1991).

The habitat requirements of juveniles of many fishery species are poorly known. Direct observations by divers and submersibles have contributed much of what we do know. Huge gaps remain in knowledge of the early juvenile stages of reef fishery species, even for species the general biology of which is well understood, such as the queen conch (Stoner and Sandt, 1991).

Juveniles may live in different areas from adults for a number of reasons. The most important include risk of predation, avoidance of intraspecific competition, and differences in food or other requirements between juveniles and adults. The latter differences perhaps evolved as a response to the former selection pressures. Predation risk and competition will be examined later in this chapter.

Crypsis is one important reason why early-stage juveniles are overlooked. For example, many species of epinepheline grouper are rarely seen as young juveniles, even where common as adults. Thresher (1984) suggested that they settle cryptically into rubble or deep within the interstices of reefs. Alternatively, they may settle into different habitats, moving onto the reef later in life (e.g. Boulon, 1990). Kaufman *et al.* (1992) recorded two fish species settling cryptically in the Caribbean, including a squirrelfish (Holocentridae) and the flying gurnard (Dactylopteridae). To make things more complex, a cryptic juvenile phase may not always be a fixed characteristic of a species. For example, adults of the massive humpheaded wrasse, *Cheilinus undulatus* are common in the Red Sea and in Palau. However, in five years of observations in the Red Sea I never observed a juvenile smaller than 25 cm long, whilst in Palau, juveniles below this size were common in the open (pers. obs.).

A common pattern observed for species important to fisheries is increase in the number of large individuals with increasing depth. A number of explanations have been offered for this pattern, including differential mortality or growth with depth, or movement to deeper water with increasing size. Fish in shallow water may be subject to higher rates of fishing mortality. Alternatively, Longhurst and Pauly (1987) have suggested that cool, deep water allows fish to devote a higher proportion of their food ration to growth compared with warmer, shallow water, where basal metabolism is higher. In the tropics certainly, shallow waters have a much higher availability of shelter compared to the deep, and may be favoured by juveniles seeking to reduce predation risk. Productivity is also higher in the shallows, probably leading to a greater availability of small prey for young fishes. Different processes are likely to predominate for different species and areas, but together they have the effect of creating a widespread pattern.

For most species, even those for which juvenile habitats have been identified, our understanding of how important those habitats are is still qualitative. This is especially true for species that are present in more than one kind of habitat as juveniles. For example, juvenile Nassau groupers, *Epinephelus striatus*, can be found in seagrass beds, patch reefs or areas of continuous fringing reef (Grover *et al.*, 1992; Tucker *et al.*, 1993; Beets and Hixon, in press). It is unclear what the relative contribution of settlement into each habitat type is to overall replenishment of populations. In a rare example of a study quantifying settlement among habitats, Shulman and Ogden (1987) showed that 95% of French grunts, *Haemulon flavolineatum*, settled into seagrass beds compared with 5% settling directly to reefs. Similar observations for species of importance to fisheries will be very worthwhile.

Seagrass beds and mangroves as nursery habitats

It has been widely claimed that seagrass beds and mangroves are important nursery areas for species caught in reef fisheries (UNEP, 1985). This belief appears to have grown out of observations of juveniles of some reef species within these habitats (Austin, 1971; Jones and Chase, 1975; Lal *et al.*, 1984; Boulon, 1990). However, the presence of juvenile fishes alone is not a good indication that a habitat has a nursery function for reef species. It is necessary to demonstrate that juveniles settling in mangroves or seagrass beds eventually move to reef habitats and that this movement is important in their life histories. This linkage has only occasionally been directly observed, although for species that are not found elsewhere as juveniles the importance of a habitat as a nursery is usually assumed.

Arguments have often surfaced as to how important mangroves and seagrass beds are as nurseries (Robertson and Duke, 1987; Hatcher *et al.*, 1989; J.D. Parrish, 1989), with the corollary being: how much will reef fisheries suffer if there are losses of these habitats? Most studies have failed to compare densities, or turnover rates, of juveniles in non-reef compared with reef habitats. However, recent work suggests several general points.

Firstly, mangroves do act as nurseries for a number of commercially important species, especially penaeid shrimps (Staples *et al.*, 1985; Robertson and Duke, 1987; Sasekumar *et al.*, 1992). These species often support important fisheries but few of them are reef-associated. Quinn and Kojis (1985) found that mangrove fish assemblages in Papua New Guinea were dominated by three species of commercially important pony fishes (Leiognathidae). However, species also found in association with reefs constituted less than 1% of total numbers in samples. Thollot (1992) studied similarity of fish faunas between mangroves and nearby reefs in New Caledonia. He found only 13% of species (43) in common between these habitats. Only 13 reef associated species were present in mangroves as juveniles, but the majority were from commercially important families including snappers (Lutjanidae), jacks (Carangidae), surgeonfishes

(Acanthuridae), rabbitfishes (Siganidae) and emperors (Lethrinidae). However, most were also able to use alternative habitats as nursery areas. Caution is necessary in interpreting these results because many juvenile fishes shelter among the aerial roots of mangroves, making them extremely difficult to sample. This may bias samples toward species found at the margins of mangrove stands and over soft substrata. If juveniles of reef-associated species shelter more deeply among roots, they may be undersampled.

Secondly, seagrass beds, which in general occur in closer proximity to reefs than mangroves, support more reef-associated species as juveniles, and seem to be important nursery habitats for a number of commercially important species, such as snappers, grunts (Haemulidae) and groupers (Serranidae) (Ogden and Gladfelter, 1983; Bouchon-Navaro *et al.*, 1992; Louis *et al.*, 1992; Van der Velde *et al.*, 1992). Baelde (1990) found that seagrass beds close to mangroves in Guadeloupe supported a different assemblage of fishes from beds close to coral reefs. The latter had more juveniles of species that migrate to and from reefs to seagrass as adults, such as grunts and squirrelfishes, than did beds close to mangroves.

Thirdly, both seagrass beds and mangroves are apparently more important as nursery areas for reef species in the western Atlantic than in the Indo-Pacific (Ogden and Gladfelter, 1983; Birkeland and Amesbury, 1987; Ogden, 1988; Van der Velde *et al.*, 1992). The reason for this difference is unknown. One possible explanation has recently been offered by Bellwood (1994). He noted that all of the 14 species of western Atlantic parrotfishes (Scaridae) inhabited seagrass beds as juveniles, adults or both. By contrast, most Indo–Pacific parrotfishes are associated with reefs throughout their lives. Western Atlantic reefs are believed to have been affected more strongly by seawater temperature drops during glaciations than those of the Indo–Pacific, so restricting reef area and resulting in a major faunal turnover. Bellwood (1994) suggested that most of the species that made it through this evolutionary crisis were also capable of living in non-reef habitats.

Many commercially important species settle in large numbers to Caribbean seagrass beds, for example French grunts. While there they feed on microinvertebrates and plankton (McFarland *et al.*, 1985). They move to reefs after approximately 1 month of development, having reached 2–3 cm long (McFarland *et al.*, 1979). Similarly, yellowtail snapper, *Ocyurus chrysurus*, can undergo juvenile development in seagrass beds, moving to reefs at a similar size to that of French grunts.

J.D. Parrish (1989) has proposed that settlement into seagrass beds and mangroves may have evolved as a strategy to gain access to reefs. He suggests that reefs are a prime habitat for many species but are difficult targets for settling larvae to hit. Non-reef habitats may thus act as an initial step towards reaching reefs and can provide additional recruits to reefs through migration, when direct settlement has been low. In this way such habitats may provide a buffer to maintain recruitment levels to reefs, even in poor years. Unfortunately, this idea is probably untestable.

Muehlstein and Beets (1992) have argued that losses of seagrass habitat in the Caribbean, due to disease and development pressures, could create a bottleneck in the life histories of species that use them as nursery grounds. If their nursery function is important, such losses could have important effects on fisheries. Similarly, losses of mangroves may affect reef fisheries under some circumstances, and could have major effects on non-reef fisheries. Declines in the area of seagrass beds and mangroves have been documented worldwide (Shepherd *et al.*, 1989; Muehlstein and Beets, 1992; Sheppard *et al.*, 1992; Walker and McComb, 1992). Declining catches of reef fishes, especially in the Caribbean, may be partly due to losses of nursery habitat, although data to demonstrate this are sparse (Sadovy, 1989). Alternatively, declines may be due to heavy fishing pressure, as populations of target species have been depleted even in areas of the Caribbean not subject to habitat losses (Vicente, pers. comm.).

4.3 FORAGING MIGRATIONS TO ADJACENT HABITATS

The importance of off-reef habitats may extend beyond the juvenile stage. Many species of reef fishes forage in habitats adjacent to reefs, especially seagrass beds. Most shelter on the reef by day and move out across foraging grounds by night. In the Caribbean, grunts are the main group of species to undertake such migrations (Ogden and Zieman, 1977; Ogden and Quinn, 1984; Meyer *et al.*, 1983).

Indo–Pacific fishes do not undertake feeding migrations to the extent that Caribbean species do. The major commuter guild, grunts, is virtually absent from the Indo–Pacific (J.D. Parrish, 1989). However, adjacent habitats still provide an important food source for species such as some emperors, jacks and goatfishes (Mullidae) (Brouns and Heijs, 1985; Blaber *et al.*, 1992; Holland *et al.*, 1993). Most species that forage away from reefs come from two trophic groups: benthic invertebrate feeders, which take advantage of rich invertebrate communities present in sediments, and piscivores, which typically prey on the many juvenile fishes present in seagrass beds. Blaber *et al.* (1992) noted that piscivorous predators dominated the larger fishes present in seagrass beds off northern Australia, and that sharks (Carcharhinidae) and jacks would move in to shallow seagrass areas at night to hunt.

In terms of fisheries, foraging migrations have the important effect of concentrating biomass into a small area (while fishes are on the reef) but

	POST-RECRUITMENT COMPETITION INTENSE	POST-RECRUITMENT COMPETITION WEAK
RECRUITMENT MODIFIED BY PR PROCESSES	**1** **Competition model** **(Smith and Tyler)**	**2** **Predation/ disturbance models** **(e.g. Talbot *et al.*)**
RECRUITMENT NOT MODIFIED BY PR PROCESSES	**3** **Lottery model** **(Sale)**	**4** **Recruitment limitation model** **(e.g. Doherty)**

Fig. 4.1 Jones' (1991) classification of models of community structure for coral reef fishes. PR, post-recruitment. Further details of models can be found in chapters in Sale (1991a). Redrawn from Jones (1991).

yields actually depend on a much larger effective area than the reef itself (Polunin, Chapter 5). Roberts and Polunin (1994) found that the standing stock of fishes in the central part of the Hol Chan Marine Reserve in Belize reached a massive 340 g m^{-2} compared with values of 25–50 g m^{-2} more typical of reefs elsewhere. Eighty-nine per cent of the biomass within the reserve consisted of predatory species and was supported by foraging over the extensive seagrass beds behind the barrier reef.

4.4 MODELS OF POPULATION REGULATION AND COMMUNITY STRUCTURE

The past 20 years have seen periods of vigorous debate and several paradigm shifts in the way we view reef fish community organization. Jones (1991) has summarized the models developed to explain how reef fish communities are assembled (Fig. 4.1) on the basis of two criteria: (1) whether or not populations on reefs are at carrying capacity, and (2) whether or not post-settlement processes are important in modifying patterns set at settlement. In describing the models I follow Jones' (1991) terminology.

The history of reef ecology is like a compressed version of terrestrial ecology. First came description of patterns, then experimentation to try to establish causal relationships. Early fish ecologists tried to fit what they saw into the theoretical framework developed by terrestrial ecologists, which at the time suggested that animal communities should exist in some kind of equilibrium, organized by the deterministic process of competition. The first of the models, the competition model, recognized the distinct zonation patterns of fishes on reefs, and together with other, less well-defined differences among species, argued that reef communities were (1) at carrying capacity, (2) equilibrial, and (3) assembled by competition among species, with coexistence mediated by fine resource partitioning.

Sale (1974, 1977) drew attention to variation in fish assemblage structure at very small scales, and pointed out that the bipartite life history of reef fishes might lead to a very different set of rules governing the assembly of communities from those operating for most terrestrial animals. One of the main problems with the competition model which Sale identified was that rather than the fine niche differentiation expected, there were often broad overlaps in habitat and food use by groups of similar species. Sale (1974, 1977) proposed the lottery model to account for coexistence. Communities were still postulated to be at carrying capacity with strong competition among species, but species abundances were non-equilibrial and determined by the filling of free space by 'luck of the draw' in settlement from the plankton.

Sale's model was challenged vigorously, especially by those arguing that reef fish communities were more ordered than his lottery hypothesis would allow (e.g. Gladfelter and Gladfelter, 1978; C.L. Smith, 1978; Anderson *et al.*, 1981; Gladfelter and Johnson, 1983). An alternative view was put forward based on predation or disturbance maintaining populations below carrying capacity, thus preventing resources from becoming limiting (Talbot *et al.*, 1978; Bohnsack and Talbot, 1980). This model tried to account for evidence for niche differentiation among species but also for the non-equilibrial population dynamics described by Sale (1974, 1977).

The controversy stimulated some excellent experimental work and, by showing the potential importance of the larval stage of reef fishes, for the first time concentrated efforts on processes of settlement and recruitment. Both the competition and lottery models assumed that rates of settlement were sufficient to maintain saturated communities at carrying capacity. Certainly, the immediate impression given by the abundance of life on reefs is that there can be virtually no unused resources. It was therefore surprising to find in detailed studies of settlement on cleared patch reefs, that recolonization could be slow and that reefs often failed to recover their former abundance of fishes even over periods of many months (Smith, 1973; Williams, 1980; Doherty, 1981, 1983). At about the same time, Victor (1983, 1986c) found that the post-settlement population dynamics of a Caribbean wrasse was determined primarily by settlement from the plankton. Population sizes of this species seemed to depend largely on levels of settlement. These observations led to the development of the recruitment limitation hypothesis, which states that populations are generally held below carrying capacity by low rates of settlement and may be non-equilibrial (Victor, 1983; Doherty, 1991).

At about the same time as this model was being developed for reef fishes, interest in larval supply as a determinant of population sizes was burgeoning in other fields of marine science and was dubbed 'supply-side ecology' (Lewin, 1986). But, in another example of the separate paths trodden by fishery biologists and ecologists, an appreciation of the importance of variability in recruitment has been an integral part of fishery science since the 19th century.

The present synthesis

Early on in the debate about how fish communities were structured, there was a polarization into different schools of thought, based on the simple models proposed. Time and again the natural world has shown that complexity is almost the only thing that can be assumed. Simple models may have great heuristic value, but they always fail because exceptions abound. Understanding of reef fish biology has grown since the mid 1980s to the point where multiple determinants of community structure are being accommodated. This is a result of a greater appreciation of the importance of scale in ecological processes and the notion that outcomes of interactions are not predetermined exclusively by biology but are contingent on local circumstances.

Warner and Hughes (1988) summarize the implications of recent population models for understanding of reef fish community structure. They suggest that dichotomies between density-dependent and density-independent processes or equilibrium versus non-equilibrium are flawed. Mathematical models suggest that even communities organized by competition need not be equilibrial, while demonstrations of stable population levels need not imply density-dependent processes.

A modern view of fish community structure lies somewhere in the middle of Fig. 4.1, overlapping all of the hypotheses (see chapters in Sale, 1991a). For example: settlement may sometimes be sufficient to reach levels such that fishes compete; even at low population densities competition may be locally important; at small scales, stochastic dynamics may predominate whereas at larger scales there is a much greater predictability in community structure. Which factor dominates may depend for instance on species, trophic level, biogeography or level of exploitation. The relative importance of settlement variability, competition and predation in regulating population and community structure are examined in more detail below.

4.5 SETTLEMENT VARIABILITY AND STOCK REPLENISHMENT

The rate at which stocks are replenished by new settlement is of critical importance to fisheries. The spatial scale significant to fisheries is generally much larger than that addressed by ecologists. Doherty and Williams (1988) have reviewed the rapidly expanding literature relating to replenishment of fish populations on reefs. The majority relates to small species such as damselfishes (Pomacentridae) and wrasses (Labridae), but the limited data available on commercially important species suggest that the main findings will apply equally to a wide spectrum of larger exploited species. Doherty and Williams' (1988) main conclusions were (1) that settlement is generally seasonal but often takes place over extended periods (e.g. Fig. 4.2), (2) within seasons, settlement takes place in pulses which do not appear to be related to environmental variables or to be synchronized across years, (3) single large pulses of settlement may dominate year-class strength at local scales, (4) episodes of settlement are often multispecific, (5) settlement pulses may be consistent over large areas (tens of kilometres), and (6) there is often considerable spatial variation in settlement strength among reefs which is consistent across years.

For most species of reef fishes and many invertebrates, settlement variation among years leads to considerable variation in year-class strength. Although there are often multispecies pulses of settlers, differences in year-class strength are rarely synchronized among species across years (Eckert, 1984; Doherty and Williams, 1988). Settlement variability among years causes enormous difficulties to managers of single-species temperate fisheries. Most stock assessment models assume equilibrium conditions and consistent recruitment to the fishery. While these problems are also common to reef fisheries, their multispecific nature provides some protection of catches. Settlement variability may alter the species composition of catches from year to year, but will have little effect on overall yields except in areas where only a handful of species dominate catches.

For long-lived species, catches may differ little among years despite quite wide variation in year-class size. This can occur where one or a few strong year classes are always present in the population to support the fishery, despite poor settlement in some years. Many commercially important reef fishes can potentially reach old ages, reproducing over long stretches of their life spans. For example, the giant Caribbean jewfish, *Epinephelus itajara*, can reach ages of up to 37 years (Bullock *et al.*, 1992). Other

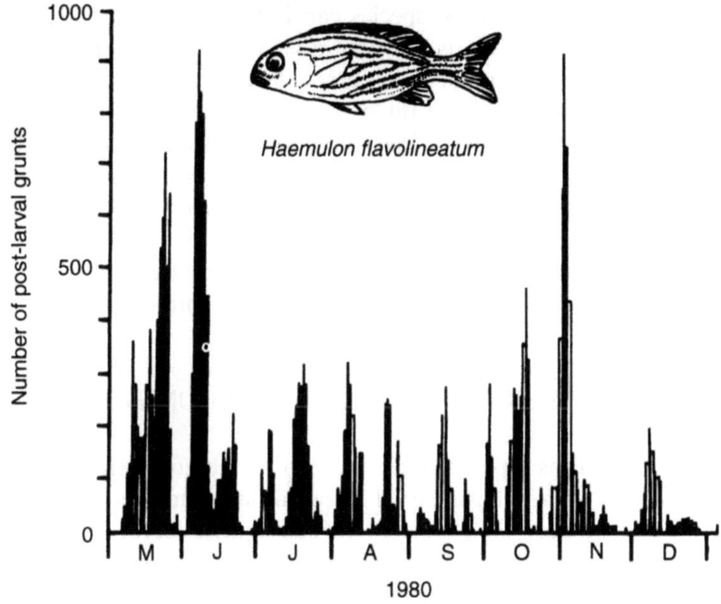

Fig. 4.2 Numbers of recently settled grunts observed on 20 patch reefs in St Croix, US Virgin Islands. Filled columns indicate actual census numbers while open columns indicate breaks between censuses and have been approximated using running averages from adjacent censuses. Settlement shows pronounced semilunar periodicity (15 day cycle) with weaker peaks at weekly intervals. Biennial peaks in settlement (May–June and October–November) correspond with peaks in the annual gonadosomatic index of many Caribbean fishes. Redrawn from McFarland *et al.* (1985).

species of grouper typically reach ages between 10 and 25 years (Sadovy *et al.*, 1992), snappers between 5 and 25 (Druzhinin, 1970; Manooch, 1982, 1987; Wright *et al.*, 1986) and grunts between 5 and 15 (Manooch, 1982; Darcy, 1983a,b). Long reproductively active life spans can buffer catches against the unpredictability of settlement over shorter time scales. Warner and Chesson (1985) have developed a model of coexistence based on this life-history feature, which they call the 'storage effect'. Their model suggests that if settlement of a species is low relative to the size of the population present on a reef, which could be the case in long-lived species, then settlement-limited populations may still appear to fluctuate around some equilibrium level. A second important prediction of their model is that even minor variations in mortality rates can greatly reduce correlations between settlement and population size (Warner and Hughes, 1988).

A long reproductively active life span provides some insurance against occasional or short-term settlement failure. However, under more intense fishing pressure even long-lived species can become heavily depleted if low settlement is prolonged. This is the case if fecundity is low or if spawning fish are particularly vulnerable to capture. For example, the US fishery for jewfish has been closed as a result of overexploitation of stocks. Episodic settlement can increase the vulnerability of stocks to overexploitation. Some species may settle in high numbers but very infrequently. For example, giant clam (Tridacnidae) populations typically show very low levels of settlement with occasional pulses (Braley, 1988). Robertson (1988) reported a massive settlement pulse of the triggerfish *Balistes vetula* in Panama, some 50–100 times greater than background levels.

4.6 STOCK–RECRUITMENT RELATIONSHIPS

Reef fishery organisms typically exist in open populations. The almost universal possession of a pelagic larval dispersal stage means that local reproductive output may be unrelated to subsequent larval input as new settlers (Boehlert, Chapter 3). This feature renders predictions of fisheries recruitment based on stock size uncertain, at least at the scales useful to tropical fisheries management. The nature of the relationship between stock size and fisheries recruitment has challenged fishery biologists for many years. Gulland (1983a) went so far as to call it 'the most serious scientific problem facing those concerned with fishery management'. It is important to management because the degree of dependence of replenishment on stock size (and on age structure of the stock, which affects fecundity) is critical to determining yields.

The problem of variable fisheries recruitment from the parent stock greatly complicates calculation of sustainable fishing levels. During good years, recruitment levels may be maintained by a small parental stock, whilst in poor years, even large stocks may be insufficient to adequately replenish populations. The manager's dilemma is whether to set fishing levels such that stock size remains large enough to produce sufficient recruitment in an average year, or in a poor year. In either case, the manager must know the shape of the relationship between stock and recruitment, and its degree of variability, before informed decisions are possible.

Figure 4.3 shows several hypothetical relationships between stock and fisheries recruitment. However, as Pauly (1984a) noted, about the only thing that is really known about the relationship is that it must pass through the origin! Two models have been widely used, both of which assume that mortality of recruits is density dependent. The Ricker (1954) model was originally developed for Pacific salmonid fishes and assumes that at high stock densities, some factor, such as competition, predation or

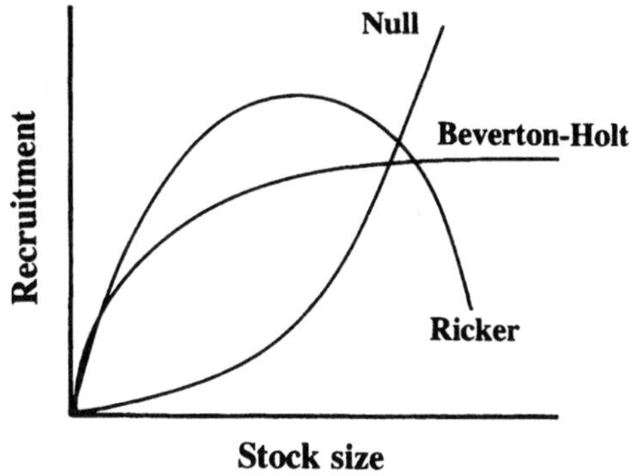

Fig. 4.3 Three possible relationships between stock size and fisheries recruitment for marine organisms. The Ricker model assumes that mortality of eggs, larvae or juveniles is strongly density dependent, while the Beverton–Holt model assumes density-dependent limitation of recruitment, possibly imposed by a ceiling in available resources. The null model assumes an increasing representation of older age classes with increasing stock size and no density dependence.

disease, may actually reduce numbers of recruits entering the population. By contrast, the Beverton and Holt (1957) model, while still predicting limitation of recruits at high stock sizes, follows an asymptotic form. Unfortunately, neither model has proven of much use in practice (Hall, 1988) because recruitment seems to vary almost independently of stock size for many species (Pitcher and Hart, 1982). The most important prerequisite for curve-fitting seems to be a fertile imagination!

Recruitment to a fishery depends on both pre- and post-settlement processes (up to the point of first capture). From the fishery-based analyses of stock and recruitment performed to date for temperate and some tropical stocks, it appears that processes operating on larvae and young fishes effectively uncouple recruitment from stock size. However, part of the high variability found in stock–recruitment relationships may be an artefact of the data. Measurement errors in estimates of stock size and subsequent recruitment may obscure what may in reality be somewhat tighter relationships (Hilborn and Walters, 1992).

Although stock–recruitment relations have been derived for a number of tropical stocks (e.g. Murphy, 1982), I am unaware of any for true reef-associated species. However, it would seem that the nature of the relationship could be derived quite easily for that subset of coral reefs where stock sizes can be well defined: isolated oceanic atolls.

Intriguingly, some recent studies have implied that there may be a relatively good temporal relationship between production of eggs and settlement into the reef population (Robertson *et al.*, 1988, 1993; Hunte von Herbing and Hunte, 1991), suggesting that production and settlement are more closely linked than previously supposed. Robertson *et al.* (1993) noted that intermonthly variation in settlement strength was on average 1.5 to 3.0 times greater (and up to 4–20 times greater) than variation in production, but interannual variation in settlement was similar to that in production. They concluded that pelagic processes introduced considerable variation into settlement of these species over the short term, but did not greatly affect year-class strength (at least within their study period of 3–10 years).

An important finding about stock replenishment is that there may be consistent among-reef patterns in settlement strength. That is, some reefs consistently receive lower densities of settlers than others. For example, on the Great Barrier Reef, where this phenomenon was first noticed, One Tree Reef consistently received lower inputs of new settlers of the damselfish *Pomacentrus wardi* than did Heron or Wistari reefs, only 10–15 km away (Doherty and Fowler, 1994a). An interesting question arises from these observations: to what extent do patterns reflect differences in production of larvae by local populations versus oceanographic processes modifying patterns of larval delivery to reefs? The answer will be of much significance to management of reef fishery stocks.

4.7 IMPORTANCE OF COMPETITION

To what extent population sizes on reefs are limited by competition after settlement has long been an important question. As evidence accumulated in the early 1980s to show that low settlement was a major factor limiting population sizes of fishes, so also did evidence mount to suggest that competition rarely appeared to limit population sizes. Its perceived influence as a force structuring communities therefore diminished. More recently, however, competition has been demonstrated in a growing number of studies. As Jones (1991) has pointed out, life after settlement is important, especially as 99% of the potential life span of a fish follows settlement, although only a tiny proportion live to enjoy it.

Failure to detect competition within or among species can arise for a number of reasons other than its absence. Studies typically have examined dynamics at very small scales and for limited periods, and so might miss important features of the life history where competition may occur. Detailed studies of damselfishes have shown that competition is usually manifested, not by limitation of population size, but by reductions in growth rate or reproductive output (Jones, 1991). Thus it may be missed in studies where the only currency measured is abundance. For example,

Robertson (1984) found that competition between territorial damselfishes in the Caribbean resulted in diminished fat reserves and smaller average size of individuals of one species in the presence of a larger competitor.

Most of the argument surrounding whether communities of fishes on reefs are equilibrial or non-equilibrial has been based entirely on species abundance data. Overall abundance is strongly influenced by settlement variability. Counts separated by only days can differ by one or two orders of magnitude due to settlement pulses. However, conclusions regarding stability of communities may be very different if populations are measured in terms of numbers of mature fishes, or of biomass of each species present. There have been no attempts to do this of which I am aware. It would be well worth re-examining this question using such data.

Even where populations of fishes on reefs appear to be undersaturated through low settlement, competition may still be important if there is variation in quality of habitat or potential mates. Wellington and Victor (1988) found that populations of an eastern Pacific herbivorous damselfish competed for space in shallow reef areas even though there was space available in deeper water near by. Shallow areas were preferred because they could sustain higher rates of algal production, which could presumably be translated into faster growth and higher reproductive output. Although most studies have concentrated on species that are fairly sedentary, such as damselfishes, the larger species important to fisheries are typically much more mobile. For them, there is likely to be substantial post-settlement movement, and concentration of fishes into higher-quality areas of habitat. The potential for both intra- and interspecific competition is considerable.

Evidence of competition is best for small fishes, which are most amenable to experimentation, but it has also been demonstrated in populations of commercially important fishes. Following the 1983 mass mortality of the herbivorous sea urchin *Diadema antillarum* in the Caribbean, there were increases in population sizes of herbivorous parrotfishes and surgeonfishes (Carpenter, 1990; Robertson, 1991b). In Panama, adult populations of two surgeonfish species increased by 160% and 250% (Robertson, 1991b), while in St Croix, densities of parrotfishes and surgeonfishes combined increased by around 300% (Carpenter, 1990). Hay and Taylor (1985) demonstrated similar responses by experimentally removing *Diadema* from sections of reef in the Virgin Islands. Parrotfish and surgeonfish populations and grazing rates increased after several months in the cleared areas.

Populations have not always responded to changes in the availability of resources. There was no apparent response by populations of herbivorous damselfishes to increases in availability of algae following the El Niño coral kill in the eastern Pacific (Wellington and Victor, 1985) or after crown of thorns starfish, *Acanthaster planci*, outbreaks on the Great Barrier Reef (Doherty, 1991). However, caution is needed in interpreting findings like these because a lack of response over the short term may indicate short-term settlement limitation only. The possibility cannot be dismissed, for example, that recruitment limitation is due to limited egg production by the resident population (Hixon, 1991; and see above). If this were so, then populations may respond only slowly to increases in available resources.

Evidence to determine how frequent competition is among fishes or other reef organisms is still inadequate. We have no clear idea to what extent it limits reproductive output or growth in biomass at the population level. However, recent work has put it back on the scientific agenda as a subject worth further study.

4.8 MORTALITY

Otolith methods

The study of age and growth of tropical reef fishes has been revolutionized by the discovery of growth bands in otoliths, especially daily growth bands (Thresher, 1988b). We are now gaining much more precise information about mortality rates and patterns by inference from age-frequency distributions in populations determined using otolith analysis. The width of growth increments appears to be a useful proxy measure for somatic growth (Radtke, 1990). Otoliths have also been used to demonstrate seasonality of growth in tropical fishes (Longhurst and Pauly, 1987). The transformation of demographic analysis by otolith methods extends far beyond mere ageing and growth. Analyses of trace elements within growth rings can help to reveal the habitats occupied by a fish during life (Radtke, 1988), for example, or in future may help define the natal sites of individuals. A review of methods of otolith analysis is beyond the scope of this chapter, but readers can find further information in Brothers (1984, 1987) and Campana and Jones (1992).

Mortality rates and patterns

Reefs are environments characterized by very high predation rates: Sudekum *et al.* (1991) estimated that two species of jacks alone consumed over 400 kg ha^{-1} $year^{-1}$ of fish on French Frigate Shoals in the Hawaiian Islands. A high proportion of fish biomass can be piscivores (Talbot, 1965; Goldman and Talbot, 1976; Polunin, Chapter 5). In addition to specialist piscivores, many species will feed opportunistically on fishes (Hixon,

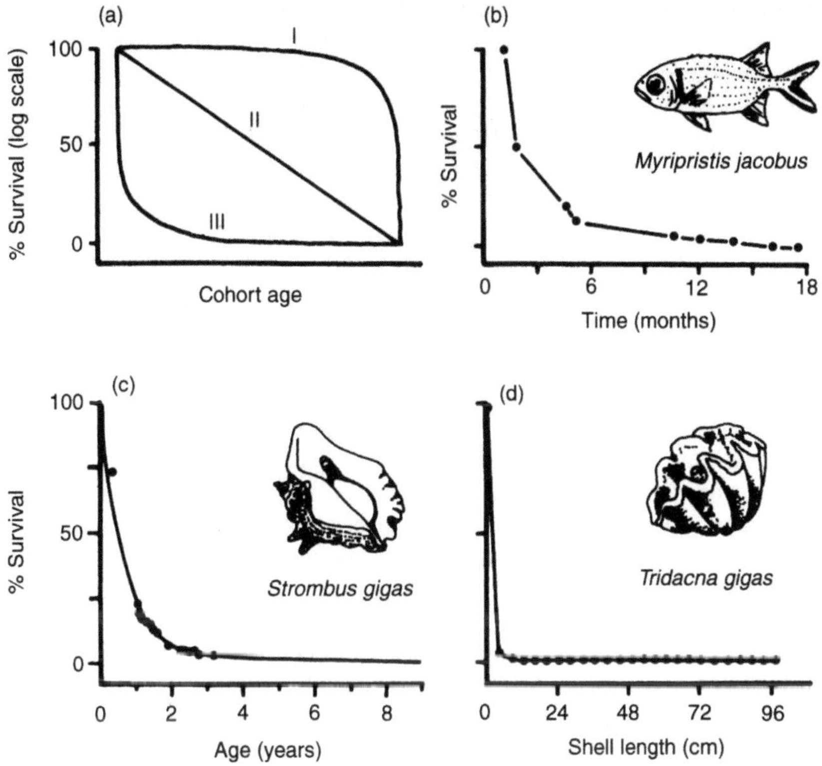

Fig. 4.4 (a) Three possible forms of mortality curve. Reef organisms appear to be characterized by the type III mortality curve. (b) Survivorship of a cohort of squirrelfish settling on artificial reefs in St Thomas, US Virgin Islands; redrawn from Hixon and Beets (1993). (c) Survivorship of queen conch in Puerto Rico based on age-specific mortality data in Appeldoorn (1988). (d) Survivorship of giant clams on Michaelmas Reef, Great Barrier Reef, based on age-specific mortality data in Pearson and Munro (1991).

1991). Norris and Parrish (1988) found that 52 of 126 fish species from reefs of the north-western Hawaiian Islands had fish remains in their guts and most of these prey were juveniles. For a fish or invertebrate settling from the plankton, the reef is a hostile environment. Kaufman *et al.* (1992) called settlement 'a suicide drop onto the reef'!

Figure 4.4 shows several possible shapes of the post-settlement survivorship for reef organisms. It appears that for most species, the type III mortality curve provides the closest approximation to actual patterns, regardless of taxon, although possible exceptions include toxic species (which do not generally support fisheries). High post-settlement mortality appears to be the norm. The small size at which most reef fishes and other organisms settle makes them especially vulnerable to predation. As they grow, they eventually may become large enough that the number of predators capable of handling them diminishes. Ralston (1976) has suggested that rapid early growth in the butterflyfish *Chaetodon miliaris* is an adaptation to high early mortality. However, the evidence for this is scant and rapid growth might also be an adaptation to early reproduction.

Early mortality rates have been measured directly for small species of reef fishes settling on the Great Barrier Reef (Doherty and Sale, 1985; Meekan, 1988; Sale and Ferrell, 1988). Losses of 10% to 30% were common during the first few days after settlement but rates slowed considerably in the following weeks. Relatively sedentary damselfishes and wrasse, which constituted the majority of the species observed in these studies, may be good models for mortality in only a subset of the fishes important to reef fisheries. Mortality rates appear to vary depending on behavioural and settlement characteristics. Species that settle in large numbers, are fairly mobile, or form schools, may be subject to higher mortality rates than solitary, sedentary fishes. Eckert (1987) found that abundant schooling species of wrasse had higher average annual mortality than less common species with a stronger association with the substratum. Shulman and Ogden (1987) found extremely high mortality rates in the French grunt, which forms large aggregations after settlement into Caribbean seagrass beds. Mortality exceeded 99% in the first 12 months after settlement. While this rate is much higher than those determined for smaller species it may be typical for species that settle in large numbers. Similar findings have been reported for damselfishes (Doherty and Sale, 1985; Williams, quoted in Medley *et al.*, 1993) although such a relationship has not always been clear (Sale and Ferrell, 1988).

Data for early mortality in commercially important species are hard to find. However, it is likely that species within families such as the fusiliers (Caesionidae), parrotfishes and surgeonfishes will experience higher mortality rates than species within families such as the groupers or emperors. The former tend to be more mobile, even from a very early age, and therefore may be more vulnerable to predation.

In addition to effects on population sizes of reef fishes, as Shulman and Ogden (1987) pointed out, high rates of mortality after settlement may substantially affect the community structure of fishes on reefs, modifying relative abundance patterns set by settlement (Hixon, 1991).

Factors affecting mortality

Probably the two principal factors determining mortality rates of reef organisms are the numbers of predators present and the ability of prey to escape. Both will differ among species for any given area of reef. However, one of the most important determinants of predator avoidance appears to be the availability of refugia inaccessible to predators, 'enemy-free space' in the terminology of Jeffries and Lawton (1984). Very few data are available on other sources of mortality, such as parasitism or disease, but these are likely to be of lesser importance than predation. However, it is clear that there may be wide variations in natural mortality among areas. For example, mortality estimates for the white grunt, *Haemulon plumieri*, varied by a factor of five between different areas of the western Atlantic (Darcy, 1983a). A ten-fold difference in mortality rates of an angelfish (Pomacanthidae) has even been documented among zones of the same reef (Aldenhoven, 1986).

Numerous studies have demonstrated a link between the physical structure of reef habitat and the diversity of fishes present (Risk, 1972; Gladfelter and Gladfelter, 1978; Luckhurst and Luckhurst, 1978; Carpenter *et al.*, 1981; Roberts and Ormond, 1987). A typical relationship shows an increase in the number of species present with increasing structural complexity, a relationship which has been well established for terrestrial habitats (e.g. MacArthur and MacArthur, 1961; Lawton, 1983). The relationship appears to be due to the combined effects of structural complexity *per se*, and of an increasing variety of microhabitats and resources, with which structural complexity is generally correlated. Roberts and Ormond (1987) attempted to distinguish between these two elements, using an index of substratum diversity (variety of benthic habitat types) which did not incorporate physical structure. They found a strong positive correlation between substratum diversity and species diversity of fishes on Red Sea reefs (Fig. 4.5).

It has been harder to demonstrate a relationship between structural complexity and the abundance of fishes in correlative studies (Risk, 1972; Luckhurst and Luckhurst, 1978; Carpenter *et al.*, 1981). However, such studies have usually not measured the availability of refuge holes directly, but simply assumed that these tend to increase with increasing complexity of the habitat. Roberts and Ormond (1987) measured refuge hole availability directly on Red Sea reefs and found, using a stepwise multiple regression, that holes of three different size classes explained 77% of the variance in fish abundance on reefs.

It seems likely that the most important effect of habitat structure is in reducing loss rates of small juveniles. Experimental evidence has shown, particularly for small species such as tube-blennies (Chaenopsidae) and gobies (Gobiidae), that availability of refuges may regulate population sizes (Behrents, 1987; Clarke, 1989; Buchheim and Hixon, 1992), presumably as a result of removal by predators of individuals lacking refuges. The majority of diurnally active fishes seek refuge within holes in the reef at night, regardless of whether or not they use them by day (Smith and

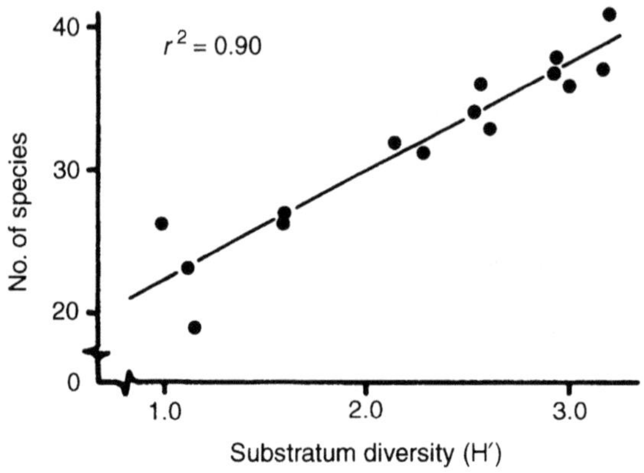

Fig. 4.5 Relationship between diversity of substratum components and the number of fish species present on Red Sea reefs. Substratum diversity was calculated from percentage cover data for different substratum types on 10 × 1 m transects at five depths at four sites using the Shannon–Weiner index (H′). Fishes were counted on 200 × 5 m transects at four depths at the same four sites. Methods are described in Roberts and Ormond (1987) and the figure is redrawn from Sheppard *et al.* (1992).

Tyler, 1972). Shulman (1985) showed how Caribbean surgeonfishes and squirrelfishes (Holocentridae) actively seek out and defend refuge holes at dusk. This behaviour is probably an adaptation to night-time predation risk.

Experiments by Hixon and Beets (1989, 1993), using artificial reefs placed over seagrass beds near St Thomas in the US Virgin Islands, have also demonstrated the importance of habitat structure and refuge hole availability. They used reefs with different numbers and sizes of holes to show that (1) reefs with refuges supported more prey fishes than those without, (2) prey fish were more abundant on reefs with holes close to their body size (which would presumably function more effectively as refuges), and (3) predators were important in controlling the maximum abundance and species richness of prey, but not in determining average prey abundance or species richness (Fig. 4.6).

The importance of refuges from predators is certainly not restricted to fishes. Braley (1987), for example, showed a very close association between giant clams and branching acroporid corals on the Great Barrier Reef. While large clams are invulnerable to almost all predation mortality, small clams form a significant component of the diet of triggerfishes (Balistidae). Braley (1987) attributed the microhabitat distribution pattern observed to selection of habitats by juvenile clams. However, differential

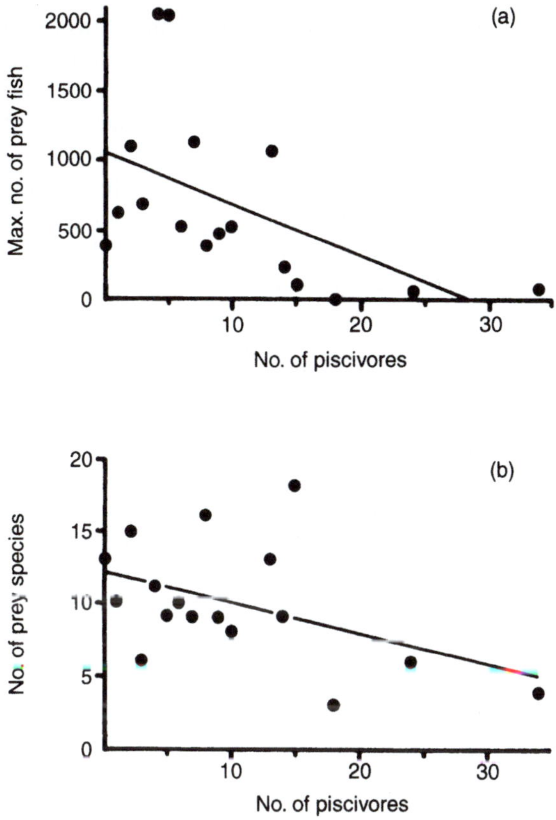

Fig. 4.6 (a) Maximum number of potential prey fish ever observed at each abundance of reef-associated piscivorous fishes on artificial reefs in St Thomas, US Virgin Islands. Each point represents a different reef ($P = 0.01$, $r^2 = 0.35$, $n = 17$). (b) Number of prey species corresponding to each point in graph (a) ($P = 0.02$, $r^2 = 0.31$, $n = 17$). Redrawn from Hixon and Beets (1993).

mortality among microhabitat types would also explain the patterns. Stoner and Waite (1990) showed that juvenile queen conch in the Bahamas were more proficient at selecting particular seagrass habitats than adults, suggesting strong selection pressure for habitat choice in small conch. They suggested that selection was through habitat-specific mortality patterns.

The addition of refuges from predators has become a central component of fishery enhancement for spiny lobsters, *Panulirus argus*, in Cuba and Mexico (Miller, 1989). The use of 'casitas', low-lying shelters placed on the seabed which simulate lobster dens, is primarily intended to aggregate lobsters, exploiting their shelter-seeking behaviour. However, Eggleston *et al.* (1992) have shown that survival of small juveniles in seagrass beds is significantly enhanced by casitas.

The abundance of predators is the second major factor affecting mortality rates of reef organisms. One probable role of nursery habitats away from the reef is to allow organisms to undergo early development in an environment where predators are less abundant than they are on reefs. Blaber *et al.* (1992) found that densities of juvenile fishes were highest in areas of seagrass with the highest density of shoots in northern Australia and attributed this to the combined effects of greater food availability and shelter from predators. In the Caribbean, Shulman (1985) found higher loss rates of experimentally tethered fishes in seagrass beds with decreasing distance from the reef.

Correlations between abundance of predators and that of prey fishes have been difficult to detect. One reason for this is that whilst higher densities of predators should reduce prey abundance, they may be confounded with other factors, such as reef productivity, settlement rates of prey and so forth. In areas with higher predator densities, prey standing stock could remain similar to that in areas with lower abundance of predators if turnover rates were higher. Such effects would blur the expected relationship. Bohnsack (1982), for example, could not find any effect on prey fishes of higher predator density within a marine fishery reserve on the Florida Keys. In the northern Red Sea, Roberts and Polunin (1992) found an inverse relationship between abundance of a small surgeonfish, *Acanthurus nigrofuscus*, and numbers of piscivorous predators. However, other typical prey species showed either no effect or a positive relationship.

Even in more 'clean' experimental settings, relationships between predator and prey abundance are weak. In the artificial reef experiments of Hixon and Beets (1993), there was no detectable relationship between average numbers of piscivores and abundance of their prey. Predation seemed important only in setting upper limits to prey numbers, with prey abundance usually fluctuating at lower levels. In the first year of an experiment on the Great Barrier Reef, Caley (1993) found that, during periods of settlement, species richness and numbers of fishes present on artificial patch reefs were generally greater on reefs where predator abundance had been reduced. By the second year, there were also inverse relationships between numbers and species richness of older, resident fishes and piscivore densities, suggesting a time-lagged effect of predation through effects on survivorship of recruits.

Competition may also influence mortality rates, although experimental work on small species has generally failed to demonstrate density-dependent mortality (Jones, 1991). However, Robertson (1988) found that mortality rates of settlers of *Balistes vetula* increased following the massive settlement pulse on Panamanian reefs. His data do not differentiate between possible sources of mortality. However, the main source was probably predation, and higher rates could have been due to predator switching, saturation of available refuges for juvenile triggerfishes, or a combination of the two.

From the foregoing discussion it is obvious that fish communities on reefs are profoundly affected by the nature of their physical environment. Physical factors interact in complex ways with others to determine the numbers and mix of species present. Although effects of habitat are complex in detail, a simple consequence of reducing habitat complexity is a loss in numbers of individuals, and number of species fishes (Sano *et al.*, 1984a, 1987; Bouchon-Navaro *et al.*, 1985; Dawson Shepherd *et al.*, 1992).

4.9 BIOGEOGRAPHIC DIFFERENCES IN COMMUNITY STRUCTURE

Throughout this chapter the emphasis has been on processes acting within reefs. These are important in regulating population and community structure throughout the geographic range of reefs, although with some regional variations. However, the players within reef communities vary enormously throughout the tropics due to biogeographical differences in species composition and differences in reef environment (Harmelin-Vivien, 1989b; Roberts, 1991; Thresher, 1991; Roberts *et al.*, 1992; Sheppard *et al.*, 1992). For example, over short distances (tens of kilometres), major faunal shifts have been documented in species composition, community and trophic structure of assemblages on reefs in areas such as the Red Sea (Roberts, 1991; Roberts *et al.*, 1992; Sheppard *et al.*, 1992) and on cross-shelf gradients within the Great Barrier Reef (Williams and Hatcher, 1983; Russ, 1984b; Wilkinson and Cheshire, 1988). At larger scales, for example Caribbean vs. Pacific Ocean, assemblage composition is almost completely different. Such differences among areas explain some of the marked regional variations in the nature of reef fisheries.

4.10 IMPLICATIONS AND FUTURE RESEARCH DIRECTIONS

Detailed *in situ* studies of the ecology of reef organisms have been under way only for the last 25 years or so. It is hardly surprising, therefore, that there are many gaps in our understanding of settlement and post-settlement processes. Perhaps the greatest deficiency has been a lack of attention to fishery species. For sound practical reasons, the emphasis has been on small species. Although studies of damselfishes, blennies and wrasses will be of great help to managing aquarium fisheries, these families may be inadequate models for processes affecting species that are eaten by people. Food species differ in important ways including larger body size, greater mobility, lower dependence on refuge holes, and generally a lower abundance.

Detailed ecological studies are needed to supplement the 'guts and gonads' approaches which have so far dominated research into food fishery species. These will pose difficult logistic challenges but are critical to shed light on important processes determining, for example, stock size and mortality rates. Such species will probably rarely be amenable to the multifactorial experimental ideal which has been adopted as the way forward in fish ecology (Hixon, 1991; Jones, 1991). Studies will have to make increased use of observation and experiments on natural areas of continuous reef (e.g. manipulation of access to resources) rather than the population-level manipulations on small, isolated patch reefs popular with those studying smaller fishes. An approach which has also proven useful is to take advantage of natural perturbations (Doherty, 1991) such as the die-off of *Diadema* sea urchins in the Caribbean (e.g. Carpenter, 1990; Robertson, 1991b).

Present knowledge of settlement and post-settlement processes provides a basis for speculation as to the implications of these processes for fisheries. For example, we are only just beginning to appreciate the abilities of species to select habitats at settlement using sometimes very precise cues. One important implication of larval habitat selection is that disruption of settlement cues or sites, through human modification of reef habitats, may have important effects on the settlement process, thereby interfering with later recruitment to reef populations. We still know very little about mechanisms of habitat selection.

Similarly, the importance of nursery habitats to reef fisheries production is still not known in any real quantitative way. Nursery habitats appear to be more important in the Caribbean than in the Indo–Pacific. The widespread losses of nursery habitats from pollution and development may therefore have greater implications for fisheries in the Caribbean compared with the Indo–Pacific. Determining the importance of nursery habitats will require more sophisticated studies than have been conducted to date. These must, at the least, attempt to determine the proportion of populations of reef species that pass through a nursery habitat before reaching the reef.

Post-settlement movement, both within and among habitats, is a key process which has been neglected. Such movements have important implications for demography and are critical to certain predicted functions of marine fishery reserves (Bohnsack, Chapter 11). Reef fish ecologists have gone to great lengths to eliminate movement in experimental studies. It is time now to study movement for its own sake, but the few attempts made to date have encountered severe logistic difficulties, especially in tracking small animals. New tagging and tracking methods will help to resolve these problems.

Recruitment is central to fisheries biology. We are further advanced in the understanding of settlement variability and its ramifications for recruitment to fisheries. However, the degree of local coupling of stock and recruitment remains unknown. There have been tantalizing results from both ecological and genetic studies which suggest that stock size and fisheries recruitment may be more closely related than experience from temperate fisheries might lead us to expect. Certainly stock and recruitment are closely intertwined in oceanic atoll settings. Populations on isolated reefs are predicted to be more vulnerable to recruitment overfishing (in which insufficient eggs and larvae are produced by the parent stock to replace itself) than those in continental or dense archipelagic settings where larval inputs from elsewhere may buffer against depletion.

There are strong indications that settlement limitation is at least common for reef fish stocks (Doherty and Williams, 1988). If it predominates, then an important implication for fisheries is that stocks may be vulnerable to recruitment overfishing. Studies of the relative importance of pre- and post-settlement processes to the mortality of fishery species will help to determine if this is so.

The quality of reef habitats is clearly of much importance to fishery stocks. One of the best-established ecological relationships is between the structural complexity of a reef habitat and the number of species and abundance of animals it supports. It is of much significance to reef fisheries that a general effect of pollution and other human impacts on reefs is a loss of habitat complexity. The relationship of habitat degradation to fisheries production has not been established. What, for example, is the effect of a 20% loss of coral cover on fisheries production? Given the present high rates of loss of coralline habitats, effects of reef degradation are a research imperative.

Adaptive management of fisheries, the testing of management regimes on stock isolates, has received much attention recently (Hilborn and Walters, 1992). Manipulative approaches to determining the outcome of management regimes could be especially useful for reef fisheries. Many of the limitations of adaptive management could be overcome in the reef context. Harvesting pressure can be closely controlled and spatially replicated. Stocks can be assessed independently of fishing gear used. Adaptive management approaches may be especially appropriate if reef fish populations turn out to be generally self-recruiting on moderately small spatial scales. Experimental approaches to management have already begun. For example, total and partial closures to fishing will provide much-needed information on population levels under reduced harvests. Such experiments may eventually help us determine the effects of fishing precisely enough that such populations can be managed effectively.

Although we know much more about the ecology of species such as are harvested for aquaria than about food fishes, we are not yet making good use of this understanding. Despite the very high value of aquarium fisheries worldwide, management has barely been attempted to date. Information on aquarium species has still not been incorporated into fisheries models. This in itself is a fertile area for research.

Our present knowledge of reef organisms is at least sufficient to know that important controls on population size and community structure may operate at many stages in the life history. The history of reef science suggests that neglect of one in favour of others is folly.

ACKNOWLEDGEMENTS

My thanks to Mark Hixon, Julian Caley, Yvonne Sadovy, Julie Hawkins and Nick Polunin for their many suggestions for improvement of this chapter.

Chapter five

Trophodynamics of reef fisheries productivity

Nicholas V.C. Polunin

SUMMARY

The fish biomass of a few studied reefs varies from 2 to 24 tC km^{-2}. Fish productivity of undisturbed tropical reefs is based on carbon sources of reefs and adjacent systems. The major primary producers of reefs are the microalgae of rocky habitats, with dominant areal coverage and high community net primary productivity (NPP). Shallow subtidal flats and reef margins support an especially high NPP. The exogenous carbon inputs to reefs derive from the plankton and other littoral ecosystems. Large external inputs are expected in areas subject to run-off and upwelling. Grazers constitute up to a quarter of the fish biomass and consume much of the microalgal NPP. There are both nocturnal and diurnal planktivores on reefs, and a Great Barrier Reef study measured import of plankton by fish to a reef front of 3 kgC m^{-1} $year^{-1}$. Invertebrate feeders constitute the major part of the reef fish biomass and can evidently support food consumption rates in excess of 20 tC km^{-2} $year^{-1}$. Piscivores contribute 2–54% to the fish biomass of the few Indo–Pacific reefs studied and piscivory apparently exceeds 6 tC km^{-2} $year^{-1}$ on many tropical reefs. Process-orientated and biomass-based (e.g. ECOPATH) models of reef fish productivity have been synthesized. Reef fishes are major reservoirs of nutrient elements such as nitrogen. Detrital fluxes are substantial, but there is little information on pathways. Total fish productivity can exceed

Reef Fisheries. Edited by Nicholas V.C. Polunin and Callum M. Roberts.
Published in 1996 by Chapman & Hall, London. ISBN 0 412 60110 9.

20 t fresh weight km^{-2} $year^{-1}$. There is no information on how reefs may systematically vary in fish productivity, but productivity is expected to be greater (1) in nutrient-rich waters than in nutrient-poor waters, (2) on small compared with large, reefs, (3) on reefs with adjacent productive communities compared with those without such communities, and (4) on reefs where a high proportion of the fish biomass consists of young individuals compared with those with older populations.

5.1 INTRODUCTION

Fish productivity relies both on numerical turnover of the individuals in a population, for which I shall use the word 'replenishment', and their growth. Productivity is sustained through replenishment combined with adequate growth of surviving individuals. Replenishment depends on spawning (Sadovy, Chapter 2), dispersal and mortality of eggs and larvae (Boehlert, Chapter 3), and survivorship of animals during and after settlement (Roberts, Chapter 4). Ageing techniques have developed (Pannella, 1980), and estimates of body growth rate are increasing in number (Loubens, 1978; Pauly, 1980b; Buesa, 1987), but we know little about relative roles of source materials from which reef fish biomass is ultimately derived, and little too about the ecological pathways by which those sources are dissipated and taken up by secondary producers. Investigation of these fluxes, referred to as trophodynamics, draws our attention to processes at the scale of the ecosystem, involving many species and large areas. Biological studies addressing replenishment have tended to be reductionist. Often our understanding has been based on analyses of particular species in small areas of reef over short periods, even if such local events have been compared across larger spans of time and space.

At present, understanding of marine fisheries population dynamics relies heavily on single-population models. Yet consideration of the multispecies and ecosystem contexts of fishing, including that on reefs, is very important for management. In situations that are biologically complex, there is also a need for alternative estimates of potential yield from exploitation; these may be derived from trophodynamic information. For reefs, estimates of sustainable yield have remained elusive because of their complexity, lack of data and the simplicity of accessible approaches (Russ, 1991).

The purpose of this chapter is to offer a trophodynamic view of relatively undisturbed tropical reefs and evaluate what this might tell us about fishery composition, sustainable yields, and their variation in time and space. The trophodynamic basis of fish production can be quantified in various ways. Units can include fresh weight, organic carbon, nitrogen or energy. Although early ecosystem studies tended to use units of energy, more recent work has focused on organic carbon because it is on this that most estimates of primary productivity are based and respiration can be easily related to it. The present summary will use both carbon and wet-weight units, however, because many fluxes are quantified in carbon terms, while catches and fishery production are estimated in units of wet-weight. Occasional reference to nitrogen-based fluxes will be made as well.

I will begin by considering the primary materials supporting food chains on reefs, both from within and outside the system. I will then focus on the pathways influenced by large consumer organisms. The final step will be to bring together data on major fluxes as a basis for predicting modes and levels of fish production in relatively undisturbed systems. However precise ecologists might like to be about what does or does not constitute a reef, the fishery definition is more hazy. Fishery data are typically imprecise with respect to habitat. In any case, fish production from a reef-associated fishery will be partly based on off-reef processes. This account is relevant to tropical coral-based hard-ground habitats in general which may act as foci for fishing activities (Munro, Chapter 1).

5.2 WHAT ARE THE ULTIMATE SOURCES OF CARBON FOR FISH PRODUCTIVITY AROUND TROPICAL REEFS?

Photosynthesis is the ultimate source of all organic carbon for fish production. This primary productivity is most likely to be from within the reef system, but primary producers from outside the system cannot be discounted as sources of organic carbon, as well as of other elements.

Organic carbon sources of reefs

There is much variability among estimates of reef primary productivity. Some of this arises from differences among studies with respect to technique and chosen habitat. Large-scale studies of reef flats of a few mid-latitude reefs (>10° both north and south), using similar techniques, show a consistency in total photosynthesis (gross primary productivity, GPP; Kinsey, 1985). The differences among habitats are attributable to effects of water movement, substratum irregularity (rugosity), and coral and algal cover. Reef flats and margins in general can be expected to sustain a GPP in the range 5–10 gC $m^{-2}day^{-1}$, while sand and rubble zones might have a GPP of 1 gC $m^{-2}day^{-1}$ (Kinsey, 1985). On reefs with steep slopes, most GPP is therefore derived from level hard substrata in shallow water (<5 m deep). Reefs with gentle peripheral gradients, however, could have a high proportion of the GPP contributed by perimeter upper slopes.

The main primary producers on reefs are the benthic microalgae, macroalgae, symbiotic microalgae of corals and other symbiont-bearing invertebrates, and seagrasses. Phytoplankton have not been considered significant primary producers over the reef (but see below). The zooxanthellae living within the tissues of hard corals make a substantial contribution to primary productivity in zones that are rich in corals, because of their density (e.g. $>10^6$ cells cm^{-2} of live-coral surface) and the high rugosity of the surfaces on which they live (often $>10\,m^2m^{-2}$), as well as their own photosynthetic potential. Estimates of GPP range from 0.5 to 10 gC $m^{-2}day^{-1}$, but zones of high coral cover typically constitute only a small proportion of entire tropical reef systems (e.g. Odum and Odum, 1955), and so their contribution to total reef GPP is small.

The benthic microalgae referred to are mostly unicellular and filamentous forms growing over and within the surface layer of reef rock and rubble. Epilithic algae either have a calcareous matrix and are mostly encrusting (crustose algae) or form a thin lawn (turf algae). In spite of living within the reef substratum, endolithic microalgae may also be important primary producers. Microalgal hard substrata often have high rugosity (up to >3 m^2m^{-2}), and the small algae maintain intense photosynthesis (GPP per unit surface area occupied ranges from <1 to 3 gC $m^{-2}day^{-1}$ for the corallines and 1 to 12 gC $m^{-2}day^{-1}$ for turfs). Benthic microalgae are major contributors to total GPP in most reef zones (Klumpp and McKinnon, 1989). Corallines tend to occupy narrow, shallow peripheral zones with strong wave action. Turf algae cover most of the surface of shallow reefs (commonly $>90\%$) and are thus the major contributors to reef GPP (Polunin and Klumpp, 1992b). Their importance extends to nitrogen because turf-algal communities contain high densities of nitrogen-fixing forms. Peripheral zones of reefs with high algal and coral cover and strong water exchange commonly support a GPP of 20 gC m^{-2} day^{-1} (Kinsey, 1985). The microalgal productivity of soft substrata is poorly known. Although sandy habitats may support low levels of GPP, on the other hand such areas can be extensive. Their contribution to system GPP may therefore be high.

Macroalgae, including calcareous forms such as *Halimeda*, maintain high rates of photosynthesis (GPP per unit area of algal bed ranges from 2 to 40 gC $m^{-2}day^{-1}$), but they are only widespread and abundant on reefs with high nutrient inputs and little grazing (Littler and Littler, 1984). They are therefore not considered major contributors to GPP on most tropical reefs. Seagrass beds are often closely associated with reefs and are highly productive (GPP ranges from 3 to 16 gC $m^{-2}day^{-1}$). They may thus make substantial contributions to the total GPP of entire reef systems.

Organic carbon sources outside reefs

Ocean surface waters in the tropics have generally low productivity, but waters subject to upwelling and coastal run-off can be much more productive than pelagic areas (Berger, 1989). The water column overlying tropical reefs can still be more productive than that of adjacent neritic waters, but most estimates of GPP are well below 0.5 gC $m^{-2}day^{-1}$ (Sorokin, 1990). More than 90% of the instantaneous GPP of a reef will typically be derived from the reef and associated benthic habitats, especially adjacent to high islands (Doty and Oguri, 1956; Rougerie and Wauthy, 1986).

Reefs are static structures, however, and the unproductive waters into which they project are continually moving. Reefs therefore have access to substantial open-water productivity. On the Great Barrier Reef of Australia, more than 80% of total continental-shelf GPP is planktonic in the southern region where the shelf is wide. Even in the north where it is narrow, planktonic sources contribute 33–49% of total GPP (Furnas and Mitchell, 1988). There is commonly a four-fold difference in radiocarbon measures of open-sea primary productivity between coastal waters and barren oceanic gyres (Berger, 1989). Thus, particularly in inshore continental waters, shallow benthic habitats such as reefs must not always be considered the dominant sources of carbon for fisheries. Outside sources can clearly be important for reefs, and while this significance is rarely estimated, its input may be in living (plankton), or dead (detrital), forms.

In coastal waters, detrital organic matter from land, the plankton and fringing marine plant communities is potentially abundant. There may be both passive advection of particulate and dissolved detrital carbon onto reefs, and active transport of such material onto reefs via fishes which shelter on reefs but feed in adjacent habitats. Other things being equal, there is a greater potential for nourishment of inshore reefs than offshore reefs by external carbon sources, and this inshore nourishment will be enhanced around large land masses (Doty and Oguri, 1956).

A few data from inshore waters do indicate that the influx of fine-particulate (particle size <1 mm) oceanic organic carbon to reefs can be substantial in relation to total GPP (Ducklow, 1990). There are high bacterial densities in reef waters (e.g. 1–10 mg bacterial C m^{-3}) and potential productivity of this microbial community is large (e.g. 20%) in relation to total GPP in one reef lagoon (Ducklow, 1990). There are also abundant phytoplankton and zooplankton in waters around reefs. These communities, however, are distinct in composition from their open-sea counterparts (Sorokin, 1990). It is likely, therefore, that the reef zooplankton and water-column micro-organisms are supported at least as much by reef GPP and carbon from adjacent productive coastal sources, such as

(a)

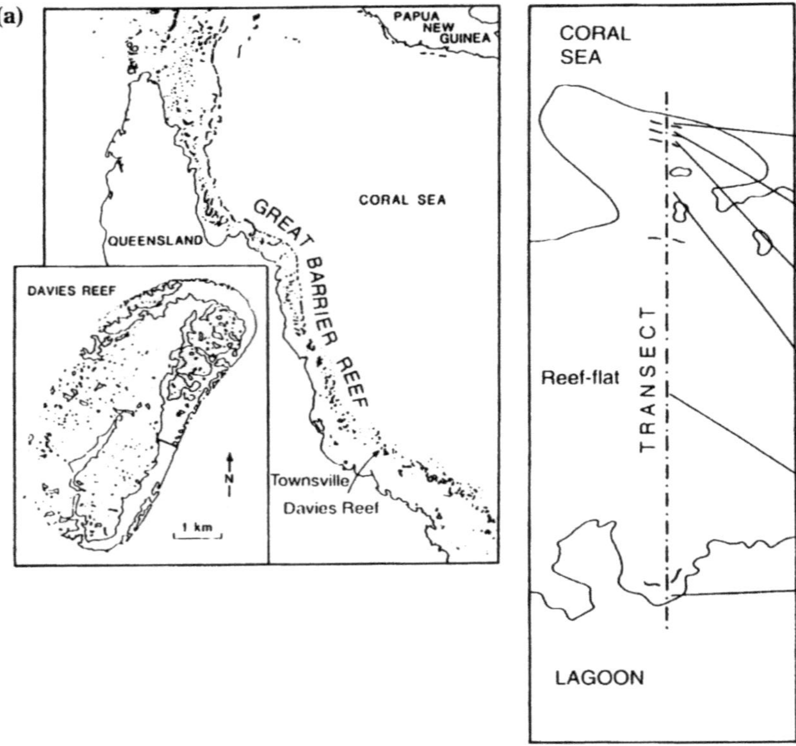

Zone	Dominant algal community	Prominent grazers	GPP/R ratio	Per cent reef tract
A	Crustose coralline	Parrotfishes	1·96	2·44
B	Thick turf	Surgeonfishes Large damsel-fishes	2·08	2·37
C	Thick turf	Large damsel-fishes	2·08	2·37
D	Fine turf	Small damsel-fishes	2·08	19·12
E	Fine turf	Small damsel-fishes blennies	1·90	59·90
F	Fine and Thick turf	Large and small damselfishes blennies	1·79	1·71
Sand	Unicellular	None	0·63	12·09

seagrass beds and mangroves, as they are by neritic or pelagic inputs (Ducklow, 1990).

Net primary productivity

GPP does not directly measure the availability of organic carbon for growth because much of it may be dissipated by respiration. Net primary productivity (NPP) should be calculated by subtracting primary producer respiration from GPP, but in practice respiration includes that of small animals and micro-organisms which are not necessarily photosynthetic. Typically, most reef GPP is thought to be dissipated within the system, so that at the level of whole reefs, NPP is often zero (Kinsey, 1985). The NPP of different primary producer groups, however, offers a measure of how important they are likely to be in supporting the food webs which maintain fisheries around reefs.

The highest levels of NPP are provided by turf algae and macroalgae. Coralline algae, corals and other organisms bearing symbiotic algae are much less productive. The areal coverage of algal turfs (Klumpp and

(b)

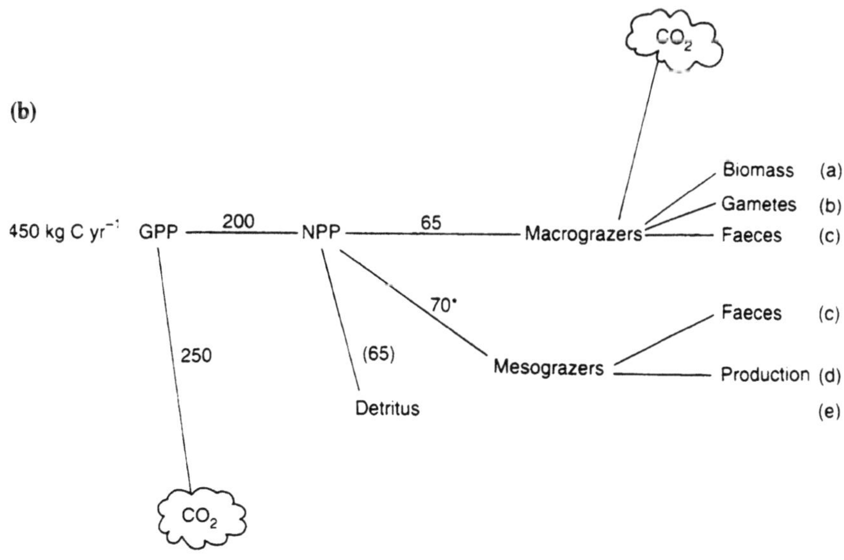

Fig. 5.1 Trophodynamic model of fish productivity at Davies Reef, Great Barrier Reef. (a) Location, transect zones, principal primary producers and grazers, and productivity across a 410 m long windward tract. (b) Fate of reef primary productivity to macrograzers (herbivorous fishes and invertebrates with body size >1 cm), mesograzers (herbivorous invertebrates <1 cm >200 μm) and detritus. From Polunin and Klumpp (1992b) by permission of Oxford University Press.

McKinnon, 1992), which is evidently determined by high grazing intensity (Lewis, 1986), means that these are the major contributors of carbon to food webs at the scale of whole reefs (Polunin and Klumpp, 1992b). The turf-algal NPP is highly seasonal, however, being greater in summer than in winter, and at 10 m depth may be half what it is on shallow flats (Klumpp and McKinnon, 1992). Macroalgae are favoured in some sites, under conditions of low grazing intensity and high nutrient inputs (Littler and Littler, 1984), and may thus be locally important to fish productivity. Their productivity and use is a neglected area of tropical reef ecology.

Reef margins, particularly those facing the open sea, exhibit a high NPP (Fig. 5.1). By contrast, back-reef lagoonal waters and sandy habitats in general support very low NPP, except where macroalgae or seagrasses are common. Reef flats will often be intermediate in NPP, but because they occupy a large proportion of the total area, they are major potential contributors to food webs (Polunin and Klumpp, 1992b). Reef flats and margins thus constitute major ultimate sources of organic carbon for fishery productivity. Sandy and deep-water habitats are likely often to be sinks in this regard, although their large areas may support more fishery productivity than is usually recognized.

The significance of this low productivity of sandy habitats to fisheries will depend on the target species involved. The principal prey of many demersal carnivores, for example, may be in lagoonal sediments even though the ultimate source of organic carbon for these prey may be the NPP of adjacent reef margins. The magnitude of community GPP and respiration in source and sink zones, their relative sizes and the flux of carbon between them are evidently the major determinants of the low overall NPP of whole reef systems. The main reasons for this disparity between rocky source and sedimentary sink zones seem to be that algae cannot be completely grazed down on hard substrata and some loss of algal carbon from productive reef zones is inevitable with water movement. This exported carbon tends to accumulate in areas of low water movement and high sedimentation. Corals do not grow well in such areas and thus fine-sandy substrata predominate.

5.3 HOW DO THE CARBON SOURCES SUPPORT FISH PRODUCTIVITY ON TROPICAL REEFS?

The biomass of fishes on margins of mid- to high-latitude (12–32°N or S) tropical reefs has been estimated by a variety of techniques, including explosives and rotenone collections, and modelling. Estimated biomass has varied by a factor of 10, from 26 to 238 t wet wt km^{-2} (approximately 2–

24 tC km^{-2} if wet weight is 10% carbon; Vinogradov, 1953) at eight widely separated sites (Goldman and Talbot, 1976; Williams and Hatcher, 1983; Polovina, 1984; Arias-Gonzalez, 1993; Opitz, 1993). The range is also high within areas, for example 17–195 t wet wt km^{-2} among reef zones at One Tree Island (Goldman and Talbot, 1976) and 92–237 t wet wt km^{-2} among reefs off Townsville (Williams and Hatcher, 1983), on the Great Barrier Reef. The data are too limited to indicate whether any trends exist, for example of fish biomass with respect to latitude.

Studies of the gut contents of fishes and other large consumers constitute the main body of information on the food webs which support them (J.D. Parrish, 1989). These investigations include those of Hiatt and Strasburg (1960), Hobson (1974) and Sano *et al.* (1984b) in the Pacific, Vivien (1973) in the western Indian Ocean and Randall (1967) in the Caribbean. Feeding habits of tropical reef animals are very varied, but among fishes a basic categorization for the purposes of this account is into herbivores, planktivores, benthic invertebrate-feeders and piscivores.

Herbivory

There are many functional groups both of herbivores and of algae on tropical reefs (Jones, 1968; Steneck, 1988; Glynn, 1990). Consumption of algae, for which I will use the general word grazing, constitutes a dominant flux of organic carbon. Major macrograzers (body size >1 cm) on reefs include damselfishes (Pomacentridae), parrotfishes (Scaridae), surgeonfishes (Acanthuridae), rabbitfishes (Siganidae), blennies (Blenniidae), chub (Kyphosidae), gastropod molluscs and sea urchins. Predominance of sea urchins in some areas is thought to be a result of predator removal by fishing (Hay, 1984). Mesograzers (body size >0.2 mm <1 cm), including amphipods and harpacticoid copepod crustaceans, may also be important consumers of the NPP.

Grazers constitute 16–24% of the total fish biomass on small reefs or around the margins of large reefs (Randall, 1963a; Talbot, 1965; Goldman and Talbot, 1976; Harmelin-Vivien, 1981; Williams and Hatcher, 1983; Parrish *et al.*, 1985). Individual grazers vary greatly in their weight-specific ingestion rate, but single species may graze a high proportion of NPP on reefs, as indicated by a 20% figure for the sea-urchin *Diadema antillarum* (Hawkins and Lewis, 1982) and 30–70% for three fishes (Chartock, 1983; Polunin; 1988; Klumpp and Polunin, 1989). Grazing is seasonal, at least at the high-latitude sites studied, and a strong correlation with temperature has been indicated (Polunin and Brothers, 1989). Macrograzing is typically high on outer reef margins (Hatcher, 1981; Russ, 1987; Polunin and Klumpp, 1992a), and may be reduced on level inner reef flats which lack shelter and become tidally exposed

(Polunin and Klumpp, 1992b). Grazers in many cases partition feeding habitat on reefs apparently according to processes such as competition, in relation to factors such as tidal exposure and wave action (Klumpp and Polunin, 1990). Availability of suitable algal foods and susceptibility to predators will also influence grazing. Relief and actual shelter, such as boulder tracts, can locally increase herbivorous-fish abundance. Where fish grazing is locally at a low level, other herbivores such as molluscs may become important (Harmelin-Vivien, 1981).

There have been few attempts to measure levels of total grazing on tropical reefs in relation to NPP. Across a windward reef tract of the central Great Barrier Reef, one-third of the NPP is consumed by macrograzers, nearly all fishes, and another one-third by mesograzers (body size >200 μm <1 cm; Polunin and Klumpp, 1992b). On a fringing reef crest in the Virgin Islands (Caribbean) 80–100% of the NPP was taken by herbivorous sea urchins and fishes, and little by smaller grazers (Carpenter, 1986).

The primary producers of tropical reefs have relatively low concentrations of nitrogen and phosphorus, and are characterized by high grazer consumption rates (90–400 gC gC^{-1} $year^{-1}$; Klumpp and Polunin 1989; Polunin and Klumpp, 1992a). The efficiency with which reef-algal diets are absorbed across herbivore gut-walls is high in some cases, but grazer populations are often dominated by older, slow-growing individuals which channel this uptake into body growth very inefficiently (Table 5.1). This carbon may be lost as carbon dioxide through respiration, or it may be channelled into reproductive output. On the one hand, these grazers are operating at high ambient temperature and therefore respire much of their carbon intake (Scholander *et al.*, 1953). On the other hand, many species also spawn repeatedly during their reproductive season (Robertson, 1991a) and these will therefore translocate much of their carbon intake to reproductive effort. A substantial carbon flux through reef grazers is apparently channelled into large carnivores via those species feeding on released eggs (Polunin and Brothers, 1989).

Although the NPP of sandy habitats appears low, fishes such as mullet (Mugilidae) and some gobies (Gobiidae) can be abundant around reefs and specialize in feeding on unicellular algae and detritus in sand. Few large species graze directly on seagrass (Harmelin-Vivien, 1983). Macroalgae appear to be common only where they are little grazed (Lewis, 1986), although some rabbitfishes, unicornfishes (Acanthuridae) and chub concentrate on them. Any productivity of such plants in the vicinity of a reef is likely to be exported as detritus or exploited by microbial decomposers and detritivores as a basis of reef fish productivity. Conversely, there may be significant import from adjacent coastal waters to reefs where such large plants occur.

Table 5.1 Indicative dietary absorption (organic matter or carbon basis) and gross growth efficiencies (carbon or energy) of particular grazers

Species	Efficiency (%)	Source
Absorption		
Diadema antillarum	26–72	Hawkins (1981)
	−126–+13	Hawkins (1981)
Acanthurus guttatus	11	Chartock (1983)
A. olivaceus	16	Chartock (1983)
A. triostegus	14	Chartock (1983)
Ctenochaetus striatus	20	Nelson and Wilkins (1988)
Plectroglyphidodon lacrymatus	58–78	Polunin (1988)
Stegastes lividus	72	Lassuy (1984)
Holacanthus bermudensis	75	Menzel (1959)
Gross growth		
Diadema antillarum	6–15	Hawkins and Lewis (1982)
Plectroglyphidodon lacrymatus	0.2	Polunin and Brothers (1989)
Scarus frenatus	0.8	Polunin and Klumpp, unpubl. data
S. sordidus	0.1	Polunin and Klumpp, unpubl. data

Planktonic inputs

A variety of planktivores occur on tropical reefs (Davis and Birdsong, 1973), with distinct nocturnal and diurnal assemblages. The nocturnal community includes some soldierfishes (Holocentridae), bigeyes (Priacanthidae), cardinalfishes (Apogonidae) and many sessile filter-feeding invertebrates such as corals and bivalve molluscs. They are common in back-reef areas as well as on exposed reef fronts and rely heavily on prey that shelter in the reef by day. In many areas, feeding on larger plankton by day is dominated by fishes, including damselfishes, fusiliers (Caesionidae), silversides (Atherinidae), anchovies (Engraulidae), half-beaks (Hemiramphidae) and scads (Carangidae). These diurnally active species exploit mostly plankton from the adjacent open sea (Sorokin, 1990) and are only abundant at reef margins (Harmelin-Vivien, 1981). Plankton-feeding fishes have been estimated to make up 8–70% of the total fish biomass at Pacific reef margins (Goldman and Talbot, 1976; Williams and Hatcher, 1983; Parrish *et al.*, 1985).

To my knowledge, the input to reefs of neritic plankton has only been estimated for the windward edge of Davies Reef on the central Great Barrier Reef. There diurnal planktivorous fishes are abundant and constitute what has been described as a 'wall of mouths' (Fig. 5.2); their consumption was calculated to be 3 kgC m^{-1} of reef front $year^{-1}$ (Hamner *et al.*, 1988), which is substantial for the reef front. Some of this carbon uptake is evidently passed on to piscivores that feed along the reef front. In terms of nitrogen and other essential elements less available than carbon, this input must be especially important. Stable isotope analyses are a way of differentiating between reef-derived and planktonic source materials in reef food chains, but the results so far are equivocal. On the one hand, growth of a temperate reef predator that feeds in open water is evidently supported by planktonic inputs (Thomas and Cahoon, 1993). Conversely at Bermuda the nitrogen in some carnivorous reef fishes comes largely from nitrogen-fixing algae (Schoeninger and DeNiro, 1984). Planktonic inputs to reef fish productivity are all the more significant when their higher nitrogen content is considered together with the particular susceptibility of pelagic prey fishes to large piscivores (Hobson, 1968). During spawning seasons, plankton feeders such as damselfishes and fusiliers may also be major beneficiaries of eggs released into open water, but such intake represents a recycling of material within the reef. There are no estimates of ingestion rates of eggs by reef planktivores (e.g. Pauly, 1989).

Planktonic carbon is also passed on to a variety of species, including some herbivores, which readily eat the faeces of planktivores (Robertson, 1982), and this intake must further contribute to production of predators. The fate of small plankton, pelagic micro-organisms and incoming detritus has scarcely been studied. Bacterial production supported by dissolved and particulate detritus could alone be equivalent to 20% of the GPP at reef fronts (Ducklow, 1990). There are in addition detrital inputs from the defecation of invertebrate-feeding fishes that forage in adjacent seagrass beds at night and spend the day over reefs (Meyer *et al.*, 1983).

Invertebrate feeding

The majority of fish species on tropical reefs are benthic invertebrate-feeders and these are probably the greatest contributors to the fish biomass on most reef margins (Randall, 1963a; Talbot, 1965; Goldman and Talbot, 1976; Williams and Hatcher, 1983; J.D. Parrish, 1989). The soldierfishes, snappers (Lutjanidae), emperors (Lethrinidae), and sweetlips and grunts (Haemulidae) are mostly nocturnal foragers which seek large (body size > 1 cm) mobile prey.

Fishes that feed mostly on large mobile invertebrates during the day include sandperches (Mugiloididae), gobies (Gobiidae), wrasses (Labridae), groupers (Serranidae), scorpionfishes (Scorpaenidae), goatfishes (Mullidae), threadfin bream (Nemipteridae), emperors (Lethrinidae), triggerfishes

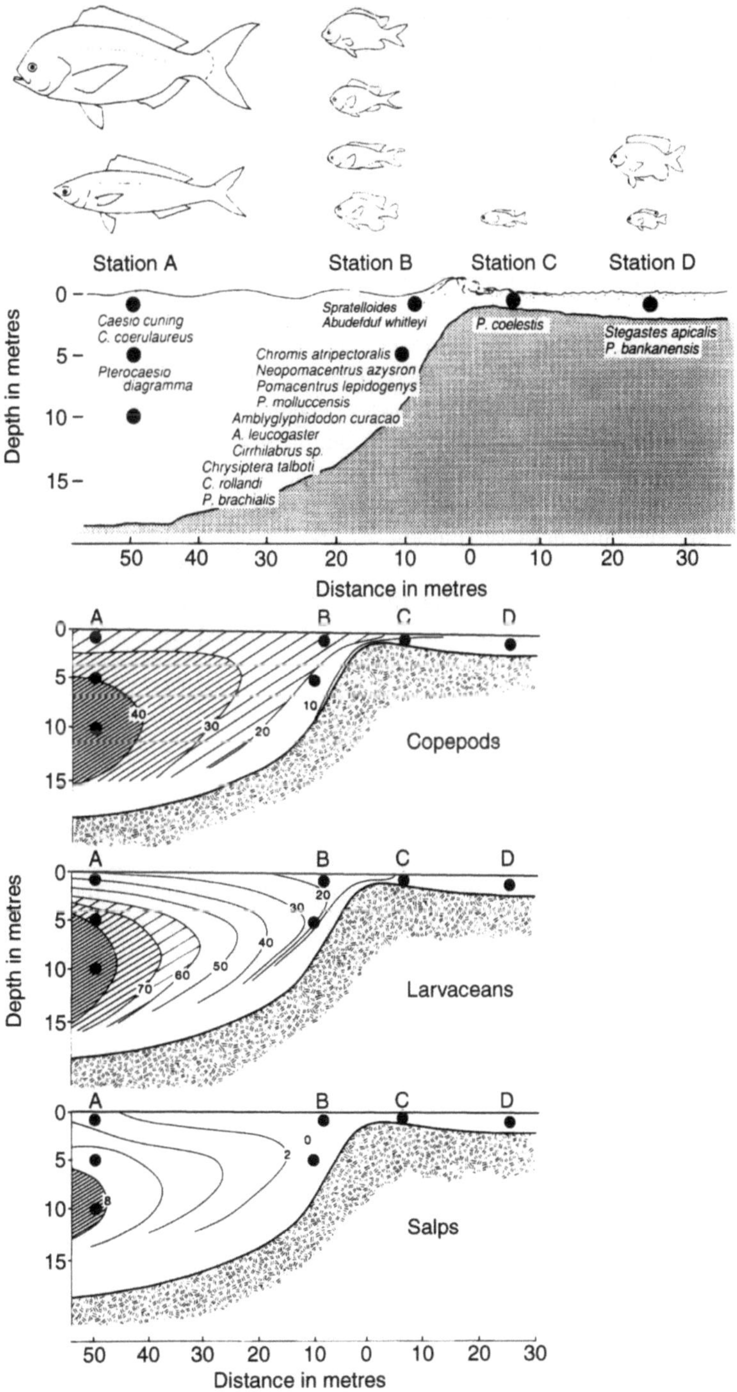
Station A
Station B
Station C
Station D
Depth in metres
Caesio cuning
C. coerulaureus
Pterocaesio diagramma
Spratelloides
Abudefduf whitleyi
Chromis atripectoralis
Neopomacentrus azysron
Pomacentrus lepidogenys
P. molluccensis
Amblyglyphidodon curacao
A. leucogaster
Cirrhilabrus sp.
Chrysiptera talboti
C. rollandi
P. brachialis
P. coelestis
Stegastes apicalis
P. bankanensis
Distance in metres
Copepods
Larvaceans
Salps
Depth in metres
Distance in metres

(Balistidae) and porcupinefishes (Diodontidae). Important mobile invertebrate prey include shrimps, stomatopods, crabs, sea urchins, gastropods and polychaetes. On two well-studied reefs, in the Indian Ocean and Pacific, crabs have been found to be the single most important dietary item (Harmelin-Vivien, 1981; Parrish *et al.*, 1985).

Few fishes of tropical reefs feed on corals, sponges, ascidians and other sessile invertebrates. Those that do include butterflyfishes (Chaetodontidae), pufferfishes (Tetraodontidae), filefishes (Monacanthidae), boxfishes (Ostraciidae) and triggerfishes (Balistidae). Most species are active only by day.

There are big spatial differences in the trophodynamic importance of invertebrate feeding (Harmelin-Vivien, 1981), but there is little information on the energetics of individual species. Feeding rates derived empirically for a few invertebrate-feeders suggest biomass-specific ingestion of 4–8 g wet wt g wet wt^{-1} $year^{-1}$ for three large snappers and three grunts (Pauly, 1989), but a multiple regression predicting ingestion in terms of temperature, body size and activity (Palomares and Pauly, 1989) indicates higher values of 8–15 g wet wt g wet wt^{-1} $year^{-1}$ for actual invertebrate-feeders (Arias-Gonzalez, 1993). Relating this to the likely biomass of such fishes on tropical reefs suggests that they will often ingest invertebrates at over 20 tC km^{-2} $year^{-1}$.

Piscivory

Most piscivores are fishes. Large piscivores include requiem sharks (Carcharhinidae), dogfishes (Triakidae), jacks (Carangidae), barracudas (Sphyraenidae), tunas (Scombridae), garfishes (Belonidae), groupers, conger eels (Congridae), moray eels (Muraenidae) and some snappers. There are also many smaller piscivores, including snake eels (Ophichthidae), hawkfishes (Cirrhitidae), lizardfishes (Synodontidae), cornetfishes (Fistulariidae) and trumpetfishes (Aulostomidae).

The contribution of piscivores to total fish biomass is variable, but can be large. Estimates range from 2% to 54% among reef margins in the Indo–Pacific (Goldman and Talbot, 1976; Harmelin-Vivien, 1981; Williams and Hatcher, 1983; Parrish *et al.*, 1985), and 1–26% among zones within an Indian Ocean site (Harmelin-Vivien, 1981).

Fig. 5.2 The 'wall of mouths' at Davies Reef, Great Barrier Reef. (top) The diurnal planktivorous fishes and their positions while feeding in the water column and across the reef. (bottom) Reduction in densities of copepods, larvaceans and salps as water masses encounter and cross the 'wall'. From Hamner *et al.* (1988), with permission of *Bulletin of Marine Science* and the authors.

Feeding rates of a few large piscivores have been calculated (Cortes and Gruber, 1990; Sudekum *et al.*, 1991) and annual ingestion is 14–19 g wet wt g wet wt^{-1} $year^{-1}$ in the jack *Caranx melampygus* (body size 2–10 kg), 10–20 g wet wt g wet wt^{-1} $year^{-1}$ in *C. ignobilis* (2–40 kg) and 6 g wet wt g wet wt^{-1} $year^{-1}$ in the shark *Negaprion brevirostris* (25 kg). Fish prey can be predominantly derived from a small number of families, such as goatfishes, wrasses, cardinalfishes and soldierfishes in the north-west Hawaiian Islands (Norris and Parrish, 1988), but there are insufficient data to indicate whether geographical trends exist in piscivore diet composition.

Ingestion rates will be greater at higher ambient temperatures, rising perhaps 1.4-fold for a 10°C warming (Pauly, 1989); greater at smaller body sizes, for example 43 g wet wt g wet wt^{-1} $year^{-1}$ in a 12 g lizardfish, *Synodus englemani* (Sweatman, 1984); and greater at higher activity levels, for example 8–32 g wet wt g wet wt^{-1} $year^{-1}$ in various pelagic species (Pauly, 1989). Based on the multiple regression of Palomares and Pauly (1989), mean values of 7–15 g wet wt g wet wt^{-1} $year^{-1}$ were considered indicative for a reef at 17°S in the Pacific (Arias-Gonzalez, 1993). Estimates of the large-scale magnitude of piscivory on reefs have rarely been made. At French Frigate Shoals in the north-western Hawaiian Islands, jacks and sharks alone are estimated to maintain a high prey consumption rate of >6 tC km^{-2} $year^{-1}$ (Sudekum *et al.*, 1991). On two shallow reefs of Moorea, French Polynesia, total piscivory is thought to be more than twice that amount (Arias-Gonzalez, 1993). Piscivory and other forms of predation are clearly very important on tropical reefs (Hixon, 1991), but we have few data on rates and impacts.

5.4 WHAT IS THE ECOSYSTEM BASIS OF FISHERY PRODUCTIVITY?

Prediction of reef fish productivity from trophodynamic first principles is constrained by limited data. There are few detailed studies of the energetics of particular species, whether at the top or bottom of food chains around tropical reefs (e.g. Polunin, 1988; Sudekum *et al.*, 1991). Even in food-web studies there are important data lacking at both ends of the body-size spectrum. On the one hand there are very few data on the micro-organisms, mesofauna, and indeed on the young of most fishes, all of which on a weight-specific basis may be very important in productivity terms. On the other hand, large predators can clearly have substantial trophodynamic impacts, yet we know relatively little of their densities and detailed distributions. Detritus is clearly very important also, yet as ever, it is poorly quantified in food-web studies (Alongi, 1988; Ducklow, 1990).

It is with respect to little-studied ecosystem processes that modelling becomes a useful approach to understanding reef fish productivity. None of the many types of models releases us from the constraint of limited data. It does, however, provide an opportunity to collate information from diverse studies, describe the context in which fish is produced, and interpolate rates of processes such as defecation, decomposition and invertebrate feeding, which we know are very important, but have yet to measure extensively.

Modelling

Models of the biological basis of fish productivity have to date been either biomass-centred or process-orientated. A process-orientated trophodynamic model was developed for Davies Reef, Great Barrier Reef by Polunin and Klumpp (1992b) using data on benthic NPP, grazing and planktonic inputs for a 410 m long tract of reef which was divided into six hard-substratum zones and sandy habitat. The model postulated six food chains based on grazing and three on planktivory, and suggested that the shortest chains, and potentially those most productive of fish, were those through which least carbon passed (Fig. 5.3). The main reason for this was that direct use of somatic production by predators was thought unlikely, while most carbon might in fact be allocated to gamete production. This production would not be directly accessible to those predators, but rather would be fed upon by specialists which would constitute an extra link in the food chain. Approximately one-third of the NPP was macrograzed, one-third mesograzed and the remainder converted to detritus. The available data (Hamner *et al.*, 1988) suggest a small planktonic input to fish productivity in carbon terms at this site.

ECOPATH is a biomass-based compartmental modelling software originally developed by Polovina (1984) and used to describe the fluxes of materials through French Frigate Shoals, Hawaii. It has since been elaborated (ECOPATH II, Christensen and Pauly, 1992a) and applied to tropical reefs of the central Caribbean (Opitz, 1993) and of Moorea, French Polynesia (Arias-Gonzalez, 1993). A more tentative application has been to a reef at Bolinao, Philippines (Aliño *et al.*, 1993).

The basis of the ECOPATH model is that functional groups are identified, and for each both biomass and biomass-specific processes such as ingestion are known, and steady state with respect to biomass is assumed. Thus production by each group is assumed to be balanced by the outputs, namely respiration, natural mortality (predation, disease etc.) and export. If we accept this, then for a group which we might designate i: production by (i) – predation on (i) – non-predatory losses from (i) – export from $(i) = 0$.

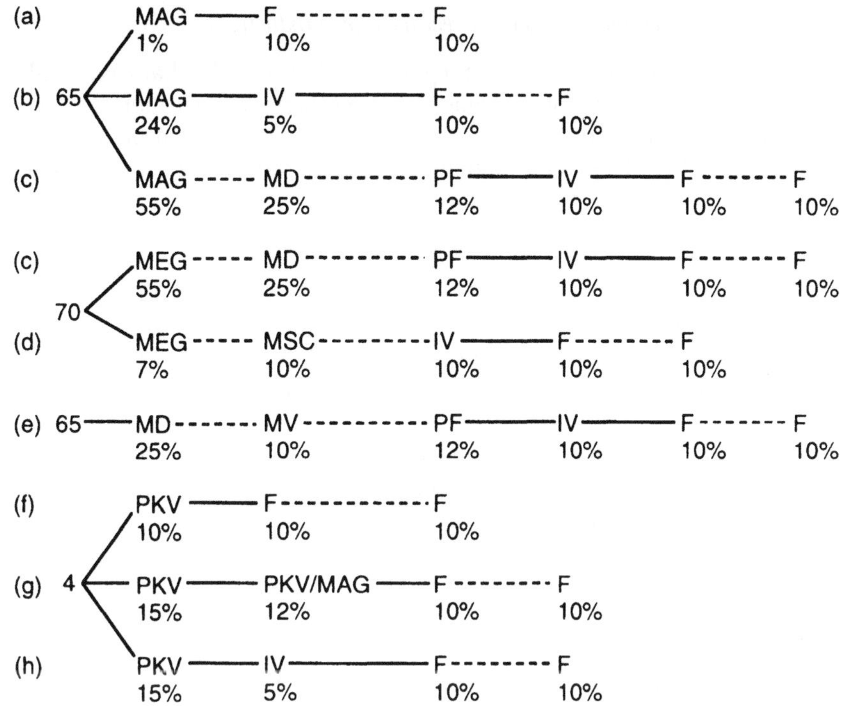

Fishery resource targeted	Yield from reef tract (t km^{-2} tract)	Yield from whole of Davies Reef (t km^{-2} reef)
	A	B
MAG	13.9	4
IV	22.3-48.4	7-16
F	0.5-6.7	0-2
PKV	8.8	3
MAG + IV	36.2-62.3	11-20
MAG + IV + PKV	45.0-71.1	14-23

Fig. 5.3 Principal food chains considered to support fish productivity at Davies Reef, Great Barrier Reef. (top) The six food chains based on algal productivity (chains a–e correspond to the a–e of Fig. 5.1, bottom) and three chains based on planktivory. MAG macrograzer (body size > 1 cm); F, piscivore; IV, invertebrate feeder; MD, microbial decomposer; PF, particle feeder; MEG, mesograzer (body size < 1 cm > 200 μm); MSC, mesocarnivore; MV, microbivore; PKV, planktivore. The short but unproductive food chains referred to are (a) and (f), where predators use the somatic production of consumers, as opposed to the gamete production at (b) and (g). (bottom) Estimated fish productivity, which varies according to which groups are hypothetically targeted to produce the 'yield'. Column A is calculated from organic carbon yield converted to total dry weight by multiplying by a factor of 2.5 and to fresh weight by a further factor of 3.33. Column B is derived from A by multiplying by 0.33, the proportion of the total area that is reef. From Polunin and Klumpp (1992b), by permission of Oxford University Press and the authors.

'Export' may be contributed to by emigration and by harvest, while 'non-predatory losses' consist of mortality from disease, starvation, parasitism, injury etc. The proportion of production that is exported or predated is the 'ecotrophic efficiency'. Mathematically the above can be expressed as:

$$P_i - B_iM2_i - P_i(1 - EE_i) - EX_i = 0 \,, \tag{5.1}$$

where: B_i is biomass of (i); P_i is production of new tissue by (i); $M2_i$ is predation mortality of (i); EE_i is ecotrophic efficiency of (i) = proportion of productivity predated or 'exported'; $(1 - EE_i)$ is the 'other mortality' (MO) of (i) = $P - (EX +$ predation$)$ = disease, starvation, and EX_i is export of (i) = catches + emigration.

If we know the productivity per unit of biomass (PB), then productivity is the product (* denotes multiplication) of biomass and the productivity per unit biomass (PB), or

$$P_i = B_i * PB_i \,. \tag{5.2}$$

If we know the combined biomass of predators (j) feeding on (i), ($\Sigma_j B_j$), their consumption per unit biomass (QB_j) and the proportion of prey (i) in the average predator's diet (DC_{ji}), then predation on (i),

$$B_iM2_i = \Sigma_j B_j * QB_j * DC_{ji.} \tag{5.3}$$

We can also say that non-predatory losses on (i),

$$P_i(1 - EE_i) = B_i * PB_i(1 - EE_i). \tag{5.4}$$

Re-assembling Equation 5.1 from Equations 5.2–5.4 we can say that:

$$(B_i * PB_i) - (\Sigma_j B_j * QB_j * DC_{ji}) - (B_i * PB_i(1 - EE_i)) - EX_i = 0. \tag{5.5}$$

This leads to the equation:

$$(B_i * PB_i * EE_i) -(\Sigma_j B_j * QB_j * DC_{ji}) - EX_i = 0. \tag{5.6}$$

Thus by applying Equation 5.6 we calculate for group (1):

$$\begin{aligned}(B_1 * PB_1 * EE_1) - (B_1 * QB_1 * DC_{11}) - (B_2 * QB_2 * DC_{21})\\ - \ldots - (B_n * QB_n * DC_{n1}) - EX_1 = 0,\end{aligned} \tag{5.7}$$

and for the nth group we have:

$$\begin{aligned}(B_n * PB_n * EE_n) - (B_1 * QB_1 * DC_{1n}) - (B_2 * QB_2 * DC_{2n})\\ - \ldots - (B_n * QB_n * DC_{nn}) - EX_n = 0.\end{aligned} \tag{5.8}$$

The function of ECOPATH is to run such simultaneous equations and balance the biomass fluxes across all groups for given input values and rates.

The information for comprehensive runs involving tropical reefs comes from many sources. For consumer animals *PB* is derived from total mortality,

$$Z_i = PB_i. \tag{5.9}$$

Estimates of mortality can come from fishery catch curves or a multiple regression which predicts mortality on the basis of body size, ambient temperature and growth parameters (Pauly, 1980b). Values of *QB* may come from empirical data, or from other multiple regressions predicting food consumption from body size, ambient temperature and other measures (Cammen, 1980; Pauly, 1989). Biomass information may come from sampling, or from abundance and body size estimates (Peters, 1983). Dietary data will rely heavily on food-web studies, while some other fluxes such as primary productivity and detritus production are derived from investigation of population energetics or system trophodynamics. Catch data are derived from field studies, but there are few data on emigration, the other aspect of export. It is clear in all of this that average values of parameters will have to be chosen, in spite of the range of species and sizes of individuals often making up groups. Averaging is also necessary across some spatial and temporal variations in the environment.

Reef processes in general

The ECOPATH models of Polovina (1984) for French Frigate Shoals (FFS) off Hawaii, of Arias-Gonzalez (1993) for two reefs at Moorea in French Polynesia and of Opitz (1993) for a generalized reef of the central Caribbean have all employed very high ecotrophic efficiencies (Equation 5.1). An illustration of this feature is given in Fig. 5.4, and the high values reflect the low system NPP of many tropical reefs, which has been alluded to above.

These models also indicate that fishes constitute only a small part (3–10%) of the total live reef biomass, which excludes detritus. Because algae make a large contribution to total biomass, however, the fish contribution to combined consumer biomass, both animal and microbial, is large (16–42%). This has a number of consequences. It underlines the trophodynamic importance of fishes on tropical reefs and indicates likely ecosystem-wide effects of changes in their biomass with exploitation. When nutrient elements such as nitrogen are considered, the role of fishes becomes still more significant; substantial amounts of nitrogen are stored in fish biomass and are removed with harvesting.

In spite of the prominence of grazing, much algal carbon is converted to detritus, while the high levels of grazing, combined with relatively low dietary absorption by grazers (Table 5.1), mean that flows to detritus are further enhanced. Much of the reef fish productivity turns out to be

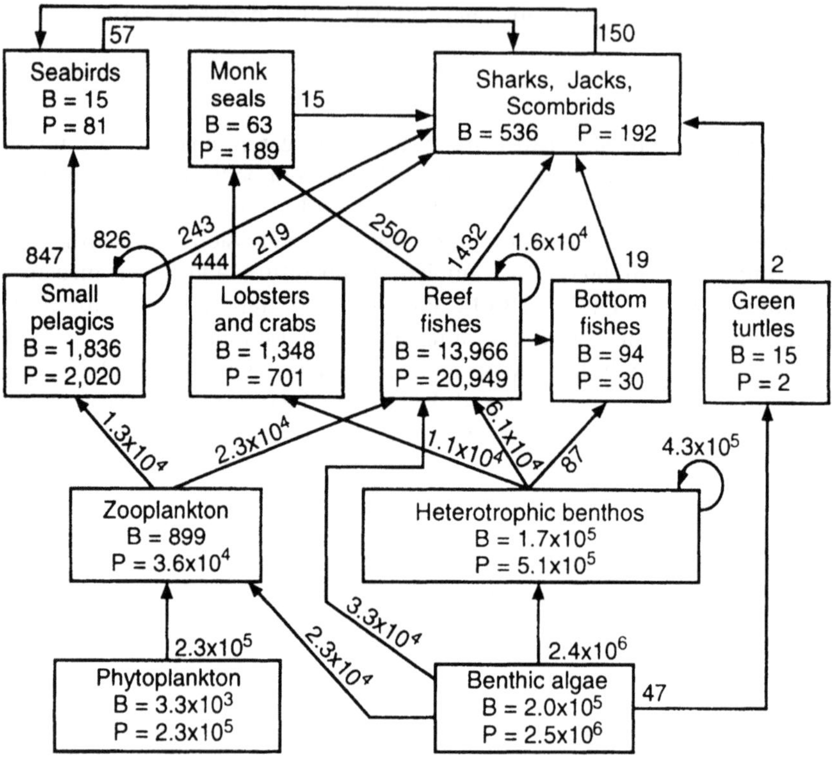

Fig. 5.4 Simple trophodynamic model of French Frigate Shoals, Hawaiian Islands, in units of kg fresh weight km^{-2}, based on 10 consumer groups and two primary producer groups, where *B* is biomass and *P* is productivity. The high estimated EEs, ecotrophic efficiencies, can be seen by comparing the *P* value with the total output flux from each group. From Polovina (1984), by permission of Springer Verlag and the author.

based upon detrital fluxes, which are poorly quantified (Alongi, 1988; Wiebe, 1988). The material involved, being ultimately derived from algae which are low in nitrogen and phosphorus, is of a quality so poor that only decomposer micro-organisms can make extensive use of it. The microbial ecology of reefs is little studied (Ducklow, 1990). Microbial–detrital material is used by filter feeders such as bivalves, copepods and ascidians in the water column, and by deposit feeders such as brittle stars, crabs and many polychaetes. From this observation, the manifestly large role of invertebrate-feeding in the fish productivity of tropical reefs is implied.

Fish productivity and effects of environment

Estimates from trophodynamic models of fish productivity of reefs in the Pacific vary from 14–23 $t\,km^{-2}\,year^{-1}$ for Davies Reef (18°S, Polunin and Klumpp, 1992b), through 26–30 $t\,km^{-2}\,year^{-1}$ for Moorea (17°S, Arias-Gonzalez, 1993) to 35 $t\,km^{-2}\,year^{-1}$ for French Frigate Shoals (24°N, Polovina, 1984). Not all of this productivity will be exploitable under normal circumstances. For example, many individuals will be too small to catch. In addition, many strategies of exploitation are conceivable. In spite of appearances, targeting grazers alone might not maximize productivity (Polunin and Klumpp, 1992b). Conversely, removal of large piscivores can be expected to increase availability of prey species to exploitation (Polovina, 1984). There are also indications that there may be alternative states of the system; under certain conditions, sea urchins apparently exclude parrotfishes (for example, McClanahan, 1994) with likely food-chain consequences.

Although rather little information is available on food-chain efficiencies and processing of detritus (Alongi, 1988), greater supply of on-reef (benthic-algal NPP), or off-reef (plankton or detritus), sources of carbon can be expected to stimulate fish productivity. Within bounds, many factors are potentially conducive to enhanced reef NPP, including space, nutrients, depth and grazing (e.g. Grigg *et al.*, 1984). Many factors may also enhance plankton and detrital inputs to reefs, including proximity to habitats with high system NPP, advection and nutrients (Mann, 1993). Work in the Pacific indicates that on open ocean reefs, long-term changes in population processes may be driven by offshore processes such as upwelling (Polovina *et al.*, 1994).

5.5 WHAT DETERMINES FISH PRODUCTIVITY AND COMPOSITION?

Modelling has helped to synthesize information on the trophodynamic basis of fish production. It is clear that additional data are needed for more accurate pictures of reef functioning. What seems most needed is an analysis of how fishery yields are built up from the trophodynamic processes of particular reefs. In the meantime, however, it is worthwhile trying to establish what factors may determine large-scale variation in fish productivity. Production of fish being a multistage process, it follows that a host of factors might as a whole determine productivity. Rather than attempt to provide a catalogue of possible factors about which almost nothing is known, I shall rather pick a few trends which might be tested in future work.

Reef location

Nutrient availability is a likely determinant of fish productivity, and major sources of inorganic nutrients in tropical waters are the land and upwelling. It is not known what factors might affect nitrogen fixation on reefs, but external supply of ions such as nitrate and phosphate are expected, up to a point, to stimulate primary productivity, whether planktonic or reef-based. Consequently, fish productivity should typically be greater on reefs close to large land masses than those in nutrient-depleted oceanic waters. Inshore reef fisheries should be greater next to large land masses (e.g. continents) than next to small land masses (e.g. atolls). Reefs in regions of upwelling will have greater fish productivity, although the relative roles of algal and planktonic inputs cannot be anticipated at this stage. There are apparently no data with which to test this idea (Dalzell, Chapter 7).

Reef shape and size

Reef margins are productive in terms of NPP and they also receive nutrient inputs from outside the system. Other things being equal, the fish productivity of small reefs should therefore be greater than that of larger reefs because of the greater contribution per unit area of reef margin. There is positive, although limited, circumstantial evidence to support this prediction (Arias-Gonzalez, 1993). It is also conceivable that for reefs of equal size, reefs with corrugated margins will produce more fish than those with smoother margins. In either case, if planktivorous fishes and reef-front piscivores in general are limited by outside inputs, then composition of the overall stock will be different, having a higher proportion of such species.

Reef depth and latitude

Because reef NPP declines rapidly with depth, deep reefs will tend to have a higher proportion of planktivores and reef-front piscivores than shallow reefs. There are indications that this may be the case for some reefs (Haight *et al.*, 1993). In the absence of substantial exogenous inputs such reefs will also have lower fish productivity in oceanic regions than in upwelling regions. There do not appear to be data available to test this prediction. Most reefs studied are at comparatively high latitudes. It is not clear what effects latitudinal changes might have on fish productivity.

Adjacent systems

Detrital inputs from adjacent habitats with a high export of carbon, such as mangroves and seagrass beds, can be expected to stimulate fish productivity on reefs, other things being equal. This idea will be difficult to test in practice because areas with well-developed mangrove and seagrass will also often have confounding characteristics such as turbidity, siltation and high nutrient levels.

Population restructuring

The indication that grazer populations may be inefficient at transferring algal productivity up food chains on reefs depends on two things: high reproductive effort and a high proportion of the biomass held in old individuals with slow body growth. Increase in mortality rates, such as through fishing, will lower the average age and favour faster-growing individuals, but it may increase prey availability to large predators still further because more of the prey productivity will be available directly to them. With 'growth overfishing', overall fish productivity at a site may increase substantially, not only because grazers are growing faster but also because food-chain transfer is more efficient. It is important that the energetic basis of this argument be developed further and that we know more about the age structure of populations, particularly those low in reef food chains, in order to define better the relative importance of numerical replenishment and individual body growth.

ACKNOWLEDGEMENTS

I thank the British Overseas Development Administration and Natural Environment Research Council for funding, and David Klumpp, Richard Grigg, Tony Hawkins and Villy Christensen for helpful reviews of this chapter.

Chapter six

Geography and human ecology of reef fisheries

Kenneth Ruddle

SUMMARY

Throughout the tropics, reef fish stocks are exploited mainly by individual households and communities to fulfil subsistence requirements and often to yield a small saleable surplus. Small-scale commercial fisheries are also common. As a consequence, the social and economic organization of reef fisheries varies considerably, precluding useful generalization here. On the other hand, harvesting in reef fisheries exhibits similar adaptations to the physical characteristics of site and cyclical change within it. Gear types are also relatively similar and simple, with the notable exception of parts of the Pacific Basin and South East Asia. A brief discussion of these similarities forms the core of this chapter. Another shared feature is that almost everywhere reef fisheries are under intensifying pressure from external forces, especially urban and commercial growth. These pressures and others lead to more intensive fishing and many contribute to the weakening of local systems that hitherto controlled it. The geography and human ecology of reef fisheries, despite some unifying similarities, is a complex and locally detailed subject, which can be only cursorily summarized here. Because reef fisheries are fast changing, and in many cases are becoming extinct, to make systematic and comprehensive geographical and human ecological studies of them is an increasingly urgent task.

Reef Fisheries. Edited by Nicholas V.C. Polunin and Callum M. Roberts.
Published in 1996 by Chapman & Hall, London. ISBN 0 412 60110 9.

6.1 INTRODUCTION

Reef fisheries are the domain of subsistence and small-scale commercial fishers, *par excellence* (Munro, Chapter 1). In many locations, and especially in the Pacific islands, such fishers employ ingenious gear types. Fishing may be minutely adapted to a complex and cyclically changing biological and physical environment. It is often based on sophisticated bodies of local knowledge. Catches may provision households and communities, satisfy local markets, fulfil an increasing urban demand for everyday food species and a growing international demand for speciality products. Such fisheries are characterized by the use of multiple gear types with multispecies targets. This complexity is heightened by the social and economic characteristics of the communities that rely on reef resource systems.

This chapter briefly describes the main characteristics of the microgeography and human ecology of coral reef fisheries, but the task is made difficult by the fragmentary and anecdotal nature of the existing literature and the rapid changes in fisheries caused by modernization. So the continued existence of characteristics described even a few short years ago cannot be assumed.

6.2 A SUMMARY CLASSIFICATION OF TROPICAL REEF FISHERIES

On a global basis, reef fishing patterns may be classified into two main types depending on human population density or remoteness of the reefs. Two other minor variants include those where only selected species are harvested, and those where ciguatera poisoning is a problem (Munro, 1984b).

Areas of high human population density

Reefs in such areas are usually massively overexploited by subsistence and small-scale commercial fisheries, as a consequence of a large local demand for food fish, low operating costs and lack of alternative employment. This situation is typical of South and South East Asia and the Caribbean (Munro, 1983a).

In South East Asia reef fisheries supply some 10–25% of the fish protein consumed by coastal zone dwellers (Langham and Mathias, 1977; Gomez, 1980) and constitute the only source of animal protein for communities living close by. In the Philippines, reef fisheries have been estimated to supply some 9% of the national fish catch (Murdy and Ferraris, 1980), but inclusion of subsistence and other unreported catches indicates the contribution may be 25% (Carpenter, 1977; Gomez, 1980; McManus, 1988). Similarly, reef fisheries of Sabah are estimated to supply some 25% of the total marine fish harvest (Mathias and Langham, 1978). An estimated 30% of the State of Trengganu marine fish catch, in Peninsular Malaysia, is derived from reefs (De Silva and Rahaman, 1982).

Areas of low human population density or remote locations

Underexploited fisheries are found in areas of low human population density such as parts of eastern Indonesia, much of Melanesia, and the outer islands of most Pacific island nations. Demand is so low that it is generally uneconomical to harvest reef fish other than for household consumption or small local markets. Satisfying distant markets is generally infeasible owing to a lack of both physical infrastructure and a distribution system. This is the case in much of the Pacific Basin, as in the Yasawa Islands of Fiji (Lal and Slatter, 1982), and in areas such as East Africa and the Sudanese coast of the Red Sea (Kedidi, 1984a). In the Tigak Islands of northern Papua New Guinea, for example, production declines with increasing distance from the Government Fish Purchasing Centre at Kavieng (Dalzell and Wright, 1990).

Most small-scale fishers are part-timers engaged in other activities, especially agriculture or wage labour (Lal and Slatter, 1982; Espeut, 1992). Occupational plurality is the main reason for low reef fisheries production in Papua New Guinea (Lock, 1986a). There, apart from a lack of financial motivation in the face of competing opportunities, especially in agriculture, development of small-scale fisheries is constrained by the distance of markets from landing sites and a perception that fishing is financially risky (Frielink, 1983).

Thus income-earning opportunities outside the fisheries sector can have a major impact on reef fisheries production. In the Tigak Islands, for example, reef fishing effort increases when copra prices are low and declines when they are high. Vegetable production and royalties from logging operations may also affect fishing effort in the same manner (Dalzell and Wright, 1990).

By contrast, the impact of market demand as an incentive driving development of reef fisheries is well illustrated by the thriving fishery in the vicinity of Port Moresby. Fishing there has become the main economic activity of some families in five villages west of the city (Lock, 1986a). Fishers from these villages exploit the local barrier and fringing reefs to supply fresh fish to the Port Moresby market.

In some areas, reef fisheries are underexploited owing to food preferences, a good example being the Maldives. Although that nation is composed entirely of reefs, reef fish have traditionally been very little exploited. Tuna (especially skipjack tuna) is both the favourite food of Maldivians and the export mainstay, although reef fish is eaten when the tuna fishing is poor (Anderson, 1992; Anderson *et al.*, 1992).

Reef fisheries based on selective market demand

Many fisheries only harvest top predators, such as groupers (Serranidae), snappers (Lutjanidae) and other large fishes. This typifies the fisheries of the Great Barrier Reef of Australia, and parts of the eastern and southern coasts of the USA (Munro, 1984b). They exist wherever customers are wealthy enough to pay the price, and are therefore a widespread characteristic of small-scale commercial fishing superimposed on subsistence fisheries. Because this chapter is focused on developing countries, this category will not be discussed further.

Reef fisheries constrained by ciguatera poisoning

Ciguatera is a form of fish poisoning which is derived from natural dinoflagellate toxins and transmitted up the grazing food chains of reefs. It is thought to become especially concentrated in the viscera of piscivorous reef fishes, to which most cases of poisoning are related. Ciguatera poisoning locally depresses demand for reef fish in many of the eastern Caribbean islands and in parts of the Pacific Basin (Munro, 1984b). Although such localized areas are known and avoided by local fishers, the ciguatera problem deters market intermediaries, such that commercial demand is generally depressed and fisheries may be left underexploited.

6.3 ELEMENTS OF REEF FISHERIES

Harvesting

An intimate knowledge of the complex physical and biological environments of tropical reefs underlies both the fishing technologies employed and the targeting of fishes by species. The Torres Strait Islanders, for example, denote coralline environments and fish habitats by finely subdivided taxonomies fitted together in a coherent framework which is at once both elegantly sophisticated and eminently practical. As Nietschmann (1989) wrote of the Torres Straits Islanders, '... the sea is knowable, useable and predictable. Marine plants and animals and sea and weather conditions are not thought of as being "out there" in some sort of undefinable, geographic mush, undifferentiated with respect to species, season or site'. The timing, location and conditions of occurrence

Table 6.1 Principal characteristics of reef fisheries in Yaeyama Archipelago, Okinawa Prefecture, Japan. After Ruddle (1987a)

	Inshore zone	Offshore zone	Open sea zone
Fishery	Subsistence	Small-scale artisanal	Large-scale commercial
Water depth (m)	2–12	Up to 60	Up to 90–120
Principal habitats	Lagoon, reef flat, surf zone, reef channel, beach, shoreline, estuary, mangrove swamp	Seaward slopes of barrier and fringing reefs	Surface layers and bottom
Techniques employed	Netting, spearing, trapping, collecting, squid trolling	Hand lining, spearing, gill netting, fish driving, lift netting, squid trolling	Trolling, hand lining, bottom longlining
Main harvest	Unicornfishes, jacks, silver biddies, garfish, wrasse, emperors, snappers, damselfishes, parrotfishes, groupers, rabbitfishes, seabream, octopus, squid, cuttlefish, sea urchin, shellfish, seaweeds	Reef species (as in inshore zone), and especially jacks in mid-water, bottom-dwellers, turtles	Bonito, spanish mackerel, wahoo, squid (at surface), emperors, snappers, seabream, groupers (on bottom)

of marine phenomena are regarded as entirely predictable. In their fisheries, decision-making islanders refer to a taxonomy of more than 80 distinct terms to describe different tides and tidal conditions, using a '... lunar-based system that keeps track of the four daily tides, changes in height and current speed, time of occurrence and duration, seasonal shift (due to changes in earth, sun and moon orbits), water clarity, surface conditions and associated movements of fish, sea turtles and dugongs' (Nietschmann, 1989).

The wide variety of habitats harvested by a range of gears which are minutely adapted to site and fish behaviour has also been described for the Yaeyama Archipelago of south-western Okinawa Prefecture, Japan (Table 6.1) (Ruddle, 1987a; Ruddle and Akimichi, 1989).

Physical characteristics of the site

Adaptation to tidal condition

Tidal condition is the principal factor governing accessibility to reef fishing spots, and hence activity patterns. Fishing is suppressed during high tide by deeper water and stronger currents. As Hill (1978) noted of American Samoans, despite the availability of different fishing methods, fishers' '... ability to harvest marine foods from the fringing reefs is strongly challenged by one meter of sea water'.

In American Samoa, as elsewhere, the relative importance of each fishing technique varies by tidal height and period. At low tide, reef gathering, done mainly by women and children (see below), assumes overwhelming importance, constituting about 55% and 27% respectively of the fishing effort. Diving and baited hook-and-line fishing are also used at low tide, becoming more important with increased tidal height. Both constitute 46% and 37% of techniques used at mid-tide and high tide, respectively. With increasing water depth, gathering declines in importance, to only 1.4% of effort and diving, baited hooks and use of outrigger canoes better adapted to deeper water conditions become more important. Gill netting, trapping and the use of cast nets are better suited to mid-tide conditions.

Use of different reef zones also varies tidally, with fishing activities moving seawards as water levels decline. At low water, reef margins, together with the reef flat, are worked by gatherers. At high tide, activities concentrate along the shoreline, and to a lesser extent the mid-reef, reef channels and front and inner reef zones. Mid-water is a transitional phase.

Fish behaviour in response to changing water levels also affects the zonation of gear use. During high water many species move shorewards from deeper water to the reef platform, where they can be harvested by hook and line, cast netting and diving. As the tide recedes, gill nets are set in or close to reef passages to catch fish returning to deeper waters. Fishers follow their targets into deeper waters during periods of low tide, and harvest them using lines from the exposed reef margins, by diving along the reef front and channels, and by cast netting and gathering over much of the reef.

Adaptation to reef zonation

Use of fishing techniques also varies according to reef zonation and degree of exposure to the open sea. Efforts naturally focus on the most productive zones and fishing spots. Thus, for example, gatherers concentrate on the outer and mid-sections of fringing reefs, seeking shellfish, octopus and echinoderms. Gill netting and trapping are usually focused on reef passages, outer slopes and the reef margin. Hook-and-line fishers and cast netters usually target mid-water and bottom species along the shore area and the reef margin.

Techniques used vary according to degree of exposure to onshore winds and oceanic swells, with those used on exposed coasts demonstrating a greater sensitivity to the diurnal tidal cycle than those used on embayed coasts. Along exposed coasts the predominant techniques are gathering and diving, with gill nets and traps set in sheltered locations and reef passages. Other techniques are infrequently used. By contrast, on embayed coasts baited hook-and-line fishing, reef gathering and diving assume almost equal importance. Other techniques are used relatively more frequently on embayed than on exposed coasts.

Along exposed coasts, most fishing is in the mid- and outer reef and reef margin zones, with the shoreline of relatively minor importance. By contrast, the various zones are used more evenly on sheltered coasts, with activities concentrated in the shore, mid-reef, reef margin, and outer reef zones.

Adaptation to the diel cycle

Composition of fishing labour, sites worked, gear used and target species all vary during the diel cycle. For example, in American Samoa both men and women fish by day, but night-time fishing is primarily a male task. This reflects mainly the overall household division of labour, but women also target mainly sessile invertebrates by gathering in daytime, whereas men seek nocturnally active lobsters and fish that must be taken by spearing. Cast nets are rarely used at night, owing to difficulty of detecting targets and increased likelihood of damage on coral. Artificial light sources are used in night-time reef gathering and diving, and sometimes in line fishing (Hill, 1978).

Adaptation to seasonal change

The principal seasonal change affecting fishing is change in prevailing wind direction between onshore and offshore. Partly because of increased difficulty caused by turbulent waters, but mostly owing to the presence of a seasonal upwelling on coasts with offshore winds, fishers shift the location of their activities, if not constrained by management systems that restrict entry (Ruddle, Chapter 12). Fishers may also adapt the location of activities to exploit periodic spawning aggregations, as for example in the reef channels of Marovo Lagoon, Solomon Islands (Johannes and Hviding, 1987; Johannes, 1989).

Social and cultural limitations on harvesting

Harvesting site and gear type used are determined not only by physical and biological criteria, but not uncommonly also by exclusive fishing rights territories and associated management systems (Ruddle, 1994a,b; Ruddle, Chapter 12) and other social and cultural controls (Ruddle, 1988a, 1994a,b; Ruddle *et al.*, 1992). The resultant situation is often complex. For example, an individual fisher of Yap, in the Federated States of Micronesia (FSM), was traditionally enmeshed in a complex set of rights and obligations. As Falanruw (1992) described it:

> Where he fished was determined by both the location and status of his village as well as the marine resources of the estate. How he fished was also determined by his social status, and [so] the methods available to him. When he fished was dictated not only by seasonal marine phenomena, but the needs of leaders, and experience of fishing specialists.... Then, depending upon the fishing method, all or some of the catch may have been used to fulfil obligations to higher ranking villages, or to his own trustee who may in turn have contributed the catch to fulfil their obligation to others. If the catch was for family use, the best of our fisherman's catch would be presented to the head of his estate. Finally, within his own immediate family, he gives his wife the fish of her choice. In return he was provided with the finest of his wife's produce from the land and obtains the favor of his elders who will someday provide his inheritance.

On Yap the rights to use fishing gear differed by social class (Table 6.2). Some members of the lower classes were thus restricted to using simple gear, such as extraction sticks and stupefacients, in the riverine and tidal pool habitats to which they had access rights. Women and children could collect invertebrates near the shore and use a hook and line inside the reef. Hook and line was also open to other individuals. Methods that

Table 6.2 Basic categories of fishing gear–fishing technique used in coral reef fishing by user category in Yap, Federated States of Micronesia. User categories: LC, 'Lower classes'; W + C, women and children; OM, old men; IND, individual; COM, community; SPE, special. After Falanruw (1992)

Technique	LC	W + C	OM	IND	COM	SPE
Extraction sticks (*tholom*)	X					
Stupefacients	X					
Gathering near shore		X				
Hook and line within reef		X	X	X		
'Butterfly' net (*k'ef*)				X		
'Butterfly' net (*yerao*)					X	
Push net (*manago*)				X		
Small trap (*yinup*)				X		
Bamboo trap (*sagel*)				X		
Stone trap (*ach*)				X		
Stone trap and leaf sweep (*ruwol*)					X	
Large nets (many methods)					X	
Deep-water deep net from canoe (*athing*)						X
Hand-netting of flying fish from canoe (*magal gog*)						X
Over-reef trolling (*wayrik*)						X

employed mostly smaller gear were also available to individuals, including 'butterfly' nets (*k'ef*), push nets (*manago*), small fish traps (*yinup*), bamboo fish traps (*sagel*) and stone fish traps (*ach*) (Falanruw, 1992). Rights to larger gears, and those which required a group to operate them, were vested in communities. These included large 'butterfly' nets (*yerao*), stone fish traps and leaf sweeps (*ruwol*), and many methods of net fishing

Table 6.3 Ranking of selected habitats by prestige level, Yap island, Federated States of Micronesia. After Falanruw (1992)

Habitat	High prestige	Medium prestige	Low or no prestige
Open sea	X		
Lagoonal holes	X		
General lagoon area		X	
Mangrove channels		X	
Mangrove and coastal fringe			X
Rivers			X

(Falanruw, 1992). Prestigious techniques that required special rights were reserved for the upper classes. These included net fishing from canoes in deep water (*athing*), using hand nets from canoes to catch flying fish (*magal gog*) and trolling beyond the reef (*wayrik*) (Falanruw, 1992).

Access to specific habitats was also controlled (Falanruw, 1992). In general, the prestige of a habitat increased seawards (Table 6.3). Inshore habitats were least prestigious and limited to the lowest social ranks, whereas access rights to the open ocean were limited to the upper classes. Although there were exceptions, in general the lower classes had rights to use only rivers and tide pools. Women and children were limited to reef gathering and collecting in mangroves.

Fishing gear and techniques

In most tropical reef fisheries, gear types are now simple, with the notable exception of South East Asia and the Pacific islands. Many Pacific island fisheries are characterized by a bewildering variety of gear types and techniques adapted to habitat type, tidal conditions and materials available for fabrication (Stair, 1897; Beasley, 1928; Anell, 1955). On Yap, for example, some 125 named fishing methods have been identified, although not precisely defined (Falanruw, 1992). Many traditional gears and construction materials were abandoned following westernization and modernization (Alexander, 1902; Smith 1947a; Titcomb, 1952; Kennedy, 1962; Hosaka, 1973).

The most important gear types worldwide are hook and line (usually hand lines), traps (pots), and gill nets (Munro and Williams, 1985). This is so in the reef fisheries of Kenya (Nzioka, 1990), Red Sea (Kedidi, 1984a), South Sulawesi, Indonesia (Sạwyer, 1992), Belize (Espeut, 1992) and Jamaica (Espeut, 1992). Many fishing techniques have persisted largely unchanged for centuries, as in Jamaica, where artisanal fishing technology has altered little since at least the early 19th century (Espeut, 1992). Other important methods include reef gathering, spearing, fish weirs or fixed traps, drive-in nets, stupefacients, poisons and blasting. Often, combinations of these techniques are used.

Traps are characteristic of the Caribbean and are limited mostly to individually buoyed traps or trap lines (Munro, 1983a). Two basic types can be distinguished, the traditional and ubiquitous 'Antillean' type of wooden-frame galvanized-wire trap, of which there are three basic designs (Munro *et al.*, 1971), and the less common spiny lobster (mainly *Panulirus argus*) trap (Munro, 1983a). Hook-and-line fishing is also important in the Caribbean. Netting is less widespread, and limited mainly to seine nets used in back-reef shallows, bays and harbours, and some drive-in gill netting. Because the capital outlay for gear is low, spear fishing is of increasing importance among impoverished fishers. On the south coast of Jamaica, 54% of fishers studied use traps, 48% gill nets, 29% hand lines, and less than 5% use beach seines. However, there is considerable local variation, and hand lines are more commonly used than gill nets in southern Jamaica (Espeut, 1992). In Belize, hook and line is the commonest method, followed by trapping (Espeut, 1992). Dynamiting and fish poisoning are a problem in some areas (Munro, 1983a; Espeut, 1992).

In the Tigak Islands of northern Papua New Guinea, net fishing with gill nets and beach seines accounts for about 35% by weight of the total catch. This is followed in importance by miscellaneous techniques and combinations of techniques (31%), hand lining (22%), spearing (6%), and trolling (6%) (Wright and Richards, 1985).

The overexploited reef fisheries of densely populated regions of South East Asia employ a wider range of gear types, for example at Cape Bolinao, Pangasinan, Philippines (McManus *et al.*, 1992). One of the few types of industrial-scale reef fisheries also occurs in the Philippines, based on the *muro-ami* drive-in set net. The gear consists of a large bag, held open by the current, and two wing nets to guide the fishes toward the bag. It is set over reefs at a depth of 13–30 m. Swimmers hold rope scare lines, with plastic strips tied at intervals and weighted with stones weighing 3–5 kg. They converge on the net and scare the fishes into it (Fig. 10.11). Each fishing unit employs 200–400 young boys as swimmers. Although condemned on environmental and welfare grounds (Castañeda and Sy, 1983; Corpuz *et al.*, 1983a; Jennings and Lock, Chapter 8; McManus, Chapter 10), this is an effective harvesting technique. It is executed by 25 Philippine-based vessels, which range throughout the Sulu and South China seas and support an estimated 15 000 people.

Basic gear types and fishing techniques

Reef gathering

Reef gathering is perhaps the most widespread reef fishing method. It is practised predominantly by women and children, mostly on reef flats. However, in many locations, as at densely populated Cape Bolinao, men have become gatherers. This reflects relatively large cash returns, low capital requirements and a lack of alternative employment opportunities (Acosta and Recksiek, 1989; McManus *et al.*, 1992). Gathering may be a major source of animal protein for household use (see below). In many locations, the nutritional and economic importance of hand-gathered products may exceed those of the other fisheries. The principal foods gathered from reefs are octopus, sea cucumber, sea urchin, various crustaceans, finfish, seaweed and shellfish.

Despite its importance, there have been few investigations of reef gathering. One study of the Lingayen Gulf in the Philippines estimated invertebrate production to be 25 $t\,km^{-2}\,year^{-1}$ within the 2 m isobath. Of this harvest, 52% of the landings, or 13 $t\,km^{-2}\,year^{-1}$, were from hand gathering in shallow water (Lopez, 1986).

Reef gathering is of major household economic as well as nutritional importance. In many parts of South East Asia, for example, and especially in the Philippines, shellcraft industries depend on the production of reef gatherers. Suppliers to this industry use specialized gathering implements that include rakes, push nets and net–rake combinations (McManus, 1988). Mostly, however, gathering is by hand, or with an extraction stick for probing holes and crevices. Sticks are commonly employed to extract octopus, in particular. In the Pacific islands, extraction sticks were formerly made from the aerial roots of mangrove trees (Falanruw, 1992; Rochers, 1992). A hooked piece of metal or wire is often now used (Rochers, 1992). Extraction sticks were sometimes used together with stupefacients, as on Yap (Falanruw, 1992).

Stupefacients

A range of stupefacients, from liquid detergents to natural plant derivatives, is used at Cape Bolinao, Philippines. Sodium cyanide is the main poison and is used to take both food and aquarium fish (McManus *et al.*, 1992). In the Pacific Basin a large range of botanical materials was traditionally used in fishing (Eldredge, 1987). However, in some areas they have been banned by traditional authorities, as in the Nenema zone of New Caledonia (Teulières, 1990, 1991). The most widespread of stupefacient was *Derris* sp., which in some places, as on Yap (Falanruw, 1992), was especially cultivated for fishing.

Hook and line

Hook-and-line fishing, although widespread, is highly varied in detail. Traditionally, in the Pacific islands, hooks were made of sticks, coconut shell, turtle shell, pearl shell and hardwoods. These have now largely been replaced by steel hooks, which may be used either baited or unbaited. The technique also includes pole fishing, drop-stone fishing, drifting lines, and trolling within the reef and the open sea beyond. In the Philippines, a multi-hook hand line, *kawil-moderno*, is used for reef species (Fig. 6.1).

Net fishing

Net fishing is both common and highly varied. Size, proportions and mesh size of nets vary greatly, according to target species, fishing technique and

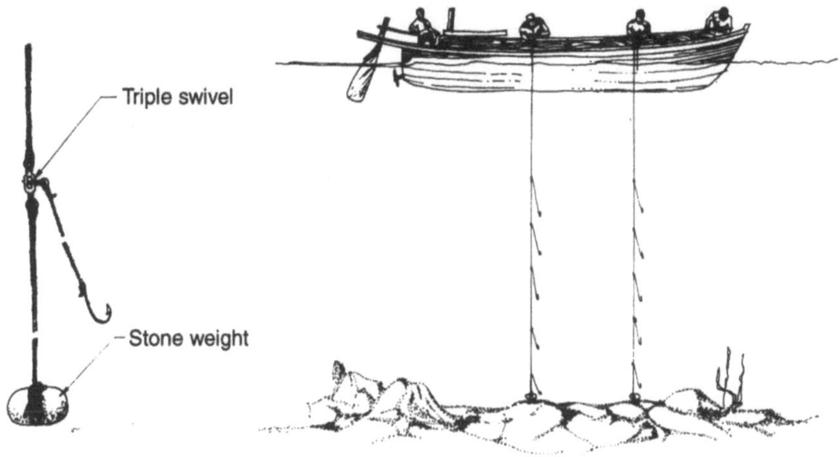

Fig. 6.1 The *kawil-moderno*, a multiple hand line used from boats in the Philippines for catching reef-associated fishes. After Umali (1950).

habitat. On Kosrae, FSM, for example, nine distinct types of nets were employed, each adapted to a particular fishing technique, habitat, tidal condition, and the number of persons in a fishing group (Safert, 1919). Nets are used both as stand-alone techniques and in conjunction with other gear. With other gear, such as leaf sweeps, they are used to drive fish into traps, to surround and concentrate fish for either spearing or scooping, or for capture in the net itself. They may be hand-held and mobile, fixed, or used from a boat.

Whereas many small nets are operated by individual fishers, cooperative net fishing is widespread. In villages near Port Moresby, Papua New Guinea, for example, several cooperative netting techniques are employed. In the surrounding net (*lekeleke*) method, six to ten nets are linked and used to surround a coral head or fish school. After closing, men enter the net to spear fishes. The targets are mainly reef fishes, especially surgeonfishes (Acanthuridae) and emperors (Lethrinidae). In another surrounding net (*haita*), stupefacients may be used in addition to spearing. In barrier netting (*aratore*), linked nets are set around small inlets at high tide. Trapped fish, mainly mullet (Mugilidae), mojarras (Gerridae) and rabbitfishes (Siganidae), are collected at low tide. In drive-in gill netting (*kwadikwadi*), several nets are linked and set in line or slightly curved. Schools of reef fishes, especially emperors, but also jacks (Carangidae) and mullet, are then driven into the net (Lock, 1986a). In the Philippines, a trammel net, *trasmaliyo*, is often operated from boats in reef waters; fishes are scared towards the net by men in the boat beating the water surface with oars and sticks (Fig. 6.2).

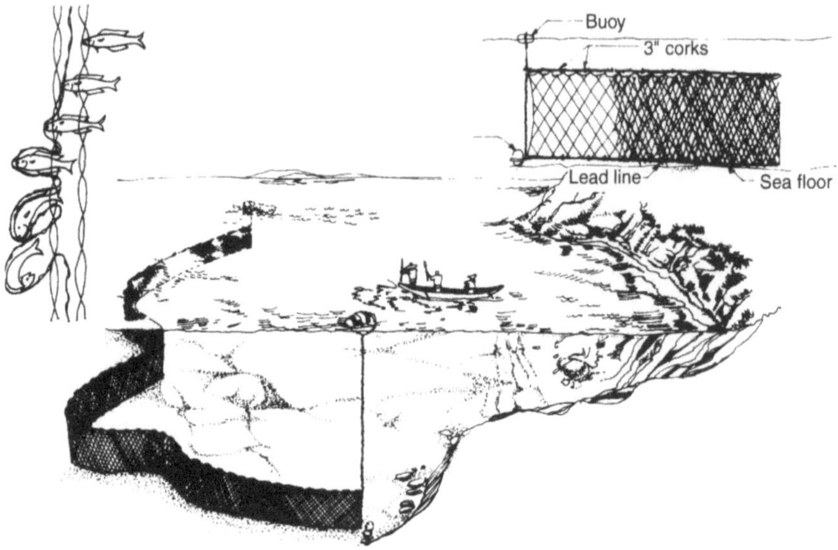

Fig. 6.2 A trammel net, *trasmaliyo* operated around reefs in the Philippines from boats; men in the boats scare fishes towards the net by beating the water surface with oars and sticks. After Umali (1950).

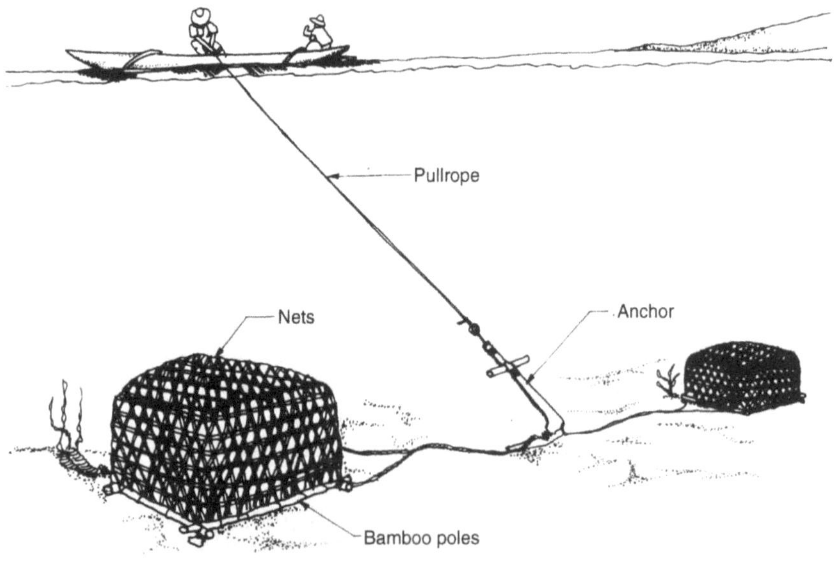

Fig. 6.3 *Bubo*, a type of fish pot used for catching reef fishes throughout the Philippines. After Smith *et al.* (1980).

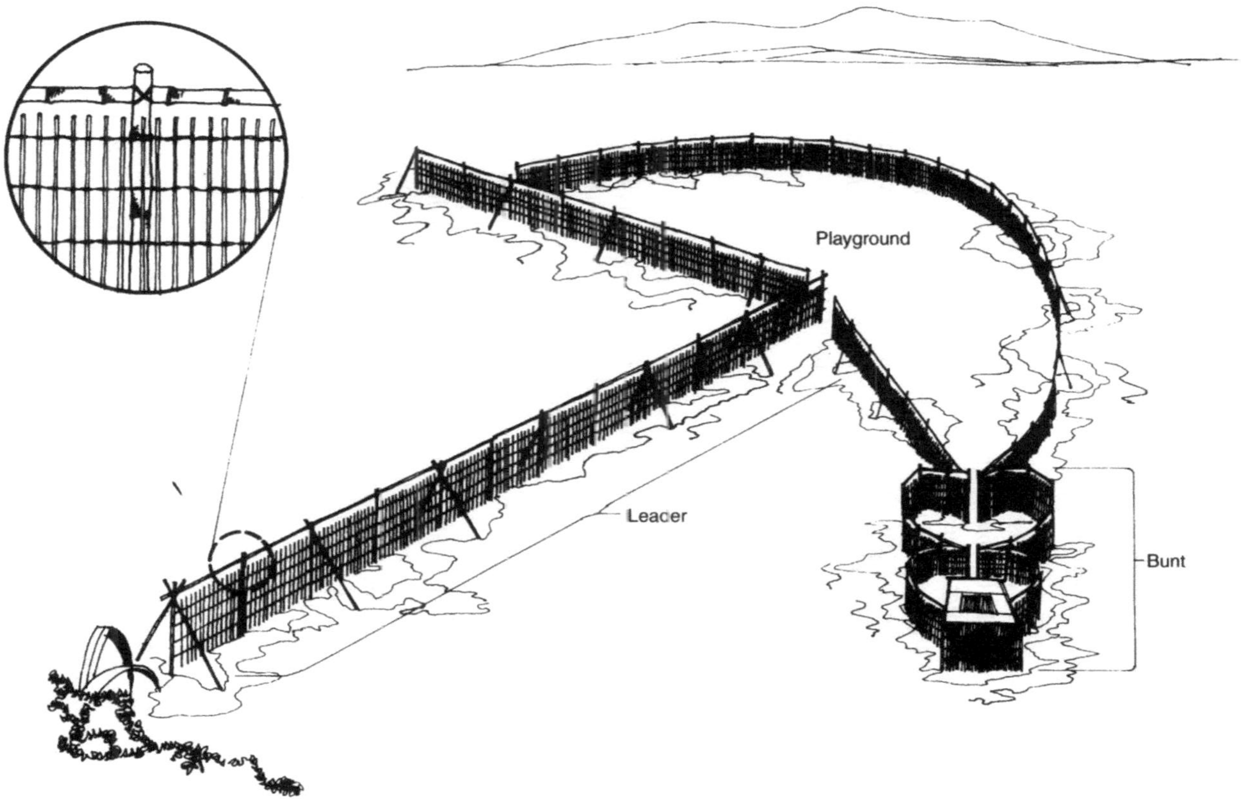

Fig. 6.4 *Baklad*, a fish corral used in the Philippines. After Smith *et al.* (1980).

Fish trapping

Fish trapping is another common technique. In much of the Pacific Basin small traps are set in coral areas, usually by individual fishers, especially older men. Large traps are placed in deep water and others are used in conjunction with stone weirs. In Philippine waters, small unbaited woven-bamboo traps fitted with horse-neck funnels, called *bubo*, are widely used (Fig. 6.3).

Large 'fish corrals', consisting of arrow-shaped fences, are widely used in South East Asia. In the *baklad*, a device used in the Philippines, a leader fence helps to conduct fishes into the main part of the corral, the playground, whence the catch gathers in the 'bunt' (Fig. 6.4). Such traps are often placed in the migration path of fishes, especially the rabbitfish *Siganus fuscescens*, in seagrass beds. In the Philippines the sites of such corrals may be leased to investors (McManus *et al.*, 1992).

Blast fishing

Blast fishing is a widespread, although generally illegal, method of reef exploitation employed worldwide. It has been reported in at least 16 countries in the Indian Ocean region, and is particularly prevalent in South East Asia (UNEP/IUCN, 1988a), as well as from 11 countries in the central and western Pacific. In that region it is particularly prevalent in Micronesia (UNEP/IUCN, 1988b). Although it is commonly known as dynamite fishing, in reality a wide range of explosive devices is used, ranging from home-made bombs to dynamite. At Cape Bolinao, the commonest device is a bottle filled with sodium azide and layers of pebbles. Catches are high and investment relatively low (McManus *et al.*, 1992). Blast fishing is regarded as a major problem in Indonesia (Soegiarto and Polunin, 1981; White, 1984), Malaysia (De Silva, 1984; White, 1984), the Philippines (Gomez, 1980; Yap and Gomez, 1985) and Thailand (Chansang, 1984).

Spear fishing

Spear fishing is both a traditional technique and, in many locations, a technique recently adopted because of low gear costs. The gear used often consists of locally made wooden spear guns and sharpened metal rods powered by large rubber strips. At Cape Bolinao, both breath-holding and compressor-assisted spearing are done. Rabbitfishes are speared at night within the reef (McManus *et al.*, 1992).

Two types of spear fishing are conducted near Port Moresby, Papua New Guinea. Underwater spearing (*korosi*) is done with simple home-made

Fig. 6.5 *Salubang*, a multiprong spear employed at night in Antique Province of the Philippines to catch fishes and marine turtles, often in conjunction with a portable lamp. After Umali (1950).

spear guns, both by day and by night, and targets large reef fishes, especially surgeonfishes, sweetlips (Haemulidae) and groupers (Serranidae) (Lock, 1986a). Surface spearing at night (*kere*) employs a simple multi-pronged spear and takes surface-swimming species, which are almost entirely needlefishes (Belonidae). The *salubang* is used from boats in Antique Province of the Philippines in conjunction with a lamp to spear fishes and marine turtles at night (Fig. 6.5).

Harvesting labour

Harvesting of reef fish is performed by men and women, but their roles are usually strictly divided. In general, men fish from canoes in the lagoon or open sea and use a wide range of gear types, whereas women fish on the reef and virtually never use canoes or elaborate and complex gear. In many societies, and particularly in the Pacific Basin, women are ritually prohibited from participating in certain fisheries, notably those that involve the harvesting of such prestigious species as bonito, tuna or turtles. However, this is but a general rule, because women do participate in some lagoon and open sea fishing, in addition to the nutritionally important but low-prestige gathering (Table 6.4). On Kosrae, FSM, women are the principal fishers (Rochers, 1992).

Unfortunately, most information on the role of women in fisheries is

Table 6.4 Involvement of Pacific islander women in reef fishing other than gathering

Type of involvement	Region	Source
No involvement	Melanesia	
	Solomon Islands	Ivens (1930)
	Micronesia	
	FSM	
	Kapingamarangi	Buck (1950)
	Ulithi Atoll	Lessa (1966)
	Polynesia	
	French Polynesia	
	Mangareva	Buck (1938)
	Marquesas	Handy (1923)
	NIUE	Loeb (1926)
Minimal involvement	Melanesia	
	Papua New Guinea	
	Manam	Wedgewood (1934)
	Solomon Islands	
	Ontong Java	Bayliss-Smith (1977)
	Tikopia	Firth (1965)
	Polynesia	
	French Polynesia	
	Raroia	Danielsson (1956)
	Tokelau	MacGregor (1937)
	Tonga	
	Niuatoputapu	Dye (1983)
	Western Samoa	Lockwood (1971)
Involved in some types of fishing	Melanesia	
	Papua New Guinea	
	Ponam	Carrier (1982)
	East New Britain	Schoeffel (1983)
	Micronesia	
	FSM	
	Caroline Islands	Alkire (1977)
	Chuuk	Murai (1954)
	Kiribati	Catala (1957), Zann (1985)
	Tuvalu	Zann (1985)
	Polynesia	
	American Samoa	
	Manua	Mead (1969)
	Cook Islands	
	Mangaia	Buck (1934)
	Pukapuka	Beaglehole and Beaglehole (1938)
	Tongareva	Buck (1932)

Table 6.4 *Continued*

Type of involvement	Region	Source
	French Polynesia	
	Society Islands	Handy (1932)
	Tubuai	Aitken (1930)
	Wallis and Futuna	
	Uvea Island	Burrows (1937)
	Western Samoa	Buck (1930), Mead (1969)
Involved in general fishing (within reef)	Melanesia	
	Fiji	
	Lau Islands	Hocart (1929), Thompson (1940)
	Moala	Sahlins (1962)
	Viwa Atoll, Yasawa Islands	Lal and Slatter (1982)
Unrestricted involvement	Micronesia	
	FSM	
	Kosrae	Rochers (1992)
	Mariana Islands	Thompson (1945)

anecdotal, with few studies of their productivity, and even fewer of their fisheries and marine environmental knowledge (Ruddle, 1994a,b,c, in press). What little quantitative data there are have been collected for Oceania, where the contribution of women and children to total fish catches ranges from 16% to 50% (Chapman, 1987). In Tonga, women's reef gathering at Tongatapu provides an estimated 77 t (1978) of shellfish per year (McCoy, 1980).

Women play important, although varied and often changing, roles in the reef fisheries of South East Asia. At Taka Bone Rate Atoll, off South Sulawesi, Indonesia, for example, the economic role of women varies among islands. On Tarupa Kecil women collect clams, octopus and other edible species from the reef flat at low tide. On Pasitalu Timur they gather small clams, holothurians and molluscs. On Jinato, women reef-gather and assist in processing the men's catch. On Latondu Besar they collect and dry octopus. By contrast, on Pasitalu Tengah, children do the gathering (Alder and Wicaksono, 1992). Women are also commonly involved in making and repairing gear and the production of shell-craft items.

An indication of the role of women in fisheries production can be derived from data on five villages in various parts of Fiji (Lal and Slatter, 1982). Fishery activities absorb from 21% to 71% of female monthly labour input to economically productive pursuits (Table 6.5). Within the

Table 6.5 Female labour input (h $month^{-1}$) to reef fisheries and other economically productive activities in five Fijian villages. Time spent marketing is not included with labour input. Calculated from data in Lal and Slatter (1982)

Activity	Najia	Namara	Navatuyaba	Votua	Nasawana
Finfishing	24	24	8	0	39
Shell gathering	24	45	17	48	0
Crustacean gathering	14	32	33	29	8
Fish processing	24	30	0	0	28
Total fisheries related	86	131	58	77	75
Agriculture	32	40	72	27	18
Domestic duties	90	120	210	120	120
Others	26	24	0	0	64
Total non-fisheries related	148	184	282	147	202
% Fisheries related	58	71	21	52	37

Table 6.6 Distribution (%) of female labour input to reef fisheries by task in five Fijian villages. Time spent marketing is not included with labour input. Calculated from data in Lal and Slatter (1982)

Activity	Najia	Namara	Navatuyaba	Votua	Nasawana
Finfishing	28	18	14	0	52
Shell gathering	28	34	29	62	0
Crustacean gathering	16	24	57	38	11
Fish processing	28	23	0	0	37
Total fisheries production	72	76	100	100	63

fishing sector of the village economies, 72–100% of female labour input to fishing is devoted to fish production, with processing absorbing 0–37% (Table 6.6). Their contribution to total household cash income ranges from 30%, in Namara Village, on Wayalailai Island in the Yasawa Islands, to 96% in Navatuyaba Village, on Viti Levu. The contribution of women in other villages is 41% at Naibalebale Village, on Viwa Atoll in the Yasawa Islands, 72%, at Najia Village, also on Viwa Atoll, and 73% at Votua Village, in Ba Province, north-western Viti Levu.

At Naibalebale, 71% of women fish, by diving and spearing and by the use of hand nets, but by only minimal use of lines. There, fishing by men and women combined contributes 82% of cash income. In most households less than 25% of the shellfish and crustaceans harvested by the women is used for family subsistence, most being sold (Lal and Slatter, 1982). In Najia Village, in all households with women fishing commercially, their income was higher than men's, ranging from 55% to 100% of total household income (Lal and Slatter, 1982).

Processing

The bulk of the world's reef fish harvest is either consumed directly or marketed fresh. In general, there is very little post-harvest processing. However, processing is done in locations remote from markets, as in Fiji (Lal and Slatter, 1982) and many parts of Indonesia (Alder and Wicaksono, 1992). Fish processing is almost universally the task of women, as it is in Fiji (Table 6.6).

6.4 CATCH DISTRIBUTION AND MARKETING

Catches are distributed within households and communities and marketed. However, for most of the world's reef fisheries, marketing is of minor importance. In many areas, and relatively well documented for some islands of the Pacific, elaborate distribution rules govern post-harvest access rights to the landed catch. Such rules include: (1) those to provision the family and community; (2) those required as subsequent and continual repayment for the acquisition of fishing rights; and (3) those enmeshed in general community sharing and reciprocity, and norms concerning equity and fairness (Ruddle, 1994a,b; Ruddle, Chapter 12).

Provisioning

In Kiribati, fishers are obliged by custom to provision their nuclear family, and offspring have to feed their elders. Sharing the catch within a family is culturally very important; disinclination or failure to do so could lead to disinheritance, as enshrined in the *Native Lands Ordinance* (1956) (Teiwaki, 1988). Similarly, in Tokelau catches were distributed through the village *inati* system, through which all residents received equal shares (Hooper, 1985, 1990).

Repayment for fishing rights

On Yap the transfer of fishing rights often obligated the grantee to share with the grantor the catch from the area transferred (Anon., 1987a; Falanruw, 1992). Many of the rights to fishing methods used by individuals carried the obligation to contribute the first catch to the overseer of the fishing area or method, to the trustees or to the village. When individual or special methods were used in the fishing territory of a higher-status village, either the first catch or a portion of the catch had to be given to that village as tribute (Falanruw, 1992).

Such obligations applied mostly to various forms of net fishing. In a survey of 36 instances of such rules, 72% applied to netting, 8% to line fishing, 6% to the use of stupefacients, and slightly less than 3% each to spear fishing, trapping, the use of weirs, probing with a stick and unspecified techniques (Anon., 1987a). The allowed techniques were specified.

The amount of catch to be shared with the rights grantor varies considerably among cases, from 100% to 1–2%, with further rules specifying frequency, species and sometimes size of fish to be shared. Occasionally, where the use of more than one technique is permitted, those details are specified for each technique. Where 100% of the catch was specified, this is qualified by such riders as 'when needed', 'of larger fish', 'of catch once a year' or 'of first trip'. Some such detailed rules specify the percentage of one particular species (e.g. '50% of humphead parrot fish', *Bolbometopon muricatum*). Often they simply specify a certain percentage of any 'preferred species' or of any 'fancy species', a modern form of the traditional tribute.

Not all fishes are suitable for use as tribute or the satisfaction of obligations entitled in rights. Those preferred for tribute comprised fishes taken mostly in the open sea using upper-class techniques and which, in former times, were enmeshed in elaborate ritual prior to a fishing trip. These include green turtle, *Chelonia mydas*, blackfin needlefish, *Tylosurus acus*, marlin, *Makaira mazara*, wahoo, *Acanthocybium solandri*, yellowfin tuna, *Thunnus albacares*, skipjack tuna, *Katsuwonus pelamis*, and humphead wrasse, *Cheilinus undulatus*. Various species which were mostly from the reef are a multipurpose category that can be used as tribute, although not preferred. These include rabbitfishes, *Siganus* spp., unicorn fishes, *Naso tuberosus*, seagrass parrotfish, *Leptoscarus vaigiensis*, butterflyfishes, *Heniochus* spp., rainbow runner, *Elegatis bipinnulata*, flying fish, *Cypselurus cyanopterus* and mahi-mahi, *Coryphaena hippurus* (Falanruw, 1992).

Community sharing and reciprocity

In Kiribati, the concepts embodied in ownership of stone fish traps demonstrate an intricate pattern of reciprocal sharing and obligation, both in construction and division of harvest. The catch is divided into three lots: 75–85% goes to the owners of the trap and of the coral blocks from which it is constructed. The remainder is given to those who assisted with harvest. In purse-seining within the lagoon, fish (or cash proceeds) are distributed equally among the fishers when they are from different households. Canoe and net owners receive a larger share for use of their equipment. Reciprocal sharing with other households is practised if catches are large.

Distribution of the harvest is fundamental in ensuring intra-group harmony and the stability of the traditional management system, especially if distribution is from higher-status persons, with species or other special access rights (see above) to the community at large. Distribution systems can be complex in terms of the categories of persons involved, as well as geographically extensive.

Ulithi Atoll, FSM, provides an example of such complexity. Such valuable species as turtles are presented there as tribute to the paramount chief, who slaughters and distributes them in a closely specified way. Some parts are given to the women in the menstrual house on Mogmog Island. They distribute what they do not need to women on other islands and to the heads of the two highest-ranked lineages on Mogmog Island. In turn, they distribute some to the heads of lesser lineages (Ushijima, 1932). Stranded whales and other fish caught by specific gear on specific days are distributed in a similarly ramified fashion.

Women also have distribution rights because canoes, although owned by lineages, are overseen by the women of a lineage. This is because canoe hulls are made from mahogany logs obtained from Yap Island, in exchange for cloth made by the women. Further, because postmarital residence is patrilocal, women are scattered throughout the various matrilineages on an island. Consequently, food distribution reaches all parts of all islands (Ushijima, 1982).

Distribution in the form of reciprocal exchange of goods also occurs among the islands of Ulithi Atoll (Fig. 12.1). For example, ecologically favoured Falalap Island provisions the rest of Ulithi with taro, breadfruit, sweet potato and banana. However, Falalap lacks fishing grounds, and must receive fish from other islands. Fishing rights areas are expansive on islands in Mangejang District, where, however, vegetable cultivation is precluded by the absence of a freshwater lens. People of Falalap and Mangejang thus exchange vegetables and fish (Ushijima, 1982).

Influence of urbanization

Urbanization is changing the nature of reef fishing throughout the tropics (Ruddle, 1994a). For example, although the reef fishery of Western Samoa has traditionally been a subsistence activity, selling in the fish market at Apia, the capital, has become a major source of income for village fishers since the early 1970s (Helm, 1992). Fish is marketed there by individual fishers or their family members, who typically bring some 5–30 kg to market at any one time. This is usually the catch of one person or of a small group of fishers using the same harvesting technique. From two to ten species are represented in the catch sold. Composition by weight of marketed fish by fishing technique in 1986 and 1987 was netting (24%, 40%), spearing (48%, 31%), trapping (12%, 16%) and hook and lining (16%, 13%) (Helm, 1992).

The bulk of the reef fish harvest of Jamaica is marketed fresh through a well-organized and highly structured yet informal system. This is well suited to distributing the small catches landed at over 200 points throughout the island. All fish not required for subsistence by the fishers is sold either directly to wholesale and retail 'higglers' (small-scale traders) and the general public. Wholesale higglers are more important and in turn sell to restaurants, fish processors, and retail sidewalk higglers, market higglers or itinerant higglers (Espeut, 1992).

Reef fishing behaviour can also be market driven, in addition to being determined by physical factors. In reef fishing villages west of Port Moresby, Papua New Guinea, for example, although low tides are the best time for fishing, fishing activity is dictated mainly by the market. Daytime fishing ceases at 1500 h to enable the catch to reach Port Moresby for the late-afternoon shopping period (Lock, 1986a). In those villages, women are responsible for cleaning and selling the catch. Some is retained for household consumption and some given away. Some fish is sold within the fishing communities. Fish not sold at the urban market is taken home to be eaten, smoked or given away (Josephides, 1982).

In the Maldives, reef fish has not traditionally been eaten, except when tuna fishing is poor. Thus the harvesting and marketing of reef fish, other than to fulfil emergency needs, is relatively recent. There are three main markets for reef fish (Anderson *et al.*, 1992). First, there is the fish market at Malé, the national capital, where fresh fish are purchased mainly by restaurants, foreign residents and nearby tourist resorts. Second, the 70 tourist resorts located mostly on Malé and Alifu atolls purchase significant quantities of reef fish; resorts distant from Malé usually purchase fish on a contract basis from artisanal fishers (Van der Knaap *et al.*, 1988). Third, low-value, salt-dried reef fish is exported to Sri Lanka. This market, which takes from 50% to 90% of the reef fish caught, and which accounts for 5.7–9.4% of all marine product exports (Brown *et al.*, 1989), is supplied mainly by fishers inhabiting the outer atolls. These people cannot market their catch fresh in Malé or the resorts (Anderson *et al.* 1992).

ACKNOWLEDGEMENT

I am grateful to John McManus for arranging the illustrations.

Chapter seven

Catch rates, selectivity and yields of reef fishing

Paul Dalzell

SUMMARY

Fishing on tropical reefs is the preserve mainly of small-scale artisanal fishers, where increases in human power, rather than machine power, are used to generate large volumes of reef fish landings. The main forms of fishing gears deployed on reefs and in reef lagoons are handlines, traps, gill nets, seine nets and spears. Trawls can be deployed on the soft-bottom substrata adjacent to coral reefs but they catch mainly non-reef-associated species. The selectivity of nets, hooks, traps and spears is reviewed using, as far as possible, examples from reef fisheries. Although there is substantial variation, all gears catch a relatively wide range of species and selectivity is primarily size-based. However, species selectivity also results from the interaction of fish behaviour and gear characteristics. Reef fishes are also captured alive for restaurants and aquaria, and small pelagic fishes in coral reef lagoons are captured live for bait for pole-and-line tuna fishing. Observed yields from tropical reef fisheries range from around 0.2 $t\,km^{-2}\,year^{-1}$ in Papua New Guinea to over 40 $t\,km^{-2}\,year^{-1}$ from American Samoa. Yields in excess of 5 $t\,km^{-2}\,year^{-1}$ are probably sustainable in the long term, although the upper limit for sustainable harvests from reef fisheries has yet to be accurately determined and will, in any case, vary among areas.

Reef Fisheries. Edited by Nicholas V.C. Polunin and Callum M. Roberts.
Published in 1996 by Chapman & Hall, London. ISBN 0 412 60110 9.

7.1 INTRODUCTION

All fishing gears are selective to some degree. Fishing gears may select by size, by species or a combination of both. Some gears, such as seine nets, may take a wide variety of both sizes and species, with selection acting only against small specimens. Others, such as gill nets, may also take a wide variety of species, but be very size selective, taking only a relatively narrow range of lengths, depending on mesh size. Some gears are designed only to select for particular species. For example, in Melanesia kite fishing has been designed specifically to catch longtoms (Belonidae), the teeth of which are snared in a spider-web lure towed from a *Pandanus* leaf kite (Hulo, 1984).

In this chapter, I describe the catch rates, catch composition and selectivity of the gears used to catch fish and invertebrates on tropical reefs. Understanding fishing gear selectivity is particularly important for stock assessment and management because it will influence catch composition and the size and age frequencies of the target species in the catch. This in turn may introduce bias into the computation of life-history parameters such as growth and mortality rates (Appeldoorn, Chapter 9). Effective management of a fishery may be dependent not only on limiting effort but also on measures such as setting mesh sizes, hook sizes or even banning certain gears based on knowledge of selection effects. Finally, I review yields from reef fisheries around the world with respect to the sustainability of exploitation.

7.2 MODERN FISHING STRATEGIES AND TECHNIQUES

The rugged topography of reefs limits the use of nets to static gill nets, seines or trammel nets that can be set and retrieved without major damage. Shallow water can also limit the size of vessel able to operate near reefs. Reef fisheries tend to be characterized by small-scale, artisanal (i.e. non-mechanized) gears that include hand lines, seine nets, gill nets, traps (fixed and moveable) and spears (Ruddle, Chapter 6).

Although mainly confined to continental shelves and large river estuaries, trawl fishing is practicable on many slopes and soft bottoms adjacent to reefs. Examples can be found on the Mozambique shelf (Saetre and Paula e Silva, 1979), the Lingayen Gulf of the Philippines (Yazon and McManus, 1987), the Australian Great Barrier Reef (Dredge, 1988) and the north-west shelf of Australia (Edwards, 1983; Sainsbury, 1984). Experimental trawling has also been conducted on the demersal stocks of the Mahé Plateau in the Indian Ocean (Marchal *et al.*, 1981; Künzel *et al.*, 1983), the Chesterfield Islands and Landsdowne Bank between New Caledonia and Australia (Kulbicki *et al.*, 1990), and in the lagoon of New Caledonia (Kulbicki and Wantiez, 1990; Wantiez and Kulbicki, 1991). Such fisheries may take species such as groupers (Serranidae), snappers (Lutjanidae), emperors (Lethrinidae) and threadfin breams (Nemipteridae) that move between tropical reefs and the demersal environment beyond.

7.3 PERFORMANCE AND SELECTIVITY OF FISHING GEARS

Trawls

The selectivity of fishing gear is the interaction of the dimensions and materials used to construct a particular gear, the morphology of the fishes caught and the life history of the targeted species. For a fishing net, the most obvious dimension is mesh size or distance between two knots. In pliant fishing nets where the diamond-shaped mesh may alter its dimensions, depending on deployment, the mesh size is expressed as stretched mesh, i.e. the maximum distance between two knots.

For trawls and seines it is usual to assume that the size composition of fish entering the mouth of the net is the same as that in the immediate vicinity of the gear. The selectivity of the gear is thus a factor of escapement through the mesh of the smaller size classes. For a given mesh size the proportion of fish escaping in each successive size class decreases until

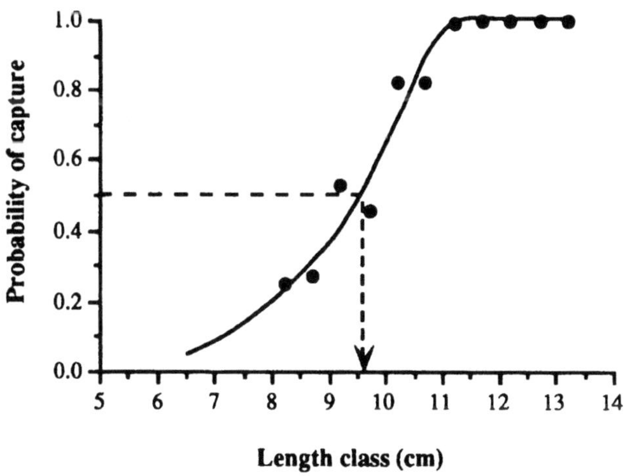

Fig. 7.1 Selection curve for the goatfish, *Upeneus sulphureus*, caught by trawl fishing with 2.0 cm mesh net in the Samar Sea, Philippines. Arrow marks size at which half the fish are retained in the trawl. Based on data in Silvestre *et al.* (1986).

all fish are retained. The selection curve of such gear is usually sigmoidal (S-shaped) and the average minimum capture length (L_c) is defined as the size at which 50% of the fish are retained by the gear (Fig. 7.1). In general, L_c is considered to be proportional to mesh size; the proportionality constant is called the selection factor (S.F.). Where this is known, the selection factor can be used to estimate L_c from

$$L_c = \text{S.F.} \times \text{mesh size.} \qquad (7.1)$$

Pauly (1984a) states that in general the selection factor of fishes is related to their overall shape. Slender fish have high selection factors whilst bulky fish have low selection factors.

Trawl net selection has been investigated by numerous workers. Selection is dependent primarily on the mesh size of the net and particularly the cod end. Experiments to determine mesh selectivity with trawls can be conducted by attaching a fine-meshed bag to the trawl cod end and comparing the size frequencies of fish retained by the cod-end with those retained in the cover. Longhurst (1960) and Silvestre *et al.* (1986) studied mesh selection for trawls catching tropical demersal species. Other factors may also influence the selectivity and catching power of trawls, such as width and height of the trawl opening and the power and speed of the vessel towing it. The introduction of higher-opening trawls and more powerful boats to the trawl fishery in Manila Bay in the Philippines increased the vulnerability of small pelagic fishes (such as mackerels, scads and anchovies) to demersal gears. This led to improved catch rates at a time when demersal stocks were declining through overfishing (Caces-Borja, 1975).

Trawling for demersal fishes commenced on Australia's north-west shelf in the 1950s (Sainsbury, 1987). Although less numerous than on the eastern coast of Australia, fringing reefs and shoals are found throughout the shelf area. Catches of trawlers comprise mainly threadfin breams, lizardfishes (Synodontidae), emperors and snappers, most of which are also caught on reefs. Experimental trawling on the Mahé Plateau in the Seychelles and on the slopes of the Chesterfield Islands and Landsdowne Banks in the Pacific produced catches dominated by species commonly associated with coral reefs (Marchal *et al.*, 1981; Künzel *et al.*, 1983; Kulbicki *et al.*, 1990). These included snappers, lizardfishes, goatfishes, triggerfishes and threadfin breams. Strongly reef-associated species such as parrotfishes (Scaridae), surgeonfishes (Acanthuridae) and boxfishes (Ostraciidae) were also included in the catch.

On the eastern coast of Australia, trawlers fish for prawns on the soft bottoms of shallow estuarine areas of the Great Barrier Reef lagoon (Dredge, 1988). Species composition of the trawl finfish bycatch is very different from that of adjacent reefs, although a small number of species common to both are taken, including jacks (Carangidae), threadfin breams and groupers. Similarly, trawling on the soft bottoms of tropical estuaries adjacent to reefs in New Caledonia, the Philippines and Papua New Guinea produced catches which contained few reef fishes (Coates *et al.*, 1984; Quinn and Kojis, 1985; Silvestre, 1990; Wantiez and Kulbicki, 1991; Wantiez, 1992). Wantiez (1992) has suggested, however, that the low densities of reef fishes in the soft bottom habitat may still represent a sizeable biomass when the entire habitat area is considered. Soft bottom habitats may act as recruitment reservoirs for reefs and extend the trophic zones for certain reef carnivores.

Given the relative unimportance of trawl fisheries around reefs, it is not surprising that little work has been done on the trawl catchability of adjacent habitats. Kulbicki and Wantiez (1990) conducted underwater visual census observations (by diving) on soft bottom habitats fished with a shrimp trawl net in the New Caledonia lagoon. They found that the trawl retained only about 10% of the biomass in the path of the net. Similar studies on trawl catches combined with underwater observations have been made for North Atlantic demersal species by Uzmann *et al.* (1977), who showed that estimates of fish density made from a submersible were eight times greater than estimates from otter trawl catches.

Gill nets and trammel nets

Gill nets are very size selective because capture relies on fish being trapped within the mesh. Selectivity depends mainly on fish size and shape and mesh size, but is also affected by the thickness, material and colour of net twine, hanging ratio (the number of meshes in a given length of net) and method of fishing. The theoretical curve for a given mesh size of gill nets is the bell-shaped normal distribution (Holt, 1963) (e.g. Fig. 7.2). The left slope of a gill net selectivity curve represents small fish wedged bodily in the mesh. The right slope represents larger fish mainly tangled by head parts. Baranov (1948) proposed a rule of thumb for gill nets which states that few fish are caught the lengths of which differ from the optimum capture length by more than 20%. A detailed review of gill net selectivity is given by Hamley (1975).

The selection curve of gill nets can be estimated as described above when the fish are captured using two gill nets of different mesh sizes and comparing the size-frequency distributions. A plot of the logarithm of the ratio of catches versus length should be a straight line, from which the optimum capture length and the standard deviation of the selectivity curves can be generated (Gulland, 1983b). As some fish may become easily tangled, the curves can be very skewed or multimodal. Where the curve is only slightly asymmetrical with some positive kurtosis, then this

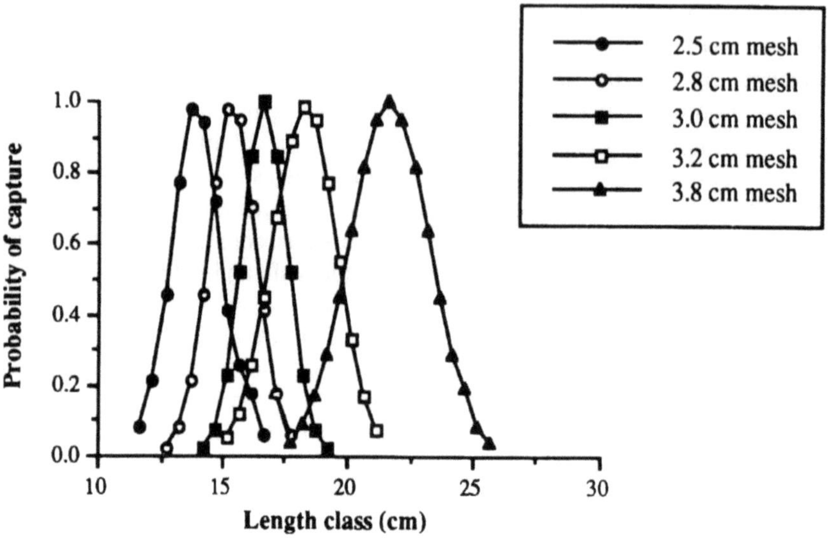

Fig. 7.2 Gill net selection curves for catches of the sardine, *Amblygaster sirm*, in Sri Lanka, with five different mesh sizes. Based on data in Dayaratne (1988).

method can still be used where the logarithm of the ratio of catches is plotted against the logarithm of the lengths.

The simplest forms of gill net selection curves are usually observed with mackerel and herring-like fish that are fusiform and have smooth profiles without spines or large fins that promote tangling. A good example is given by Dayaratne (1988), who investigated selectivity in the Sri Lankan gill net fishery for the sardine *Amblygaster sirm* and found that the selectivity at different mesh sizes could be described by the bell-shaped normal distribution (Fig. 7.2). Similar results were found by Ehrhardt and Die (1988) for Spanish mackerel, *Scomberomorus maculatus*, in southern Florida. Like sardine, Spanish mackerel lack protuberances and spines, being caught by snagging of the gills or wedging of the body. Owing to this latter form of retention, the selectivity of gill nets changes with the onset of the spawning season, where progressive gonad development increases fish girth.

Gill nets may be set at the surface, in mid-water or on the sea bottom and may be used to fish passively; alternatively fishers may locate schools of fish and drive them into the path of the net (see also below). They may also be used to surround schools of fish, particularly pelagic species such as sardines (Clupeidae), scads (Carangidae) and half-beaks (Hemiramphidae), which then snag themselves in the net as they try to escape.

Unlike the gill net, which snags the fish, the trammel or entangling net consists of a small-meshed net sandwiched between two layers of large-mesh net. When a fish swims through the larger outer meshes it encounters and pushes against the loose interior net so that a pocket is formed around it (Fig. 6.2). To be successful, the meshes of the two outer nets must fit exactly one upon the other so that the pocket can be formed on the impact of a fish. Selectivity is influenced by the same factors as those for gill nets; however, mesh selection is a function of two nets rather than one. Losanes *et al.* (1992a,b,c) made detailed comparative studies of catching efficiency of gill and trammel nets. Selection curves of trammel nets were wider than those of gill nets and were bimodal, because trammel nets catch fish using the property of gill nets plus an additional entangling property of the larger nets. Losanes *et al.* (1992c) suggested that selectivity of a trammel net could be estimated by comparing it with a gill net of the same mesh size as the middle net.

Gobert (1992) compared the selectivity of trammel nets with catches from wire-mesh traps in a Martinique reef fishery. Trammel nets were more species selective than fish traps, targeting mainly spiny lobster or crayfish (Panuliridae). They also caught finfish at much greater sizes than did the most commonly used mesh sizes employed to build fish traps. Acosta (1993) has investigated the selectivity of gill nets and trammel nets catching fishes on reefs and amongst mangroves in Puerto Rico. He found significant interactions between mesh size and hanging ratio for both gears. The best catches with gill nets were obtained with a combination of a high hanging ratio with large meshes or low hanging ratio with small meshes. For trammel nets, the largest catches were obtained with low hanging ratios with large meshes or high hanging ratios with small meshes. The selectivity curves of gill nets and trammel nets were unimodal, with optimum length increasing as mesh size increased.

Beverton and Holt (1957) suggested that with static gears such as gill nets, there would be a saturation effect, with decreased rate of fish being retained by the gear with increasing soak time. The catch (*C*) is proportional to the time soaked (*t*) and a measure of the abundance or availability (*A*) of the targeted species. The relationship takes the form:

$$C = C_{\infty}(1 - e^{-At}) \qquad (7.2)$$

where C_{∞} is a theoretical asymptote of the catch curve. The same basic equation has been used to describe long-line catches of temperate-water fishes (Gulland, 1955) and catches of reef fishes in wire-mesh traps (Munro, 1974).

Acosta (1993), however, found that over a 24 h period, the saturation effect of decreased catch with soak time did not occur. The efficiency of gill nets was variable over time but differences were not significant. Changes in trammel net catches were negligible. Losanes *et al.* (1992a) found that catching efficiencies of gill nets and trammel nets were greatest at the

Table 7.1 Summary of catch rate (CPUE, catch per unit effort) and catch composition data for gill net and trammel net fishing on reefs in the Caribbean, Indian Ocean and South Pacific regions (n.a. not available)

Location	Net length (m)	Mesh size (cm)	Target stock	CPUE		Principal catch components	Source
				Range	Mean		
Kiribati	n.a.	5.7–12.7	Reef and lagoon species	5.0–96.0 kg/trip	43.4 kg/trip	Albulidae, Carangidae, Mugilidae, Mullidae	Anon. (1989)
Solomon Islands	n.a.	5–15	Reef and lagoon species	0.26–0.90 kg/100 m net-h	0.46 kg/100 m net-h	Sharks, Chanidae, Carangidae, Mugilidae	Blaber *et al.* (1990)
Cook Islands	90–230	4.5–5.0	Small pelagics and reef fish	0.14–18.04 kg/10 m of net	2.2 kg/10 m of net	Carangidae, Priacanthidae, Mullidae, Caesionidae	Chapman and Cusack (1989)
Fiji (Rabi Island)	150	1.9–7.6	Reef and lagoon species	15–26 kg/set	18.9 kg/set	Lethrinidae, Lutjanidae, Mugilidae, Holocentridae	Anon. (1983a)
Fiji (Rotuma)	229	7.6	Reef and lagoon species	10.0–60.0 kg/set	31.8 kg/set	Mugilidae, Carangidae, Lutjanidae, Lethrinidae	Anon. (1983b)
Papua New Guinea (Port Moresby)	n.a.	5.0–12.7	Reef and lagoon species	n.a.	2.0 kg/man-h	Lethrinidae, Lutjanidae, Carangidae, Scombridae	Lock (1986a)
Papua New Guinea (Rabaul)	35–100	3.8	Small pelagics	0.7–6.7 kg/set	3.0 kg/set	Carangidae, Clupeidae	Dalzell (1993a)
Tonga	100–1200	5.0	Reef and lagoon species	5.6–7.2 kg/set	6.0 kg/set	Acanthuridae, Labridae, Siganidae, Lethrinidae	Halapua (1982)
Guam	n.a.	n.a.	Reef and lagoon species	0.67–12.24 kg/set	4.24 kg/set	Acanthuridae, Mullidae, Scaridae, Labridae	Katnik (1982)
Seychelles	50	5.7–6.4	Small pelagics	38–75 kg/set	55.7 kg/set	Scombridae, Caesionidae, Carangidae	de Moussac (1987)
Puerto Rico (gillnets)	50	7.6–12.7	Reef fish	0.16–3.90 kg/set	1.03 kg/set	Sharks, Haemulidae, Scaridae	Acosta (1993)
Puerto Rico (trammel nets)	50	7.6–12.7 (inner net)	Reef fish	0.10–4.25 kg/set	1.16 kg/set	Sharks, Haemulidae, Scaridae	Acosta (1993)
Martinique (trammel nets)	300–1000	8.0 (inner net)	Reef fish and spiny lobster	1.60–4.62 kg/100 m net	3.43 kg/100 m net	Lobsters, Haemulidae, Lutjanidae, Serranidae	Gobert (1989, 1992)

shortest and longest soak times, based on experimental observations with a captive population of rainbow trout, *Oncorhynchus mykiss*, and with soak times ranging between 0.5 and 5 h. Minns and Hurley (1988) showed with freshwater species in Lake Ontario that catch per unit of effort (CPUE) was proportional to length of soak in some species and inversely proportional to soak in others. These results were attributed to species-specific differences in activity patterns and net saturation effects. It was concluded that use of gill net catches as indices of fish abundance must be validated species by species.

Table 7.1 shows examples of catch rates and catch composition for gill net and trammel net fishing in lagoons and on coralline shelves. Reporting of catch rates of gill nets in the literature was varied and included expression of effort as trips, sets, sets of unit net lengths, sets of unit net length and time, and man-hours. It was not possible to standardize between these as readily as with other gears, so units of effort are quoted for each example in Table 7.1. Catch rates are modest, with average catch per set (the most common CPUE) ranging from 3.0 to 55.7 kg. Target species were not only reef and lagoon species such as snappers, groupers and surgeonfishes, but also small pelagic fishes such as Indian mackerel, *Rastrelliger kanagurta*, scads, *Selar* spp., and sardines.

Seines

Seine nets do not snare or snag fish, but act as a barrier into which fish can be concentrated then hauled. Seines can be deployed from the beach, hence 'beach-seines', or on the reef or lagoon floor where fish can be driven into them. This is known as drive-in-net fishing. Selection is mainly for smaller fishes, and the selection curve is sigmoidal rather than bell-shaped like that for gill nets.

Table 7.2 gives examples of catch rates and catch composition from drive-in-net fishing and beach seining. As with gill net fishing, catch rates are reported as per set or per man-hour. The methods of drive-in-net fishing are broadly similar to *muro-ami* (Ruddle, Chapter 6; McManus, Chapter 10). A large horizontal scare line may be employed to sweep over the coral, in place of many individual vertical scare lines used by *muro-ami* fishers. This method is used in the islands of Micronesia and is described in detail by Smith and Dalzell (1993). In some instances no scare lines are employed, and instead a group of fishers form a semicircle which converges on the net mouth. The fish are alarmed by repeated splashing of the water, which drives them into the net.

Catch rates of drive-in-net fishing are modest, apart from *muro-ami*. However, the *muro-ami* catch is generated by about 500 fishers as opposed to between 2 and 10 fishers in the other examples. Each set of the *muro-*

Table 7.2 Summary of catch rate (CPUE, catch per unit effort) and catch composition data for drive-in-net and beach seine fishing on reef stocks in the Indian Ocean and South Pacific regions (n.a. not available)

Location	Net length (m)	Mesh size (cm)	Target stock	CPUE		Principal catch components	Source
				Range	Mean		
Woleai (Micronesia)	35	4.5	Reef and lagoon species	8.1–129.4 kg/set	42.7 kg/set	Acanthuridae, Scaridae, Siganidae, Lethrinidae	Smith and Dalzell (1993)
Philippines	240	2.5	Reef and lagoon species	n.a.	1900 kg/set or 3.8 kg/man-h	Caesionidae, Acanthuridae	Corpuz *et al.* (1985)
Papua New Guinea (Port Moresby)	n.a.	5.0–12.7	Reef and lagoon species	1.41–4.95 kg/man-h	2.52 kg/man-h	Lethrinidae, Carangidae, Mugilidae, Siganidae	Lock (1986a)
Papua New Guinea (Tigak Is)	100	6.3–7.5	Reef and lagoon species	n.a.	3.9 kg/man-h	Mugilidae, Chanidae, Carangidae, Scaridae	Wright and Richards (1985)
Palmerston Atoll (Cook Islands)	14–480	2.3	Reef and lagoon species	0.41 kg/set	13.1 kg/set	Scaridae	Anon. (1988)
Nauru	n.a.	n.a.	Reef and lagoon species	1.1–8.0 kg/h	3.9 kg/h	Kyphosidae, Mugilidae, Acanthuridae, Lutjanidae	Dalzell and Debao (1994)
Guam	140–280	n.a.	Reef and lagoon species	0.09–0.46 kg/man-h	0.25 kg/man-h	Siganidae, Acanthuridae, Labridae, Lethrinidae	Katnik (1982), Amesbury *et al.* (1986)
Seychelles (beach seine)	n.a.	n.a.	Small pelagics and reef species	8.5–565.3 kg/set	159.0 kg/set	Scombridae, Caesionidae	de Moussac (1987)
Papua New Guinea (Rabaul) (beach seine)	200	2.5	Small pelagic species	n.a.	350 kg/set	Carangidae, Clupeidae	Dalzell (1993a)
Philippines	230–300	2.5	Small pelagic species	7.4–20.7 kg/set	14.2 kg/set	Exocoetidae, Hemiramphidae, Belonidae	Dalzell (1993b)

ami takes about an hour, and this represents an average catch rate of 3.8 kg man-h^{-1}, similar to that of the small-scale drive-in-net fishing used in Papua New Guinea. Catches of drive-in nets and beach seines include mainly reef species and small pelagic fishes. Fusiliers (Caesionidae) form about 80% of *muro-ami* catches, with surgeonfishes making up most of the balance (Corpuz *et al.*, 1985; Dalzell *et al.*, 1990). The principal reef species taken by drive-in-net fishing are surgeonfishes, parrotfishes, mullets (Mugilidae) and rabbitfishes (Siganidae), whilst the two examples of beach seining fisheries target scads, fusiliers and mackerels.

Hook and line

Hand lines are probably the most commonly used gear to catch reef fishes, and select mainly for predatory species, although herbivorous fishes, such as surgeonfishes and rabbitfishes, occasionally take baited hooks (Katnik, 1982; Wright and Richards, 1985; Dalzell, pers. obs.). Table 7.3 summarizes rates and composition of catches from hand lines. Catches were typically dominated by snappers, groupers, emperors (grunts, Haemulidae, in the Caribbean) and jacks. Hand-line fishing on shallow (<60 m depth) reef species typically yields average catch rates of between 0.5 and 2.0 kg line-h^{-1}, although the maximum reported was 4.4 kg line-h^{-1} in the Seychelles. Catches of fishes from the deeper slopes beyond the shallow reefs are considerably higher. Dalzell and Preston (1992) reported on hand line fishing targeting deep slope fishes from between 80 and 400 m from most of the countries and territories of the South Pacific. Average catch rates from high islands and atolls were 6.0 and 7.7 kg line-h^{-1} respectively, on mainly virgin stocks. Average catch rates from exploited deep slope stocks in Papua New Guinea, Vanuatu and Tonga ranged between 3 and 5.0 kg line-h^{-1}. Very high catch rates of between 12.0 and 15.0 kg line-h^{-1} were reported from the relatively shallow (40 m) hand-line fishery on the Norfolk Ridge, 10–30 km from Norfolk Island (Grant, 1981). Almost 90% of the catch of this fishery is the emperor, *Lethrinus chrysostomus*, with the balance formed by groupers and jacks. Despite these remarkably high catch rates, there is evidence from an earlier survey in 1959 (Van Pel, 1959) that average catch rates were once even higher (23 kg line-h^{-1}).

Longlining of reef fishes is less common than hand-line fishing, but may be effective for snapper and grouper stocks on deep reef slopes. Table 7.4 gives examples of catch rates and composition from longlining in reef areas. Average CPUE ranged between 4.6 and 124.6 kg per 100 hooks deployed. The exceptionally high catch rates reported from Espiritu Santo in Vanuatu are the result of a limited number of sets on virgin deep slope stocks, conducted under ideal fishing conditions. Catches in all areas were

Table 7.3 Summary of catch rate (CPUE, catch per unit effort) and catch composition data for hand line fishing on reef and associated stocks in the Caribbean, Indian Ocean and South Pacific regions (n.a. not available)

Location	Target stock	CPUE (kg line-h^{-1})		Principal catch components	Source
		Range	Mean		
Jamaica (Pedro Bank)	Shallow reef and deep slope species	1.0–3.7	2.0	Lutjanidae, Carangidae, Serranidae, Haemulidae	Munro (1983c)
Papua New Guinea (Port Moresby)	Shallow reef species	0.68–4.45	2.46	Lutjanidae, Carangidae, Serranidae, Lethrinidae	Lock (1986a, c)
Papua New Guinea (Tigak Is)	Shallow reef species	n.a.	1.2	Lutjanidae, Carangidae, Serranidae, Lethrinidae	Wright and Richards (1985)
Guam	Shallow reef species	0.03–2.04	0.55	Carangidae, Lethrinidae, Acanthuridae, Siganidae	Katnik (1982)
Belau	Shallow reef species	2.8–7.32	5.1	Lutjanidae, Lethrinidae, Serranidae	Anon. (1990, 1991)
Norfolk Island	Shallow reef and slope species	11.8–15.0	13.6	Lethrinidae, Serranidae, Carangidae, Labridae	Grant (1981)
Maldives	Shallow reef species	0–8.5	1.8	Lutjanidae, Lethrinidae, Carangidae, sharks	Van der Knaap *et al.* (1991)
Philippines (Cape Bolinao)	Shallow reef species	0.32–0.94	0.59	Lethrinidae, Lutjanidae, Priacanthidae	Acosta and Recksiek (1989)
Maldives	Shallow reef species	1.75–3.33	2.4	Lutjanidae, Serranidae, Carangidae, Lethrinidae	Anderson *et al. (1992)*
Seychelles	Shallow reef species	0.55–19.07	4.38	Carangidae, Scombridae, Lutjanidae, Lethrinidae	de Moussac (1987)
Fiji (Ba)	Shallow reef species	0.14–12.12	2.27	Lethrinidae, Lutjanidae, Carangidae, Serranidae	J. Anderson, MRAG, London (pers. comm.)
Nauru	Shallow reef and deep slope species	0.75–7.2	3.0	Lutjanidae, Serranidae, Carangidae, Holocentridae	Dalzell and Debao (1994)
Chuuk Outer Banks (Micronesia)	Shallow reef and deep slope species	1.31–4.57	2.30	Lutjanidae, Carangidae, Lethrinidae, Scombridae	Diplock and Dalzell (1991)
Yap	Shallow reef and deep slope species	0.97–3.1	1.67	n.a.	Anon. (1987b)
Tropical Pacific atolls	Deep slope species	0.4–19.0	7.7	Lutjanidae, Serranidae, Carangidae, Lethrinidae	Dalzell and Preston (1992)
Tropical Pacific high islands	Deep slope species	2.2–13.2	6.0	Lutjanidae, Serranidae, Carangidae, Lethrinidae	Dalzell and Preston (1992)

Table 7.4 Summary of catch rate (CPUE, catch per unit effort) and catch composition data for longline fishing on deep and shallow reef and associated stocks in the Caribbean, Indian Ocean and South Pacific regions (n.a. not available)

Location	Target stock	CPUE (kg/100 hooks)		Principal catch components	Source
		Range	Mean		
New Caledonia	Shallow reef species	3.0–12.2	8.2	Lethrinidae, Serranidae, Labridae, Lutjanidae	Kulbicki *et al.* (1987); Kulbicki and Grandperrin (1988)
Chuck Outer Banks	Deep slope species	1.6–12.3	6.8	Lutjanidae, Carangidae, Lethrinidae, Carangidae	Diplock and Dalzell (1991)
Maldives	Shallow reef and deep slope species	3.0–24.0	10.3	Lutjanidae, Lethrinidae, Serranidae, sharks	Van Der Knaap *et al.* (1991)
Maldives	Shallow reef and deep slope species	9.8–24.0	16.8	Lutjanidae, Serranidae, Lethrinidae, sharks	Anderson *et al.* (1992)
Sri Lanka	Shallow reef and deep slope species	0–24.8	4.55	Lethrinidae, Serranidae, Lutjanidae, Carangidae	Pajot and Weerasoorriya (1980)
Sri Lanka	Shallow reef and deep slope species	1.55–14.4	6.4	Lethrinidae, Lutjanidae, Serranidae, Carangidae	Anon. (1982)
Sri Lanka	Shallow reef and deep slope species	4.0–11.1	6.8	Lethrinidae, Lutjanidae, Serranidae, Carangidae	Weerasoorriya *et al.* (1985)
Indonesia (Sumatra)	Shallow reef species	n.a.	9.0	Serranidae, Muraenidae, Haemulidae, Lutjanidae	Kunzmann (1988)
Tonga	Deep slope species	n.a.	11.0	Lutjanidae, Serranidae, Carangidae, sharks	Mead (1987)
Fiji	Deep slope species	12.5–29.2	19.1	Lutjanidae, Serranidae	Walton (pers. comm.)
Vanuatu (Paama and Espiritu Santo)	Deep slope species	10–92.5	33.7	Lutjanidae, Serranidae, sharks	Fusimalohi and Preston (1983)
Vanuatu (Espiritu Santo)	Deep slope species	15–389	124.6	Lutjanidae, Serranidae, Lethrinidae, Carangidae	SPC (unpublished data)
Martinique	Shallow reef and deep slope species	4.0–27.0	10.8	Lutjanidae, Serranidae, Muraenidae, sharks	Gobert (1989)
Gulf of Mexico	Shallow reef species	15–60	n.a.	Lutjanidae, Serranidae, Haemulidae, Sparidae	Tashiro and Coleman (1977)

dominated by snappers, groupers, emperors (or grunts in the Caribbean) and jacks.

Hook-and-line gear can also be trolled from a vessel to catch pelagic species and reef fish that venture into the pelagic zone to take lures. Selectivity of trolled gears is influenced by hook size, lure type, vessel speed and sea condition. Table 7.5 summarizes catch rates and composition from troll fishing adjacent to reefs. Large scombrid fishes, mainly tunas, Spanish mackerel and wahoo, dominate catches. However, catches include reef species such as groupers, jacks, barracudas, snappers and emperors.

Hook selectivity depends on size. With conventional 'J' hooks, size is measured as the distance between the point of the hook and the shank (Fig. 7.3(a)). Other measures of size include the product of length and width of the hook (Fig. 7.3(b)) used by Ralston (1982) for tuna circle hooks employed for handline fishing on groupers and snappers in Hawaii. The results of fishing experiments with different hook sizes are contradictory and partly inconclusive. Koike *et al.* (1968) and Kanda *et al.* (1978) have suggested that catches of spiny goby, *Acanthogobius flavimanus*, and mackerel, *Scomber japonicus*, demonstrate a tendency towards an optimum catchable length which is dependent on hook size. Others have reported sigmoidal selection curves, analogous to those for trawl fisheries (Chatwin, 1958; McCracken, 1963; Saetersdal, 1963).

Ralston (1982) found that selectivity was strongest against smaller fish for deep slope snappers and groupers in Hawaii. Ralston (1990) conducted a further study in the Mariana Islands to determine the most appropriate model describing hook selectivity. The two models tested were the logistic (S-shaped) curve for trawls and seines, or the normal (bell-shaped) curve for gill nets. Neither model adequately described hook selectivity. Small hooks caught substantially more small fish, while large hooks were more effective in capturing larger size classes.

Bertrand (1988) studied hook selectivity in the hand-line fishery of the Saya de Malha Banks in the Indian Ocean, catches from which comprise mainly emperors. For a given population under varying fishing conditions there was no effect on selectivity caused by modifying baits and hook sizes. Further, he suggested that the high abundance of larger fishes in the catch relative to smaller fishes indicated competition for bait. Cortez-Zaragosa *et al.* (1989) and Tandog-Edralin *et al.* (1990) studied hook selectivity for Philippine yellowfin and skipjack tunas caught by hand-line fishers. In both instances all hook sizes captured a wide range of lengths, but optimum capture length increased with hook size.

Lokkeborg and Åsmund (1992) reviewed species and size selectivity of longline fishing. The same selection effects apply as with hooks used for hand lining but they also suggest that bait size has an influence on size selectivity. This stems from the relationship between predator and prey size

Table 7.5 Summary of catch rate (CPUE, catch per unit effort) and catch composition data for troll fishing on reef and associated stocks in the Caribbean, Indian Ocean and South Pacific regions

Location	CPUE (kg line-h^{-1}) Range	Mean	Principal pelagic species in catch	Principal reef components	Source
Papua New Guinea	3.7–6.9	4.9	*Euthynnus affinis, Auxis thazard, Scomberomorus commerson*	*Caranx ignobilis, Sphyraena barracuda, Epinephelus fuscoguttatus, Symphorus nematophorus, Lethrinus* spp.	Anon. (1984), Wright and Richards (1985), Dalzell and Wright (1986), Lock (1986a)
Palau	3.7–12.6	8.2	*Katsuwonus pelamis, Thunnus albacares, Acanthocybium solandri*	*Caranx ignobilis, Sphyraena* spp.	Anon. (1990, 1991)
Fiji	1.6–16.2	5.7	*S. commerson, Grammatorcynus bicarinatus, Gymnosarda unicolor, E affinis*	*Sphyraena qenie, S. barracuda, Plectropomus areolatus, Plectropomus* spp., *Caranx ignobilis, Caranx* spp.	Chapman and Lewis (1982), Lewis *et al.* (1983)
Maldives	0–7.5	0.93	*E. affinis, A. solandri, A. thazard, Katsuwonus pelamis, G. unicolor*	*S. barracuda, Aprion virescens*	Anderson *et al.* (1992)
Tuvalu	0.5–7.0	2.7	*A. solandri, K. pelamis, T. albacares, Istiophorus platypterus*	*A. virescens, Aphareus furcatus, S. barracuda, Caranx* spp.	Chapman and Cusack (1988)
New Caledonia	0–7.3	4.0	*S. commerson, E affinis, A. solandri, T. albacares*	*C. ignobilis, S. barracuda, A. virescens, L. bohar, Epinephelus* spp.	Chapman and Cusack (1989)

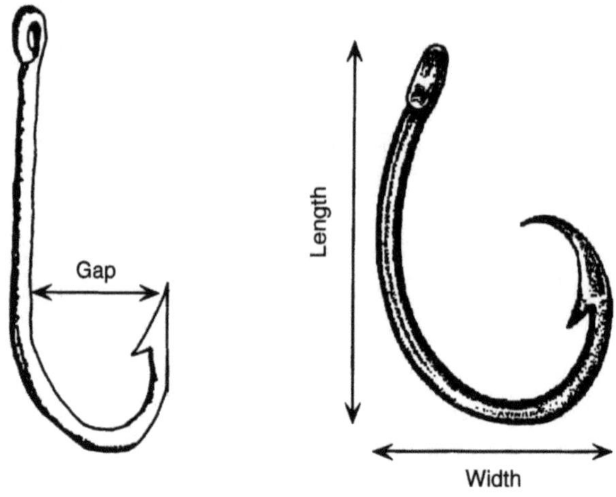

Fig. 7.3 Types of hook employed for line fishing and the dimensions used to measure hook size.

which has been observed in the field for a number of species (Hart, 1986). As Lokkeborg and Åsmund (1992) point out, the behaviour of a fish responding to a baited hook is similar to that of a foraging fish, and bait size will affect the size of fish caught.

Hook spacing on the main line can also affect fishing power and selectivity of longlines. Hamley and Skud (1978) and Skud (1978) reported increases in Pacific halibut, *Hippoglossus stenolepis*, catches when the interval between hooks was increased from 2.8 to 5.5 m; the mean size of the captured fish also increased. The authors suggested that smaller halibut have smaller feeding ranges and therefore, when hooks are widely spaced, the probability of encountering a hook is reduced for small fish. Further, the ratio of halibut in catches increased with increasing hook interval, suggesting that they may be more successful than other species in competing for available baits. Catch rate generated by varying the intervals between hooks was used by Eggers *et al.* (1982) as a method for computing area fished by longline.

Polovina (1986), in a hand-line fishing experiment on a seamount in the Mariana Islands, demonstrated species selectivity resulting from competition for baits. An intensive fishing experiment, to estimate biomass (see below), led to a general decline in the CPUE with cumulative catch, including declines in CPUE of the snappers *Pristipomoides zonatus* and *Etelis carbunculus*. However, the catch rates of the snapper *P. auricilla* increased over the same time period. *P. zonatus* and *P. auricilla* were found in the same depth range whilst *E. carbunculus* was caught at greater depths. Polovina (1986) suggested that if *P. zonatus* is more aggressive than *P. auricilla* in competing for bait, then the initial catchability of *P. auricilla* will be low, but will rise as the population of *P. zonatus* is reduced.

Kulbicki (1988b) used a combination of fishing and underwater visual census to investigate species selectivity of longlines used in a New Caledonia reef lagoon. Total catch rates were proportional to biomass as determined from the underwater census counts. The same was true for most of the major species groups in the catch apart from emperors of the genus *Lethrinus*, where longline CPUEs were much higher than for species found at similar densities such as wrasses (Labridae), groupers (Serranidae) or even other emperors (*Gymnocranius* spp.). Kulbicki concluded that *Lethrinus* spp. were either especially vulnerable to longline gear or largely underestimated by underwater visual census techniques. A similar study conducted at Johnston Atoll (North Pacific), where hand-line catch rates of deep slope species were compared with abundance estimates from a submersible, also concluded that there is a direct relationship between observed fish density and hand-line CPUE (Ralston *et al.*, 1986).

Traps

Portable fish traps are common in the Caribbean, East Africa, Arabian Gulf and South East Asia. While used in some parts of Micronesia, however, they are uncommon elsewhere in the South Pacific. Figure 7.4 shows the types of traps commonly deployed on reefs in the Caribbean islands. Fish traps, like seine nets, are relatively unselective. Selectivity of traps built from chicken wire depends partly on mesh size. Chicken wire mesh is either rectangular or hexagonal in shape, and is rigid, unlike a net mesh that alters in size with the disposition of the net. Hexagonal wire mesh size is measured as the distance between the parallel sides of the mesh, although this is not the same as the maximum aperture of the mesh, which is the maximum distance between knots. Rectangular wire mesh size is expressed as the lengths of the sides forming the rectangle. Munro (1983c) suggested that the selection effect of rigid wire-mesh traps is a function of the maximum aperture of the mesh and maximum body depth of the fish. Other factors such as the depth at which the trap is set, trap shape and use of bait may affect selectivity and catching power.

Examples of the various designs of fish traps used on reefs and coralline shelves are given in Smith (1947b), Munro (1983c), Quinn *et al.* (1985) and de Moussac (1986b). The dynamics of fish and invertebrate catches in portable traps set on shallow reefs have been investigated in detail by Munro *et al.* (1971), Munro (1974), Luckhurst and Ward (1987), Bohnsack *et al.* (1989), C.R. Davies (1989) and Dalzell and Aini (1992).

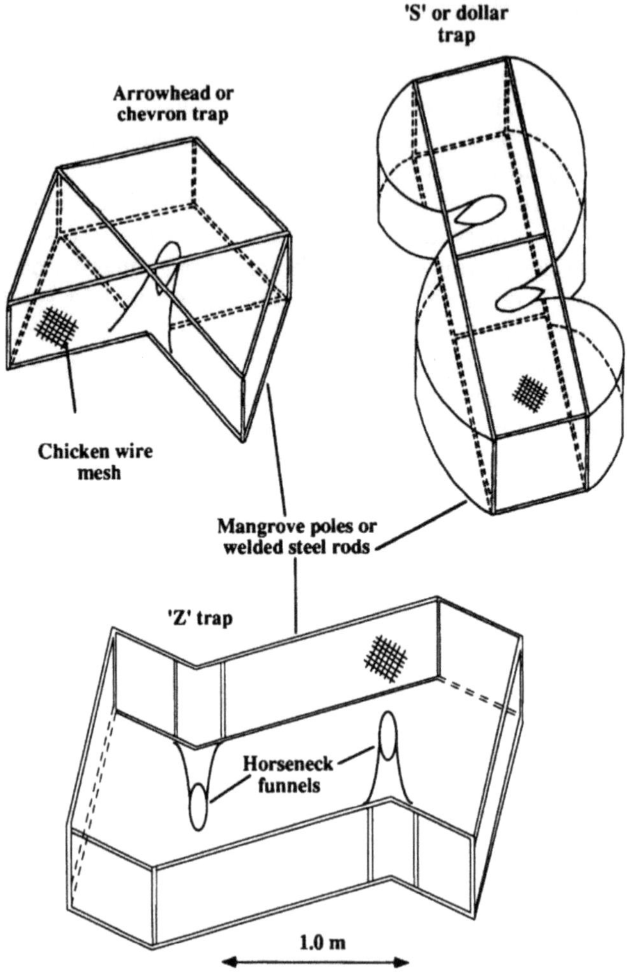

Fig. 7.4 Examples of wire-mesh traps used to catch reef fish in the Caribbean islands. Adapted from Dalzell and Aini (1989).

Munro *et al.* (1971) and Munro (1974) observed the build-up of fishes in unbaited traps on Jamaican reefs and showed that catches tend towards an asymptote (page 167 and Equation 7.2) as the entry or ingress is balanced by the escapement rate or egress from the trap. Catches with baited and unbaited traps on shallow reefs in all areas maximized at soak times between 4 and 7 days (Munro, 1983c; Luckhurst and Ward, 1987; Felfoldy-Fergusson, 1988; C.R. Davies, 1989; Van Der Knaap *et al.*, 1991; Dalzell and Aini, 1992).

Table 7.6 summarizes catch rates from different types of trap set on shallow reefs. Traps set close to the reef catch mainly herbivorous species such as surgeonfishes, rabbitfishes and parrotfishes, and small scavengers such as emperors, grunts and groupers. The main difference between shallow reef catches in the Caribbean and similar fishing in the Indo-Pacific region, is the propensity for Caribbean spiny lobster (Panuliridae) and crabs (Majiidae) to enter traps. These crustaceans eventually form up to 20% of the catch, greatly increasing its value (Munro, 1983c). Tropical Indo-Pacific spiny lobsters do not enter fish traps as readily as their Caribbean congeners. However, species in Hawaii and Western Australia readily enter traps and form the basis of local fisheries. Felfoldy-Fergusson (1988) also found that slipper lobsters (Scyllaridae) around Tongatapu in the southern islands of Tonga would enter fish traps.

Munro (1974) found that while bait initially increased the rate of ingress, ingress declined when bait was exhausted and catch stabilized at levels similar to those of unbaited traps. Bait is extremely important, however, in catches of predatory species (snappers, emperors, groupers and jacks) in deep water beyond the slope where catch rates are maximized in hours rather than days (Table 7.6). Whitelaw *et al.* (1991) reported that a large quantity of bait (4.0 kg) was necessary for traps to be effective and that pilchard, *Sardinops neopilchardus*, was better than less oily fish such as emperor, *Lethrinus choerorynchus*. Wolf and Chislett (1974) also found that large quantities of bait, between 2.2 and 11.4 kg per trap, were required for traps to fish effectively for snappers and groupers on the demersal shelves of Caribbean islands. The most effective baits were scads, Spanish mackerel and herring.

Selectivity of fish traps has been investigated in the Caribbean. Following observations on traps constructed from three mesh sizes, Olsen *et al.* (1978) concluded that rigid wire mesh is size selective, both within and among species. Bohnsack *et al.* (1989) observed catches in traps constructed from ten different rectangular mesh sizes, ranging from 1.3 × 1.3 cm to 7.6 × 15.2 cm, and a 3.75 cm hexagonal mesh. The hexagonal mesh caught the greatest numbers and weight of fish, with declining catch rates with smaller or larger rectangular mesh. There were relationships between mesh shape and size and individual retention for snappers, groupers, jacks, porgies (Sparidae) and surgeonfishes.

Based on observations from trap catches, Munro (1983c) developed a simple linear model for determining the mean retention length (L_c) of Caribbean reef fish caught by a particular trap mesh size. The model takes the form:

$$L_c = \mathrm{d}D + \mathrm{v} \qquad (7.3)$$

where v and d are constants and D is the maximum aperture of the wire mesh. Hartsuijker and Nicholson (1981) proposed, however, that average recruitment size of fish to a trap of a given mesh size was more likely to be

Table 7.6 Summary of mean catch rate (catch per trap) and catch composition data for trap fishing on reef and associated stocks in the Caribbean, Indian Ocean and South Pacific regions (n.a. not available)

Location	Trap design*	Depth set (m)	Target stock	Mean catch per trap (kg)	Average soak time (days)	Principal catch components	Source
Papua New Guina (Tigak Is)	Arrowhead	5–20	Shallow reef species	2.3	5	Acanthuridae, Lethrinidae, Serranidae, Scaridae	Dalzell and Aini (1992)
Papua New Guinea (Tigak Is)	S	3–13	Shallow reef	0.8	5	Acanthuridae, Lethrinidae, Scaridae, Serranidae	Dalzell and Aini (1992)
Vanuatu	Z	100–430	Deep slope	3.1	1.2	Lutjanidae, Lethrinidae, Serranidae, Carangidae	Guerin and Cillauren (1989)
New Caledonia	Z, round, rectangular	80–320	Deep slope	8.6	0.4	Lutjanidae, Serranidae, Lethrinidae	Desurmont (1985)
New Caledonia	Z	10–50	Shallow reef	4.5	0.6	Serranidae, Lutjanidae, Lethrinidae	Kulbicki and Mou-Tham (1987)
Tonga	Z	10–35	Shallow reef	n.a.	4	Mullidae, Lethrinidae, Lutjanidae, Theraponidae	Felfoldy-Fergusson (1988)
Australia (Great Barrier Reef)	Z	3–14	Shallow reef	n.a.	5	Siganidae, Chaetodontidae, Scaridae, Lutjanidae	C.R. Davis (1989)
Australia (NW shelf)	O,S,Z	7–20	Shallow reef	9.7	0.2	Lethrinidae, Serranidae, Lutjanidae	Whitelaw *et al.* (1991)
Philippines (Cape Bolinao)	Rectangular	1–5	Shallow reef	0.1	1	Siganidae, Labridae, Scaridae, Serranidae	Acosta and Recksiek (1989)
Philippines (Tawi-Tawi Is)	Round and rectangular	5.5–9	Shallow reef	0.6	2	Chaetodontidae, Scaridae, Nemipteridae, Siganidae	Tahil (1984)
Maldives	Arrowhead	1–60	Shallow reef	1.9	5	Lutjanidae, Lethrinidae, Serranidae	Van Der Knaap *et al.* (1991)

India (Tamil Nadu)	Heart-shaped	< 30	Shallow reef	0.48	1	Lethrinidae, Siganidae, Scaridae	Mohan (1985)
Sri Lanka	Various	20–200	Shallow reef and deep slope	0.7	0.5	Lethrinidae, Serranidae, Balistidae, Lutjanidae	Hammerman (1986)
Seychelles	Arrowhead	10–40	Shallow reef	4.6	2	Siganidae, Lethrinidae, Serranidae, Scaridae	de Moussac (1987)
Kenya/Tanzania	Arrowhead	< 30	Shallow reef	3.3	2	Siganidae, Scaridae, Labridae, Mullidae	de Moussac (1986b)
South Florida	Z	9–73	Shallow reef	5.6	10	Lutjanidae, Serranidae, Haemulidae, Sparidae	Sutherland and Harper (1983)
South Florida	Rectangular	15–18	Shallow reef	4.7	38	Serranidae, Haemulidae, Pomacanthidae, Lutjanidae	Taylor and McMichael (1983)
South Florida	Rectangular	7–40	Shallow reef	2.3	7	Serranidae, Balistidae, Haemulidae, Lutjanidae	Bohnsack *et al.* (1989)
Jamaica (inshore)	Z and S	1.5–29.5	Shallow reef	2.6	16	Scaridae, Haemulidae, Acanthuridae, Crustaceans	Munro (1983c)
Jamaica (offshore)	Dollar and hexagonal traps	15–30	Shallow reef	21.0	2	Balistidae, Serranidae, Haemulidae, Acanthuridae	Munro (1983c)
Caribbean banks and seamounts	O, Z and D	40–200	Shallow reef and deep slope	16.0	1	Lutjanidae, Serranidae, Carangidae	Wolf and Chislett (1974)
Puerto Rico	S, Z, arrowhead	40–250	Shallow reef and deep slope	3.6	2.5	Lutjanidae, Serranidae, Carangidae, Haemulidae	Stevenson and Stuart-Sharkey (1980)

*See also Fig. 7.4. An O trap is a circular trap of Australian design, shaped rather like a bass drum (fig. 2 in Wolf and Chislett, 1974).

a function of the ecology and behaviour of the species in question than mesh size. They argued that changes in foraging patterns and home range size with age probably determine the size at which a species becomes susceptible to traps. The size at recruitment may be larger than that predicted by the equation, as fish may be able to squeeze through the mesh.

Ward (1988) showed Munro's (1983c) model accurately predicted average retention lengths of some species but not others. The predicted L_c for surgeonfishes was much smaller than the minimum size observed in the catch, a result similar to that of Dalzell (1989) for dollar and arrowhead traps in Papua New Guinea. Dalzell concluded that the observed selection effects for these species were a reflection of recruitment; as they increase in size they move from nursery areas on the reef flat and mangroves, onto the reef slope, where they become vulnerable to capture by traps.

Trap shape and construction materials also affect selectivity and catch rates. On deep reef slopes of New Caledonia, Desurmont (1989) found that Z and rectangular traps with straight, rather than curved or horse-neck, funnels produced greater catches than circular traps. An interesting bycatch of traps set deeper than 300 m in New Caledonia were nautilus, *Nautilus macrophthalmus*, which are valuable in the ornamental shell trade. In the Philippines, Tahil (1984) found that traditionally manufactured bamboo traps were less efficient than chicken wire or plastic mesh traps. Overall, rectangular traps caught more than circular traps. In the Caribbean, Munro *et al.* (1971) found that S-shaped dollar traps captured 25% more fish than traditional Z traps and suggested that the continuous curve of the dollar trap guides fishes to the funnel entrance.

Differences were found between the performance of wire and bamboo traps in Jamaica by Shaul and Reifsteck (1991). During the mid 1980s, Jamaican fishers reverted to constructing traps from traditional materials such as plaited bamboo as this was cheaper than chicken wire. There was a significant difference in catch composition: bamboo traps caught more goatfishes (Mullidae) whilst wire-mesh traps caught more breams and parrotfishes. Luckhurst and Ward (1987) showed that escapement from traps with straight funnels was far higher than from those with horse-neck or curved funnels.

Eggers *et al.* (1982) devised a method by which the area fished by a trap can be found by recording the catches of pairs of traps set at increasing distances from each other. Davies (1989) applied this to catches of Z traps on the Great Barrier Reef and found that the effective area fished for a small snapper, *Lutjanus carponotatus*, was 143 m^2 and for the rabbitfish, *Siganus doliatus*, 1256 m^2. Miller and Hunte (1987) used a combination of catch rate and underwater visual census observations to estimate the effective fishing area of arrowhead traps set on reefs in Barbados. Effective area fished ranged between 148 and 346 m^2 for five taxa of fishes (parrotfishes, surgeonfishes, squirrelfishes, groupers and angelfishes). Recksiek *et al.* (1991) used the same methodology for arrowhead traps set in Puerto Rico and found that effective area fished was respectively 25–90 m^2, 24 m^2 and 93 m^2 for the parrotfishes *Sparisoma aurofrenatum* and *S. viride*, and surgeonfish *Acanthurus bahianus*.

Little work has been carried out into the selection and dynamics of stationary traps or fish corrals on reefs. At their simplest, these traps are V-shaped stone and stick enclosures with an entrance that faces the shore, as for example in Papua New Guinea (Hulo, 1984) and Cook Islands (Baquie, 1977). More complex structures may comprise a series of leaders or barriers which guide the fish into several interconnecting chambers. The chambers terminate in a single catching chamber where the fish may be netted or speared (Fig. 6.4). The more complex structures are found in South East Asia (Umali, 1950), French Polynesia (Grand, 1985), Guam (Amesbury *et al.*, 1986), Tonga (Halapua, 1982) and Palau (Johannes, 1981). Fixed barrier traps take advantage of the tidal foraging migrations of reef fishes such as parrotfishes, surgeonfishes and goatfishes, which feed on the reef flat and follow the receding tide into deeper water. When they encounter a fence they swim along this and concentrate in a chamber or net where they can be caught.

Escapement rates from fixed traps made up of rigid wire mesh have not been estimated. If fixed traps are built from chicken wire as in French Polynesia, Guam and Tonga, then selection effects similar to those exhibited by portable traps are likely, and the average length at first capture will be a function of maximum mesh aperture and fish body depth. Fixed traps made from fishing net, as in Palau, or with a portion of fishing net forming the apex of the trap, as in the Cook Islands, will also have selection effects dependent on the interaction of mesh size and fish dimensions. Traps made from tightly packed stones, or from bamboo strips, where there are no sizeable gaps, will retain all but the smallest fishes.

The operation of fish corrals (or *parcs*) at Tikehau Atoll in French Polynesia has been studied by Caillart and Morize (1985). Over 30 species were captured, principal among them being the snappers *Lutjanus gibbus* and *L. fulvus*, the unicornfish *Naso brevirostris*, the big-eye scad *Selar crumenophthalmus*, and the goatfish *Mulloidichthys flavolineatus*. These six species together formed two-thirds of the total catch. Catches were seasonal, with the greatest production in the austral summer from November to February and with little to no production between April and September. Average catch rates ranged from 380 to 580 kg $parc^{-1}$ day^{-1} during the fishing season.

Baquie (1977) describes some aspects of the fish corrals (*pa-ika*) set on the reef flat at Rarotonga in the Cook Islands. These consisted of V-shaped barriers of stone that directed fish into a net set at the apex of the V. The principal target of the *pa-ika* is the big-eye scad, although some lagoon species and other pelagic fishes are also caught. Woodland (1960) found that newly established fish corrals in tropical north-eastern Australia catch between 230 and 320 kg of fish per week, but that this would eventually decline to around 90 kg per week after about 3 months. Depending on location, fish corrals caught a mixture of estuarine and reef species including barramundi, *Lates calcarifer*, rabbitfishes, groupers, snappers, jacks and Spanish mackerels.

Fixed traps (*kesokes*) in Palau catch mainly rabbitfishes, with the balance made up of emperors, wrasses (Labridae), parrotfishes and surgeonfishes (Kitalong and Dalzell, 1994). Estimates of the size at first capture of the dominant species and of the optimum mesh size for the terminal portion of the net, based on yield-per-recruit analysis, suggested that 7.6 cm nets be retained in the fishery. This would select for the optimum average capture length in a majority of the dominant species.

Spears

Spear fishing is conducted both above water and below. Fishers may target fishes from land or boat using spears and arrows, or by diving with hand spears and spear guns. The development of masks, fins, scuba gear and spear guns has meant the fishing power of the fishers has greatly increased.

Data collected from spear fishing on temperate rocky reefs in Australia show that catch rates vary among spear fishers due to experience (Lincoln-Smith *et al.*, 1989). Less skilled divers could not fish as deep as skilled fishers, nor were they as familiar with the habitats and behaviour of the species. CPUE was also inversely correlated with wave height.

As fishers actively hunt fish, there is more scope for fishers to directly influence gear selectivity. A good example is shown by the length frequencies of the small surgeonfish *Acanthurus nigrofuscus* at Woleai Atoll in Micronesia, caught by drive-in-net fishing and spearing (Fig. 7.5) (Smith and Dalzell, 1993). The fishing occurred at similar locations on the lagoon back reefs over a period of 1 month. Speared fish were larger on average than drive-in-net catches (mean 14.4 vs. 11.3 cm).

Few published observations exist on the catch rates of fishers using spears. All but one of the examples in Table 7.7 are from fisheries in the South Pacific. Catches range between 0.4 and 5.7 kg man-h^{-1}, although a much higher rate of 8.5 kg man-h^{-1} is reported from Palau. Catches mostly comprise strongly reef-associated species such as surgeonfishes, parrotfishes, squirrelfishes and groupers.

Little is known about surface spear fishing. In Papua New Guinea and

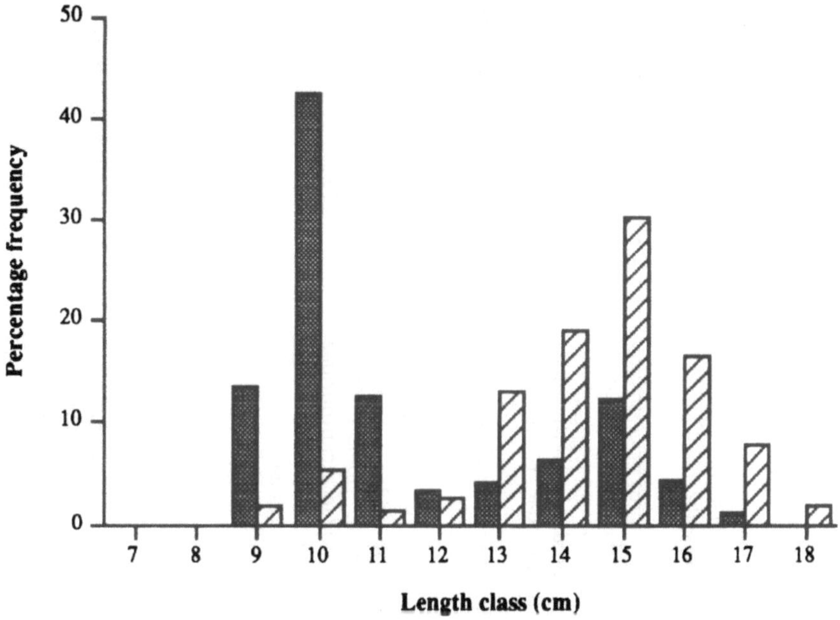

Fig. 7.5 Length–frequency distributions of the surgeonfish *Acanthurus nigrofuscus* caught by drive-in-net fishing (shaded columns) and spear fishing (hatched columns) at Woleai Atoll, Micronesia. Based on data in Smith and Dalzell (1993).

elsewhere in the Pacific, multiprong spears are thrown amongst dense schools of small pelagic species such as the herring, *Herklotsichthys quadrimaculatus*, and scads (*Selar* spp.). These may be eaten or more commonly are used for bait to catch larger species. Surface spear fishing is practised at night in southern Papua New Guinea for longtoms (Belonidae) which are attracted to lights mounted in canoes (Lock, 1986a; Fig. 6.5). The tendency of longtoms to jump from the water when pursuing prey or when startled, has led to many injuries, some fatal, to fishers in the southern Papuan region due to impalement by their sharp beaks (Barss, 1982).

Other methods

A variety of other fishing activities is conducted on reefs (Ruddle, Chapter 6), although there are few data on catch rates. Cast-net fishing, for example, takes place in shallow waters bordering reefs and mangrove. Cast nets on Guam range in diameter from 2.5 to 3.7 m with mesh sizes between 0.5 and 10 cm, depending on the target species (Amesbury *et al.*, 1986). Catches were 0.26–1.92 kg man-h^{-1} with a mean of 0.69 kg man-h^{-1}, comprising rabbitfishes, surgeonfishes, jacks and goatfishes

Table 7.7 Summary of catch rates (CPUE, catch per unit effort) and catch composition from spear-fishing on South Pacific reefs (n.a. not available)

Location	Target stock	CPUE (kg man-h^{-1})		Principal catch component	Source
		Range	Mean		
Philippines (Cape Bolinao)	Shallow reef species	0.9–2.0	1.3	Siganidae, Mullidae, Labridae, Scaridae	Acosta and Recksiek (1989)
Papua New Guinea (Port Moresby)	Reef and lagoon species and large pelagics	1.2–3.6	2.4	Serranidae, Acanthuridae, Scombridae, Haemulidae	Lock (1986a)
Papua New Guinea (Kavieng)	Reef and lagoon species	n.a.	2.4	Scaridae, Serranidae, Lutjanidae, Haemulidae	Wright and Richards (1985)
Palau	Reef and lagoon species	7.4–9.6	8.5	Scaridae, Serranidae, Acanthuridae, Lethrinidae	Anon. (1990, 1991)
Nauru	Reef and reef slope species	0.2–4.3	2.0	Lutjanidae, Holocentridae, Serranidae, Acanthuridae	Dalzell and Debao (1994)
Guam	Reef and lagoon species	0.08–1.14	0.4	Scaridae, Kyphosidae, Siganidae, Acanthuridae	Katnik (1982)
Woleai (Micronesia)	Reef and lagoon species	0.55–2.04	1.2	Acanthuridae, Scaridae, Balistidae, Labridae	Smith and Dalzell (1993)
Tonga	Reef and lagoon species	1.2–1.7	1.4	Acanthuridae, Scaridae, Holocentridae, Serranidae	Halapua (1982)
Fiji (Dravuni)	Reef and lagoon species	0.81–1.6	1.2	Serranidae, Acanthuridae, Lutjanidae, Carangidae	Emery and Winterbottom (1991)
Fiji (Ba)	Reef and lagoon species	0.12–5.7	1.5	Lethrinidae, Lutjanidae, Serranidae, Scombridae	J. Anderson, MRAG, London (pers. comm.)

(Katnick, 1982). In Nauru, an average of 2.8 kg man-h^{-1} was recorded from cast-net fishing on the reef flat, with a catch made up of surgeonfishes, chub (Kyphosidae), parrotfishes and other small reef species (Dalzell and Debao, 1994). Similar species form the main targets of cast-net fishing in Palau (Johannes, 1981).

7.4 LIVE FISH CAPTURE

The preceding sections consider gears used to capture reef fishes for food. Fishes are also captured live for the aquarium and restaurant trades, particularly in East and South East Asia. Small schooling pelagic species, such as anchovies, sprats, sardines and fusiliers, are also captured from reef lagoons and held alive on board pole-and-line tuna vessels. Fishes are either aggregated at night by light attraction and captured with lift nets, or caught in the daytime with surround nets. The baitfish are loaded into wells set in the hulls of the pole-and-line vessels. During tuna fishing the live baitfishes are scattered on the sea to attract tuna schools within range of the pole-and-line fishers, who catch the tuna using lures fitted with a barbless hook.

The selectivity of the gear used to catch baitfishes is similar to that of seines and trawls, although there may be some other influences due to the differing species' response to light attraction when fishing at night. Capture of small pelagic fishes from reef areas, once widespread, has declined sharply, as pole-and-line tuna fishing has been largely replaced by purse seining. Live-bait fisheries in lagoons still persist in the Maldives, Solomon Islands, Kiribati, Fiji and Hawaii. Reviews and biological studies of these and other bait fisheries in the Indo–Pacific region are contained in Shomura (1977), Conand (1988), Dalzell and Lewis (1989) and Blaber and Copeland (1990). Catches from tuna bait fisheries are generally modest, usually less than 200 t $year^{-1}$, although the Papua New Guinea bait fishery generated catches of between 1000 and 1500 t $year^{-1}$, and the Solomon Islands bait fishery now catches an estimated 2500 t $year^{-1}$ (Nichols and Rawlinson, 1990). The latter is 20% of the estimated total coastal fish production (Dalzell and Adams, 1994).

The aquarium fish trade in the Pacific and elsewhere has been reviewed by Axelrod (1971), Conroy (1975) and Pyle (1992). According to Whitehead *et al.* (1986), the aquarium fish industry is one of the world's most valuable fisheries, with annual retail sales of fish and equipment exceeding US $4 billion. At least 25 countries are involved throughout the world. Whitehead *et al.* (1986) suggest that the Philippines, Hawaii and Caribbean account for about 96% of international trade, although Pyle (1992) considers this an overestimate.

A large number of species are of interest to the aquarium trade. Pyle (1992) lists ten families (Pomacanthidae, Chaetodontidae, Acanthuridae, Labridae, Serranidae, Pomacentridae, Balistidae, Cirrhitidae, Blenniidae and Gobiidae) as the most important. Over 240 species are listed in export figures by the Palau Division of Marine Resources (Anon., 1992a). The most commonly listed are the damselfishes (Pomacentridae) *Chrysiptera cyanea*, *Chromis albipectoralis* and *Dascyllus aruanus*, although these have little unit value (US$ 0.20 and US$ 0.30 per fish). Comparatively rare species such as the small surgeonfish, *Ctenochaetus tominiensis*, may be worth as much as US$ 25.00 per fish.

Aquarium fishes are captured in hand nets, barrier nets and traps. The barrier net is essentially a small drive-in net used by divers to catch a wide range of species. In some countries, such as the Philippines, Indonesia and Mexico, chemicals such as cyanide have been widely used to stun marine aquarium fishes for capture. This has led to a decline in quality and reputation of exports from countries where this practice is widespread.

Certain species of reef fishes are in demand as live food for restaurants in South East Asian countries. Groupers such as *Plectropomus* spp. and *Epinephelus tauvina* are much favoured. In Palau, groupers and Napoleon wrasse, *Cheilinus undulatus*, were caught by hand lines to supply the live fish market in Hong Kong (Johannes, 1991; Richards, 1993). Fishing was discontinued, however, when one of the two regular spawning aggregations of groupers, situated near Koror, Palau's capital, was fished so heavily it failed to occur again.

Live reef fishes and invertebrates for consumption are exported from Papua New Guinea to Hong Kong (Richards, 1993). Like Palau, the main exports are groupers (*Plectropomus* spp., *Epinephelus polyphekadion* and *E. malabaricus*) and Napoleon wrasse, which are caught from the reefs of the Hermit Islands in the north. Other species exported live from Papua New Guinea to Hong Kong include stonefish, *Synanceia verrucosa*, sought for its meat and venom for traditional medicines, and crabs and lobsters.

7.5 REEF FISHERY YIELDS

Yields of finfishes and other marine organisms have been reviewed by Munro and Williams (1985) and Russ (1984c, 1991). Initially, it was thought that the sustainable yield from tropical reef fisheries was 4–5 $t\,km^{-2}\,year^{-1}$ (Stevenson and Marshall, 1974), based mainly on data from the Caribbean, and questionable data from one Pacific atoll (A. Smith, SPREP, Apia, pers. comm.). Evidence suggesting that yields may be much higher (15–30 $t\,km^{-2}\,year^{-1}$) was regarded as anomalous (Marshall,

1980). However, more recent data make it clear that yields well in excess of 5 $t\,km^{-2}\,year^{-1}$ are possible, particularly from reefs in South East Asia and the Pacific. Apart from normal fishing activities in these fisheries, significant amounts of shellfish are gathered from the reef at low tide (see below). Known yields from reef fisheries are summarized in Table 7.8.

As Russ (1991) has pointed out, some of the confusion that has arisen with respect to estimating reef yields is due to different definitions of the area of reef being fished. Some workers have included only actively growing hermatypic reef to depths ranging between 8 and 60 m (Russ, 1991; Munro, Chapter 1). Others have estimated yields based on reef area and on shallow lagoon area, which included tidal mangrove forests, seagrass beds and sand flats. Some workers have even included the adjacent shelf area to a depth of 200 m in their estimation of reef yields.

The species to be included in reef harvests also creates problems when comparing yields among reefs. Catches may include scombrid fishes, such as tunas and mackerels, or snappers, groupers and other deep slope species which have been caught away from the reef. Reef yields also include shellfish and other invertebrates collected from inshore reefs at low tide, which at some locations may account for a significant fraction of the total harvest (Ruddle, Chapter 6).

The observed yield is primarily a function of fishing effort, and this varies greatly among reefs of the world. The South Pacific provides striking contrasts between levels of exploitation. The coast of Papua New Guinea has extensive reefs but fish yields are likely to be in the vicinity of 0.5 $t\,km^{-2}\,year^{-1}$, as for example in the Tigak Islands (Wright and Richards, 1985). Even on reefs considered to be heavily fished such as those near Port Moresby (Lock, 1986b) and on the west coast of Manus (Chapau and Lokani, 1986), yields ranged from 2.8 to 6.4 $t\,km^{-2}\,year^{-1}$. The overall yield is limited by fishing activity and in many cases also by a tradition of obtaining most food from the land. By comparison, the smaller Polynesian and Micronesian islands have far higher reef fishery yields due to the necessity of procuring protein from the sea. Even with increased urbanization, the demand for fresh fish and shellfish remains high and countries such as Samoa, Nauru and Niue and have yields at or in excess of 5.0 $t\,km^{-2}\,year^{-1}$ (Table 7.8).

Geographic variation in yield has been related to variations in density of fishers for small-scale reef fisheries in Jamaica (Munro and Thompson, 1983), Samoa (Munro, 1984a) and southern Papua New Guinea (Lock, 1986b). A similar approach has been applied to near-shore artisanal fisheries in the Philippines that include reef fisheries (Fox, 1986). The result of these approaches has been a variant on the Schaefer–Fox equilibrium production model where, instead of a time series of catch–effort data couplets being used to generate maximum sustainable yield (MSY) for single-species

Table 7.8 Estimated yields from reef fisheries in the Indian Ocean, Caribbean and South Pacific (n.a. not available)

Location	Habitat type	Area fished (km^2)	Maximum depth fished (m)	Yield ($t\ km^{-2}\ year^{-1}$)	Source
Papua New Guinea					
Kavieng	Fringing reefs and patch reefs	207.7	30	0.42	Wright and Richards (1985)
Port Moresby		116	40	5.0	Lock (1986b)
Total	All coral reefs	39 940	30	0.21	Dalzell and Wright (1986)
Jamaica	Coralline shelf	3 422.0	200	1.2–4.3	Munro (1977)
American Samoa	Fringing reef	3.0	40	8.6–44.0	Wass (1982)
Western Samoa	Fringing reef	300.0	40	11.4	Zann *et al.* (1991)
Tarawa (Kiribati)	Atoll reef and lagoon	459.0	30	7.2	Mees *et al.* (1988)
Ontong Java (Slomon Is)	Atoll reef and lagoon	122.0	n.a.	0.6	Bayliss-Smith (MS)
Nauru	Fringing reef and reef slope	7.5	100	4.5	Dalzell and Debao (1994)
Niue	Fringing reef and reef slope	6.2	60	9.3	Dalzell *et al.* (1993)
Philippines					
Apo	Fringing reef	1.5	20–60	11.4–24.9	Alcala (1981), Alcala and Luchavez (1981),
Sumilon	Fringing reef	0.5	40	14.0–36.9	Alcala and Gomez (1985), Lopez (1986),
Panilacan	Fringing reef	1.8	29	10.7	Savina and White (1986), Bellwood (1988),
Hulao-hulao	Barrier reef	0.5	15	5.2	Alcala and Russ (1990)
Lingayen Gulf (invertebrates only)	Fringing reef	70.0	2	25.3	
Total	All coral reef	27 044	20	5.9	Based on catch statistics in BFAR (1988) and reef area from Murdy and Ferraris (1980)
Puerto Rico	Coralline shelf	5 300	100	0.1–1.4	Weiler and Suarez-Caabro (1980)
Tanzania coast	Coralline shelf	12 160	n.a.	0.8–5.7	Wijkstrom (1974)
Gulf of Suez (Red Sea)	Coralline shelf and reefs	n.a.	30–60	0.31	Sanders and Kedidi (1984)
Red Sea coast of Saudi Arabia	Coralline shelf and reefs	n.a.	30–60	0.4	Kedidi (1984b)
Mauritius	Fringing reefs and barrier reef lagoons	243.0	n.a.	3.96–7.5	Munbodh *et al.* (1988)
Palau	Fringing and barrier reefs	450.0	20	1.7–3.0	Kitalong and Dalzell (1994)
Maldives (North Male)	Atoll fringing and patch reefs	552.0	45	0.95	Van der Knaap *et al.* (1991)
	Fringing reefs				
Fiji (Koro and Lakeba)	Fringing reefs	8.4	n.a.	5.0	Bayliss-Smith (MS)
Fiji (Yanuca, Dravuni, Moala, Totoya, Navatu)		n.a.	n.a.	0.3–10.2	Jennings and Polunin (1995)
Ifaluk atoll	Atoll reefs and lagoon	5.0	n.a.	5.1	Stevenson and Marshall (1974), based on observations by Alkire (1965)

stocks, the geographic variation in effort and catch is used to generate MSY for a multispecies assemblage of reef fishes (Appeldoorn, Chapter 9).

7.6 LIMITATIONS OF YIELDS AND THEIR PERCEPTION

The most urgent question for fisheries biologists is what are the limits of sustainability for yields from reef fisheries? It should be stressed that like any fishery, reef fisheries are dynamic. There may, for example, be gross changes in fishing effort with time. The archaeological record shows that Pacific island reef resources have been exploited over long periods, sometimes greater than 10 000 years (Butler, 1988; Kirch *et al.*, 1991; White *et al.*, 1991; contributions in Carroll, 1975). Further, some islands had much larger populations prior to European contact than at present. Populations fell because of exposure to unfamiliar diseases and, in some cases, conscription for work elsewhere. For example, Palau is thought to have had a population in excess of 45 000 prior to the arrival of Europeans, in contrast to the present population of 16 000 (Semper, 1873). Similar severe declines in population following European contact have occurred in Vanuatu (McArthur, 1981), the Caroline Islands (Connell, 1983) and the Marquesas (Rallu, 1990).

More recently, the Pacific Micronesian Islands were occupied by a collective population of more than 200 000 soldiers during World War II. Cut off from their supply lines towards the end of hostilities, the Japanese heavily fished reef stocks and resorted to explosive fishing in some instances. Smith (1947b) records the general scarcity of fish in locations such as Kosrae, Chuuk Lagoon and Pohnpei immediately following WW II, while Johannes (1991) reports that Palauan chiefs introduced traditional conservation measures at about the same time, due to severe depletion of stocks by the large war-time Japanese population on Palau. It is important, therefore, not only to quantify the catch rates, selectivity and standing stocks prevailing in reef fisheries, but to be aware of the history of these fisheries and the information that this can provide towards assessing sustainability and management of stocks.

Expansion of fishing effort, by targeting snappers and groupers on the deep reef slope, was once thought to offer additional commercial potential for fishers in coralline regions, diverting fishing pressure from shallow areas. However, the deep slope fish assemblage comprises apex predators that have long life spans and what appear to be low recruitment rates (contributions in Polovina and Ralston, 1987). Catch rates may decline in less than a year to as little as one-half to one-third of those on virgin stocks, and catches on some small seamounts can decline to zero (Ikehara *et al.*, 1970; Nath and Sesewa, 1990), suggesting that most of the fishable biomass has been captured. Current commercial fisheries development initiatives in the South Pacific and other tropical seas are aimed towards fisheries for high-value, large pelagic species, where resource limitation is not yet a serious problem. Reef fisheries will continue, however, to be an important source of protein for many inhabitants of the coralline regions, particularly where agriculture is limited.

ACKNOWLEDGEMENTS

I am grateful for the many helpful comments and suggestions made by Nick Polunin, Callum Roberts, Michel Kulbicki, Andy Richards, Tim Adams and two anonymous referees, all of whom greatly improved this chapter.

Chapter eight

Population and ecosystem effects of reef fishing

Simon Jennings and John M. Lock

SUMMARY

Fishing activities affect the population structure, growth, reproduction and distribution of target species and have indirect effects on non-target fish or invertebrate populations and their reef habitats. We review the multifarious impacts of fishing and conclude that existing knowledge of fishing effects is remarkably primitive given the burgeoning literature on reef fishery science and management. In particular, fishing effects are widely treated as synonymous with overfishing and thus many studies of fishing effects have been based upon examination of localized catastrophic events rather than changes that occur within sustainable fisheries subject to different fishing intensities or cropping regimes. An improved understanding of fishing effects in sustainable fisheries may assist the development of a range of new monitoring, assessment and management methods. These would provide an alternative to methods based on conventional population analyses within reef fisheries of limited economic importance which are unlikely to be selected for rigorous scientific study. Initial development of the fishing effects science could be achieved at moderate cost by refining existing procedures for collection of catch and effort data and by treating observed combinations of gear, effort and catch composition as a series of experimental manipulations.

Reef Fisheries. Edited by Nicholas V.C. Polunin and Callum M. Roberts.
Published in 1996 by Chapman & Hall, London. ISBN 0 412 60110 9.

8.1 INTRODUCTION

Fishing activities have direct and indirect effects on reef fish populations and their ecosystems. Such effects are significant because they lead to changes in the size and composition of yield from the fishery and alter subsequent fish production processes. In theory, knowledge of fishing effects should help the fishery manager to predict how a fishery responds to fishing and to select a management strategy that maintains the favoured type and quantity of yield. However, despite current concerns for the sustainability of reef fisheries and a burgeoning literature suggesting approaches for their management, there are remarkably few studies of fishing effects.

In this chapter we review the effects of fishing on the population structure, growth, reproduction and distribution of target species and consider indirect fishing effects on non-target fish and invertebrate populations and their reef habitats. These divisions serve to structure the review, but they are primarily artificial because responses to reef fishing are not mutually exclusive and have repercussions throughout the reef ecosystem. Russ (1991) produced a fine review of the effects of reef fishing, providing an introduction to the terminology associated with fishing effects and describing a number of seminal studies in considerable detail. In our review we inevitably refer to many of the same studies. However, our discussion of individual studies is relatively brief, particularly when they describe effects which are intuitively obvious (e.g. that the removal of fish by fishing leads to an immediate reduction in the number of fish remaining in the water) but which may not be reliably recorded due to problems with methodology or the scale on which the study is conducted. We attempt to describe the general effects of fishing whilst focusing on the efficacy of techniques for examining fishing effects and the potential management applications of the fishing effects science.

8.2 EFFECTS OF FISHING ON TARGET SPECIES

Numerical abundance and biomass

Successful fishing introduces an additional source of mortality within a fish stock and therefore causes decreases in numerical abundance and biomass. The initial responses to fishing have been documented when reserves or previously unfished areas were opened or became accessible to the fishery. It is frequently suggested that the most likely detectable effects of fishing pressure are declines in the abundance or biomass of species selectively targeted by the fishery (Russ, 1985; Russ and Alcala, 1989).

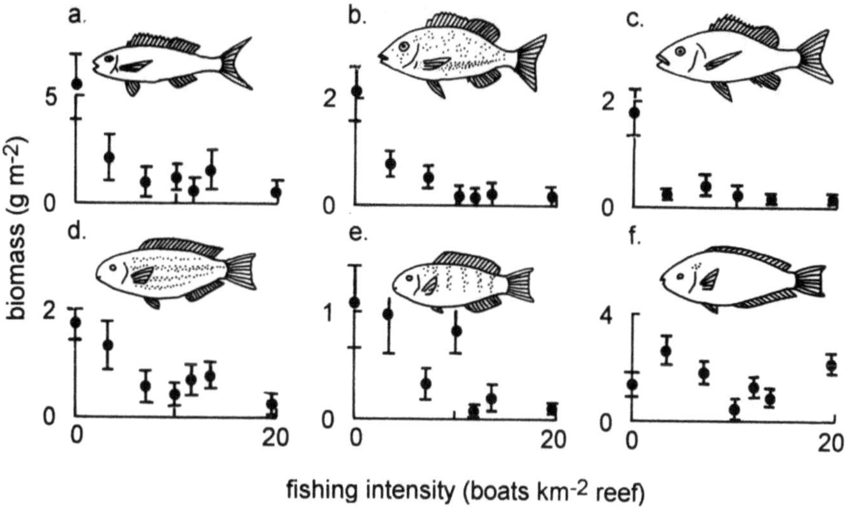

Fig. 8.1 Relationship between mean biomass (+ SEM, $n = 48$) as estimated using underwater visual census, and an index of fishing intensity for species targeted by the Seychelles reef fishery: (a) *Aprion virescens*; (b) *Lethrinus obsoletus*; (c) *Lutjanus bohar*; (d) *Scarus frenatus*; (e) *Scarus ghobban*; (f) *Scarus niger*. Redrawn with modifications and additions from Jennings *et al.* (in press b). Note differing vertical scales.

Several studies have documented such effects (Bohnsack, 1982; Ayling and Ayling, 1986a; Polovina, 1986; Alcala, 1988; Samoilys, 1988; Beinnsen, 1989; Russ and Alcala, 1989; Roberts and Polunin, 1992; Polunin and Roberts, 1993; McClanahan, 1994; Watson and Ormond, 1994; Jennings *et al.*, in press b,c; Fig. 8.1). However, few accurate measurements of fishing mortality have been produced because natural and fishing mortality are effectively inseparable without knowledge of population structure (Munro and Williams, 1985; Russ, 1991).

The effects of mortality are rapidly reflected in the fishery because many reef fishes are relatively site attached and there is limited replacement through adult immigration (Munro and Williams, 1985). Fishers often witness an initial decline in catch per unit effort (CPUE) of target species (Smith and Dalzell, 1993) followed, at high levels of exploitation, by a decline in total catch (Gaut and Munro, 1983). Thus during the development of the Hawaiian line fishery from 1959 to 1978, the total yield remained at similar levels whilst CPUE fell (Ralston and Polovina, 1982). In the Caribbean, Appeldoorn *et al.* (1992) recorded temporal declines in CPUE for several target species and Koslow *et al.* (1988) documented an

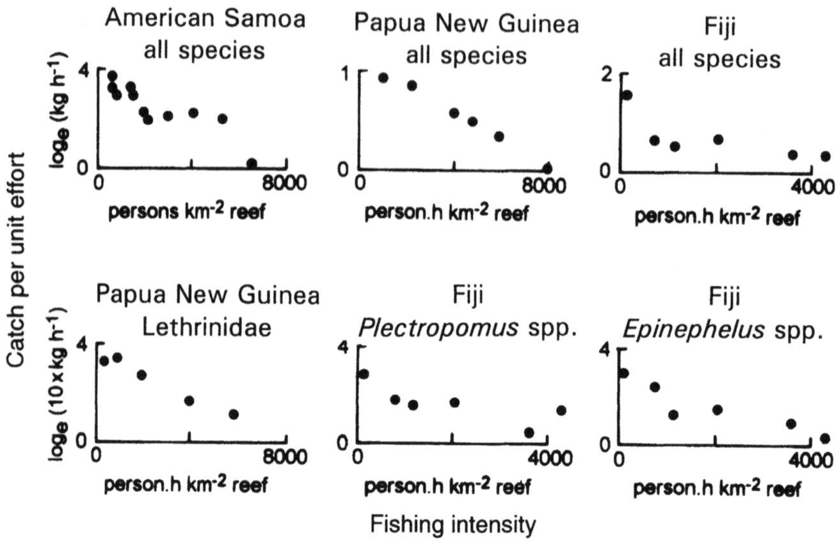

Fig. 8.2 Relationships between catch per unit effort and various indices of fishing intensity in reef fisheries. American Samoa, all species combined: redrawn from Munro and Williams' (1985) treatment of data in Wass (1982). Papua New Guinea, all species combined: calculated from Lock (1986c). Fiji, all species combined: redrawn from Jennings and Polunin (in press a). Papua New Guinea, Lethrinidae (emperors): calculated from Lock (1986c). Fiji, *Plectropomus* spp. and *Epinephelus* spp. (day spear fishery): plotted from data in Jennings and Polunin (in press a). Note differing vertical scales, and that American Samoa data are not expressed in person-hours.

82% long-term decline in CPUE at SE Pedro Bank in Jamaica where fishing effort had increased by 100%, and a 33% decline in CPUE at Port Royal Cays where fishing pressure increased by over 30% between the periods 1969–1973 and 1981–1986. However, there was no decline at SW Pedro Bank, which was initially unexploited and latterly fished with approximately 10% of the effort at SE Pedro Bank.

Similarly, spatial comparisons among grounds subject to different fishing intensities indicate that CPUE is lower at higher fishing intensities (e.g. Wass, 1982; Fig. 8.2). Kawaguchi (1974) noted that exploited banks within the operating range of the Jamaican nearshore canoe fishery yielded a mean CPUE of 0.45 kg line-h^{-1} whereas small lightly exploited or unexploited banks yielded 3 kg line-h^{-1}. Moreover, Lock (1986b) reported maximum catch rates which ranged from over 4 kg line-h^{-1} on the less accessible areas of the Port Moresby barrier reef to less than 1 kg line-h^{-1} in areas close to the main fishing centre. As fishing intensity increases, so the exploitable stock biomass is further reduced and increases in effort may no longer maintain the total catch. Thus Munro (1983c) and Gaut and Munro (1983) recorded lower CPUE on heavily exploited Jamaican Banks but noted a decline in total catch in the most heavily fished areas and Alcala and Russ (1990) reported a decrease in yield following intensive fishing in the Philippines.

In the longer term, at certain levels of exploitation, population models suggest that a stock may adjust to the new source of mortality, and compensatory changes in growth and recruitment may reverse declines in CPUE or total catch. However, the practical evidence for such an effect remains equivocal.

Size

Different fishing techniques remove fish of different sizes and species. Fishers typically target a restricted size range of fishes, and thus changes in the size and age structure of populations should be expected following fishing. Decreases in size attributed to fishing have been demonstrated following the opening of reserve areas or the development of a new fishery (Craik, 1981a,b; Bohnsack, 1982; Aiken 1983a,b; Gaut and Munro, 1983; Munro, 1983a; Reeson, 1983a,b; Thompson and Munro, 1983b,c; Wyatt, 1983; Ferry and Kohler, 1987; Koslow *et al.*, 1988; Luchavez and Alcala, 1988; Beinssen, 1989; Dalzell and Pauly, 1990) or through comparison of neighbouring sites subjected to different levels of fishing intensity (Bell, 1983; Russ, 1985; Ayling and Ayling, 1986a; Ferry and Kohler, 1987; Beinssen, 1988, 1989; McClanahan and Muthiga, 1988; Samoilys, 1988; Buxton and Smale, 1989; Polunin and Roberts, 1993; Watson and Ormond, 1994). Whilst it is clear that fishing affects the size structure of populations, the results from these studies must always be interpreted with caution as they may be biased by immigration, emigration and, in the longer term, recruitment and growth fluctuations. In addition, there are size-specific spatial and temporal variations in the distribution of fishes, and fishing or other survey methods may not produce samples representative of the population. For example, Kingsford (1992) showed that coral trout, *Plectropomus leopardus*, of different sizes fed on different parts of the reef and it has been reported that the mean size of some line-caught grunts (Haemulidae) (Gaut and Munro, 1983) and triggerfishes (Balistidae) (Aiken, 1983b) increased with depth of capture on Pedro Bank, Jamaica.

Size-selective fishing may affect the genetic structure of fish populations and lead to decreases in heterozygosity (Smith *et al.*, 1991) and growth rates (e.g. Law and Grey, 1989; Sutherland, 1990; Law, 1991). Although there are no detailed studies of these effects in reef fisheries, it is possible that artisanal reef fishers who crop a range of life-history stages would cause less marked genetic effects than commercial fishers who selectively target larger fishes.

Sex ratios and reproductive output

A change in population size structure may have profound effects upon species that undergo size-related sex changes (Sadovy, Chapter 2). Many commercially important reef fishes are protogynous hermaphrodites (Robertson and Warner, 1978; Warner and Robertson, 1978; Young and Martin, 1982; Thresher, 1984; Choat and Randall, 1986; Sadovy and Shapiro, 1987; Shapiro, 1987a) and thus, for species in which sex change is predominantly controlled by endogenous factors, the proportion of male fish would be expected to fall in response to fishing that targets the larger fish in the population (Munro and Williams, 1985; Bannerot *et al.*, 1987). For example, Thompson and Munro (1983c) noted that the male:female ratios of *Epinephelus striatus* and *Mycteroperca venenosa* were 1 : 0.72 and 1 : 0.85 on the lightly exploited offshore banks, compared with 1 : 5.6 and 1 : 6.0 at heavily fished Port Royal Cays. Similar effects could be expected for many other grouper species which mature as females and become males at sizes well in excess of the size of first capture (Thompson and Munro, 1983c; Shapiro, 1987a).

Most reported sex ratios are from fishery samples, in which sex ratio may vary markedly due to interactions between the size class retained and the size at which fish mature or change sex. However, comparisons using similar gears at different sites probably give useful information for comparative purposes if it is assumed that size-specific catchability is equivalent for both sexes. The extent to which apparent changes in sex ratio may be attributed to fishing effects is not clear. For example, Thompson and Munro (1983c) noted that snapper sex ratios in catches varied according to depth and type of fishing. *Lutjanus buccanella* male:female sex ratios of 1 : 2.1 were recorded in catches from traps set at 40–100 m whereas a ratio of 1 : 0.7 was recorded from line catches at 40–280 m. Thresher (1984) reviews several examples of male and female fishes aggregating on different areas of the reef.

Size-selective fishing is also expected to alter the sex ratios of gonochoristic fishes when males and females mature at significantly different sizes. There are no long-term analyses of the relationships between size and age of maturity in reef fishes. However, evidence from detailed studies of temperate species suggests that, within sexes, size is a more important determinant of maturity than age and that the size of maturity shows little plasticity in response to forcing factors such as temperature or growth

(Beverton, 1987; Jennings and Beverton, 1991). Munro (1983d) suggested that the male:female sex ratio of around 1:0.4 for *Pseudupeneus maculatus* was attributable to females growing more slowly and reaching a retainable size later than males. Conversely, *Mulloidichthys martinicus* male:female sex ratios were 1:1.52 offshore and 1:1.86 at Port Royal Cays, suggesting that the males grew more slowly. If changes in sex ratio are real, then their effects on reproductive processes are unclear. Evolutionary theory suggests that, in unfished populations, the ratio of male to female fish will attain an equilibrium that maximizes reproductive output (e.g. Werren and Charnov, 1978; Charnov, 1982). A reduction in the number of male fish due to fishing will disturb this equilibrium and may reduce the proportion of eggs that are fertilized. Not surprisingly, there are no direct tests which indicate when sperm production might become limiting. However, Shapiro *et al.* (in press b) do suggest that the energetic cost of sperm production in the wrasse, *Thalassoma bifasciatum*, is such that males consistently release the minimum amount of sperm needed to fertilize the egg clutch of the female.

Shifts in the size and age distributions of fish populations could have profound influences on their reproductive output. Larger reef fish of a given species will have higher relative fecundity (number of eggs per unit of somatic body mass) than smaller fish. Thus a population of a given biomass will have greater potential fecundity when composed of larger rather than smaller individuals (e.g. Bohnsack, 1990). Furthermore, when the reproductive life span of fishes is artificially curtailed by fishing, their potential reproductive output will not be realized. Within an unfished population, evolutionary logic suggests that a reduction in reproductive output would only occur naturally when it ensures that more progeny will subsequently attain maturity (Calow, 1979). Such a reduction may be forced by changes in other life history traits or may be a response to physical and biological characteristics of the environment (Jennings and Beverton, 1991). However, when the reduction in reproductive output is a direct consequence of fishing mortality, and of no evolutionary benefit, then the population will only maintain evolutionary fitness by rapid changes in reproductive strategy. These changes are expected to involve increases in reproductive output at a given size or age. They have not been documented for reef fishes but have been noted in exploited populations of spiny lobster, *Panulirus marginatus* (DeMartini *et al.*, 1992) and plaice, *Pleuronectes platessa* (Horwood *et al.*, 1986). DeMartini *et al.* (1992) recorded a 16.9% increase in size-specific fecundity after exploitation and Horwood *et al.* (1986) demonstrated that younger North Sea plaice of a given length had higher absolute fecundity.

It is difficult to determine the effects of fishing on recruitment when the effects may only be apparent at unidentified 'downstream' sites and when they will be masked by massive and largely unpredictable variations in larval survival (e.g. Doherty, 1991). If fishing does affect patterns of juvenile settlement and thus the subsequent recruitment of fish to the fishery, it is likely that the settlement patterns have a more general effect on distribution at the among-reef scale than within reefs (Doherty, 1991; Fowler *et al.*, 1992). Although most of the studies reviewed by Doherty (1991) focused on the initial settlement of non-fished species, recent studies of recruitment suggest the pattern is also common to commercial species (Appeldoorn *et al.*, 1992). The study of recruitment is further confounded by sound evidence that initial settlement patterns of juvenile fishes may be substantially modified by predation (Jones, 1991; Caley, 1993). As in temperate waters, where the search for stock–recruitment relationships has frustrated fishery biologists for several decades (Cushing, 1988; Shepherd and Cushing, 1990), it is unlikely that it will be possible to detect changes in recruitment rate that can reliably be attributed to reef fishing.

Behaviour and distribution

The behaviour and distribution of fishes will affect the observed response of a fishery to fishing. It is widely accepted that spear fishing can alter the behaviour of fishes and Randall (1987) noted fishes becoming shy and wary after harassment by divers collecting aquarium fishes. General diver disturbance may have a similar effect. For example, Harmelin-Vivien *et al.* (1985) noted lower fish abundance when consecutive visual surveys were conducted in the same area and divers involved in visual census work have frequently commented that different species respond to them in different ways (Samoilys, 1992). It is probable that the behavioural changes of fishes following spear fishing would bias abundance estimates obtained by visual census. In addition, CPUE would not provide a valid index of relative biomass because catchability would be negatively correlated with fishing effort.

Fishing disturbance may stimulate the redistribution of fish populations. When Sumilon Island Reserve in the Philippines was opened to fishing, Alcala and Russ (1990) attributed the significant increase in the abundance of parrotfishes (Scaridae) on the reef slope to their being driven from normal reef habitats to deeper areas by the use of drive nets and similar gears. Although it has not been documented for tropical species, there is evidence for the contraction and expansion of the effective range of fish populations with changes in their abundance (Beverton, 1984). It is also probable that there would be localized changes in distribution of roving carnivores following differential exploitation of their prey.

8.3 EFFECTS OF FISHING ON ECOSYSTEMS

Community structure

Fishing is expected to alter the structure of reef fish communities in a number of ways. Some individuals are removed directly because they are selected by particular gears (Dalzell, Chapter 7) or favoured for capture (page 194) whilst others are affected by changes in trophic interactions within the ecosystem. These changes may involve shifts in the biomass or size-specific biomass of different trophic groups (e.g. Fig. 8.3) or shifts in the relative dominance of populations with specific life history strategies (Adams, 1980).

Relatively few comprehensive studies have recorded changes in multi-species communities in response to fishing (Russ, 1985; Lock, 1986c; Koslow *et al.*, 1988; Russ and Alcala, 1989). Existing knowledge of fishing effects is such that it is not possible to separate direct and indirect fishing effects in these studies. Koslow *et al.* (1988) observed significant changes in the overall composition of reef fish catches at heavily exploited sites in Jamaica: the dominant trapped groups shifted from grunts and surgeon-fishes (Acanthuridae) to squirrelfishes (Holocentridae) and there was a virtual disappearance of parrotfishes (Scaridae). At another site, there was a shift from triggerfishes (Balistidae) to soldierfishes (Holocentridae) and non-balistid Tetraodontiformes. Similar changes occurred with increasing fishing effort at both exposed offshore and sheltered inshore sites, and thereby suggested a common response to exploitation. There was also a decline in the catch of primary commercial species at a lightly fished site and a shift in species composition that reflected the shift at the heavily fished sites. Moreover, there was a decline in overall catch rate and a disproportionately fast decline in the catch rate of deeper-bodied fishes such as groupers (Serranidae), snappers (Lutjanidae), parrotfishes, triggerfishes, angelfishes (Pomacanthidae) and surgeonfishes. Koslow *et al.* (1988) suggested that this rapid decline was a direct effect which resulted from their susceptibility to the gear for more of their life cycle. Russ and Alcala (1989) documented significant decreases in species richness which were probably attributable to fishing. They considered that the decreases in species richness of large piscivore communities were due to direct removal by fishers, whereas those of butterflyfishes (Chaetodontidae) were probably due to the reduction of live coral cover following habitat-destructive fishing techniques.

These studies suggest that descriptions of catch composition and catch rate may provide a crude index of the extent of exploitation in reef fisheries. In his analysis of the Port Moresby fishery, Lock (1986c) suggested three levels of exploitation effects in multispecies and multigear fisheries.

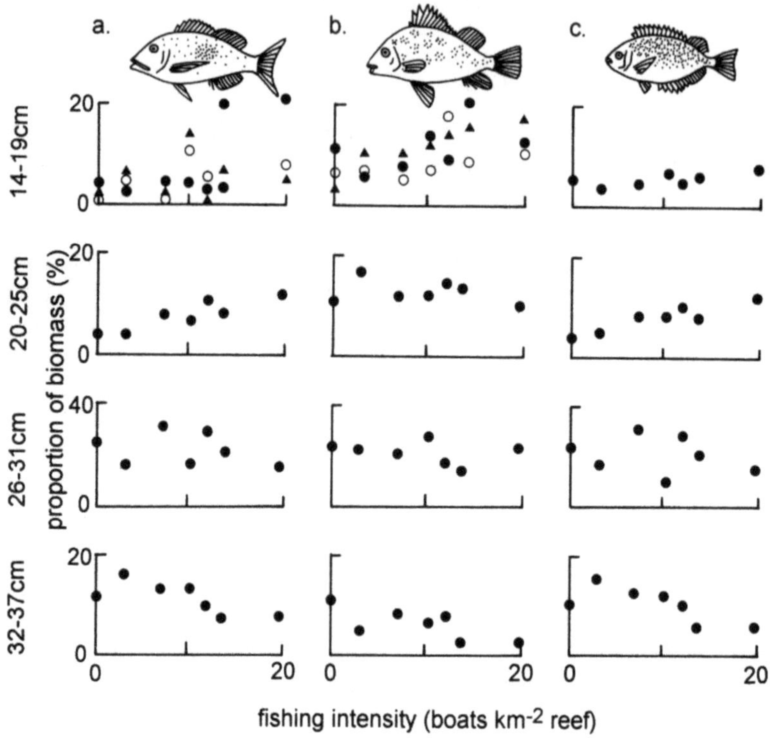

Fig. 8.3 Relationship in the Seychelles between fishing intensity and the proportion of biomass within fish of specific size classes (total length) in each of three trophic groups: (a) piscivorous and invertebrate-feeding fishes; (b) invertebrate-feeding fishes; (c) herbivorous fishes. Open circles, filled circles and filled triangles on the same plot represent means in different type of reef habitat (n = 16 for each habitat). Filled circles alone represent pooled means (n = 48) when differences between habitats were not significant (ANOVA, $P < 0.05$). Previously unpublished information based on data collected according to Jennings *et al.* (in press b). Note differing vertical scales.

Each level was characterized by catch composition and catch rate. In areas of light to moderate fishing intensity, catch rates were initially high, with large herbivores (surgeonfishes) dominating spear and net catches and large predatory groupers and snappers dominating the line catches. Catches of these species declined over a short period as the fishing pressure increased and, with fewer large fishes available, netting was adopted as the main fishing method. Small emperors (Lethrinidae) dominated catches using all fishing techniques in the moderately exploited phase. Eventually, hand lining was replaced entirely by netting and small herbivorous rabbitfishes (Siganidae) dominated the catch. These three phases are generally similar to those reported by Koslow *et al.* (1988) and are in accordance with Munro and Smith's (1984) proposed scenario for the evolution of a multispecies and multigear fishery.

Since reef fishes exhibit a range of life history tactics which are presumably shaped by natural selection to fit particular ecological demands (Stearns, 1976), it would be expected that fishing will affect fishes with different life history traits in different ways (Munro and Williams, 1985). Species with short life spans and rapid population growth, which channel a large proportion of their resources into reproductive activities, are likely to respond rapidly to fishing and may be fished sustainably at younger ages and higher levels of mortality. If fishing pressure is reduced it is predicted that these species have the capacity to recover quickly from a minimum population size caused by overfishing (Adams, 1980). Species that put a large proportion of their resources into increasing individual fitness and have low natural mortality will produce a high yield-per-recruit, but maximum yields will only be obtained with late entry to the fishery and at lower levels of fishing mortality (Adams, 1980). Fisheries based on slower-growing species are likely to be vulnerable to intensive exploitation despite naturally more stable population sizes which are buffered by numerous age classes against recruitment failure of individual cohorts. Their rate of recovery from fishing will be slow, and the multispecies system of which they are members may shift into one of the alternative stable states which have been suggested by theorists (Beddington, 1984) and possibly documented in some fisheries (Koslow *et al.*, 1988).

Fish predator–prey relationships

In reef ecosystems, piscivorous fishes are probably the most significant consumers of fish biomass (Grigg *et al.*, 1984) and, as in temperate waters (Bax, 1991; Overholtz *et al.*, 1991), piscivores may consume considerably more fish biomass than is removed by fishing. For example, the consumption of fish biomass by giant trevally, *Caranx ignoblis*, and blue trevally, *C. melampygus*, on French Frigate Shoals in the NW Hawaiian Islands is estimated to be approximately 50 $t\,km^{-2}\,year^{-1}$ (calculation from Sudekum *et al.*, 1991), well in excess of recorded or expected multispecies fishery yields from a variety of similar habitats (e.g. Munro and Williams, 1985; Russ, 1991; Jennings and Polunin, 1995; Dalzell, Chapter 7). Furthermore, there is good circumstantial evidence to indicate that piscivorous fishes consume a large proportion of fish biomass because a relatively high proportion of the species encountered in reef ecosystems are piscivores (Talbot, 1965; Goldman and Talbot, 1976; Brock *et al.*, 1979; Williams and Hatcher, 1983; Parrish *et al.*, 1986; Norris and Parrish, 1988; Hixon,

1991; Wantiez, 1992; Polunin, Chapter 5). Many of the favoured target species in reef fisheries are piscivores and it is reasonable to assume that their removal or shifts in their size distribution will cause perturbation within the reef ecosystem.

Numerous studies show disproportionately high mortality of reef fishes shortly after settlement, apparently due to severe predation on new recruits (e.g. Shulman and Ogden, 1987; Hixon, 1991). For example, Shulman *et al.* (1983) studied the fish assemblages on a series of artificial reefs and showed that the early survival of grunts was negatively correlated with the presence of piscivores and Hixon and Beets (1989) demonstrated that there was a negative relationship between the number of resident piscivores and the maximum number of prey fishes. It is likely that in many reef systems a reduction in piscivore density would affect community structure by decreasing the post-settlement mortality of favoured prey. However, with the exception of Caley (1993), controlled attempts to manipulate piscivore densities in order to observe the response of prey have been hampered by difficulties with removing piscivores and preventing the immigration of new piscivores to the study area (Stimson *et al.*, 1982; Kulbicki, 1988a; Shpigel and Fishelson, 1991) and by experimental artefacts and other extraneous factors affecting prey abundance (Lassig, 1982; Thresher, 1983b; Doherty and Sale, 1985; Jones *et al.*, 1988; Schroeder, 1989). In addition, with the exception of those by DeCrosta (1984); Sweatman (1984) and Sudekum *et al.* (1991), there are virtually no quantitative studies of the food consumption rates or favoured feeding strategies of major piscivores (Polunin, Chapter 5).

The best examples of significant and sustained reductions in the abundance of piscivorous fishes are found in fisheries (page 194). As such, fisheries may be useful systems in which to examine predator–prey relationships. There is some evidence from the study of catch rates and catch composition in reef fisheries that predator removal may lead to increases in the abundance of prey fish. In Lock's (1986c) study, for example, the increase in surround-net catches of lethrinids may have been a response to the removal of larger predatory lutjanids and serranids by spearing and handlining. However, Jennings *et al.* (in press b) censused 134 species of reef fishes in seven Seychelles fishing grounds subject to different fishing intensities and provided no evidence for a clear release of prey species despite significant differences in the abundance of predators (Fig. 8.4). Their data are too variable and based on too little replication to indicate that prey release is not occurring (Jennings *et al.*, in press b) but they do suggest that any increases in prey populations are less marked and less consistent than the decreases in predator populations. In enclosed lake fisheries, there are examples of clear interactions between predators and their prey (Zaret and Paine, 1973; LeCren *et al.*, 1977; Eggers *et al.*,

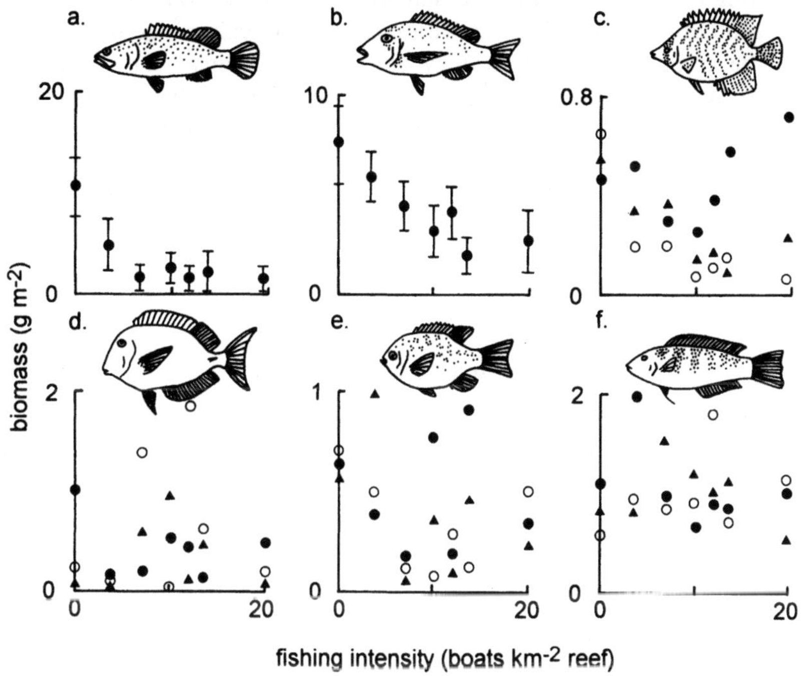

Fig. 8.4 Relationship between biomass and fishing intensity for six trophic groups on Seychelles reefs: (a) piscivores (mean ± 95% CI, $n = 48$); (b) invertebrate feeders and piscivores (mean ± 95% CI, $n = 48$); (c) corallivores with a maximum size < 20 cm total length; (d) detritivores < 20 cm; (e) omnivores < 20 cm; (f) invertebrate feeders < 20 cm. Open circles, filled circles and filled triangles on the same plot indicate means on three types of reef habitat, $n = 16$). Adapted from Jennings *et al.* (in press b). Note differing vertical scales.

1978; Marten, 1979a,b; Wanink, 1991; He and Wright, 1992; Kitchell, 1992; Persson *et al.*, 1992; Tonn *et al.*, 1992) but these systems lack the complexity of trophic linkages which characterize reef ecosystems and the diversity of piscivores is considerably lower (Jones, 1982; Ligtvoet and Witte, 1991; Kitchell, 1992).

It is tempting to suggest that we should seek effects which may be manipulated by judicious fishing strategies in order to confer desirable attributes upon the fishery. Grigg *et al.* (1984) and Munro and Williams (1985) have suggested that actively fishing piscivorous species may lead to increases in the potential or actual catches of prey. However, on the basis of existing evidence it is unlikely that such a strategy would provide sufficient additional yield to be effective unless a very wide range of predators could be cropped, using a strategy akin to the complex age-, size- and situation-related feeding strategies of the piscivorous fishes. A fishing strategy of this type is likely to be impracticable because it would require a multiplicity of gears and techniques and because it would be necessary to crop small predators to an extent that their maximum yield would not be attained as adults. More benefit may be derived from intensive cropping of prey species on the basis that they are typically more productive and faster growing than species at higher trophic levels.

Algal and invertebrate communities

Herbivorous fishes have a substantial impact on the distribution and abundance of reef algae and their invertebrate fauna (Randall, 1974; Potts, 1977; Brock, 1979; Hay, 1981, 1986, 1991; Hixon, 1982; Carpenter, 1986; Lewis, 1986; Steneck, 1988; Choat, 1991). Algae tend to dominate the benthic community in the absence of herbivores and therefore if fishing causes a direct or indirect reduction in the abundance of herbivores an outgrowth of algae might be expected. It has been suggested that algal growth can block the settlement of corals and that the grazing of algae by herbivorous fishes may clear space for coral settlement and enhance survival and growth of young coral colonies (Randall, 1974; Potts, 1977; Brock, 1979). In reality, studies in a range of locations suggest that when herbivorous fish densities are lowered by fishing, invertebrates compete for algal resources more successfully and they come to dominate the grazing community (Cook, 1980; McClanahan, 1989). However, when the effects of intensive exploitation on herbivorous fish populations are coupled with a decrease in the abundance of invertebrate herbivores, the resulting increase in algal biomass may have marked influence on the development of coral reefs. For example, the mass mortality of the algal-feeding sea urchin, *Diadema antillarum*, in the Caribbean (Bak *et al.*, 1984; Lessios *et al.*, 1984a,b; Carpenter, 1985, 1988, 1990; Hunte *et al.*, 1986; Hughes *et al.*, 1987 a,b; Lessios, 1988) was followed, in intensively fished regions, by significant increases in algal cover and significant decreases in coral cover (Hughes *et al.*, 1987b; Done, 1992; Hughes, 1994). Such development of the algal community is known to inhibit the recruitment and growth of corals (Birkeland, 1977; Potts, 1977; Bak and Engel, 1979; Sammarco, 1980; Lewis, 1986).

Relationships between predatory fish and their invertebrate prey have been explored by comparing herbivore communities at a series of sites subject to different fishing intensities (Carpenter, 1984; McClanahan and

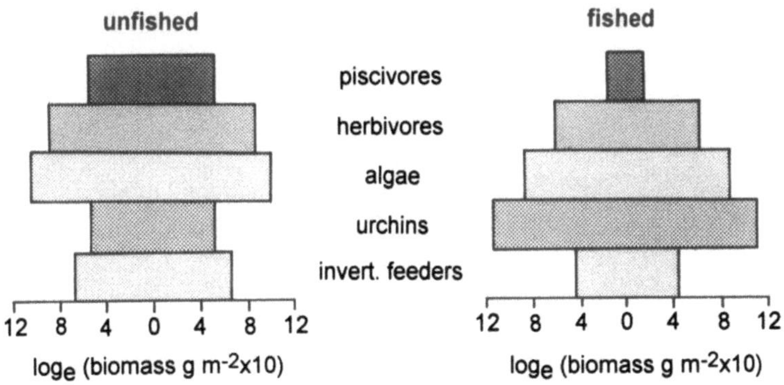

Fig. 8.5 Estimated biomass of piscivorous fishes, herbivorous fishes, algae, sea urchins and invertebrate-feeding fishes on fished and unfished Kenyan reefs. Adapted from McClanahan (1990).

Muthiga, 1988; McClanahan and Shafir, 1990). In Kenya, the more heavily exploited reef lagoons were characterized by denser populations of larger sea urchins, fewer and smaller fish and less coral cover (Fig. 8.5). Predator removal through fishing appeared to result in the ecological release of sea urchins and the competitive exclusion of weaker competitors (McClanahan, 1990; Fig. 8.6). Predation on sea urchins in unfished areas was attributed to the triggerfishes, *Balistapus undulatus* and *Rhinecanthus aculeatus* (McClanahan and Shafir, 1990). Similar increases in sea urchin populations which may have been attributable to predator removal were documented by Carpenter (1984), Hay (1984) and Muthiga and McClanahan (1987). Clearly, a reduction in predator density may cause competitive exclusion within closely related guilds of unfished organisms and a subsequent decrease in species diversity. There is frequently an inverse relationship between sea urchin and herbivorous fish density (Hay and Taylor, 1985; Foster, 1987; Morrison, 1988; Carpenter, 1990; McClanahan and Shafir, 1990; Robertson, 1991b) and McClanahan (1992) developed a biomass-based energetic model to describe the competition for resources between sea urchins and herbivorous fishes. It suggests that sea urchins tolerate low algal biomass due to their low consumption and respiration rates. This would allow them to persist at low levels of algal biomass and productivity, out-competing herbivorous fishes such as parrotfishes and surgeonfishes and reaching maximum biomass levels an order of magnitude higher than that of herbivorous fishes. The ecological release of *Echinometra mathaei* reported by McClanahan and Muthiga (1988) led to an increase in substrate bioerosion, a loss of topographic

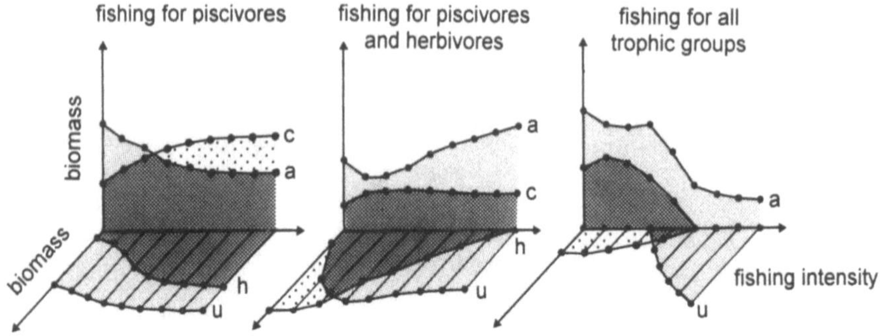

Fig. 8.6 Relationships between the biomass of coral (c), algae (a), herbivorous fishes (h), and sea urchins (u), after reefs have been subjected to different fishing regimes for a period of 30 years. Adapted from the results of a model developed by McClanahan (1990).

complexity and an associated decrease in reef fish biomass and productivity. McClanahan (1992) suggested that the persistence of herbivorous fishes on many reefs may be dependent on the abundance of sea urchin predators which maintain sea urchins at a level where their low gross production makes them ineffective competitors with herbivorous fishes. A number of species, in particular triggerfishes, which may be the targets of reef fisheries, prey on sea urchins (Hiatt and Strasburg, 1960; Randall, 1967; Hoffman and Robertson, 1983; Reinthall *et al.*, 1984; McClanahan and Shafir, 1990; McClanahan, in press).

Glynn *et al.* (1979), Glynn and Wellington (1983) and Macintyre *et al.* (1992) have suggested that sea urchins may have been responsible for impeding horizontal reef progression in the Galápagos Archipelago. Coral-eating sea urchins and their wrasse, triggerfish and pufferfish (Tetraodontidae and Diodontidae) predators co-occur on Galápagos reefs and on those fringing the eastern Pacific mainland. However, sea urchin predation appears to be low in Galápagos, resulting in high densities of sea urchins. Birkeland (1989) and Bak (1990) also concluded that sea urchins had profound influences on reef development.

In Caribbean reef systems, the area-specific rates of bioerosion attributed to sea urchins (Ogden, 1977; Stearn and Scoffin, 1977; Scoffin *et al.*, 1980; Bak *et al.*, 1984) are frequently an order of magnitude higher than those attributed to parrotfishes (Gygi, 1969, 1975; Ogden, 1977; Stearn and Scoffin, 1977; Frydl and Stearn, 1978; Scoffin *et al.*, 1980; but see Bruggeman *et al.*, 1994 a,b). However, the rates of parrotfish bioerosion on the few Indo–Pacific reefs to have been studied (Bellwood, 1995) may equal or exceed the bioerosion rates of urchins (Russo, 1980; Bak, 1990, 1994). In the Caribbean, parrotfish bioerosion is largely attributable to the activity of one excavating species, whereas in the Indo-Pacific there are 16 excavators in three genera (Bellwood, 1995). Clearly, spatial variations in the abundances of scraping and excavating parrotfishes at different localities (Williams, 1983; Russ, 1984a,b; Jennings *et al.*, 1994) will have marked affects on the total rates of bioerosion. The ecological release of urchins may lead to greater increases in rates of bioerosion at localities where scrapers previously dominated the fish fauna.

It has been suggested that crown of thorns starfish, *Acanthaster planci*, populations may proliferate following intensive fishing of their predators. However, the existing evidence for such associations is weak. Ormond *et al.* (1991) reviewed fish–starfish interactions and demonstrated an inverse relationship between the density of starfish and their predators. They concluded that starfish population outbreaks could have resulted from the removal of fishes such as emperors and triggerfishes which prey upon juvenile starfish. However, Sweatman (1995) studied predation on juvenile *Acanthaster* in one location and suggested that the predation rates which he observed would be too low to regulate *Acanthaster* populations. Further studies on larger temporal and spatial scales are needed to determine the indirect impacts of fishing on *Acanthaster* populations. If intensive fishing can lead to *Acanthaster* outbreaks, then there is good evidence to suggest that this would reduce fish production from the reef ecosystem. Increases in *Acanthaster* density can lead to marked decreases in reef structural complexity (Sano *et al.*, 1987) which, in turn, may reduce the availability of suitable habitat for reef fish communities (Williams, 1986). In Panama, Glynn (1973) found that 27 *A. planci* ha^{-1} consumed 15% of annual coral growth on a patch reef and estimated that 250 ha^{-1} would prevent any coral growth. Densities in excess of 250 ha^{-1} have been widely reported (Goreau *et al.*, 1972; Endean and Stablum, 1973; Nishihira and Yamazato, 1974; Birkeland, 1982; Zann *et al.*, 1990) and damage to living coral has frequently been attributed to *A. planci* (Goreau *et al.*, 1972; Endean and Stablum, 1973; Laxton, 1974; Nishihira and Yamazato, 1974; Faure, 1989; De Vantier and Deacon, 1990). Following *Acanthaster*-associated reef damage, Bouchon-Navaro *et al.* (1985) showed that the density of butterflyfishes decreased significantly and Sano *et al.* (1984a) also noted decreases in the abundance of resident reef fishes at a Japanese site. The rate of reef recovery following *Acanthaster* damage has been the subject of considerable debate (e.g. Done, 1987, 1988; Done *et al.*, 1988; Endean *et al.*, 1988). Moran (1990) considered the available evidence and concluded that it would take 12–15 years for a new coral community based on the fast-growing *Acropora* to develop, but 50 or more years for massive coral development.

Reef habitats

Some fishing techniques have direct effects upon reef habitats. Drive netting (Carpenter and Alcala, 1977; Gomez *et al.*, 1987), trapping (Munro *et al.*, 1987) and explosives (Alcala and Gomez, 1987; Munro *et al.*, 1987; Saila *et al.*, 1993) cause physical damage to coral. Poisons have the potential to cause chemical damage to coral and non-target fishes and invertebrates (Rubec, 1986; Eldredge, 1987; McAllister, 1988; Pyle, 1993). *Kayakas* and *muro-ami* drive-netting techniques (Carpenter and Alcala, 1977; Polunin, 1983; Gomez *et al.*, 1987; Ruddle, Chapter 6; Dalzell, Chapter 7) target fusiliers (Caesionidae), surgeonfishes, wrasses, rabbitfishes and parrotfishes. While most other benthic reef-associated fishes avoid the scare lines, there is significant damage to the benthic habitat. Carpenter and Alcala (1977) calculated the damage to 1 ha of reef during a single *muro-ami* operation involving 50 fishers who each struck the bottom 50 times within 1 ha. With 50% coral cover and a 4 kg weight on each scare line, some 305.4 m^2 ha^{-1} were damaged. Alternative weight designs have been examined in order to minimize coral damage (Anon., 1985).

Blast fishing is practised in many reef fisheries (Gomez *et al.*, 1981; Polunin, 1983; Chansang, 1984; De Silva, 1984; Yap and Gomez, 1985; Alcala and Gomez, 1987; Galvez and Sadorra, 1988; Ruddle, Chapter 6). A variety of explosives are used including those removed from armaments, TNT from mining operations and locally produced mixes using sugar or charcoal with oxidizing agents such as potassium or sodium chlorate and potassium or ammonium nitrate. Completed bombs range in size from 0.5 kg in a bottle to 10 kg or more in a drum. Alcala and Gomez (1987) report that a bottle bomb exploding at or near the bottom will shatter all corals within a radius of 1.15 m, and that a gallon-sized drum will have the same effect within a radius of 5 m. A 'typical' charge will kill most marine organisms including invertebrates within a radius of 77 m, an area of 1.9 ha. Munro *et al.* (1987) report that postlarval fishes are also killed. Clearly, the method is non-selective, and yet only a small proportion of the fishes killed are collected (Alcala and Gomez, 1987).

Habitat degradation will affect fish yield, both by causing a redistribution of the exploitable fish biomass and, in severe cases, by reducing total productivity of the fishery. Russ and Alcala (1989) suggested that reduced butterflyfish abundance in a newly exploited Philippine reserve was due to a reduction in live coral cover associated with destructive fishing techniques, although they commented on the difficulty of differentiating between effects due to the direct removal of fishes and those due to habitat modification. Porter *et al.* (1977) noted that a significantly lower biomass of zooplankton was associated with rubble rather than coral habitats, and it might be expected that fish density would change in response to such changes in food supply. The abundance of reef fishes is positively correlated with habitat complexity (Risk, 1972; de Boer 1978; Luckhurst and Luckhurst, 1978; Carpenter *et al.*, 1981; Thresher 1983b; Kaufman and Ebersole, 1984; Patton *et al.*, 1985; Roberts and Ormond, 1987; Grigg, 1994; Jennings *et al.*, in press a) although the evidence for relationships between fish abundance and live coral cover in areas of similar topographic complexity is equivocal (Bell and Galzin, 1984; Sano *et al.*, 1984a; Bell *et al.*, 1985a; Bouchon-Navaro *et al.*, 1985; Roberts and Ormond, 1987; Hourigan *et al.*, 1988; Bouchon-Navaro and Bouchon, 1989; Fowler, 1990; Cox, 1994). It should be recognized that the aforementioned studies usually consider the abundance of small, relatively site-attached fishes rather than biomass of fishes that are significant in fisheries. In addition, they consider a range of habitats within regions of overall high habitat complexity rather than comparing well-developed reefs with those that have been fished destructively until they have little topographic complexity. Shulman's (1984) study of artificial reefs demonstrated that increased habitat complexity (as reefs per unit area) led to significant increases in fish abundance, but that the rate of increase decreased as the habitat became progressively more complex.

A disproportionate reduction in the biomass of commercially desirable fish is often associated with habitat-destructive fishing methods, although other species such as small wrasses may proliferate (Russ and Alcala, 1989). Habitat complexity will also influence settlement events (Jones, 1988; Connell and Jones, 1991), and blast-fishing techniques, which are almost entirely unselective, may lead to the loss of the youngest recruits.

Plant poisons extracted from species of *Derris*, *Barringtonia*, *Tephrosia* and *Wikstroemia* are extensively used for reef fishing, as are synthetic chemicals such as sodium cyanide, quinaldine, chlorine and benzocaine (Rubec, 1986; Eldredge, 1987; Pyle, 1993; Ruddle, Chapter 6; Dalzell, Chapter 7). McAllister (1988) estimated that 150 t of sodium cyanide is used annually on Philippine reefs to catch aquarium fishes. There is little knowledge of the effects of these chemicals on the various life-history stages of reef fauna and flora (Rubec, 1986; Pyle, 1993) and although concentrations that have an acute effect are quickly dispersed, the chronic effects may be significant.

8.4 CONCLUSIONS AND RESEARCH RECOMMENDATIONS

Most contributors to this book have reviewed a large contemporary literature in a well-defined field. This chapter is atypical. The existing understanding of fishing effects is undoubtedly primitive and much of the material we have presented is extracted from studies which were not initiated with the aim of assessing fishing effects. Accordingly, we wish to consider why and how the study of fishing effects should be developed. After addressing the interpretation of evidence for direct and indirect fishing effects, and the potential advantages of basing management decisions on an improved understanding of fishing effects, we suggest methods for collecting data to describe fishing effects and we discuss the importance of determining appropriate scales on which to test for fishing effects and impose management strategies.

Fishing effects and management

Empirical evidence indicates that direct effects of fishing (pages 194–200) are often reversible within the time scales (years) of relevance to fishery managers operating within contemporary political and socio-economic constraints. Indeed, the reversibility of fishing effects is often a key assumption of fisheries yield models (Jennings and Polunin, in press c). However, in considering the apparent significance of long-term changes in fish community structure it should be recognized that most 'long-term' monitoring programmes on coral reefs have been operative for periods rather shorter than the life span of a large grouper. This has precluded the examination of longer-term natural fluctuations in fish populations which are an increasing focus of study in temperate ecosystems (e.g. Soutar and Isaacs, 1974; Steele and Henderson, 1984; Shepherd and Cushing, 1990). Thus a shift in fish community structure within the monitoring window may represent a shift to a new state or a return to a former state, or it may be part of a cycle with a lower frequency than the period of monitoring. Given these possibilities it is clearly unwise to assume that all shifts in community structure are fishing effects in a naturally persistent system. Indeed, if preserving an existing state is taken as the unquestioned aim of fisheries management or conservation, the evidence from temperate ecosystems suggests that such management will fail (Jennings and Polunin, in press c).

The indirect effects of fishing, in particular those on algal and invertebrate communities (pages 201–211), have shown little tendency to be reversible within the short time scales (years rather than decades or centuries) which matter to those people who rely on the productivity of their fisheries (e.g. Done, 1992; Knowlton, 1992; Hughes, 1994). As a result, conventional yield models will provide inappropriate estimates of yield. For example, if the biomass of herbivorous fishes were reduced to a theoretical point of maximum biomass regeneration, then sea urchins may compete for algal resources more effectively and prevent the re-establishment of the herbivorous fishes. Fishing strategies that help to prevent the proliferation of invertebrate competitors are likely to be a better solution to management problems in these fisheries. McClanahan and colleagues (pages 207–208) are investigating a number of useful approaches through modelling and empirical 'fishing effects' studies.

Conventional methods of population analysis will prove difficult to apply in tropical fisheries that are not selected for rigorous scientific study. Indeed, there is increasing recognition that an examination of sustained yields from existing fisheries may provide the most realistic approach to estimating the potential yields from ecologically similar areas (Munro and Fakahau, 1993b). Despite useful developments in methods of length-frequency analysis and predictive equations for the estimation of natural mortality in reef fishes, there are relatively few independently validated measures of growth or mortality rates — two of the basic requirements for population analysis (Appeldoorn, Chapter 9). In addition, little is known about the processes that govern recruitment, the integrity of reef fish populations and the possible interactions between different species within the fished multispecies community. Much of the funding for work that addresses these processes has supported reef fish ecologists who work with small site-attached species of little importance in fisheries. Their adoption of these species has facilitated the application of many elegant and effective experimental designs to field situations (e.g. Doherty, 1991). However, many of the species that have been studied do not relocate after settlement (in contrast with most fished species) and analyses of population variability are often based on the numerical abundance of settling fishes rather than their ultimate contribution to interannual variance in population biomass. Fisheries scientists recognize that fluctuations in the mortality of early life-history stages are the primary cause of variability in fish stocks and appreciate the constraints within which they apply yield models (Appeldoorn, Chapter 9). Thus it is not surprising that the recent demonstrations of spatial and temporal variation in settlement rates of reef fishes have not been followed by a rapid rejection of yield models. These models were often developed in environments where similar variation has been known to exist since the early 1900s and where study of early life history was conducted by laboratories which were also charged with fisheries management (e.g. Heath, 1992). With a few notable exceptions, there seem to have been few positive exchanges between tropical fishery biologists and reef fish ecologists, as evidenced by the consistent disparity of reference lists within fisheries and marine ecology journals. This lack of communication also appears to hamper the constructive development of fishery-related studies of the early life history of tropical reef fishes. It contrasts with the collaborative approach in temperate waters, where fishery management programmes have often funded and driven the ecological investigation of marine fish populations (but see Frank and Leggett, 1994). Better collaboration between tropical fishery biologists and reef fish ecologists is urgently needed, particularly when an improved understanding of larval dispersal remains an essential prerequisite for defining management and assessment units in tropical reef fisheries.

For the foreseeable future, and despite a burgeoning literature on the theory of reef fishery management, it is unrealistic to expect that population analyses will provide valid predictions of the potential yield from, or a means of assessment for, those reef fisheries of limited economic importance which are unlikely to be selected for rigorous scientific study. However, a more comprehensive understanding of fishing effects may help to provide a range of alternative assessment, management and monitoring tools for use in these fisheries. With such understanding, it should be increasingly possible to respond to changes in reef ecosystems with appropriate fishing strategies. Clearly, an understanding of fishing effects will not be the panacea which prevents habitat destruction and catastrophic loss of yield in overpopulated coastal zones where the fishers struggle to alleviate hunger and poverty (e.g. Pauly *et al.*, 1989). Rather, it may help to rationalize approaches to fisheries development and management in those artisanal reef fisheries which provide a vital source of protein for coastal communities and where catches are close to the maxima recorded in ecologically similar areas (e.g. Munro and Fakahau, 1993b).

Study of fishing effects

The results of many empirical studies, albeit scattered through space rather than time, demonstrate that fishing initiates change in the reef ecosystem. The basic responses of fish communities to fishing are remarkably similar in different environments (sections 8.2 and 8.3). Initially, more extensive and rigorous empirical studies will be required to describe the range of fishing practices and management regimes that have led to acceptable yields from a variety of reef ecosystems. Valid fishing effects data could be acquired through the refinement of existing procedures for the collection of fisheries data. However, the acquisition of fishing effects data must proceed in conjunction with a broader examination of the spatial and temporal scales on which to study fishing effects and impose management strategies.

The potential for further exploitation in an established fishery, or the state of exploitation in a previously unstudied fishery, might be assessed from an examination of fish community structure. Lock (1986c) and Koslow *et al.* (1988) noted the similarity of changes in catch composition at several sites subject to fishing, and Koslow *et al.* (1988) suggested that if quantitative relationships could be derived between catch rate, catch composition and fishing effort, they may provide useful information for managers. The means by which such relationships would be derived or used is not discussed. At present there are two particular concerns: firstly, that responses to fishing have not been investigated across gradients where fishing itself has been adequately quantified, and secondly, that fishing effects are rarely studied in sustainable fisheries. Thus most existing studies of effects fall into three categories: those which compare fish populations in fished and essentially unfished areas such as reserves (e.g. Bohnsack, 1982; Polunin and Roberts, 1993), those which describe fishing effects across a generally ill-defined gradient where details of catch composition, gear-specific effort and historical changes in effort have not been accurately quantified (e.g. Lock, 1986c; Samoilys, 1988; Jennings *et al.*, in press b) and those which describe responses to catastrophic rather than sustainable fishing events (e.g. Russ and Alcala, 1989; Alcala and Russ, 1990). If the study of fishing effects is to progress, then there is an urgent requirement for accurate quantitative studies which relate specific fishing effects to a range of specific fishing regimes. (Jennings and Polunin, in press e).

The data requirements for the study of many fishing effects and for conventional population analyses are closely related. Thus the collection of catch and effort data for species- and gear-specific categories would satisfy both the increasing demand for more detailed fishery-assessment work and the demand for fishing effects studies which relate quantitative changes in species-specific CPUE to accurately defined gradients of fishing intensity. To date, many fisheries departments have been pressured to justify their receipt of funding by gathering data on tropical fisheries and instituting their typically reactive management policies. Accordingly, there has been a stronger emphasis on collecting data than on developing a conceptual framework for the subsequent collection of valid data. Thus the changes in CPUE and catch which are reported for many fisheries are insensitive measures of fishing effects because they are not based on species divisions, the origins of fish sampled are rarely clear and the measures of fishing effort may not be rigorously defined (e.g. Appeldoorn *et al.*, 1992; Jennings and Polunin, in press a,d).

When procedures for the collection of catch and effort data have been refined, and the opportunities for the study of fishing effects in existing fisheries are largely exhausted, then large-scale manipulation of fishing practices will be required (e.g. Russ, 1991). Such manipulation would fulfil two major roles. Firstly, it would allow the effects of cropping regimes which are not observed in existing fisheries to be investigated. Secondly, it would allow new management strategies to be tested. Some forms of manipulation, such as targeting species from alternative trophic groups, could be tested in existing artisanal fisheries. Other manipulations may lead to unacceptable decreases in short-term yield. Socio-economic factors would determine that these manipulations could only be tested effectively in marine reserves, lightly fished areas or at sites where the coastal population is not reliant on the fishery for food or income.

Fishery-independent assessments of biomass and numerical abundance will be required to investigate the effects of fishing on species not targeted by the fishery and to validate changes in species composition or biomass implied by data from the fishery. Such fishery-independent estimates are characteristically obtained using visual census techniques. Visual census techniques have been the subject of wide-ranging critical discussion in recent years (Sale and Douglas, 1981; Brock, 1982; Bell *et al.*, 1985b; Harmelin-Vivien *et al.*, 1985; Sanderson and Solonsky, 1986; Thresher and Gunn, 1986; McCormick and Choat, 1987; Lincoln-Smith, 1988; Greene and Alevizon, 1989; Parker, 1990; St John *et al.*, 1990; Bortone and Kimmel, 1991; Samoilys, 1992; Watson *et al.*, 1995; Jennings and Polunin, in press b) and it is widely accepted that they underestimate the numerical abundance or biomass of many species. However, the few comparisons of CPUE and visual census to have been conducted suggest that they provide very similar estimates of relative abundance (Kulbicki *et al.*, 1987, Kulbicki, 1988b).

Fishing effects data should be collected for individual species, but the scientific and philosophical approach to the analyses and interpretation of the data should be increasingly flexible and generalist. In particular, we should consider the overall production of protein rather than emphasizing the apparent proliferation or demise of individual species. Within tropical fisheries, a range of communities dominated by different groups of fishes may all yield adequate protein. On this basis, changes in fish community structure may be of less socio-economic significance to artisanal fishers with catholic tastes than to those fishers who have entered the market economy and are increasingly reliant upon the sale of relatively few highly valued species. Ironically, much fisheries development has increased economic reliance on a limited pool of species (Jennings and Polunin, in press c,d).

The significance of scale

To interpret the significance of fishing effects and to provide valid management advice, it will be necessary to determine the scales at which fishing effects operate. For this purpose, two types of fishing effect need to be examined: firstly, those which are immediate and largely unaffected by growth-rate responses in the population, and secondly, those in the longer term which may or may not involve compensatory changes in the population and are confounded by natural fluctuations in recruitment. Whilst the value of short-term studies has frequently been questioned, such studies could, at present, provide particularly useful information on the spatial scale at which fishing effects may be detected and the methodology for their detection. The removal of fish by size-selective fishing has a direct and immediate effect on population size and structure. If the effects cannot be detected at any early stage, then the unit area of study is incorrect or the methodology for detecting changes in abundance is in error. There is a voluminous literature on the methodology for assessing size and numerical abundance (Polunin *et al.*, Chapter 14) but scale has yet to be adequately addressed despite indications of significant movements of fish in and out of study areas. Many fished species make foraging or spawning migrations (e.g. J.D. Parrish, 1989; Holland *et al.*, 1993) and these migrations are likely to influence stock assessments which are often conducted on small portions of the reef. It has been suggested that coral reefs are excellent microcosms in which to address the question of fishing effects, but such approaches are questionable without an appreciation of the scale at which fishing effects operate.

On larger spatial scales and over longer time periods, the early life history of reef fishes warrants further investigation. The biomass of fishes on many reefs is expected to be maintained by the settlement and subsequent growth of larvae that are the progeny of fishes at other sites. There is clearly a need to further improve the understanding of recruitment processes of reef fishes in order to examine the transport of larvae in relation to hydrographic phenomena (e.g. Shenker *et al.*, 1993; Thorrold *et al.*, 1994a,b) and the extent to which isolated reefs are self recruiting or reliant on larvae from other sources. Such studies are particularly relevant to fisheries management issues because they will demonstrate the extent to which the effects of fishing may be collective. Whilst we do not know the scale on which we should test for fishing effects, it is equally difficult to know the scale on which to impose management strategies. The examination of connectivity between reefs will be the first step towards identifying populations of reef fishes that have sufficient temporal and spatial integrity to warrant consideration as self-perpetuating management units (Jennings and Polunin, in press e). It should be recognized that genetic techniques are unlikely to provide a useful shortcut to identifying such stock units because genetic homogeneity may be maintained by the exchange of only a small proportion of individuals per generation (Gauldie, 1991).

We conclude that rapid development of the fishing effects science could be achieved at moderate cost by refining existing procedures for collection of catch and effort data and by treating observed combinations of gear, effort and catch-composition as a series of experimental manipulations. An improved understanding of fishing effects and the scales on which they can be detected may assist with the development of a range of new monitoring, assessment and management methods which would provide a useful adjunct to the existing approaches to tropical reef fishery science.

ACKNOWLEDGEMENTS

We wish to thank all those who have discussed and criticized ideas presented in this chapter and the referees for a number of constructive suggestions. We offer our apologies to those whose relevant work we have misinterpreted or overlooked. The senior author wishes to thank the British Overseas Development Administration for funding through the Fish Management Science Programme and the Marine Studies Programme at the University of the South Pacific for providing facilities during extended periods of work in Fiji.

Chapter nine

Model and method in reef fishery assessment

Richard S. Appeldoorn

SUMMARY

Mathematical models are used extensively to estimate the size and productivity of fishery stocks and to assess the potential effects of alternative management strategies. Although mathematical modelling in general is quite advanced, models employed in the management of tropical fisheries tend to be quite simple. A principal limitation is the amount and quality of data. The lack of information is determined by the spatial and biological diversity of resources, the complexity of the fishery and the limited capacity for data collection and analysis. Nevertheless, focused sampling strategies can yield data suitable for analysis and management advice. Both length-based and otolith-based methodologies are available for estimating aspects of growth, mortality and recruitment. Length-based methods are simpler, but their many assumptions may be difficult to meet, while otolith analysis is more costly and time consuming. The most common assessment approaches are based on single-species surplus-production or yield-per-recruit models, even though these might not be appropriate in a multitrophic, multispecies context. Multispecies models either require a prohibitive amount of data or they lump all species together at the cost of predictive power. Useful approaches have been identified, these problems notwithstanding, often involving aspects of both single-species assessment and small-scale lumping. Even the simplest models provide some useful management information; if nothing else, they

Reef Fisheries. Edited by Nicholas V.C. Polunin and Callum M. Roberts.
Published in 1996 by Chapman & Hall, London. ISBN 0 412 60110 9.

provide a target against which the need for regulation or development may be gauged.

9.1 INTRODUCTION

Assessment models for reef fisheries are varied and range from simple approximations of potential yield to complex systems models. The majority, however, are simple extensions of basic models (e.g. Ricker, 1975; Gulland, 1983b; Hilborn and Walters, 1992). Pauly (1984a) and Sparre *et al.* (1989) specifically address quantitative assessment of tropical fisheries, while Saila and Roedel (1980) and Pauly and Murphy (1982) address broader problems of assessment in tropical systems. Huntsman *et al.* (1982), Polovina and Ralston (1987) and Medley *et al.* (1993) have summarized information on reef fisheries in particular. This chapter gives an overview of quantitative approaches currently or potentially useful for reef fisheries assessment. It first reviews methods for estimating abundance, proceeds to methods for modelling the dynamics of growth, mortality and recruitment, and ends with models for estimating potential yield.

9.2 STOCK ABUNDANCE

Determination of stock abundance is useful in detecting trends or perturbations and in estimating mortality, potential yield, and stock status relative to potential yield. The estimation of stock abundance is essentially a statistical undertaking, with randomization, stratification and multistage sampling (Raj, 1968; Cochran, 1977) being pertinent procedures. Stock abundance can be measured in either relative or absolute terms, using either fishery-dependent or fishery-independent data.

Traps

Traps catch a wide variety of fishes over a broad size range, can be deployed in rugose habitats, are inexpensive and easy to use, and are often employed in commercial fisheries. To estimate stock size, traps must be calibrated for catchability or effective area fished (EAF) (Miller, 1975; Eggers *et al.*, 1982).

EAF is the area wherein it is assumed that all fish(es) are caught and outside of which none is caught. To estimate EAF, one deploys traps over various intertrap distances; a decline in catch rate occurs when traps set close together interfere with one another (Sinoda and Kobayashi, 1969; Eggers *et al.*, 1982). However, practical application in reef fisheries is limited by the requirement of a large number of traps set over a fairly uniform environment in order to rule out habitat effects. Alternatively, EAF can be estimated if an independent estimate of density (*D*, fish m^{-2}) is available, with EAF (m^2 $haul^{-1}$) being defined as follows:

$$EAF = (c/f) / D \tag{9.1}$$

where c/f is catch per unit effort. Density has been variously estimated using visual (e.g. Miller, 1975) or mark–recapture (Chittleborough, 1970; Morgan, 1974) techniques.

Miller and Hunte (1987) estimated density of diurnally active, abundant, reef-resident fishes using visual census. Recksiek *et al.* (1991) explored the use of tag–recapture techniques for estimating density of more dispersed or wide-ranging species (page 224). Sampling stratification may limit the accuracy and precision of this approach, as the diversity of reef habitats across a shelf can be substantial (Williams, 1982; Kimmel, 1985), with each habitat requiring its own calibration (Recksiek *et al.*, 1991). Acosta *et al.* (1994) found that site-specific EAF estimates fluctuated over 5 years by a factor of two, but consistent differences related to habitat were observed. Furthermore, for one well-studied species, EAF increased when abundance became very low (Fig. 9.1).

Visual surveys

Underwater visual census (UVC) has long been used to estimate reef fish abundance (Brock, 1954), but is limited to non-cryptic and diurnally active fishes (Sale and Douglas, 1981; Brock, 1982). There are two main categories of UVC. In strip transects, all individuals within a specific area (fixed path width) are counted, i.e. the probability of sighting is equal to unity. In line transects, one counts all individuals observed, regardless of their distance from the observer. The probability of sighting declines with distance and the probability density function must be estimated, but a much greater area can be surveyed than with strip transects. Total density is obtained by correcting the number observed based on the probability of sighting for a given distance. The descriptive statistics for visual surveys are well established (Eberhardt, 1978; Gates, 1979), but UVCs have largely ignored this literature (Thresher and Gunn, 1986). Only recently have underlying assumptions to these methods been addressed.

Strip transects are most often employed in UVC, including linear transects along a depth contour, compass course or demarcated path, and point counts, where all fishes within a circle around the observer are counted (e.g. Bohnsack and Bannerot, 1986). The precision and accuracy of strip transects for small reef fishes is a function of transect length relative to the abundance and patchiness of the species examined (Sale and Sharp, 1983;

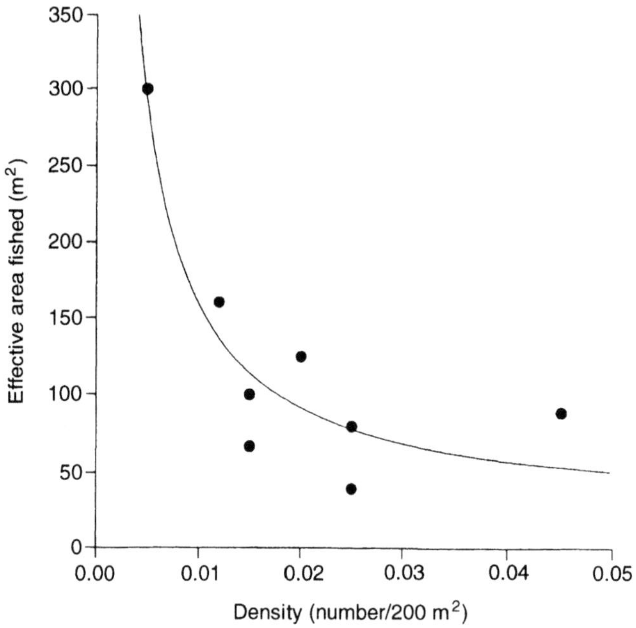

Fig. 9.1 Relationship between fish density and effective area fished by Antillean fish traps for the ocean surgeon, *Acanthurus bahianus*. Reproduced with permission from Acosta *et al.* (1994).

Fowler, 1987; McCormick and Choat, 1987; Lincoln-Smith, 1989). Generally, precision increases as transect length increases. Accuracy is largely unaffected by transect length, but abundance estimates may decrease as transect width increases for cryptic species (Sale and Sharp, 1983; Fowler, 1987) due to a decrease in the probability of sighting as transect width increases. Sale and Sharp (1983) recommend that counts from a series of transects of different widths be conducted and results extrapolated to the theoretical density at a transect width equal to zero. Their extrapolation used a linear relation between density and transect width. However, probability of sighting is more likely to decline with distance in a bell-shaped fashion (Gates, 1979; see also the data figured in Sale and Sharp, 1983). If so, strip transects are valid for estimating abundance because there exists a transect width (the top of the bell-shaped curve) where the probability of sighting should approximate unity. The appropriate transect width will vary, however, relative to the visibility of target species.

Line transects, and their point count variations, have not been extensively applied to reef fishes. Thresher and Gunn (1986) used a point count variation to estimate the abundance of jacks. Density estimates obtained were found to be quite sensitive to the estimate of distance, as this affected the estimation of the probability density function.

UVC has been increasingly applied to estimate the abundance of commercially important reef fishes, especially for small-scale comparisons of relative abundance or biomass among areas, or monitoring changes in a single area over time (Bohnsack, Chapter 11). Surveys designed to estimate absolute abundance are fewer because of the low density of larger species, particularly under high fishing pressure, and the difficulty in surveying large areas (Craik, 1982; Ayling and Ayling, 1986b; Thresher and Gunn, 1986; McCormick and Choat, 1987; Bellwood, 1988).

Stock biomass can be calculated from UVC if fish lengths are estimated during surveys. Divers need to be trained (Craik, 1982; Bell *et al.*, 1985b; Rooker and Recksiek, 1992) to estimate lengths underwater prior to the start of any survey, and periodic retraining is recommended (Bell *et al.*, 1985b). Even short training sessions significantly improve accuracy and precision of length estimates. Performance may be enhanced when training uses real fish or fish-shaped cutouts (Craik, 1982). Biomass is obtained from length estimates by using length–weight relationships, often obtainable from the literature (Abdul Nabi, 1980; Loubens, 1980a; Munro, 1983j; Wright and Richards, 1985; Mathews and Samuel, 1987, 1991; Samuel and Mathews, 1987; Bohnsack and Harper, 1988).

Nets

Gill nets and trammel nets can catch a larger number of individuals than other gears and they sample a broad range of species. However, these nets are highly size selective, and assumptions about how they work have not been adequately tested in reef environments.

Acosta (1993, 1994) and Acosta and Appeldoorn (1995) found the ability to estimate stock abundance of individual reef species by using entangling nets was limited because the number of individuals caught across nets of different mesh size was insufficient to quantify size-selectivity effects. Also, obtaining an independent estimate of abundance to calculate effective area fished was difficult. For example, at Acosta's (1993) inshore stations, one of the dominant species caught (sea bream, *Archosargus rhomboidalis*) was completely absent from concurrent UVCs.

Catch per unit effort

In the absence of absolute abundance data, estimates of relative abundance can be derived from catch-per-unit-effort data collected by either fishery-dependent or independent surveys. In the latter, gear effects can be controlled, but sample size is generally low. For example, under the US Southeast Area Monitoring and Assessment Program (SEAMAP–Caribbean), an annual fishery-independent survey using standardized fish trap and hook-and-line methods is conducted off Puerto Rico. It can estimate mean total catch with a precision of only 50% (Rothschild and Ault, 1991), but can statistically verify long-term trends. Precision around species-specific estimates is substantially less than for all species combined.

Although sampling the fishery offers the advantage of an increased sample size, it is complicated by non-random sampling and the variety of different gears used. Differences among gears would need to be standardized through calibration, or a particular gear(s) would need to be chosen as a standard, and only the standard sampled.

Fishing depletion experiments

Exploitable stock can be estimated using fishing depletion methods (Leslie or Delury methods, Ricker, 1975) if the area being fished is discrete and a closed population can be assumed. Many species of reef fishes are relatively site attached over short periods of time (Bardach, 1958; Randall, 1961, 1963b; Recksiek *et al.*, 1991), and reefs are distinct. Because some species have restricted home ranges, fishing effort should be distributed throughout the target area to ensure an equal probability of capture.

The Leslie method was used by Polovina (1986) to estimate biomass of deep-water reef fishes. He extended the method to account for variable rates of catchability among species due to competition for baited hooks. Appeldoorn and Lindeman (1985) applied Chapman's (1970) open-population modification of the Leslie method to a complex of grunt species using catch and effort data collected from the local fishery.

Tagging

The principal limitation to applying mark–recapture techniques (Ricker, 1975; Begon, 1979; Seber, 1982; Burnham *et al.*, 1987) to reef organisms is the assumption that marked and unmarked fish have an equal probability of capture. In fact, the behaviour of marked individuals may differ from that of unmarked ones, for example through injury (Randall, 1961), although this often can be avoided through careful handling and adequate testing of marking methods (Recksiek *et al.*, 1991).

The assumption of equal probability of capture requires that there be complete, random mixing between marked and unmarked individuals, yet restricted movement and non-random distribution are characteristic of many reef species. The scale on which complete mixing occurs will depend on the time between marking and recapture and the behaviour of particular species. For reef-restricted and reef-associated species (Hartsuijker and Nicholson, 1981; Parrish, 1982) short-term mixing can be expected only on the scale of individual reefs or parts of reefs. Complete mixing can be achieved by marking and releasing fish over a broad area and by restricting studies to a subsample of reefs.

There are few examples of abundance estimation using tagging techniques. Recksiek *et al.* (1991) estimated abundance for four species on a small reef using fish-trap data and Chao's (1989) estimator for sparse capture–recapture data. Beinssen (1989) estimated the abundance of coral trout on a 342 ha reef using Bailey's formulation of the Petersen method (Pauly, 1984a).

9.3 DYNAMICS

Growth, mortality and recruitment determine stock productivity, and estimating their magnitude and variability is important for assessing stock status and potential yield. Estimates of growth and mortality are required for most simple and multispecies yield models, while estimates of year-class strength are important for determining the relationship between spawning stock and recruitment, for assessing the temporal variability in yield, for understanding the underlying processes affecting recruitment and for predicting future yield.

Growth

Growth in fishes is typically described using the von Bertalanffy growth function, VBGF (Ricker, 1975):

$$L_t = L_\infty \{1 - \exp[-k(t - t_0)]\} \tag{9.2}$$

$$W_t = W_\infty \{1 - \exp[-k(t - t_0)]\}^3 \tag{9.3}$$

where L_t (W_t) is length (weight) at time t, L_∞ (W_∞) is asymptotic length (weight), k is the instantaneous growth coefficient, and t_0 is the hypothetical time at which length (weight) = 0. Growth is routinely modelled using length data, with W_∞ determined by converting L_∞ to weight using a length–weight relationship. The length–weight function should be derived from the same population from which the growth data were collected.

Growth parameters have been determined for a wide variety of reef fishes (Munro, 1983f; Munro and Williams, 1985; Manooch, 1987), but suitability of the model has been questioned for protogynous hermaphrodites (Sadovy, Chapter 2), as it is thought (Charnov, 1982) that fishes

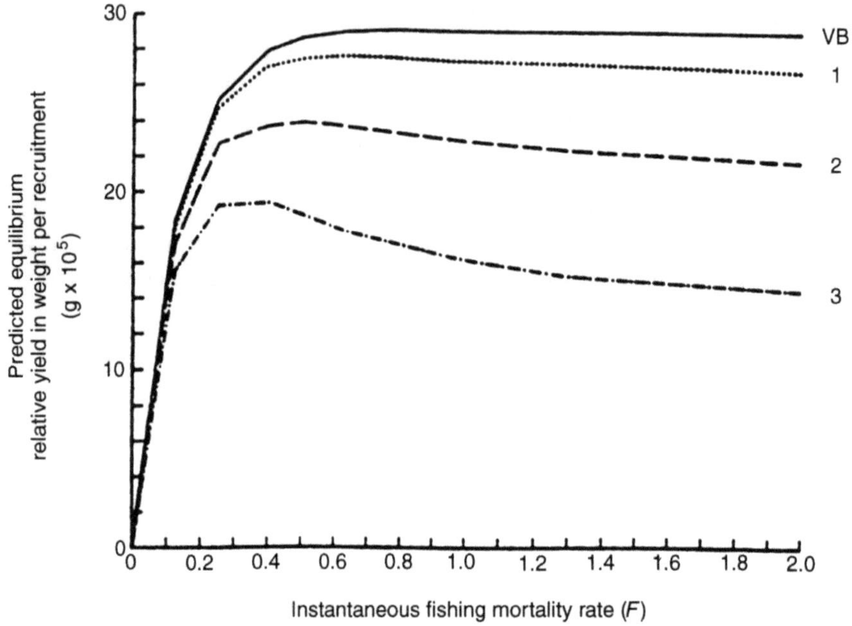

Fig. 9.2 Equilibrium yield per recruit (YPR) for gag, *Mycteroperca microlepis*, with constant recruitment under four hypothesized growth patterns: von Bertalanffy growth (VB) and slight (1), moderate (2) and large (3) growth spurts after sexual transition. $F_{Y_{max}}$ values are 0.870 for the VB fit and 0.622, 0.470 and 0.305 for growth spurt hypotheses (1), (2), (3) respectively. Modified with permission from Bannerot *et al.* (1987).

undergo a significant growth spurt at the time of sex change. Failure to account for growth spurts in yield models can result in significant over-estimation of both maximum yield and optimal effort (Fig. 9.2) (Bannerot *et al.*, 1987).

Growth analyses have shown distinct differences in the sizes of equal-age males and females of protogynous species (Moe, 1969; Nagelkerken, 1979; Manooch (cited in Bannerot *et al.*, 1987); Garratt *et al.*, 1993), and experiments have shown growth spurts under the same conditions that lead to sex change (Ross *et al.*, 1983; Ross, 1987). However, no data clearly demonstrate increased growth following change of sex (Bannerot *et al.*, 1987). Garratt *et al.* (1993) incorporated sex change and growth spurt in modelling the growth of slingers (Sparidae), assuming equal rates of growth among females, that a certain fraction of females change sex over time, and that the fraction is age dependent. Their model did not account for theories on the social control of sex change (Shapiro, 1987a; Sadovy,

Chapter 2); alternative assumptions (without a growth spurt) might equally have explained observed patterns.

Seasonal growth occurs in many reef fishes (Appeldoorn, 1987a; Longhurst and Pauly, 1987) and can be accounted for using an oscillating VBGF model (Pauly and Gaschütz, 1979, modified by Somers, 1988; Appeldoorn, 1987a). If seasonal growth variation is substantial, the basic VBGF will underestimate yield per recruit if the species is fast growing and short lived (e.g. 2 years) (Sparre, 1991), but the longer life spans of most commercially important reef species (> 10 years) result in only minor yield-per-recruit variations.

Sexually dimorphic growth in length has been reported for several species including snappers (Lutjanidae), emperors (Lethrinidae), threadfin breams (Nemipteridae) and whiptails (Pentapodidae) (Loubens, 1980a), angelfishes (Pomacanthidae) (Pauly and Ingles, 1981) and goatfishes (Mullidae) (Munro, 1983d). In all cases, growth was suitably modelled by the VBGF, but parameter values differed between sexes.

Mortality

Basic models

The basic model for mortality in fishes is of exponential decay in abundance over time:

$$N_t = N_0 \exp[-(F + M)t] \qquad (9.4)$$

where N_0 is initial abundance, N_t is abundance after time t, and F and M are the coefficients of instantaneous fishing and natural mortality, respectively (Ricker, 1975). Total mortality (Z) is the sum of F and M.

The model assumes that mortality is constant. This assumption, particularly for natural mortality, is reasonable for large, long-lived species but may not hold for smaller prey species or for young juveniles of predatory species. Mortality tends to decline with age, but by the time individuals recruit to the fishery the mortality rate has stabilized. Caddy (1991) and Appeldoorn (1988) have shown for short- and long-lived invertebrates, respectively, that natural mortality continues to decline significantly during the exploited stage, thus affecting subsequent yield estimations. The assumption of constant natural mortality during the exploited phase has not been assessed for exploited reef fishes. Doherty and Fowler (1994b) suggest that mortality declines during the first few years in two damselfishes. If size or age at first capture is relatively large, constant mortality can be assumed. However, many small species and juveniles of larger species are taken in small-mesh gear, and changes in natural mortality rate might need to be accounted for.

Caddy (1991) proposed a hyperbolic model to describe a decline in natural mortality with age:

$$M_t = M_A + (b / t) \qquad (9.5)$$

where M_t is natural mortality at time t, M_A is the lower asymptotic value of M and b is a constant. Appeldoorn (1988) used the Weibull function for the same purpose:

$$M_t = Dt^{(1 - C)} \qquad (9.6)$$

where D and C are constants.

Several studies have discussed the relationship of natural mortality rate to the abundance of predators, arguing that natural mortality in prey species will decline as predators are removed from the system (pages 244–245).

Mortality estimation

The principal methods for estimating mortality in reef fishes have been catch-curve analysis and various length-based methods. These all assume a steady-state population (i.e. constant mortality, constant or random recruitment) and an unbiased sample of the underlying population (Hoenig *et al.*, 1987; Hilborn and Walters, 1992). Length-based methods further assume constant growth from year to year, require an assumption about the timing of recruitment (e.g. continuous or discrete), and require estimates of the VBGF parameters, k and L_∞.

The assumption of a steady-state population is difficult to achieve. Recruitment patterns may appear random over the short term (Doherty and Williams, 1988; Doherty and Fowler, 1994b), but long-term climatic forcing can lead to non-random trends (Polovina *et al.*, 1994). Combining samples from several years will help to average out variations in recruitment. Many species of reef fish show evidence of offshore ontogenetic migrations, with larger mean sizes found further from shore (e.g. Smale, 1988; Rooker, 1991, 1995; Dennis, 1992; and references in Parrish, 1987; Williams, 1991). Under these circumstances, the assumption of an unbiased sample can be difficult to achieve (Morales-Nin and Ralston, 1990; Posada and Appeldoorn, in press). Density and site-specific abundance will depend upon habitat composition and the time course of migration. Observed mortality will be overestimated by migration out of the sampling range (Posada and Appeldoorn, in press). For length-based methods, an assumption of continuous recruitment would be violated where recruitment is seasonal, as occurs at higher latitudes and in some reef species, such as groupers (Serranidae), but the assumption can be approximated by pooling samples taken throughout the year. An assumption of knife-edge selection (i.e. full vulnerability to capture occurs over a short age [length] range) may be violated in reef fisheries where the size compositions of landings result from different size-selective properties of various gears (Ehrhardt and Ault, 1992; Dalzell, Chapter 7).

Catch curves (Ricker, 1975) model the linear decline in the natural logarithm of abundance by age, and total mortality is estimated from the slope of a regression through these data (Sadovy and Figuerola, 1992; Doherty and Fowler, 1994b). Length-converted catch curves (LCCCs) (van Sickle, 1977; Pauly, 1984a) assume continuous recruitment, and data need to be grouped into small length intervals (Pauly, 1984a). Strong seasonal variations in growth, especially in short-lived species, cause traditional LCCCs to overestimate mortality (Sparre, 1990). A modified LCCC based on pseudo-cohort analysis corrects this bias (Pauly, 1990a).

Assuming knife-edge selection, total mortality can be estimated from the mean length of individuals in the population (Beverton and Holt, 1956; Ssentongo and Larkin, 1973; Powell, 1979; Hoenig *et al.*, 1983; Hoenig, 1987; Weatherall *et al.*, 1987; Ehrhardt and Ault, 1992). The basic formula of Beverton and Holt (1956) is:

$$Z = \frac{k(L_\infty - \bar{L})}{\bar{L} \quad L_c} \tag{9.7}$$

where $\bar{L}$ is mean length of fish above L_c, the knife-edge selection length. The method assumes continuous recruitment and a long exploited life. Hoenig *et al.* (1983) felt an estimator based on median length (L_{med}) was more robust because it uses more information at the lower end of the length range, where length is a better predictor of age. Their equation is as follows.

$$Z = \frac{k \log_e 2}{L_{med} + \log_e [l - (L_c / L_\infty)]} \tag{9.8}$$

Several workers have found the estimator proposed by Ssentongo and Larkin (1973) to be unsuitably biased (Hoenig *et al.*, 1983; Weatherall *et al.*, 1987).

Hoenig (1987) and Ralston (1989) showed the Beverton–Holt equation to be biased when recruitment is seasonal. Ralston (1989) showed the bias to be low if the lengths of all fully recruited cohorts were included in the analysis. Hoenig (1987) developed an alternative formula for use when recruitment is seasonal:

$$Z = \log_e \frac{e^{-k}(\bar{L} - L_\infty) + (L_\infty - L_c).}{\bar{L} - \bar{L}_c} \tag{9.9}$$

Ehrhardt and Ault (1992) examined behaviour of the Beverton–Holt estimator for situations where fishing mortality declines at some larger size, such as where large fish are not susceptible to a gear or where ontogenetic migration to deeper water offers a refuge. They found a positive bias, which was greatest when life span or exploited life relative to life span was short and fishing mortality was low. Bias was typically less than 10% when F was greater than 0.3 and the exploitable life was at least 50% of the total life span. Ehrhardt and Ault (1992) developed an alternate methodology to estimate Z that removes this bias:

$$\left(\frac{L_\infty - L_\lambda}{L_\infty - L_c}\right)^{\frac{Z}{k}} = \frac{Z(L_c - \bar{L}) + k(L_\infty - \bar{L})}{Z(L_\lambda - \bar{L}) + k(L_\infty - \bar{L})} \tag{9.10}$$

where L_λ is the largest size adequately represented in the catch or largest selection size. The equation is solved iteratively.

Weatherall *et al.* (1987), modified by Pauly (1986), present a regression method to estimate L_∞ and Z/k. Given an independent estimate of k, the latter gives an estimate of Z. The estimate of Z/k is unaffected by bias in the length distribution due to offshore ontogenetic migration, although L_∞ will be underestimated under such conditions (Posada and Appeldoorn, in press).

An empirical relationship for estimating total mortality from the age of the oldest individual in a stock (t_{max}) is given by Hoenig (1983). His equation for fishes (53 species, 84 stocks; $r^2 = 0.68$) is:

$$\log_e Z = 1.46 - 1.01 \log_e t_{max} \tag{9.11}$$

and that for all organisms (79 species, 134 stocks; $r^2 = 0.82$) is:

$$\log_e Z = 1.44 - 0.982 \log_e t_{max}. \tag{9.12}$$

Because of the similarities among taxa, Hoenig suggested the second equation be used. The model assumes constant mortality within a population. Although Z can be estimated by ageing just a few of the largest fish in a sample, the estimate depends upon the sample size from which t_{max} is determined, so this increases variability. Hoenig (1983) felt this precluded the estimation of confidence limits using this method.

Empirical formulae relating natural mortality to VBGF parameters have gained wide use in tropical fisheries. Those of Pauly (1980b), applicable to all fishes, are

$$\begin{aligned}\log_{10} M = -0.2107 - 0.0824 \log_{10} W_\infty \\ + 0.6757 \log_{10} k + 0.4627 \log_{10} T\end{aligned} \tag{9.13}$$

$$\log_{10}M = -0.0066 - 0.2790 \log_{10}L_\infty + 0.6543 \log_{10}k + 0.4634 \log_{10}T \qquad (9.14)$$

where T is mean environmental temperature (C), W is weight (g) and L is total length (cm). The equations are based on data from 175 stocks of 84 fish species ranging from polar to tropical waters. The second formula is that of Ralston (1987) for snappers and groupers (15 species, 19 stocks), where using functional regression

$$M = -0.0666 + 2.52k. \qquad (9.15)$$

The approximate 95% confidence limits for Pauly's equations are 0.61–1.63 times the estimate of M (Gulland, 1984). Ralston gives the approximate 95% confidence limits of his equation as 0.24 year^{-1}. Pauly's (1980b) equations cannot account for systematic deviations, as he showed for clupeid fishes. Ralston's equation is thus felt to be a better estimator for snappers and groupers.

Recruitment

Virtual population analysis, where abundance of each recruiting year class is back calculated from abundance in the catch and known natural mortality rates, is generally impracticable for tropical reef fisheries. The stock structure is often unknown, and catch and ageing data are poor. Nevertheless, relative estimates of recruitment strength can be made on a local scale, either from surveys of newly settled individuals or from ageing studies, especially if coupled with length–frequency distributions.

Age-structured data can be used to estimate relative year-class strength through catch-curve analysis (pages 228–229), where deviations from the regression line represent variations in year-class recruitment strength. Doherty and Fowler (1994b) showed for two damselfishes on a local scale that relative year-class abundances determined by ageing are good predictors of initial year-class strength. Appeldoorn *et al.* (1992) used the catch-curve analysis of Sadovy and Figuerola (1992), the growth function of Sadovy *et al.* (1992) and annual length–frequency distributions to demonstrate coherent temporal patterns in recruitment variability for red hind, *Epinephelus guttatus*, across Puerto Rico and the US Virgin Islands.

Surveys of recently settled individuals of certain species have been conducted along the Great Barrier Reef for a number of years (Doherty and Williams, 1988), with surveys done at the end of the reproductive season. These surveys accurately reflect the abundance of settling larvae during the season (Williams *et al.*, 1994). A strong correlation between survey counts and subsequent abundance (Doherty and Fowler, 1994b) indicates that surveys might be used to predict recruitment for some species.

9.4 SINGLE-SPECIES ASSESSMENT MODELS

Assessment models attempt to estimate yield and the potential effects of fishing. Ultimately, it is necessary to ensure that a sufficient number of spawners survive for sustained reproduction and recruitment. Surplus or stock-production methods model the rate of biomass production, including recruitment, as a function of the current biomass. Yield-per-recruit (YPR) models attempt to explicitly model growth and mortality to predict yield, but do not specifically account for reproduction. More limited are approximate models that attempt only to estimate potential yield where little information is available. A number of methodologies or indices have been proposed to estimate the point where spawning stock has been reduced to a critical level (Myers *et al.*, 1994), but many of these rely on knowledge of a stock–recruitment function, or at least stock and recruitment estimates (Mace and Sissenwine, 1993), which are unlikely to be available for most reef fisheries.

Spawning potential

A potential indicator of recruitment overfishing in reef fisheries is the potential fecundity/recruit (P) or spawning-stock biomass/recruit (SSBR). SSBR can be defined in terms of the biomass of all adults, all mature females, or the eggs produced by those females; under the third definition SSBR is equal to P (Goodyear, 1993). Calculation of P is as follows (Goodyear, 1989):

$$P = \sum_{i=t_r}^{n} X_i L_i \prod_{j=0}^{i-1} e^{-(F_{ij} + M_{ij})} \tag{9.16}$$

where t_r is the age at recruitment, n is the maximum age considered, X_i is the mean fecundity at age i, L_i is the fraction of age i females that are mature, F_{ij} is the fishing mortality rate for females by age j, and M_{ij} is the natural mortality rate for females by age j.

While calculations of SSBR are relatively straightforward and analogous to YPR (e.g. Prager *et al.*, 1987; Gabriel *et al.*, 1989; Torres *et al.*, 1992), the potential of SSBR analyses has not been fully realized. In part this is due to the lack of an accepted standard against which to compare results (Mathews, 1991). Goodyear (1989) defined spawning potential ratio (SPR) as the ratio of P under fished and unfished conditions:

$$\text{SPR} = P_{\text{fished}} / P_{\text{unfished}}. \tag{9.17}$$

Comparison of SPRs with empirical data from temperate fisheries suggests average threshold SPR values of 20–30%, below which there is a significant risk of recruitment overfishing (Goodyear, 1989; Mace and Sissenwine, 1993). This has been accepted for use with tropical species (Mathews, 1991). However, Mace and Sissenwine (1993) found threshold values varied significantly among taxa. They were also inversely correlated to maximum body weight and weight at which 50% of the fish become mature, and positively correlated to natural mortality. The generally higher natural mortality relative to body size found in tropical species (Pauly, 1980b) is likely to result in relationships with markedly different parameter values. A minimum threshold SPR level of 30%, still protecting only 80% of the stocks examined, is recommended, especially in the absence of information on a stock–recruitment relationship (Mace and Sissenwine, 1993).

In fact, red drum, *Sciaenops ocellata*, in the Gulf of Mexico was found to decline even at an SPR of 28%, suggesting that a level of 30% would not be sufficiently high, and red snapper, *Lutjanus campechanus*, in the same area declined while managed at a 20% SPR (SAFMC, 1991). For the red hind, *Epinephelus guttatus*, in Puerto Rico, low recruitment over several years occurred despite an SPR $>$ 20% (Sadovy, 1993b).

Sadovy's (1993b) study is of methodological interest because the red hind is a protogynous hermaphrodite (Sadovy, Chapter 2), and the general equations for calculating P and SPR cannot be applied directly. She conducted the analysis only for females, with the rightmost term in Equation 9.16 additionally including the instantaneous rate of sex change. A catch-curve analysis on females summed the losses due to mortality and sex change. This approach assumes that at high F, the number of males needed to fertilize eggs does not become limiting.

Surplus production

Models such as those of Schaefer, Gulland–Fox, and Pella and Tomlinson (Ricker, 1975) have intrinsic appeal because their data requirements are seemingly simple, needing only estimates of catch and effort. They also report management targets in terms of effort and yield, interpretation of which is straightforward relative to fishing mortality (F) or YPR. However, a long time series of data is needed, the models are simplistic, their few assumptions are all-encompassing and therefore may be difficult to meet, and they provide little insight into underlying processes. They are therefore inflexible in suggesting management options. Schnute (1977) generalized the development of these models and put them on a stochastic basis.

Few stock-production models have been attempted for reef species. Because of high diversity in gears and landing sites, it is difficult to estimate species-specific catch and effort. Standardization of effort among gears, while feasible, is often impractical (Munro, 1983i,j). Catch is often lumped over some or all species (section 9.5). Effort may be lumped under broad categories, such as number of fishers or number of boats. Fishery-independent surveys offer greater control, but there exist few long-term studies.

Surplus-production models were originally designed for species near the highest trophic levels, where a constant carrying capacity or virgin stock biomass could be assumed. Large species, such as barracuda (Sphyraenidae), snappers, groupers and jacks (Carangidae), would meet this assumption, but many commercially important species, such as parrotfishes (Scaridae), grunts (Haemulidae), surgeonfishes (Acanthuridae), and small snappers and groupers, might not. Much of the surplus production of prey species is harvested by predators (Slobodkin, 1972; May *et al.*, 1979; Pauly, 1979). A 'prudent predator' might harvest its prey to the point of maximum biological productivity; if so, then in the absence of fishing, prey species would already be harvested at the level of maximum sustainable yield (MSY) (Slobodkin, 1972). Prey abundance may depend on the abundance of predators. What happens to prey species as a fishery expands depends on the life-history traits of those species.

Caddy and Csirke (1983) explored the first point by modelling production as a function of total mortality (Fig. 9.3). The peak of the upper curve marks the point of maximum biological production (MBP); the peak of the lower curve marks the point of MSY. The mortality at which MBP is achieved is always less than that for MSY. Caddy and Csirke (1983) show that as natural mortality (M) increases, the two curves become more disparate, i.e. fishing at F_{MSY} results in a greater degree of overfishing of total biological production. Because of generally higher values of natural mortality in tropical waters (Pauly, 1980b), reef fishes may be more susceptible than their temperate counterparts to fishing pressure.

How a species responds to increasing fishing pressure depends on its potential rate of increase, r (= 2 F_{MSY}: Caddy and Csirke, 1983). Species with higher values of r will be able to withstand better the effects of fishing (Pauly, 1979; Caddy and Csirke, 1983). Also important is the time course of exploitation relative to prey and predator species. If a fishery selectively exploits the larger, K-selected, predatory species, the stocks of more r-selected prey species may expand. A reduction in predation (natural mortality, Fig. 9.3) makes more surplus production available to the fishery. However, if fishing increases equally across all species, the reduction in predation on prey species may not be sufficient to offset the added fishery mortality. With total biological production of the prey species already overexploited, the prey stock could rapidly collapse. Pauly (1979) felt this explained the decline of ponyfishes (Leiognathidae) in the Gulf of Thailand.

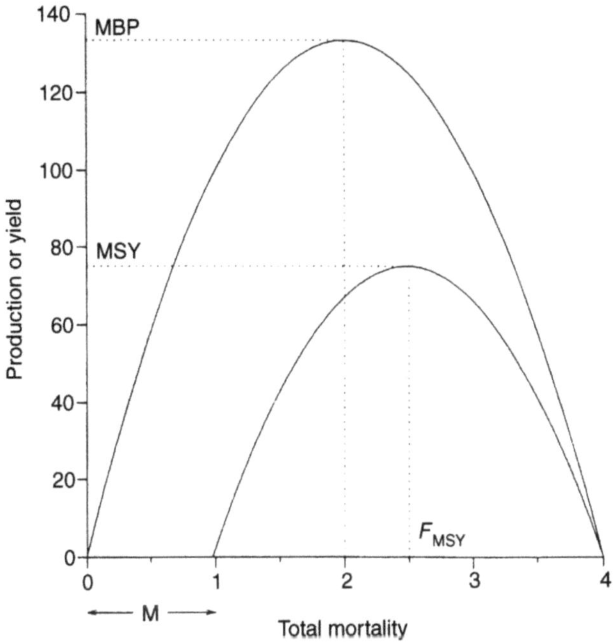

Fig. 9.3 Relationship between total mortality with fisheries yield and total biological production for the case where natural mortality (M) = 1 and intrinsic rate of increase (r) = 3. Y-axis units are arbitrary; MBP, maximum biological production; MSY, maximum sustainable yield; F_{MSY} (= 2.5), fishing mortality needed to achieve MSY. Reproduced with permission from Caddy and Csirke (1983).

Munro (1983i) suggested that, given detailed biological data obtained from other methods, surplus-production models are largely unnecessary for stock assessment. While perhaps an extreme viewpoint (Gulland and Garcia, 1983), it does reflect the difficulties in applying production models to tropical fisheries in general (Munro, 1983j) and reef fisheries in particular. Despite difficulties, single-species stock-production models are still useful. Arreguín-Sánchez (1987) modelled the Mexican fishery for red grouper, *Epinephelus morio*, using the Schaefer model, with effort expressed as annual total fleet capacity. His results (Fig. 9.4) showed effort to be way above that needed for MSY. Subsequent YPR analyses (Salazar-Ruíz and Sánchez-Chávez, 1992) corroborated these results.

The need for a long time series might be obviated with reef fishes if catch and effort data could be collected on an areal basis (Munro and Thompson, 1973; Caddy and García, 1983). The method assumes that ecological production is similar among areas, and that there will be little

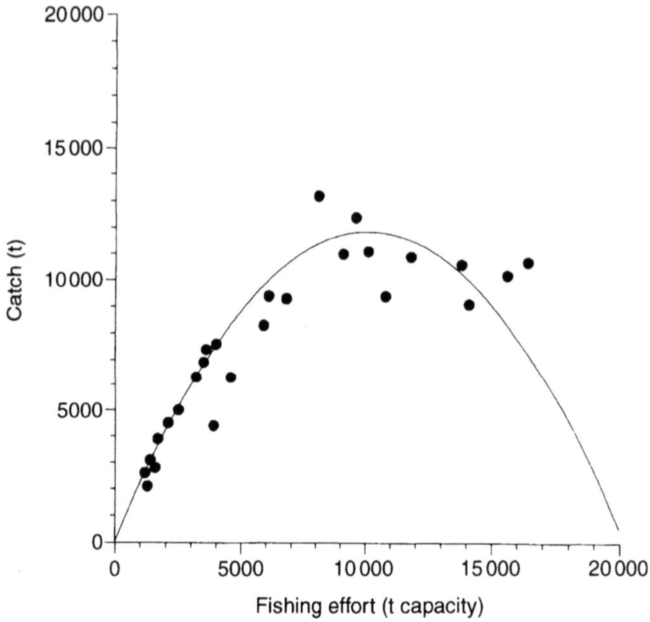

Fig. 9.4 Schaefer production model for the Mexican red grouper, *Epinephelus morio*, fishery. Axes are scaled in tonnes; fishing effort is measured in terms of fishing fleet tonnage. Reproduced with permission from Arreguín-Sánchez (1987).

short-term mixing among areas. This model is further discussed on pages 242–243.

Yield per recruit

Models of YPR assume that production of an entire stock can be modelled by summing the production of a single cohort over its life span, that is:

$$Y = \int_t F\,N\,W\,dt \qquad (9.18)$$

where Y is yield (weight), F is fishing mortality and N and W are the number and weight of individuals alive at time t.

In the YPR formulation of Beverton and Holt (1957), the equation is solved by assuming knife-edge recruitment, VBGF growth, exponential decline in abundance due to mortality (i.e. constant mortality rate), relatively long life and that recruitment is independent of stock size. The model is expressed as follows:

$$\text{YPR} = F \exp[-M(t_c - t_r)] \, W_\infty \sum_{n=0}^{3} \frac{U_n \exp[-nk\,(t_c - t_0)]}{F + M + nk} \qquad (9.19)$$

where k, W_∞ and t_0 are VBGF parameters, F and M are fishing and natural mortality rates, t_r and t_c are the ages of recruitment and selection, respectively, and $U_0 = 1$, $U_1 = -3$, $U_2 = 3$ and $U_3 = -1$. A simplified, relative version of this model (Beverton and Holt, 1964) relies on only three parameters and the additional assumptions that recruitment is constant (i.e. steady state) and continuous:

$$\text{YPR} = E(1-c)^b \sum_{n=0}^{3} \frac{U_n\,(1-c)^n}{1 + \frac{n}{b}(1-E)} \qquad (9.20)$$

where E is the relative exploitation rate (F / Z), c is the relative size at first capture (L_c / L_∞) and b is the ratio of M / k. The parameter c can be approximated from information on the maximum and minimum sizes represented in the catch, and b can be estimated from length–frequency data (e.g. Weatherall *et al.*, 1987) or taken from related taxa (e.g. Munro, 1983i: table 19.1; Munro and Williams, 1985: table 2).

Commercially important reef fishes are sufficiently long-lived often to meet the assumption of the model. Constant exponential mortality appears to fit the data for individual species. However, most reef fishes are exploited in multispecies systems that include both predators and prey. Munro (1983f) attributed higher natural mortality in unexploited areas to the abundance of predators and suggested that natural mortality be modelled as follows:

$$M = M_x + g\,P \qquad (9.21)$$

where M is natural mortality from all causes, M_x is natural mortality due to causes other than predation, P is the biomass of predators, and g is biomass-specific predation mortality. Natural mortality would decline as fishing intensity on the entire system increased. Departure from the assumption of knife-edge recruitment causes significant effects only when selection occurs over a broad age/size range relative to the range commercially exploited and only at high levels of fishing mortality (Fig. 9.5) (Pauly and Soriano, 1986; Silvestre *et al.*, 1991). While most gears have a relatively short selection range, such a situation might occur where data come from a multigear fishery, where each gear has species-specific values of age/size at capture. VBGF growth is typical, with potential problems arising only in hermaphroditic species if a growth spurt

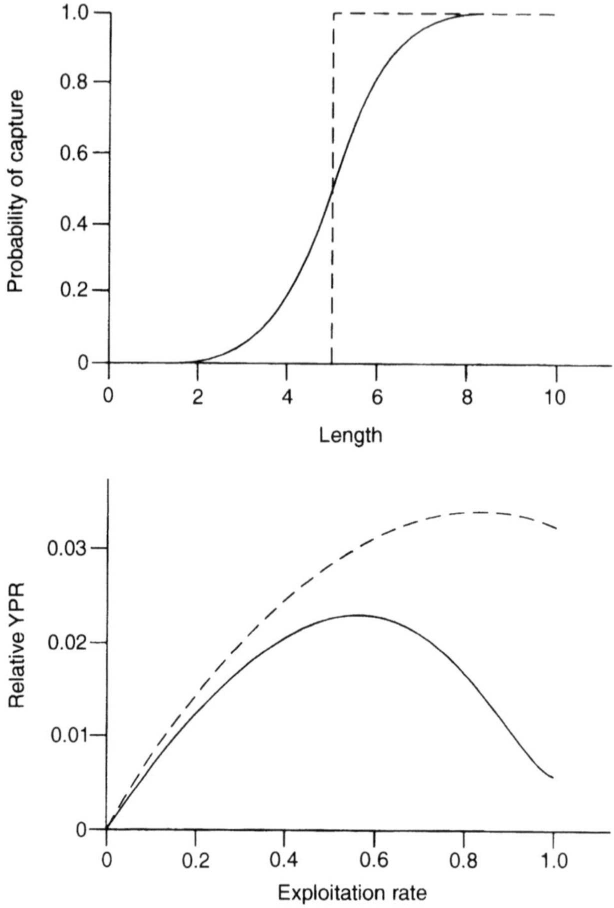

Fig. 9.5 Effect of selection range on relative yield per recruit (YPR). Top graph shows knife-edge (broken) and sigmoid (solid) selection curves. Bottom graph shows differences in estimates of relative YPR under knife-edged (broken) and sigmoid (solid) selection. Adapted from table 1 and fig. 1 in Pauly and Soriano (1986).

accompanies sex change. However, significant sexually dimorphic growth may require separate analyses for males and females. The assumption that recruitment is independent of stock size is difficult to assess without a specific stock–recruitment relationship. However, it is clear that fishing pressure can be sufficiently high to reduce recruitment (page 233). Mace and Sissenwine (1993) found that prevention of growth overfishing, using either MSY or the $F_{0.1}$ criterion (i.e. that level of F where the addition of one unit of mortality results in a 10% increase in yield relative to the amount caught by the first unit of F), would not necessarily prevent recruitment overfishing. The assumptions of the simplified YPR model are more restrictive, but by combining length data over seasons and years prior to estimating parameters, this approach can give useful results.

YPR models have become the principal tool for assessing reef fisheries (e.g. Stevenson, 1978; Huntsman *et al.*, 1983; Munro, 1983g; Matheson and Huntsman, 1984; Dennis, 1988, 1991; Acosta and Appeldoorn, 1992; Sadovy and Figuerola, 1992). The assumptions are often reasonably met (or ignored) for the basic model. Difficulties with the collection of reliable catch and effort data leave few suitable alternatives. Because YPR models take into account underlying processes of growth and mortality, they are more powerful and offer greater flexibility in management advice than surplus-production models. They can easily be extended to examine biomass per recruit or spawning stock biomass per recruit. There is also the view (Munro, 1983i,j) that estimation of the necessary parameters is no more difficult, and perhaps significantly easier, than the collection of reliable catch and effort data. Yet it is in the estimation of individual parameters that the major difficulties arise (pages 225–231).

Approximate yield models

Gulland (1971) proposed the following formula for estimating MSY:

$$\text{MSY} = \text{X}\, M\, B_0 \tag{9.22}$$

where B_0 is the virgin stock biomass and X is a multiplier. The model assumes that F_{MSY} roughly equals M. The basis for the model comes from the Beverton–Holt relative yield equation (Equation 9.20). Using the tables of Beverton and Holt (1964), Gulland found that for values of L_c / L_∞ between 0.4 and 0.7, MSY did not change significantly over the range of M / k values, and MSY approximated $0.6MB_0$ to $0.4MB_0$ for the majority of cases. In the absence of more specific information, the mid-point (0.5) could be used as an estimate of X, and this agrees with the Schaefer surplus-production model, where MSY occurs at $0.5B_0$. While this model has proven useful in some cases, it has been criticized (Francis, 1974; Beddington and Cooke, 1983; Caddy and Csirke, 1983; Kirkwood *et al.*, 1994) and MSY found generally to be overestimated (i.e. $F_{\text{MSY}} < M$), especially for tropical, short-lived species. Kirkwood *et al.* (1994) found the X parameter to range more generally between 0.2 and 0.3 and to be positively related to the relative size at first capture and the degree of density dependence between stock and recruitment.

Where estimates of virgin stock biomass are not available, Cadima (in Troadec, 1977) developed an analogous model under the same assumption of F_{MSY} equals M:

$$\text{MSY} = 0.5\ Z\ \overline{B} \tag{9.23}$$

or equivalently

$$\text{MSY} = 0.5(Y + M\ \overline{B}) \tag{9.24}$$

where $\overline{B}$ is the average *exploited* biomass and Y is yield. Cadima's model assumes that total biological production ($P = Z\ \overline{B}$; Allen, 1971) is constant, which is in contradiction to the premise of production models (Garcia *et al.*, 1989). The model is unbiased only in the absence of fishing (where it collapses to Gulland's model) or at MSY (i.e. where $F = M$). At other points it will either over- or underestimate MSY.

Garcia *et al.* (1989) offer a generalized model, under the same assumptions, that is unbiased at any level of fishing:

$$\text{MSY} = \frac{\overline{B}\ M^2}{2M - F} = \frac{\overline{B}\ \text{M}^2}{2M - \left(\frac{Y}{\overline{B}}\right)}. \tag{9.25}$$

Given the difficulty in assessing the fisheries potential of poorly studied reef-fish stocks, use of the Gulland and Garcia *et al.* models would be desirable. Estimates of biomass could be obtained from fishery-independent surveys. Nevertheless, the assumption of $F_{MSY} = M$ should not be blindly accepted (Beddington and Cooke, 1983; Garcia *et al.*, 1989). Also, the issues raised regarding production models (pages 233–235), particularly the defining of B_0 for prey species, should be kept in mind. In this sense, comparison of the Caddy and Csirke model and the Garcia *et al.* model might offer insight on how the fishery would react if the natural mortality structure were altered.

Polovina and Ralston (1986) estimated potential yield from the ratio of yield (Y) to B_0. Yield was calculated using the Beverton–Holt equation, with maximum yield obtained by varying the input parameters F and t_c. B_0 was similarly calculated by setting F equal to zero. The ratio of these two cancels the recruitment term. Given Y / B_0, potential yield can be estimated with an independent estimate of B_0 (section 9.2). This approach is subject to the same assumptions as the Beverton–Holt model (pages 336–339).

9.5 MULTISPECIES ASSESSMENT MODELS

Multispecies models allow the use of lumped-species catch, effort or abundance data, and they account for species interactions to more realistically view resource dynamics. Multispecies extensions of stock-production models vary in complexity from a single overall production model to individual production models linked through interaction terms. Individual yield-per-recruit models have been pooled to assess and manage multi-species fisheries, and complex models have been developed, based either on the Beverton–Holt model (e.g. Andersen and Ursin, 1977) or virtual population/cohort analysis (e.g. Helgason and Gislason, 1979; Pope, 1979b). In these models, species interactions are modelled through predator–prey relationships, which require complex matrices to show species-specific prey preferences. Owing to their complexity and data requirements, it is unlikely that these approaches will ever prove useful in the analysis or management of reef fisheries, although they might be of heuristic value.

Stock-production models

The simplest model of a multispecies complex is a stock-production model where production of the entire system is a function of the current total system biomass. Pauly (1979) criticized the assumptions of the single-species Schaefer model as applied to total system biomass on two points. The growth of species assemblages cannot be modelled by a single logistic curve because various species will have different growth, mortality and recruitment rates. Also, maximum biomass cannot be equal to the sum of the maximum biomass of component species. However, the total-biomass model is the integration of all processes affecting system production rather than the sum of individual models.

Theoretical support for the total-biomass approach comes from the study of ecological succession (Odum, 1968). Figure 9.6 shows net (= surplus) production to increase, then decrease, while biomass increases as a sigmoid function, over time. These are the key elements for application of a total-biomass production model; only the form of the function describing the process is needed. Because of lag effects on the response of the community to changes in exploitation rate, a function incorporating a curvilinear relationship between system catch-per-effort and effort is expected (Kirkwood, 1982).

Total-biomass production models offer limited management advice. They can determine potential yield and the effort needed to maximize biomass production, but they give no information on the required distribution of effort among species, the catch composition, or the fate of specific species. Ecological theory implies that increased fishing pressure will drive the system to an earlier successional state, and that under high fishing pressure, fundamental changes can be expected (Jennings and Lock, Chapter 8). In the Caribbean, changes associated with heavy exploitation have included a relative decline in the catch of larger predators (e.g. Nassau grouper, *Epinephelus striatus*) and an increase in small species, such as parrotfishes and surgeonfishes (Koslow *et al.*, 1988; Appeldoorn *et*

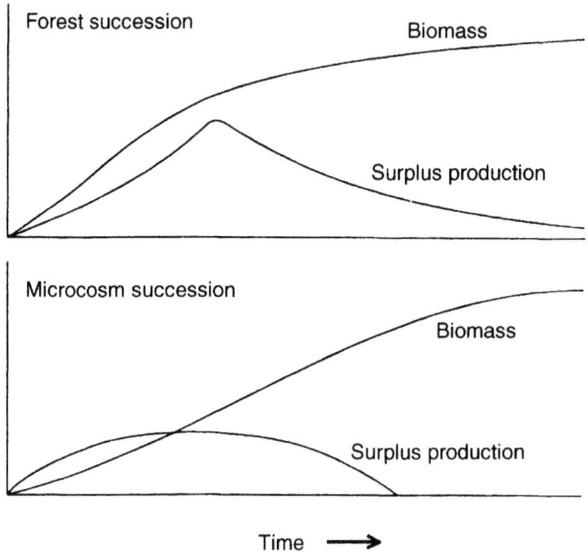

Fig. 9.6 Expected changes in biomass and surplus production during ecological succession. Derived from Odum (1968: fig. 1).

al., 1992). There was a dramatic increase in the abundances of black sea urchin, *Diadema antillarum* (Hay, 1984), and damselfish, *Stegastes planifrons* (Vicente, 1994), with a subsequent decline in the former due to an unknown pathogen (Lessios *et al.*, 1984b), transmission of which was probably aided by high density. Fishing-induced instability has led to a shift from coral dominance to algal dominance in highly affected areas (Hughes, 1994).

The predictive limitations of total-biomass models necessitate additional information if management aims to regulate particular species. Thus, Brown *et al.* (1976) applied the total-biomass Schaefer model to estimate an overall quota, but individual species were also regulated by species-specific models. The sum of MSYs from the individual models was 50% greater than the MSY from the total-biomass model.

In tropical reef areas, total-biomass models were used by Munro and Thompson (1973), Ralston and Polovina (1982) and Appeldoorn and Meyers (1993). Ralston and Polovina used the Schaefer model. The other studies employed the Gulland–Fox model using total catch and effort data collected on an areal basis, with effort being the number of boats/area and catch including all demersal and coastal pelagic species.

Koslow *et al.* (1994) argued that the results of total-biomass, area-based production models are not likely to be reliable due to violation of one or more model assumptions. Studying areas from Jamaica and Belize, they found that catches were heterogeneous among areas (possibly due to selective fishing as well as community differences), catches were not in equilibrium, and productivity possibly differed among sites.

A cluster analysis of the Hawaiian deep-water catch identified three species groups, somewhat differentiated by depth (Ralston and Polovina, 1982). Total potential yield (MSY) obtained by adding the groups was similar to that calculated by pooling all species into one analysis. The clustering approach minimized the information loss inherent in pooling data, and a greater understanding was gained of how effort could be allocated among species groups (effectively by depth) to achieve the potential overall yield.

Stock-production models have been applied to a subset of species with similar life histories (e.g. Ellis, 1969; Caddy, 1984). Appeldoorn and Lindeman (1985) presented the ecological and empirical arguments favouring this approach for reef fishes when data for related species are lumped. The species group is assumed to respond to fishing in the manner of a single species. The removal of individuals of the larger, slow-growing species will result in ecological replacement and a relative increase in abundance by the smaller species in the group. Appeldoorn and Lindeman (1985) analysed the grunts (Haemulidae) fished off south-west Puerto Rico, and predictions of catch rate closely matched limited fishery-independent data.

Species-specific production models with interspecific interactions can be constructed using Lotka–Volterra equations with added terms for fishing, e.g.

$$\frac{1}{N_t}\frac{dN_i}{dt} = a_i - b_i N_i - q_i f_i \pm c_{ij} N_j \tag{9.26}$$

where a, b and q are species-specific parameters and c is the parameter describing the strength and magnitude of interaction. Pope (1976, 1979a) and Kirkwood (1982) found that if the ratio of fishing mortalities on component species is restricted, then total effort may be used to model the fishery, and total maximum yield would be less than the sum of single-species MSYs.

While some heuristic insight has resulted from this approach (May *et al.*, 1979), its practical application is questionable due to problems in estimating model parameters. Sissenwine *et al.* (1982) could not separate true significant interactions from Type 1 errors, so any attempt to apply their results would be meaningless. Because the Sissenwine *et al.* (1982) study used the best fisheries data available, this approach will be even less promising in more complex but data-poor situations, including tropical reefs.

Yield per recruit

Stevenson (1978) and Huntsman *et al.* (1983) calculated a series of YPR models for individual species within multispecies fisheries. Rather similar behaviour of the YPR models across species suggested that uniform management measures could be applied over all species, particularly involving the regulation of age-at-capture. Such average-based system management, with some species overfished and others underfished, is valid, but only when the range of sizes (and productivities) of the fishes exploited is fairly narrow. Huntsman *et al.* (1983) analysed an offshore hand-line fishery primarily for large snappers and groupers ranging from 600 to 1300 cm in asymptotic length, while Stevenson (1978) analysed an inshore fishery dominated by small groupers, goatfishes, grunts and parrotfishes.

Munro (1983f) used YPR calculations for 24 species to calculate relative total yield and species composition for different combinations of fishing mortality and mesh size. Recruitment was based on the number of recruits needed to produce the observed average catch/effort. To account for expected declines in predators and natural mortality as fishing intensity increased, M/k declined exponentially with increasing effort. The validity of Munro's approach to modelling changes in M has not been tested, but by treating individual species, he could assess potential changes in species composition of the catch at different mesh sizes. At the mesh size giving the highest total catch (8.25 cm), the analysis showed this catch comprised larger species; smaller species either declined in yield or dropped out completely.

The analysis of related taxa presented by Appeldoorn and Lindeman (1985) was extended using the Beverton–Holt YPR approach (Appeldoorn, 1987b). M, k and W_∞ shift over time from those characteristic of smaller species to those of larger species, i.e. smaller species have higher natural mortalities and will selectively be removed from the group. W_∞ increases monotonically over time,

$$W_{\infty t} = W_{\infty S} + \left(\frac{W_{\infty L}}{0.857} - W_{\infty S} \right) \left(1 - e^{-\rho(t - t_r)} \right) \qquad (9.27)$$

where $W_{\infty S}$ and $W_{\infty L}$ are asymptotic weights of the smallest and largest species. The parameter ρ is calculated from Equation 9.27 by defining that $W_{\infty t}$ equals $W_{\infty L}$ when time t is equal to the maximum age observed. The latter is calculated from M using Equation 9.12. Corresponding changes in M and k are calculated, respectively, from Equation 9.13 and Appeldoorn's (1993) Φ equation relating k and W_∞ across species. The 0.857 is invariant. It comes from an approximate relationship between asymptotic size and maximum observed size (Pauly, 1984a, p. 26).

System models

ECOPATH models (Polovina, 1984; Christensen and Pauly, 1992b; Aliño *et al.*, 1993; Opitz, 1993) are useful for displaying how sensitive production is to the presence of higher carnivores or to the availability of different potential food groups. The models can provide only gross estimates of potential fishery productivity (Polunin, Chapter 5). Polovina (1984) suggested that at French Frigate Shoals a potentially higher yield could be achieved if apex predators were reduced.

Saila (1982) and Saila and Erzini (1987) attempt to model empirically changes in species composition based on Markov processes. A time series of catch-per-effort data by species group is required, with the number of groups being less than the number of samples. The probability that relative abundance will shift from one group to another is assumed constant, but the analysis of north-west Atlantic bottom fishes showed that this assumption does not always hold if the time series is long (Saila and Erzini, 1987). Indices of the rate of change in relative abundance are also calculated and can be used to forecast future species composition. Thus, the model can examine changes in species composition over time due to fishing or other factors. Although it has been suggested for modelling reef fisheries (Saila, 1982), only a small number of species groups could be analysed. The longer time series needed to increase the number of groups may violate assumptions of constant probabilities, especially considering the possibilities of alternative system states and the role of large ecological disturbances in tropical reefs.

In another empirical approach, Saila (1992) applied fuzzy-graph theory to analyse changes in species composition. The number of species groups is not limited, but the model is not primarily quantitative and cannot be used for forecast purposes. Instead, it produces a flow diagram indicating directions of changes among species groups and calculates the strength of those changes. The model was tested against the Gulf of Thailand trawl-survey data. For this rapidly developed fishery, most successional changes could be detected using short time-series data.

9.6 CORRELATIVE MODELS

Empirical relationships between environmental parameters and abundance or potential production have been developed for a number of temperate species (e.g. Sissenwine, 1974; Dow, 1977), which in some instances can be related to biological processes (e.g. Flowers and Saila, 1972; Crecco *et al.*, 1986). In freshwater lake fisheries, production can be estimated from temperature and morphoedaphic indices (MEIs) (Hanson and Leggett,

1982; Schlesinger and Regier, 1982). The most noted MEI (Ryder, 1965) relates production to total dissolved solids divided by depth, with the former being related to nutrients and the latter to stratification.

Stevenson and Marshall (1974), Munro (1977) and Marten and Polovina (1982) indicate that some form of predictive MEI might be developed for tropical reefs, but present data are insufficient in environmental details (Polunin, Chapter 5).

While definitive correlative models have not been developed for reef fishes, it is clear from recruitment studies (Doherty and Williams, 1988) that meso- to large-scale processes affect recruitment. Polovina *et al.* (1994) have related changes in climatic patterns and oceanic conditions to changes in the abundance of reef fishes in the north-west Hawaiian Islands. Such relationships may prove useful in predicting potential yield from routinely collected atmospheric data.

9.7 FUTURE APPROACHES

To enhance the applicability of models to the assessment and management of tropical reef fisheries requires progress in three areas. First, more and better data need to be collected, with attention given to designing efficient sampling protocols. Second, the study of fisheries dynamics must expand beyond the routine estimation of growth and mortality to include new areas. Third, current models need to be made readily available and further tested, while new, more efficient or heuristic models need to be developed.

Data acquisition

The major limitation to model application and development is the amount and quality of available data. For traditional biological models, the pertinent data are on catch, effort, species composition, length–frequency distribution and age structure, with reference given to areas fished and their characteristics (e.g. depth, distance from shore, habitat type). Models for management should also incorporate socio-economic data (McManus, Chapter 10).

Significant improvements can be made with existing resources if data collection strategies are developed with potential analyses in mind. Pulse sampling, targeting indicator species, and applying statistical surveys of catch and effort could improve sampling efficiency. Pulse sampling concentrates intensive data collection on a few species, with the species of interest rotating on a periodic basis. This is particularly useful for obtaining adequate samples for length–frequency analysis, i.e. large numbers measured over a short time period. Data collection efforts can be limited to only a subset of species. Criteria would include importance to the fishery, abundance in the catch (relative to various gears), sensitivity to exploitation or environmental perturbation, and applicability to other species (based on similar habitat or life-history parameters). Statistical sampling could replace comprehensive surveys of catch and effort and lead to improved estimations, plus calculation of variances. A starting point might be the sampling techniques originally developed for recreational fisheries.

Dynamics

To further understand and model production processes, new areas need to be emphasized: identification of critical habitats and habitat links, rates of migration and recruitment processes (Roberts, Chapter 4). All are important in the establishment and assessment of marine fishery reserves (Bohnsack, Chapter 11), while habitat maintenance and sustained recruitment are at the heart of resource conservation in general. Recruitment processes are still largely an unknown for reef fishes with respect to potential yield and stock assessment; indeed, in only a few extreme cases (e.g. isolated areas such as Bermuda) have the bounds of biological stocks been determined, and this is prerequisite for examining the relationships between spawning stock and subsequent recruitment. At present, the standard stock–recruitment models can only be applied on the basis of faith.

Modelling

Existing techniques need to be made more readily available and to emphasize estimation of variances as well as means. First attempts included software packages such as the Compleat ELEFAN (Gayanilo *et al.*, 1989), LFSA (Sparre, 1987) and FSAS (Saila *et al.*, 1988). FiSAT (Gayanilo *et al.*, 1994) is a welcome second-generation package. Nevertheless, the rapid advances in computer technology can put much more calculation-intensive applications and methodologies within the hands of the average user. Simulation studies will be useful for testing and modifying existing models with respect to their potential biases, although selected field studies will be needed to identify possible problems and corroborate results.

The development of new models will include (1) those designed to work within the data-limited context of reef fisheries, (2) more complex models designed for those few cases where more data are available, with the goal of increasing our understanding of system dynamics (e.g. the effects of species interactions) and data needs, and (3) still more complex, heuristic models. An important future trend will be the development of spatially structured models to account for differences in productivity among areas or various reef-associated habitats, critical habitats and migration effects. The use of new concepts, such as fuzzy logic and artificial intelligence (e.g. neural networks), may facilitate parameter estimation and enhance model flexibility and performance.

ACKNOWLEDGEMENTS

D. Pauly, G. Kirkwood and an anonymous reviewer critiqued the manuscript and offered valuable suggestions. Thanks are due to the Gulf and Caribbean Fisheries Institute, the United States National Marine Fisheries Service, and the Bureau Océanographique, ORSTOM, for allowing reproduction of figures. This work was supported by the University of Puerto Rico, Mayagüez.

Chapter ten

Social and economic aspects of reef fisheries and their management

John W. McManus

SUMMARY

Because fisheries management involves the regulation of human activities, it should properly be considered a social science. Unfortunately, social aspects of management have been largely neglected compared with natural science investigations of the population biology of harvested organisms.

Tropical reef fisheries suffer from the 'tragedy of the commons' wherein individuals stand to gain by exceeding equitable levels of use of a common resource. This effect is exacerbated in areas subject to escalating population pressures, an increasingly large proportion of the developing world, where alternative livelihoods are unavailable or are economically unattractive. In many areas, population expansion and poverty have led to Malthusian overfishing, where fishers initiate wholesale resource destruction in order to maintain their livelihoods in the face of declining stocks. The behaviour of fishers under these circumstances often follows the predictions of a simple bioeconomic fixed-price model. In this model, profit from catch follows a parabola with increasing fishing effort and is intersected by a cost curve which rises linearly. The point at which the curves intersect is the bionomic equilibrium point, and occurs at the level of

Reef Fisheries. Edited by Nicholas V.C. Polunin and Callum M. Roberts.
Published in 1996 by Chapman & Hall, London. ISBN 0 412 60110 9.

fishing effort at which no net profit is possible, a level at which the stock is seriously overfished. In areas with abundant unemployed labour, new entrants are attracted to a fishery until the bionomic equilibrium is reached. At a societal level, the greatest net profit from a fishery can be obtained when fishing is restricted to the level wherein the difference between the gross profit and cost curves is the greatest. This point is termed the maximum economic yield (MEY) and generally lies well to the left of the maximum gross profit point. The latter point corresponds to the maximum sustainable yield of the Schaefer model. Numerous studies of reef fisheries suggest that the MEY lies at approximately 40% of the fishing effort level of the bionomic equilibrium point. Many intensively exploited reef fisheries could thus benefit from a 60% reduction in effort, usually through a reduction in numbers of fishers.

Implementing effective management is difficult and must be based on a thorough understanding of the interacting ecological, sociocultural and economic systems. This generally requires one or more years of intensive study prior to the formulation of a strategy, and should include the affected community in a close and substantive way. I summarize some guidelines for the effective acquisition of economic and sociological data in coral reef settings. Management will rarely be effective without community participation but most management strategies will need to mix both 'top-down' and 'bottom-up' elements. Devising effective management strategies in the future will require the collaborative efforts of social and natural scientists, each aware of the basics of the others' specialities, and each contributing to assist human communities dependent on reefs to ensure the sustainability of their resources.

10.1 INTRODUCTION

Fisheries management is the act of influencing human activities so as to enhance some characteristics of a harvestable resource, such as production, economic yield, equitable access or sustainability. Thus, it is primarily an applied social science, which operates with respect to predictions and recommendations stemming from natural science investigations. Most of the emphasis in fisheries has been on the natural sciences (Appeldoorn, Chapter 9), and far less has involved social sciences such as economics, sociology, social psychology, anthropology or political science (Orbach, 1989; McManus *et al.*, 1992). This imbalance in approach has been especially problematic in reef fisheries, wherein the majority of attempts effectively to implement management plans have failed because of non-compliance by part or all of the target human populations (Johannes, 1980; White, 1984, 1986; McManus, 1988; McManus *et al.*, 1988;

Adams, Chapter 13). On the other hand, some major attempts at management have failed because of the application of economic policies without sufficient consideration of ecological effects. For example, national and international lending institutions in past decades have often supported governmental attempts to 'assist' fishers in overfished areas by providing capital for improved fishing gear (Smith *et al.*, 1983). This practice has greatly worsened overfishing and associated environmental problems. Its devastating effects on reef fisheries and coastal fisheries generally in many developing countries continue to be major concerns. Future effective management of reefs must be based on tightly integrated efforts involving both social and natural sciences.

This chapter presents a brief overview of selected social science considerations important in the management of reef harvest systems. I include discussions of reefs as common property, of the problems of overfishing related to overcrowding (Malthusian overfishing), of economic and social research approaches, and of management options and implementation, and end with selected case histories. Social science aspects of reef management are approached elsewhere in this volume (Ruddle, Chapters 6 and 12; Adams, Chapter 13). It is important to bear in mind that each social science is highly developed with respect to theory and to both quantitative and qualitative methods. The number of qualified social scientists who have worked on reef harvest management has been limited, although their pioneering work has often been exemplary. This chapter is based on a variety of projects, many of which have involved biologists and others who are not social scientists. These people have conducted the research or applied the methods out of desperation to accomplish a management objective. It is hoped that this chapter will open the door to more concerted efforts on the social science of tropical reef fisheries management.

10.2 TRAGEDY OF THE COMMON REEF: SOCIOLOGICAL MODIFICATION OF ECONOMIC GROWTH

The ocean is commonly perceived in modern societies to be in the public domain. It is a source of food and livelihood of last resort, from which nobody can be turned away. This view conflicts with concepts in many regions of coastal resources traditionally being under strict control of societies and individuals (Ruddle, Chapters 6 and 12). A number of factors have led to the weakening or obfuscation of traditional management concepts. These include population pressure, transmigration of cultural groups, colonialism, competition among economic strata, the influx of foreign ideals through the media, and other factors (Ferrer, 1991; McManus *et al.*, 1992; Ruddle, 1993, 1994b).

As coastal reefs have passed further into the public domain, they have become subject to the same problems generally expected to arise from individual, competitive use of common property. The problem was best described by Garrett Hardin (1968) as 'the tragedy of the commons' (see also Berkes, 1985). A society may benefit in theory from having a common resource which is shared by all its members. However, individuals may benefit by exceeding the equal-use level. This leads to an inherent conflict which makes management of the resource difficult.

There are many restrictions on following the preceding pattern. Most reefs are not truly open-access resources. In many areas, traditional management schemes limit new influxes of people. In others, various political groups, tongs (gangs) or influential fishing families exert social or physical pressures to restrict entry. In all but the most strictly controlled areas, this does not entirely prevent the influx of new people, but it restricts the influx to relatives or otherwise favoured individuals.

10.3 MALTHUSIAN OVERFISHING AND THE SCOPE OF ECONOMIC AND SOCIAL PROBLEMS INVOLVING REEFS

Categories of overfishing may be grouped into four major classes: growth, recruitment, ecological and Malthusian (Pauly *et al.*, 1989; Pauly, 1990b; Russ, 1991; McManus *et al.*, 1992). Growth overfishing involves the harvest of individual organisms at sizes which are suboptimal with respect to potential system yield. Recruitment overfishing is that which leads to declines in available recruits, thus reducing system yield. Ecological overfishing describes situations where harvest pressure has led to a shift in the system from economically or nutritionally useful organisms to those which are less valuable to the fishers.

Malthusian overfishing defined

Overfishing resulting from population growth and poverty tends to have broadly predictable consequences in addition to those listed above, and thus requires formal recognition. The term Malthusian overfishing has been proposed, in honour of the Rev. I.R. Malthus (1766–1834), who elucidated the inevitable conflicts which arise when exponentially growing human populations depend upon resources which increase linearly or not at all. The use of the term acknowledges the fact that fisheries and resource ecology are as much social sciences as they are physical sciences. The originally proposed definition was (Pauly *et al.*, 1989):

> *Malthusian overfishing* occurs when poor fishermen, faced with declining catches and lacking any other alternative, initiate wholesale resource

> destruction in their effort to maintain their incomes. This may involve in order of seriousness, and generally in temporal sequence: use of gears and mesh sizes not sanctioned by the government; ... use of gears not sanctioned within the fisherfolk communities and/or catching gears that destroy the resource base; ... and use of 'gears', such as dynamite or sodium cyanide that do all of the above and even endanger the fisherfolks themselves.

Malthusian overfishing generally involves growth, recruitment and ecological overfishing, as well as a variety of destructive fishing methods.

A simple bioeconomic model

It is useful to illustrate some general trends among fishing communities through the use of a simple graphical model. More advanced treatments on bioeconomic modelling can be found in Smith (1979) and Clark (1989, 1990).

The simplest bioeconomic model is the fixed-price model of Gordon (1954). This is based on the simple parabolic yield–effort model, commonly referred to as the Schaefer model (Appeldoorn, Chapter 9). If one assumes that the price per kg of fish in the market is constant at all levels of fishing effort, then the yield curve can be translated directly into a gross profit curve of the same shape (Fig. 10.1). Assuming similarly that the cost of fishing rises at a constant rate as effort increases, one can graph the cost as a line extending from the origin and intersecting the gross profit curve. The difference in value read from the vertical, value axis, between the gross profit and the cost for any level of effort, is the net profit. The point at which the profit and cost curves intersect is known as the bionomic equilibrium point, and occurs at the level of effort at which no net profit is possible (Fig. 10.1). This point is generally well to the right of the maximum gross profit point, and is usually at an effort level wherein the fishery is seriously overfished from recruitment, and often growth and ecological, standpoints.

In theory, an open-access fishery in a region of abundant, unemployed labour tends to attract new entrants until the bionomic equilibrium point is reached, and nobody (including the original fishers) can make a profit. From this, it is immediately obvious why most fishers in developing countries are poor. Blast fishing, poisoning and other forms of destructive fishing are often ways of reducing the cost line to generate a profit when net profits of any kind are scarce. These additional profits are often only temporary, however, as they merely result in a lowering of the bionomic equilibrium point and an invitation for more fishers to crowd in on the resource.

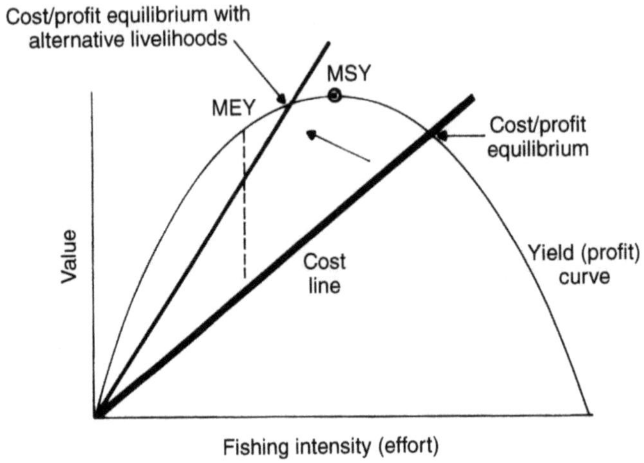

Fig. 10.1 Fixed-price model of a harvest system (MSY, maximum sustainable yield; MEY, maximum economic yield).

At a societal level, the greatest net profit from the fishery can be obtained when fishing is restricted to the level wherein the difference between the gross profit and the cost ('economic rent') is the greatest. This point, the maximum economic yield (MEY), is generally well to the left of the maximum gross profit point. The latter point corresponds to the maximum sustainable yield (MSY) of the Schaefer model, and represents the point at which the maximum harvest can be obtained from a single-species population subject to no natural variability. In that tropical reefs and similar systems encompass considerable variability and rarely involve isolated target species, a lower level of fishing is often recommended (a maximum ecological yield point) which falls somewhat to the left of the MSY. Target effort levels which account for important ecological, sociological or economic considerations are termed optimum yield (OY) points (Orbach, 1989), and these generally fall well to the left of the MSY as well. The MEY point generally is not far from OY, and, given the vagaries involved in making these models, it would normally be difficult to distinguish one from the other with any degree of accuracy.

Qualifiers on the model

There is very little in the way of economic incentive to encourage adoption of the maximum economic yield level of harvest effort. The tragedy of the commons concept leads us to predict that even if a small community were to identify that level of fishing as the most desirable, the individual fishers would be under constant pressure to improve their own incomes by overfishing (Berkes, 1985). Families would be under pressure to arrange for access by unemployed relatives from elsewhere, leading to increased local effort. Even if the entire fishery were to be controlled by a single monopolistic entity, present concepts of economic discounting predict that the entity would probably cash in on the resource in order to raise capital for investment in other enterprises with higher levels of return (Clark, 1990). Finally, one must ask why it is that economists have traditionally recommended that all such resources be harvested at a level of highest profit margin. Simply because a profit can be made by a society does not mean that the most desirable strategy is to make it (Hardin, 1991). There is a growing realization that what is most important to optimize is the standard of living, not economic production (Costanza, 1991; Potvin, 1993). The yield per fisher virtually always declines as the number of fishers increases. A society may at some scale make more profit by increasing the number of fishers on a coral reef. However, because of the decline in yield per hour of fishing, the lifestyle of the fishers and the welfare of a smaller-scale community may decline substantially before effort levels even approach the MEY.

All of these considerations are still too narrowly focused to provide rational answers to management of reefs. A manager must never think of the reef solely as a production system. Most people involved in reef management take an anthropocentric view of ecological management – that is, they feel that it is necessary to justify decisions concerning the future of the reef in terms of benefit to human beings (Eckersley, 1992). In this case, it is the use of the reef which is important, to be evaluated on short-time, single-generation, and intergenerational time scales. The latter time scales are particularly difficult to deal with in resource valuation studies. It is very difficult to know how valuable a reef species might be in the next century or two. Resource economists commonly account for the cash-in tendency by determining a value for the resource in present terms, and then decrementing ('discounting') the value over a period of a few years to estimate a total value. This practice is highly suspect, because the ecosystem and its products are becoming more rare and thus, more valuable over time (Hardin, 1991; see other articles in Costanza, 1991). There is also a growing ecocentric view, for which the preservation of nature need not be justified in terms of tangible benefits to people (Eckersley, 1992). Similar beliefs underlie many traditional philosophical systems, and conflicts or mismanagement arise when these views are not accounted for (Ruddle, Chapter 12).

Reefs provide a very wide range of uses which lead to either greater economic production or improved standards of living (Salvat, 1987; McManus *et al.*, 1992; Wells and Price, 1992). One obviously important use is for generating tourist income (Boo, 1990; Wong, 1991). Tourism is a major source of income in many developing countries, generating US$ 1–2 billion annually in some countries of South East Asia. An example of the potential importance of coral reefs in tourism economics can be seen in Australia, where the Great Barrier Reef is believed to generate, directly and indirectly, tourist incomes exceeding US$ 1 billion annually. Even very low levels of fishing can cause changes in the community structure of a reef (Munro and Williams, 1985; Russ, 1991; Jennings and Lock, Chapter 8). Effects include declines in diversity and abundance of big fishes, both of which can have adverse effects on the potential tourist value of a reef. Ongoing fishing activities tend to inhibit tourist diving, and such diving may make fishing difficult to conduct safely. Both fishing and tourism are often important to national economies. Both tend to be destructive to reefs in various degrees and require active regulation. The degree to which one use is to be favoured over the other should be the result of concerted analytical efforts.

The assumptions of the fixed-price model are rarely met completely. However, simple violations of the assumptions usually do not change the general conclusions from the model. The cost line may actually curve upwards, as in the analysis by Trinidad (1993) of pelagic fisheries. An example of applications of a variable-price model may be found in Panayotou and Jetanavanich (1987). The most serious deficiency arises when the model is used for a multispecies assemblage such as on a reef, instead of for the single-species fisheries on which it is based. When used as a management tool on reefs, it is important to consider the effects of various levels of fishing effort on the fish community (Munro and Williams, 1985; Russ, 1991), and to consider the effects of the management strategy on the biodiversity and sustainability of the resource, as well as on the value of the catch thus modified.

The 60% effort-reduction rule

In many cases it is difficult to obtain effort and yield or profit data of sufficient range to produce a reasonable fixed-price model on a system. In these cases, it is often possible to obtain a first-order approximation of the MEY level of effort without actually constructing the curves. Reefs in their natural state generally provide highly concentrated abundances of valuable fish, which are relatively easily caught by hook and line, spear or trap. These abundances vary considerably among reefs, but are generally much higher than in adjacent non-reef areas. In a cash-based economy (as opposed to barter-based), if reasonably experienced reef fishers cannot rise far above national poverty lines given reasonable hours of fishing, and there is no other obvious cause for their misfortune (e.g. poor transportation to markets), then it is likely that they are operating on an overfished reef. If there is known to be an excess of labour available, and it can be established that the near-poverty condition has persisted for a few years, then one can reasonably assume that the fishery is at or near a bionomic equilibrium.

The literature contains abundant examples wherein the MEY was estimated quantitatively to lie at approximately 40% of the effort level at bionomic equilibrium (e.g. Silvestre *et al.*, 1987; Pauly and Chua, 1988; Trinidad, 1993). In other words, the fishing effort needed to be reduced by 60%, usually in terms of fishers, but alternatively in terms of hours of fishing. It is possible to show that it is a simple property of numbers that a 60% reduction will not be far from the MEY except in extreme cases (McManus, 1992). Given a perfectly symmetrical production curve of a particular simple formulation and a straight cost curve, the MEY will always be at exactly 50% (Anderson, 1980). However, most reef production curves tend to have MSY points shifted to the left, and the 60% reduction is a more reasonable approximation in such cases (McManus, 1992).

The implementation of a fishery reserve may reduce poverty by providing higher production levels in adjacent fished areas (Alcala and Russ, 1990; Bohnsack, Chapter 11). However, the extra production would tend to attract new entrants to the fishery, thus driving the fishery back toward a bionomic equilibrium. Thus, the implementation of a reserve should be complemented by ways of limiting entry into the fishery more generally, if such mechanisms do not exist already. The same may be said of many fishery management strategies designed to optimize production, including shifts to a more optimal mesh size or suite of allowable gear types.

The 60% rule should not be used instead of an actual analysis when the data are available. However, it can serve to provide a guideline for assessing the scope of the overfishing problem on a local, or particularly a national, scale.

Global scope of the Malthusian overfishing problem

Most of the world's reefs are in developing countries, and these have predominantly high rates of population growth. More than 40 countries and island states have reported problems with the use of blasting devices on reefs. At least 15 have reported the use of poisons (UNEP/IUCN, 1988a,b,c). However, not all these reports of destructive fishing methods can be attributed to overpopulation at the coast. In Papua New Guinea, for example, where the human population tends to be high only in local pockets, the recent rise in blasting appears to be related more to a rising interest in making fast profits and a general feeling that the reefs are too vast to be seriously affected (B. Crawford, pers. comm.). Several key reef areas have been identified to suffer from population-driven reef destruction, including cases from Indonesia, the Philippines, the Maldives and Jamaica. Human population growth rates resulting in doubling times of 30 years are common. Globally, population growth rates tend to be rising in developing countries despite efforts to the contrary. Even in countries such as Thailand and Malaysia, where most of the population excess ends up in cities, the future may be expected to hold a destructive population increase along coasts and near reefs. Ultimately, it is the exponential rise of human populations in most developing countries that poses the greatest threat to reef sustainability worldwide.

10.4 ANALYTICAL AND MANAGEMENT OPTIONS

There is no single, proper way to manage a reef harvest system. Effective management must be based on a thorough understanding of the interacting ecological, sociocultural and economic conditions. Experience has shown that investigations of these systems leading to effective management generally require one or more years of intensive study prior to the formulation of a strategy (Fig. 10.2) (Cabanban and White, 1981; Alcala, 1988; Christie *et al.*, 1990; McManus *et al.*, 1992). On the other hand, there are claims that reefs are degrading on a global scale (Wells and Hanna, 1992; Wells and Price, 1992), and thus there is often a sense of urgency in planning. This section summarizes some alternatives for management strategy, data-gathering approaches and management implementation.

Management options

Reefs are complex resource systems with a broad variety of potential uses to people. Modern approaches to managing their fisheries often depart radically from the traditional approaches of limiting gear use and harvests. Fishery production may depend on limiting pollution. Restrictions on entry into the fishery may be impracticable without offering alternative livelihoods, and these may add to the pollution. The success of the alternative livelihoods may depend on training people with no education, and this lack of education may relate to time spent in childhood as fishers. Tourism may provide better incomes, but may be limited by blast fishing. Tourism itself may lead to pollution or anchor damage to corals. Tourism and mariculture may generate more total income, but may direct resources away from villagers to outside investors or immigrants. Small-scale mariculture and shellcraft may be considered to be ‘women’s work’ and not

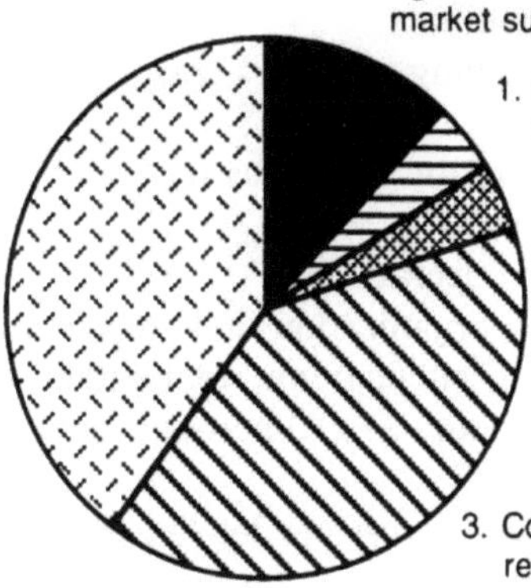

Fig. 10.2 Time invested in research activities in small-scale fishing communities in the Philippines. Data from Pomeroy (1989).

suitable for underemployed males. Women may be too overburdened with other tasks to handle part-time activities. Attractive new jobs may go to outsiders rather than to local fishers.

It is often very difficult to find rational ways to reduce the effects of reef-dependent human populations on reefs. When a promising solution is found, it often loses promise in the realization that many such populations will double in the next 30 years.

The management measures applied will vary considerably between cases. However, some idea of the scope of consideration may be gathered from the following list of options which have been proposed for various reef fisheries (Munro, 1975; L.T. McManus, 1989; J.W. McManus *et al.*, 1992; White *et al.*, 1994):

1. impose gear limitations (e.g. mesh size adjustments);
2. limit entry to the fishery (e.g. through licensing, directed taxes, prohibitions);
3. impose harvest limits or quotas;
4. impose closed seasons;
5. establish closed areas (protected areas);
6. ban destructive and dangerous gears;
7. shift resource use (e.g. from harvest to controlled tourism);
8. enhance and restore habitat (e.g. by planting mangroves);
9. control pollution and siltation (e.g. through reforestation);
10. control tourist activities and development;
11. establish organizations governing resource use (e.g. fishery and tourism councils);

12. introduce monitoring and adaptive management;
13. improve enforcement;
14. improve fish-handling facilities (e.g. reduce spoilage and health problems);
15. diversify use of marine resources (e.g. harvest more species or implement mariculture);
16. organize labour cooperatives;
17. foster community organization and empowerment;
18. support education, publicity and motivation activities;
19. reduce child labour (e.g. increase education and occupational mobility);
20. equalize work loads across gender (e.g. diversify income sources);
21. improve use of land (reduce marine harvest pressure);
22. reduce human population growth rate.

Management options are further discussed elsewhere (Adams, Chapter 13). One option which has been frequently proposed but not included above is the installation of artificial reefs. The idea of adding to fish habitat space is often attractive. However, there is a growing body of evidence that where such structures are fished, they tend to deplete local fish stocks more than they add to them (Polovina, 1989a; Bohnsack, Chapter 11). Where they are not fished, but used to 'replace' or 'restore' coralline habitats, they may be shown to be not only ineffectual in the long term, but also prohibitively expensive to install over the tens to hundreds of km^2 that reefs generally occupy (McManus, 1995). Generally, it makes little sense to attempt to restore habitats unless the causes of their demise have been removed (Munro and Balgos, 1995; Bohnsack, Chapter 11). Removing these causes and preventing future damage may constitute a far better use of limited financial resources.

General guidelines for data collection

Following an intensive study of the economics of reef and adjacent fisheries in the central Philippines, Pomeroy (1989) offered several pointers for people conducting fishery research in such areas. These suggestions are summarized below:

1. Develop a close and trusting relationship with the study population.
2. Spend time explaining why the data are being collected, including their benefits and uses.
3. Ask local leaders to assist in introducing the researcher and research goals, and to provide insights into culture and practices.
4. Accompany the fishers and fish buyers as they do their work and assist them where possible.

5. Develop ways to measure the accuracy of the data being collected.
6. Provide regular feedback on the preliminary results to the study population.
7. Keep research methodology flexible in order to adapt to changes over time.
8. Coordinate multidisciplinary survey activities so that duplication can be reduced and so that data can be used to cross-check accuracy.
9. Involve and train local individuals as research assistants.
10. Discuss ongoing research with fishers in groups regularly so that they know you are not holding back or treating anyone differently.
11. Use a methodology that permits comparisons with similar studies done elsewhere.

These procedures are basically those followed in the study of McManus *et al.* (1992) and many others (e.g. Cruz, 1993; Buhat, 1994). The practice of accompanying fishers to their fishing grounds fits well with the suggestion by Munro (1986) that generally an efficient and reliable approach to gathering important fishery parameters would be to actually help 'fish' for samples. A failure to adopt a close enough working relationship with the target community can lead to severe difficulties in the success of the study or project (Stoffle *et al.*, 1991; Stoffle and Halmo, 1992; Tan, 1993).

In addition to the procedures listed above, it is becoming increasingly clear that the participation of the community in the gathering of data may not only help with community understanding of the project, but may also capitalize on local knowledge to improve the results of the project (Renard, 1991; Stoffle and Halmo, 1992).

Gender and child-labour issues are particularly important to emphasize in both sociological and economic analyses and strategy developments. Classical studies often emphasized the analysis of male elements of the workforce. The changes that ensued in the management innovation often created unforeseen hardships for women, who generally work longer hours and are more sensitive to developmental change than men. These hardships can cut deeply into the sociocultural system, resulting in programme failures and hostility toward future innovations. Further information on the role of women in fishing villages is available elsewhere (e.g. Acheson, 1981; Yater, 1982; Oosterhout, 1987; McManus, 1989; Tungpalan *et al.*, 1991; McManus *et al.*, 1992; Ruddle, Chapters 6 and 12), but much further research is needed.

Child labour in reef fisheries is even less well studied (but see Yater, 1982). In the Philippines, children are employed in all aspects of reef harvest, from building blasting devices and gathering daily food on the reef flats to participating as breath-holding net layers at depths exceeding

Table 10.1 Core and support experts constituting a hypothetical coastal area management team. After Scura *et al.* (1992)

Expert	Main responsibility (core expert) or specific concern (support expert)
Core expert	
1. Coastal zone planner	Team leader; provides planning direction; general land and water uses; institutional and organizational arrangements
2. Resource economist	Macroeconomic policies and their implications; economic valuation; benefit–cost analysis of management options
3. Ecologist	Resource assessment analysis of ecological impacts and potential changes
4. Sociologist	Social and cultural issues; community consultation and participation
5. Environmental engineer	Technical and physical interventions; carrying capacity of coastal areas
Support expert	
6. Parks/tourism specialist	Tourism-related issues, including marine parks and nature reserves
7. Pollution expert	Environmental impacts; waste management
8. Fishery expert	Fishery resource assessment; harvesting and utilization
9. Aquaculture expert	Farming system; aquaculture planning and management
10. Forestry expert	Coastal forest planning and management

30 m in commercial *muro-ami* operations. The employment of children has long-lasting implications for their self-perceptions, as well as having a direct effect on their education and subsequent job mobility. Alternative livelihoods should be planned such that the additional income to the fisher is enough to discourage the use of children for supplemental income.

Economic data collection and analysis

Options for economic analysis of small-scale fisheries and coastal zone management have been reviewed by Stevenson *et al.* (1982), Pomeroy (1992) and others. Pomeroy's broader approach is particularly relevant as there has been a growing realization that reef fisheries management must be approached as a component of coastal zone management (Table 10.1) (Hodgson and Dixon 1988; McManus *et al.*, 1992).

A classic study in cost–benefit analysis in reef fisheries is that of Hodgson and Dixon (1988). They determined that logging on a hillside in

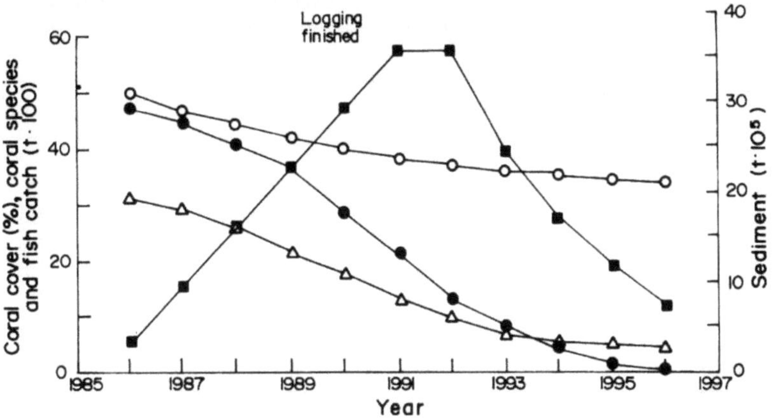

Fig. 10.3 Predicted changes in the number of coral species (filled circles), the abundance of corals (open circles) and the fish catch (triangles) if sedimentation from logging operations (filled squares) were to continue for a 10 year period at El Nido, Palawan, Philippines. After Hodgson and Dixon (1988).

Palawan, Philippines, was in direct conflict with the future use of a coralline bay for fishing and tourism. After an intensive study of sedimentation processes on the hillside and reefs, fishing patterns and productivity, and the local tourist industry, predictions were made regarding the future of the potential resource uses for the watershed system (Fig. 10.3). A conservative analysis of the potential benefits of continuing or discontinuing the logging, accompanied by convincing sensitivity analyses, showed clearly that the optimal strategy for the management of the bay and its watershed area was to eliminate logging (Table 10.2).

Sociological data gathering and analysis

As with economic analysis, the options for obtaining sociocultural data are vast, and they must be restricted to clearly defined objectives. However, it is usually helpful to supplement traditional surveys based on prepared questionnaires with investigative inquiries designed to uncover important new facts about the resource or its use patterns (Johannes, 1981, 1993). There are generally fishers who have considerable knowledge of such factors as fish migration or breeding patterns, or of potential problems in a proposed management scheme. Often the investigative team has no prior knowledge of such factors, and thus could not have included them in a prepared list of questions to ask. The methods for systematically

Table 10.2 Comparison of incomes predicted over a 10 year period with (option 1) and without (option 2) a ban on logging operations at El Nido, Palawan, Philippines. After Hodgson and Dixon (1988)

	Option 1	Option 2	Option 1 minus 2
Gross revenue			
Tourism	47 415	8 178	39 237
Fisheries	28 070	12 844	15 226
(with tuna)	(46 070)*	(21 471)	(24 599)
Logging	0	12 885	−12 885
Total	75 485	39 907	41 578
Present value (10%)			
Tourism	25 481	6 280	19 201
Fisheries	17 248	9 108	8 140
(with tuna)	(27 308)*	(15 125)	(13 183
Logging	0	9 769	−9 769
Total	42 729	25 157	17 572
Present value (15%)			
Tourism	19 511	5 591	13 920
Fisheries	14 088	7 895	6 193
(with tuna)	(23 122)*	(13 083)	(10 039)
Logging	0	8 639	−8 639
Total	(33 599	22 125	11 474

*Tuna revenues (in parentheses) are not used to calculate totals.

investigating traditional knowledge are discussed in Johannes (1981, 1993).

An excellent list of considerations to choose from for sociological surveys is included in an appendix in Stevenson *et al.* (1982). A more recent book describing methods for acquiring sociological data in coastal communities is that of Townsley (1993). Other helpful guidelines can be found in Pollnac (1988, in press).

An excellent analysis of perceptions and preferences in a fishing populace may be found in Johnson *et al.* (1989). This study focused on target species selection among recreational fishers, principally from Florida. Ordination and clustering methods were used to determine patterns in the perceived relationships among preferred target fish. Explanatory analysis took the form of belief-frame analyses, wherein series of perceptions were linked one to the other as appropriate (Fig. 10.4). The logic behind the belief-frame results was further analysed through entailment analysis, which involved a multivariate contingency analysis of paired dichotomous variables. The results of the analyses were used in producing educational material aimed at increasing the use of

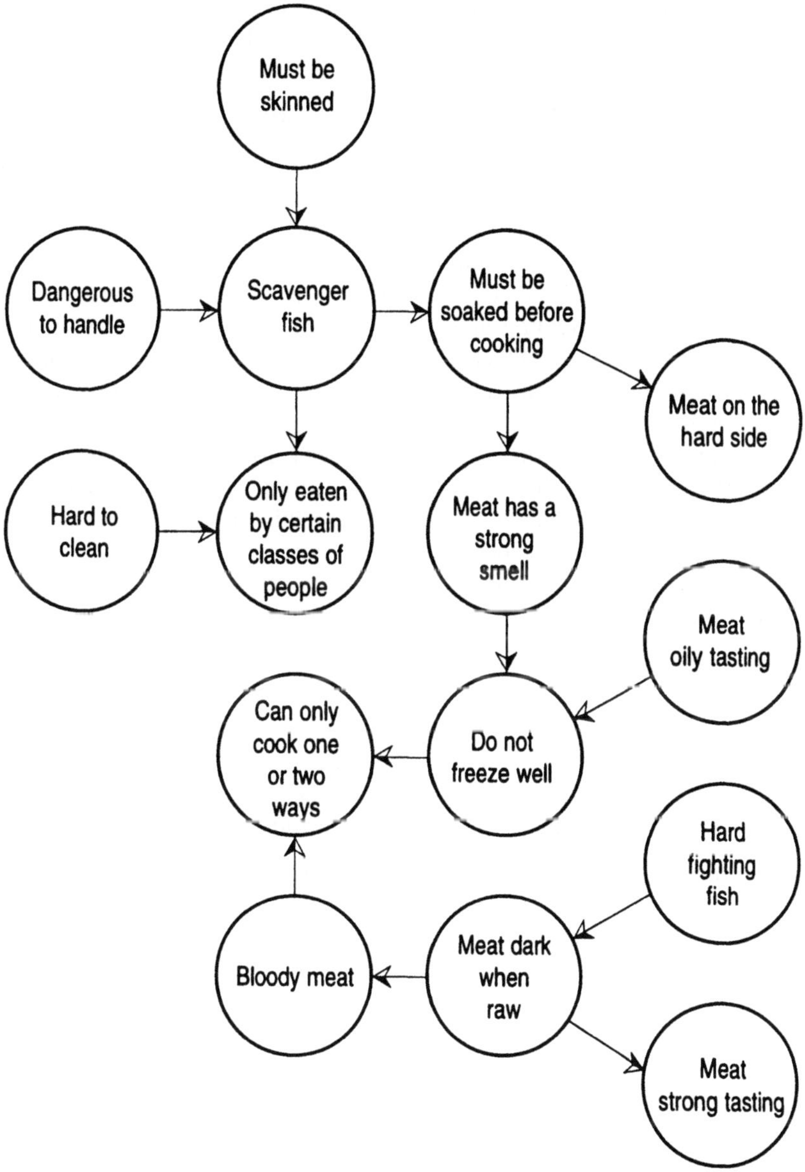

Fig. 10.4 Belief-frame analysis showing the logic behind various perceptions leading to fish preferences among Florida sport fishermen. After Johnson *et al.* (1989).

non-traditional species among recreational fishers. A similar approach would be useful for broadening the range of organisms harvested from some coral reefs, to reduce pressure on over-harvested target species.

Management implementation

There has been a growing realization that effective coastal management requires local participation (White, 1979, 1981, 1986; Johannes, 1980, 1981; Cabanban and White, 1981; Castañeda and Miclat, 1981; Gawel, 1981; De Silva, 1985; Savina and White, 1986, White *et al.*, 1986; White and Savina, 1987; Alcala, 1988; McManus *et al.*, 1988; Ferrer, 1989; Christie *et al.*, 1990; Hviding and Ruddle, 1991; Polunin, 1991; Pomeroy, 1991, in press; Renard, 1991). Community-based management promotes equity, is economically and technically efficient, is adaptive and responsive to social and environmental variations, and engenders management stability through local commitment (Renard, 1991). In particular, community management facilitates the development of attitudes such that undesirable practices become not merely illegal, but socially unacceptable as well. Community management efforts must often start with the strengthening of community-based organizations. Five elements are generally important in community development (Renard, 1991):

1. research and documentation of popular resource use and management systems;
2. definition and provision of legal instruments;
3. use of a participatory planning approach;
4. definition of clear management agreements;
5. building and developing community institutions.

The strengthening of the capabilities of the community to control and efficiently use its resources is a process of community development. Much of the emphasis in such programmes is on modifying attitudes, and some key techniques and theory have arisen from the field of social psychology. For example, a common and powerful method to overcome people's natural tendency to avoid facing important decisions (avoidance of cognitive dissonance; Festinger, 1972) is the application of directed group dynamics, wherein peer pressure is exerted to uncover underlying attitudes and permit modification to strengthen group coherence (Asch, 1972). In practical terms, this usually involves community meetings in which a well-trusted facilitator helps direct discussions toward particular ends. The 'well-trusted' requirement is often a difficult qualification to achieve, and may require many months of involvement in the community. It also generally requires a considerable knowledge of the society, including the traditional management aspects of its history. Finally, the facilitator must maintain credibility by being well informed of ecological and economic aspects of the system.

The rising emphasis on community, or bottom-up, approaches to reef resource management represents a considerable departure from previous management approaches which often concentrated on the development of plans and laws by governments and consultant teams (top-down). The more useful of the earlier approaches generally involved polls of community opinions, occasional fora for voicing concerns and sometimes efforts to accumulate and integrate traditional knowledge (e.g. Kenchington and Hudson, 1984). At least one early management handbook (Dahl, 1982) advocated community involvement in reef monitoring.

Ultimately, attempts to promote management entirely from a bottom-up approach will encounter problems, in that neither the social, economic nor ecological systems of which a reef forms a component are ever entirely closed to outside factors. Even early efforts at small-scale management on small, isolated islands often owed their successes in part to national-level reinforcement in terms of training and information resources, as well as in the enactment of supportive legislation such as fishery administrative orders (Casteñeda and Miclat, 1981). The limitation of qualified investigators and community developers in developing countries requires that efficient use be made of them. Also, many relevant ecological and resource use patterns tend to operate on large scales, and some coordination is necessary among small management units to avoid conflicting or confusing policies from developing along a given coast. It was in this light that proposals arose for developing national systems of community development specialists cross-trained in ecological management (McManus, 1988; McManus *et al.*, 1988). The necessary balance among national, provincial and local management inputs in a management scheme will tend to vary on a case-to-case basis (Wells, 1993).

In the Philippines, a national system of similar orientation was developed in the late 1980s, primarily for forest management. The Community Environment and Natural Resource Officers (CENROs) work in municipalities to encourage community development and proper resource use. The system has had credibility problems in some areas, as might be expected in attempts to manage resources with a long history of lucrative, illegal exploitation. However, the programme has had notable successes, and efforts are under way to extend the system to include coastal management.

Sri Lanka is currently experimenting with a combination of bottom-up and top-down approaches in coastal zone management (Lowry and Sadacharan, 1993). A national programme in Thailand focused efforts on relatively uncontroversial coral reef protection issues to build local and then national support for a broader programme of coastal zone management (Lemay *et al.*, 1991; Wells, 1993). A series of pilot projects funded under the ASEAN–US Coastal Resources Management Program in Thailand, Malaysia, Singapore, Indonesia, Brunei and the Philippines developed coastal management plans with varying degrees of community, provincial and national involvement (Scura *et al.*, 1992).

10.5 SPECIFIC CASE HISTORIES

Included below are brief summaries of cases illustrating sociocultural and economic factors in reef fisheries management. The first concerns an economic analysis of a reef fishery in Papua New Guinea by Lock (1986d), illustrating the role of opportunity cost in bioeconomic models. The next is an example of community development leading to village-level management of a marine reserve by Christie *et al.* (1990) and Buhat (1994). The final example illustrates the interplay of ecological, economic, social and political dimensions of commercial-level destructive reef fishing. Further examples can be found in a recent book edited by White *et al.* (1994).

Economics of reef fisheries in the Port Moresby area, Papua New Guinea

The reef fishery of Port Moresby is based primarily in four mainland villages and one island village, all within 20 km of the city (Fig. 10.5). Fishing is conducted from powered fibreglass dinghies (6 m) and outrigger or double-hulled canoes (8–10 m), or from single outrigger canoes (to 8 m) which are poled or sailed. Daytime fishing occurs from 0900 to 1500 h, allowing time to sell the catch at the end of the day in Port Moresby. Fish caught at night are sold in the morning or held on ice for the afternoon market. Approximately 15% of the catch (2 kg per family per day) is used for home consumption, and another 5% is given to school teachers, the elderly, or relatives. The remaining 80% is sold. An analysis of the labour patterns and profits indicated that most fishers earned close to the government's rural minimum wage. Excess labour influx was not a major problem, and this fishery was not in a state of Malthusian overfishing. A cost analysis is shown in Table 10.3.

The relationship between catch per unit effort and fishing intensity is shown in Fig. 10.6. It is clear that the resource availability per hour of fishing declined rapidly as fishing intensity increased in this relatively uncrowded fishery. A simple yield–effort curve based on the same regression indicates that the fishing grounds range from lightly fished to slightly overfished (Fig. 10.7). Converting this to a fixed-price curve, and adding cost lines (Fig. 10.8) shows different bionomic equilibria depending on whether costs exclude opportunity costs (TCI), include current opportunity

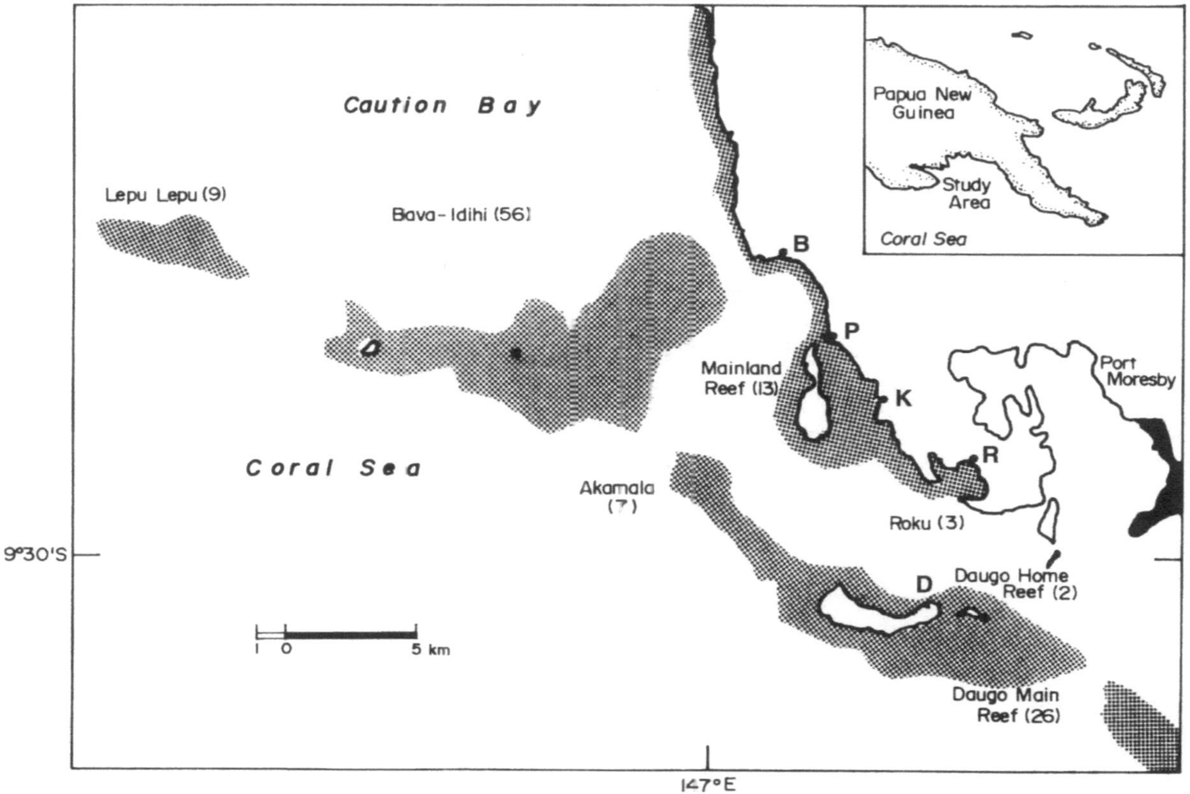

Fig. 10.5 Map of fishing groups (stippling) and villages west of Port Moresby, Papua New Guinea. Fishing villages: B, Boera; P, Porebada; K, Kouderika; R, Roku; D, Daugo; area of each fishing ground in km^2 in parentheses. After Lock (1986d).

Table 10.3 Evaluation of costs by vessel type for reef fisheries near Port Moresby, Papua New Guinea. Costs in kina, where one kina = US$0.8 approximately. After Lock (1986d)

	Daugo net	Daugo troll	Roku	Mainland engine	Mainland sail
Fixed costs					
Vessel*	645	645	60	60	40
Engine†	615	615	615	615	–
Total fixed cost	1260	1260	675	675	40
Variable costs					
Engine repair	200	200	200	200	–
Nets‡	300	–	500	300	200
Troll lines	–	100	–	–	–
Fuel					
Daily	10	15	10	10	–
Annual	1740	2445	1400	1400	–
Market charges					
Daily	8	4	6.4	6.4	3.2
Annual	1392	652	896	896	448
Fish seller's costs					
Hourly rate	0.64	0.64	0.64	0.64	0.64
Hours per annum	1044	489	840	840	420
Annual cost	668	313	538	538	269
Total variable cost	4300	3710	3354	3334	917
Total annual cost	5560	4970	4209	4009	957

*Capital cost discounted over 5 years.
†Capital cost discounted over 2 years.
‡2 year life.

costs (TC), or include opportunity costs as they would be if the alternative livelihoods were to be available only at the rural minimum wage rate (TCA). The opportunity cost here refers to the potential additional income that a fisher gives up by spending time fishing rather than working in some other, more lucrative occupation. Under conditions of high unemployment and no welfare, opportunity costs become negligible. In this case, the equilibrium is near the maximum sustainable yield (MSY) point.

As is often the case, the optimal net income, the maximum economic yield (MEY), could be achieved by maintaining each fishing ground at an effort level of approximately 40% of the bionomic equilibrium (a 60% reduction). This would mean decreasing the effort in four fishing grounds (Roku, Daugo home reef, Mainland Reef and Daugo main reef), and increasing effort in three others (Lepu-Lepu, Bava-Idihi and Akamala), as can be seen by comparing the positions of the grounds on Fig. 10.7 with the MEY_1 on Fig. 10.8. However, it is clear that for the individual fishers of the lightly fished grounds, it would be better to keep fishing pressure

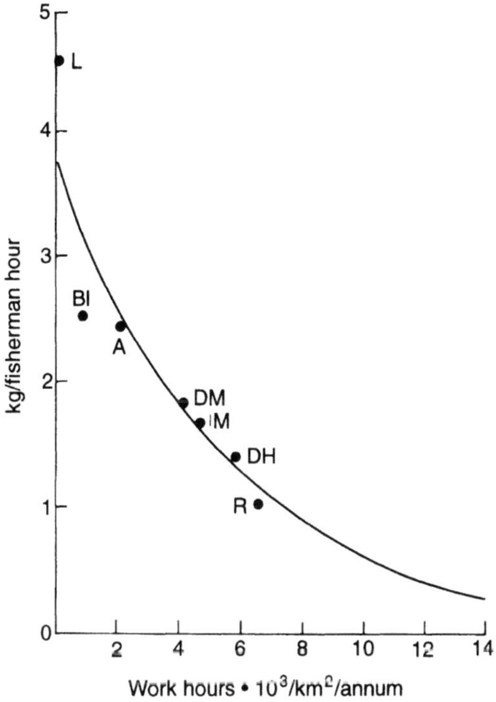

Fig. 10.6 Relationship between catch per unit effort and total effort on a set of coral reefs in Papua New Guinea. Log–linear regression, $r^2 = 0.92$. L, Lepu-Lepu; BI, Bava-Idihi; A, Akamala; DM, Daugo Home Reef; R, Roku. After Lock (1986d).

low. If the objective is to approach equity of living standards, then an analysis would have to be done of the advantages and disadvantages of being located closer to Port Moresby, and these would have to be weighed against the disadvantages of fishing more crowded grounds. Finally, such analysis would require weighting by fisher family perceptions of standard of living, and how these are likely to change in the near future.

It is clear that if alternative livelihoods were to become restricted to rural minimum wage jobs, and a labour excess were to move into fishing, the effort would increase until virtually no net profit would be possible at the bionomic equilibrium point where the TCA line intersects the profit curve (Fig. 10.8). This point would be far to the right of MSY and the fishery would be overfished. Under those conditions, the new MEY would be at roughly 40% of the new bionomic equilibrium, a point closer than the old MEY to MSY but not beyond. Restricting effort to this new MEY would be efficient in an economic sense, but would place more stress on

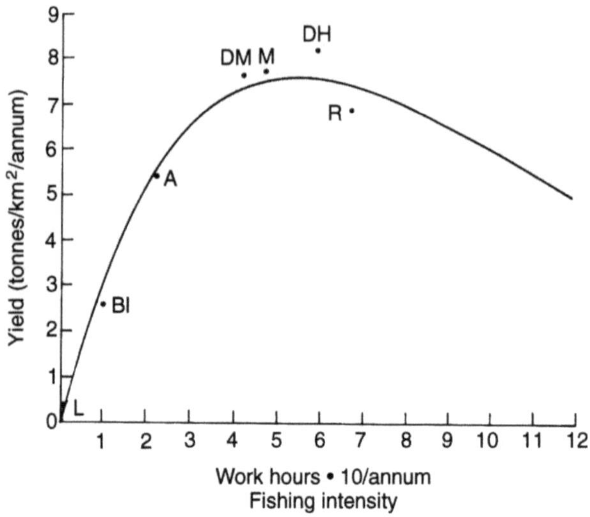

Fig. 10.7 Yield–effort curve for reefs near Port Moresby (symbols as in Fig. 10.6). After Lock (1986d).

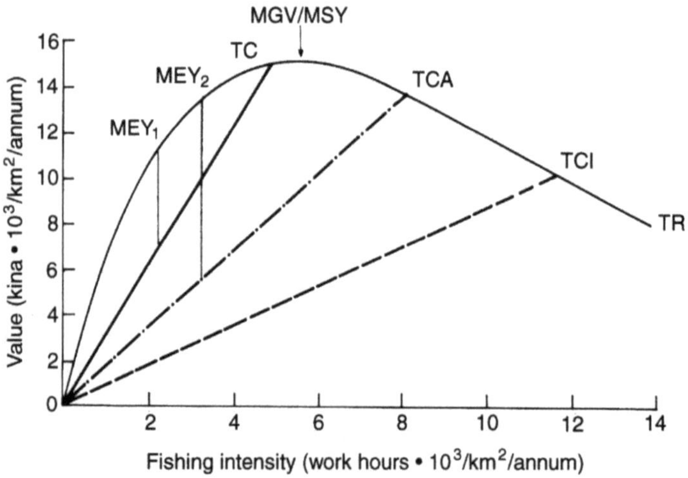

Fig. 10.8 Fixed-price model for the Port Moresby reef fisheries. Equilibria are shown for cost lines with opportunity costs omitted (TCI), included as is (TC), and included if equal to rural minimum wage levels (TCA). TR, total revenue; MEY_1 and MEY_2, maximum economic yield using the TC and TCA cost lines respectively; MGV, maximum gross value; MSY, maximum sustainable yield. One kina = US$ 0.8 approximately. After Lock (1986d) with graphically determined MEYs added.

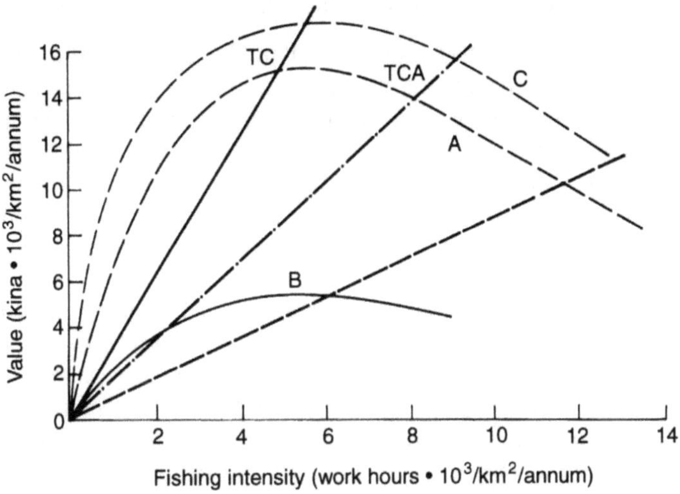

Fig. 10.9 Bioeconomic model showing the effects of a downward shift in fish prices (A to B), and an upward shift in prices (A to C). Symbols as in Fig. 10.8.

the ecosystem than forcing the system to the old MEY. The best approach would be to ensure that sufficient alternative livelihoods are always available at the higher, urban minimum wage level. In fisheries that are heavily overfished, the general objective is usually to move from an equilibrium similar to TCA to one such as TC through the generation of an improved job market outside the fishery.

A final analysis shows the effect of reducing the market prices of fish on the fishery (Fig. 10.9). At a selling price of 0.7 PNG K per kg, the fishery would no longer be viable, unless there was a concurrent drop in availability of alternative livelihoods at the higher urban minimum wage rate. An alternative scenario would be that economic difficulties lead to a decline in available food causing the price of fish to rise, as well as a decline in the job market. Both conditions could easily occur under a general state of economic decline in Papua New Guinea. In this case, we would have an upward shift in the value curve and a downward shift in the cost line, leading to a bionomic equilibrium at a higher rate of fishing intensity. This equilibrium point would then be readily filled by the unemployed, whose numbers also tend to increase during times of economic hardship. The reefs would then suffer under a high rate of fishing intensity, as well as an increased tendency toward the use of destructive gear by more desperate fishers. This clearly illustrates the direct relationship between the sustainability of harvests on ecosystems such as reefs, and the economic status of a country.

Community development in the north-western Philippines

A US Peace Corps volunteer spent a year in the small, coastal community of San Salvador Island off the Masinloc, Zambales assessing community needs. Substratum surveys on the fringing reef indicated the need for better management of the reef resources. A project was developed with the Haribon Society, a conservation organization based in Manila. Funds were obtained from the Netherlands Embassy and the Jaime Ongpin Foundation of the Philippines. The project included funds for the hiring of an experienced Filipino community organizer.

The volunteer and community organizer served as facilitators for community development meetings, and as links to national-level scientific resources. Various groups were brought in to conduct surveys and seminars. A field trip was arranged for community members to visit the successful Apo Island Marine Reserve in the south-central Philippines. During two general assembly meetings, the community drafted a resolution for the establishment of a 127 ha no-fishing sanctuary (Fig. 10.10). New regulations concerning a larger, traditional fishing reserve surrounding the island included the banning of illegal and unsound fishing practices. The community banned all collecting of aquarium fish, regardless of collecting method, because of the historical use of sodium cyanide by some collectors. The community organization presented the resolution to the Masinloc Mayor and Municipal Council. A municipal ordinance legalized the sanctuary and reserve as of July 1989. A later resolution led to the banning of a beach seine method involving scare lines with coconut leaves, because the average fish caught was only 15 cm and the method was damaging to corals. The latter resolution caused some dissent among 18 affected fishers. A Fisheries Administration Order was prepared by the national government to reinforce the municipal ordinances.

The sanctuary was marked with buoys and signs in the national language. Enforcement of the regulations has involved both local residents and the municipal government. A motor launch was loaned to the community by the Mayor of Masinloc to facilitate enforcement. Thirty-five of 39 violations in the first 8 months were by non-resident fishers, including some from the south-central Philippines. All those arrested were warned and released. On three occasions, the ordinance was not enforced immediately, but the violators were from the island and were warned by the *barangay* (village) council. They subsequently complied with the regulations. On another occasion, a violator was suspected of carrying a gun and was not detained. Incidence of blast and cyanide fishing dropped dramatically. Blast fishing decreased from frequencies of up to 3.2 blasts per day to virtually none. A sole violator of the blasting restriction was driven off before he could collect any fish.

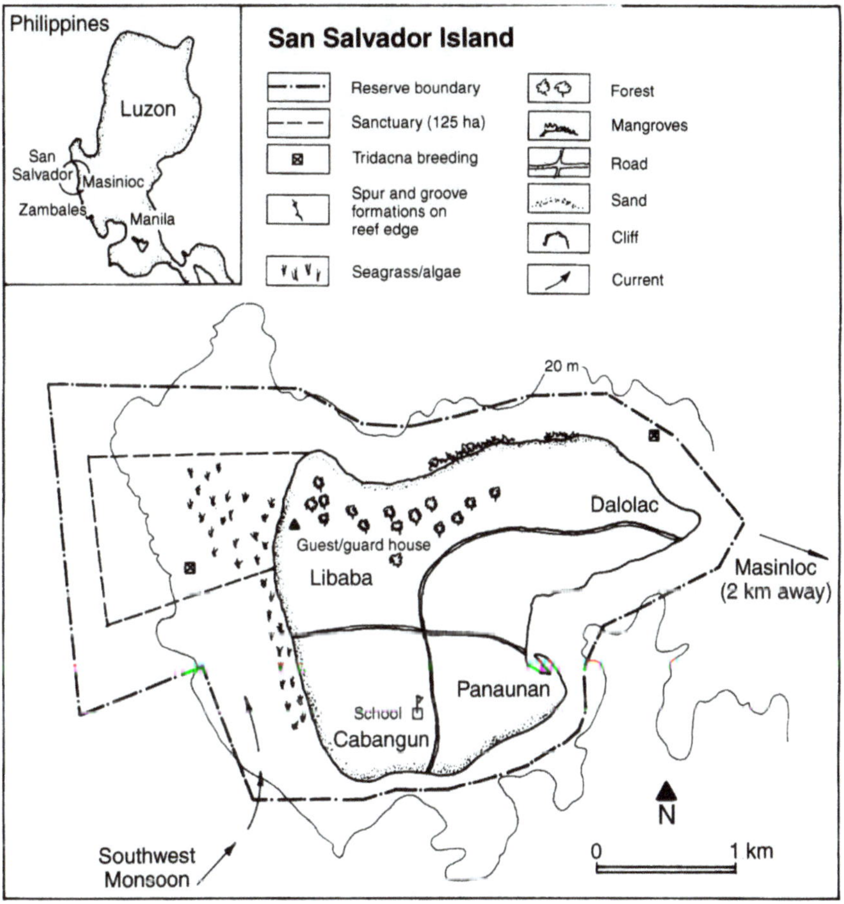

Fig. 10.10 Map of San Salvador Island, Zambales, Philippines, showing the sanctuary and reserve implemented by the community. After Christie *et al.* (1990).

Fish censuses at the island indicated substantial increases, and fishers reported greater catches. Giant clams were introduced and maintained by two volunteer caretakers so as to reseed the area. On land, community volunteers planted 2500 seedlings, of which 60% survived one to two severe dry seasons and grazing to heights of up to 5 m.

The programme at San Salvador Island did not proceed without some difficulties. The major difficulties and potential solutions are presented in Table 10.4. This case study indicates that with the proper approach, an outside facilitator can sometimes overcome a myriad of difficulties, and become a catalyst for improving the management of resources at the

Table 10.4 Issues and options for solution in the development of a community-based fishery management scheme in the Philippines. MCPSS, Marine Conservation Project for San Salvador; LTK, *Lupong Tagapangasiwa ng Kapaligiran* or Environmental Management Committee. After Christie *et al.* (1990)

Issue	Example	Potential solutions
Alienation of people adversely affected by MCPSS	Alienation of San Salvador from MCPSS whose controversial fish methods (e.g. aquarium fish gathering and '*Kunay*') had been banned	Implementation of barrier net training and alternative income project prior to passage of MCPSS ordinance. Compromise between opposing parties
	Resistance to MCPSS by corrupt government officials supporting illegal fishing	Gain support of enthusiastic higher officials
	Feeling by some San Salvador residents that MCPSS is an economic burden	Continue relevant education feedback of environmental survey results. Initiate alternative income programme affecting more people to show economic benefits from management
Lack of coordination between community leaders for MCPSS implementation	MCPSS ordinance not always enforced	Formalize LTK/*barangay* council joint enforcement body, deputize officials Supply enforcement equipment such as boat and signs
	Jealousy between villages on San Salvador as to sanctuary location	Show how sanctuary contributes to whole island ecosystem Establish small sanctuary for each village
	Struggle between LTK and *barangay* council on control and credit for MCPSS projects	Integrate two bodies for MCPSS implementation
Weakness of MCPSS management body (LTK)	Some elected LTK members lack dedication to MCPSS	Conduct more field trips to other such project sites Extend formal MCPSS for 1 year. Replace inactive LTK members through re-election
	MCPSS funds are occasionally used for personal needs	Monitor fund use Identify reliable and trusted community members

village level (Cabanban and White, 1981; Castañeda and Miclat, 1981; White and Savina, 1987; and contributors in Cruz, 1993).

Commercial-level destructive *muro-ami* fishing

Muro-ami fishing is a form of drive-in-net fishery in which swimmers use weights on vertical scare lines to drive fish into a net (Ruddle, Chapter 6). The technique is destructive to corals (Jennings and Lock, Chapter 8). The municipal version of the method is used in small-scale fishing operations on Philippine reefs involving a few tens of divers at a time (Carpenter and Alcala, 1977). The commercial *muro-ami* operation differs particularly in scale, as it involves large ships and hundreds of divers operating principally on offshore reefs in the Philippines and the South China Sea.

The commercial operation generally begins soon after a pinnacle or platform reef has been located by the ship's sonar. A group of breath-holding divers anchor a large net with a 36 m wide bag flanked by wings of approximately 100 m (Fig. 10.11; Corpuz *et al.*, 1983a). In some cases, the net may be laid in waters exceeding 30 m in depth, and the divers anchoring the net must often exit through very narrow openings, risking drowning if they are caught in the netting.

The ship circles from one net wing to the other, dropping off swimmers in a large convex arc. Each swimmer holds a float wrapped with scare line. The scare line is extended to the bottom, releasing white plastic strips tied at intervals which wave in the current. Most swimmers use simple goggles made of wood and glass, and use one or two oval paddles as foot fins. The swimmers converge on the net, diminishing the size of the arc and driving fish into the net. When the net is full, divers free it from the coral as it is pulled in.

The *muro-ami* issue is as much a sociocultural and economic problem as it is an ecological one. Until recently, many of the participants were under the age of 18 (Oosterhoot, 1987). The crews were recruited from the central Philippines and spent periods of time on a remote island north of Palawan in a fishing camp. The cemetery near the camp was extensive and said to have included boys killed in knife fights. At least one ship is known to have burned and sunk with considerable loss of lives (McManus, unpublished information). Living conditions on the ships are very cramped, and involve primarily temporary nipa and bamboo structures packed onto the upper deck, around which food is cooked on open fires. On one ship which I visited in 1980, fishers complained that they were forbidden to eat the fish they caught, and had to survive primarily on corn meal and other supplies loaded at the beginning of voyages lasting several weeks at a time. A later study indicated that the arrangements for food during the voyage varied among cruises (BWAM, 1986).

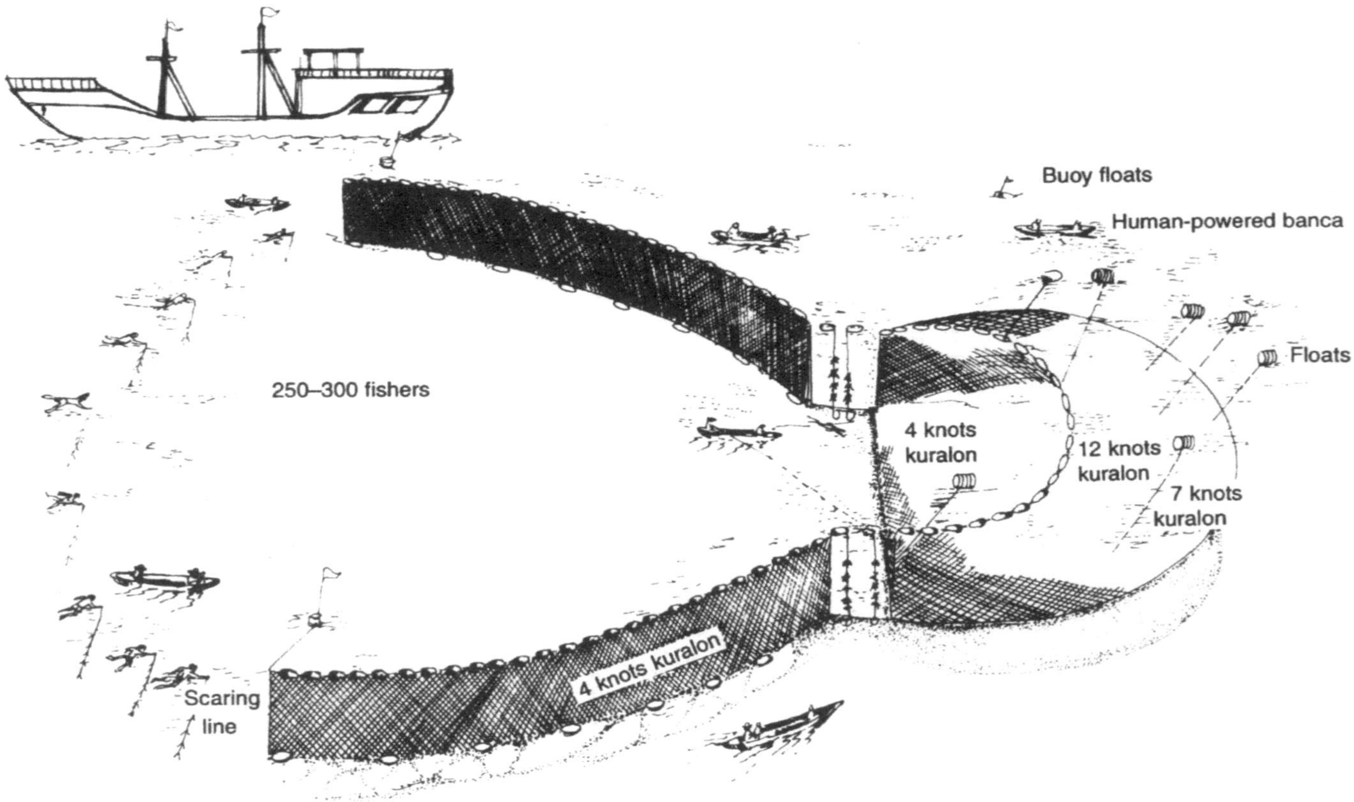

Fig. 10.11 Sketch of the commercial *muro-ami* fishing operation. After Corpuz *et al.* (1983a).

The *muro-ami* fleet is capitalized by the Frabal Fishing Corporation, a large company involved in many aspects of commercial fishing. A second company, the Abines Muro-ami Fishing Corporation, outfits the cruises and recruits the fishers. In 1982, there were 26 crews hired, including approximately 7000 fishers in groups of 200–300 (Corpuz *et al.*, 1983a). The Frabal Corporation received 70% of the total gross income. The Abines Corporation got 10% and the fishers got 20% (although this was reported as 5% and 25% respectively by a later study; BWAM, 1986). The costs of the trip came out of the latter two shares (Oosterhout, 1987). Fishers often received as little as US$ 10–30 per month. However, the areas from which they were recruited were heavily overfished and overfarmed, and more than half expressed considerable job satisfaction (BWAM, 1986). Many of the families were indebted to the Abines Corporation for borrowings during difficult times, and some who left the industry rejoined after one or two seasons, partly to avail themselves of the credit line offered.

In 1982, the Philippine Bureau of Fisheries released the results of a study indicating that anchor chain links and metal spindles caused significantly less damage to corals than the rocks in common use. The fishers were advised (with limited success) to avoid using rocks (Corpuz *et al.*, 1986a). During the mid 1980s, two internationally broadcast television specials and outcries by diving groups forced a re-evaluation of the industry. A labour investigation (BWAM, 1986) led to a reduction in the use of child labour. Recommendations from several quarters called for the total banning of the fishery. However, the economic hardship issue kept the practice alive, despite the negative ecological and sociocultural aspects. A few years later, the Bureau of Fisheries announced the successful development of a technique, *pa-aling*, wherein the scare lines were replaced with air hoses (Corpuz *et al.*, 1983b). Some of the vessels consequently adopted the new method. As this goes to press, however, a new outcry has arisen that *pa-aling* is too efficient at depauperating reefs of fish, and that it too should be banned.

The following quotes illustrate the interrelationships between *muro-ami* fishing, tourism, human rights and political dynamics as perceived by key government officials:

> It has come to my attention that this highly destructive and exploitative form of fishing is being conducted in areas of high touristic value. Kindly note that such a practice turns coral gardens into submarine deserts. If the destruction of our natural marine attractions are allowed to continue unchecked, I believe that in the long run, the viability of tourism infrastructure in nearby areas will definitely be adversely affected. – Jose A.U. Gonzalez, Minister of Tourism, Philippines, 22 July, 1986. Letter to Philippine President Corazon C. Aquino.

> It has been established in the attached position paper that 'Muro-ami' is not only a highly destructive form of fishing that causes considerable damage to our coral reefs, but also a gross and serious violation of our labour laws. This method of fishing, which utilises several thousands of young Filipino boys, aged 7 to 15, as the bulk of its labour force, has been allowed to prosper by the past administration, who were either compensated or coerced into ignoring the enforcement of the law. – General Fidel V. Ramos, Chief of Staff, Armed Forces Philippines (subsequently President of the Philippines), May 29 1986. Letter to Philippine President Corazon C. Aquino.

10.6 CONCLUSION AND OUTLOOK

During the last decade, the search for viable solutions to reef fishery problems has widened the scope of the field. Studies have increasingly emphasized the broad range of products being harvested from reefs, ranging from fish and other vertebrates to molluscs, echinoderms, sponges and seaweeds. This range of organisms is expanding as harvest pressure increases in developing countries, particularly those with high human population growth rates. The number of harvestable species is expected to grow even more rapidly as new food and drug products from the reef are identified or developed and as managers seek to lower specific harvest impacts by encouraging a broadening of the range of target organisms. The ecological, economic and sociocultural aspects of the harvesting of different groups of reef organisms are tightly linked to each other. It is often not productive to concentrate analytical efforts on any particular group, such as finfish. The field of reef fisheries is rapidly becoming subsumed within a broader field of resource ecology.

Concurrent with the increased variety of target organisms is a broadening of the scope of science that is being applied to find effective management strategies. The failures of the past have increasingly emphasized that management is more of a social science than an exercise in finding useful ecological options. Many resource economists have recently gone through similar shifts in perspective, with the increasing realization that management involves the modification of human behaviour, and that the system of capital dynamics is only one of several important factors. As population pressures rise, reef management increases in complexity. Effective management in the future will arise primarily from the collaborative efforts of practitioners in a variety of social and natural sciences, each aware of the basics of the others' specialities, and each contributing to assist human communities dependent on reefs to ensure the sustainability of their resources.

ACKNOWLEDGEMENTS

I would like to thank Ms Janet Poot for gathering materials on *muro-ami* fishing. I am indebted to Angel C. Alcala, Elmer Ferrer, Lynne Hale, Liana McManus, Joseph Padilla, Daniel Pauly, Robert Pomeroy, Garry Russ, Geronimo Silvestre, Sue Wells, Alan White and others for particularly helpful discussions. Many people made helpful comments on an early manuscript, including Brian Crawford, Ken Ruddle, Richard Pollnac, Frank Talbot, Nick Polunin, Callum Roberts and an anonymous reviewer. The first draft of this paper was completed while the author was an adjunct faculty member of the University of the Philippines Marine Science Institute (UPMSI). This is ICLARM publication number 1106 and UPMSI publication number 240.

Chapter eleven

Maintenance and recovery of reef fishery productivity

James A. Bohnsack

SUMMARY

Non-conventional methods to maintain and restore reef fishery productivity include protecting fishery habitats, hatchery releases, artificial reefs, introduction of exotic species, habitat restoration and marine fishery reserves. I conclude that it is far better to prevent overfishing and stock collapse in the first place than to have to rebuild fishery productivity later. The most important strategies to prevent loss of fishery productivity are switching to less destructive fishing methods, preventing destruction of fishery habitats and protecting some areas by establishment of marine fishery reserves. If fisheries must be rebuilt, habitat restoration and marine reserves appear to be the more promising alternatives over the long term. Except for unique circumstances, deployment of artificial reefs and release of hatchery-raised organisms have less potential for retrieving lost fishery productivity. Because of unpredictable consequences and the general inability to correct mistakes, the introduction of exotic organisms is the least favoured alternative for rebuilding fishery productivity.

11.1 INTRODUCTION

Conventional approaches to maintaining or rebuilding fishery productivity typically attempt to maintain fisheries by allowing a sufficient number of

Reef Fisheries. Edited by Nicholas V.C. Polunin and Callum M. Roberts.
Published in 1996 by Chapman & Hall, London. ISBN 0 412 60110 9.

fishes to escape capture; this is achieved either by decreasing fishing effort or by increasing the size at first capture (Appeldoorn, Chapter 9). Conventional management actions include size limits, quotas, bag limits, gear restrictions, closed seasons and limited entry. Unfortunately these often fail to maintain sustainable harvests and protect resources for various economic, social and biological reasons (Ludwig *et al.*, 1993; Rosenberg *et al.*, 1993). The result has been a decline or collapse of many fisheries (Plan Development Team, 1990; Butler *et al.*, 1993).

Tropical fisheries exist typically in poor countries. They are especially prone to failure because of inadequate biological and fisheries data, poor compliance with regulations (Adams, Chapter 13) and inadequate fishery models. Sufficient information is rarely available to evaluate the status of stocks and allow action to maintain sustainable high productivity (Appeldoorn, Chapter 9).

Below I examine what are, in a sense, non-conventional methods for recovering productivity. These include switching to less environmentally damaging or wasteful fishery practices (Jennings and Lock, Chapter 8), hatchery restocking, introductions of exotic organisms, habitat restoration, habitat alteration using artificial reefs, and use of areas permanently closed to fishing. The two manipulative approaches to enhancing populations, hatchery releases and altering habitats, have probably attracted most public and management interest. Although emphasis is on tropical fisheries, examples from other areas are used when appropriate.

11.2 CHANGE TO LESS DESTRUCTIVE FISHERY PRACTICES

Perhaps the most important first step in rebuilding depleted fisheries is to reduce or replace destructive and wasteful fishery practices. Practices particularly destructive to habitat include use of dynamite, *muro-ami* fishing, bottom-set nets, and chemicals such as bleach and cyanide (Öhman *et al.*, 1993; Ruddle, Chapter 6; Jennings and Lock, Chapter 8). In the southern US Atlantic region, roller trawls were prohibited to prevent habitat damage to sensitive reef areas. In the same region, efforts are being made to eliminate collecting live rock (reef material with attached organisms) because of habitat concerns. Fish traps have also been eliminated there because of concerns about excessive waste through mortality of juveniles, adults, and non-targeted species from both lost traps and directed fishing efforts.

Although replacing destructive fishing practices with more environmentally sensitive methods is common sense, actual implementation can be difficult because of poor education, lack of enforcement, strong cultural traditions, and political and economic factors (Ludwig *et al.*,

1993; Rosenberg *et al.*, 1993; McManus, Chapter 10). Although beyond the scope of this chapter, it is important to recognize that non-fishing activities can also destroy fisheries productivity and can potentially also be managed. For example, poor land use practices can harm fisheries by damaging and destroying habitat through pollution and by release of sediments, nutrients and sewage.

11.3 ARTIFICIAL POPULATION ENHANCEMENT (HATCHERIES)

Hatchery-raised marine species have been released extensively for well over a century. Richards and Edwards (1986) reported that almost 3.5 billion fry of various species were released in the New England area alone in 1926. Using hatcheries to restock or supplement wild populations is popular, especially with the general public, fishing interests, politicians and hatchery proponents (Martin *et al.*, 1992). However, despite a long history of releasing hatchery-reared organisms in the wild, the idea remains controversial in the scientific community (Richards and Edwards, 1986; Maccall, 1989; Martin *et al.*, 1992).

Most hatchery release programmes have been directed at temperate freshwater and anadromous species such as shad and salmon. Marine species released in the US have included cod, *Gadus morhua*, Spanish mackerel, *Scomberomorus maculatus*, pollack, *Pollachius virens*, Atlantic mackerel, *Scomber scombrus*, haddock, *Melanogrammus aeglefinus*, winter flounder, *Pseudopleuronectes americanus*, California halibut, *Paralichthys californicus*, red drum, *Sciaenops ocellatus*, and white sea bass, *Atractoscion nobilis*. Similar stocking attempts have been conducted in Europe for cod, plaice, *Pleuronectes platessa*, turbot, *Scophthalmus maximus*, and lemon sole, *Macrostomus kitt*, and in Japan for sea bream, *Chrysophrys major*. Surprisingly few hatchery release attempts have been directed at tropical species (Richards and Edwards, 1986). For most of them, there is little evidence of success (Richards and Edwards, 1986; Maccall, 1989; Martin *et al.*, 1992). The few marine attempts that have been at least partially successful have occurred in estuaries. No evidence of success exists for open water species. Below I briefly review potential benefits and obstacles to hatchery releases and review case histories involving tropical and subtropical species.

Potential uses of hatcheries

There are several reasons for releasing hatchery-reared organisms into the wild. Release of hatchery-reared organisms could:

- maintain the abundance of wild populations;
- restore depleted or endangered populations;
- increase the yields of wild stocks (Maccall, 1989);
- improve public support and funding for research;
- increase scientific knowledge of the biology and ecology of marine organisms;
- provide a mechanism to test environmental and other scientific hypotheses (Richards and Edwards, 1986);
- increase fishery economic efficiency;
- improve the genetics of wild stocks;
- introduce new productive species.

The two ostensible uses of hatchery stocking are to rebuild populations of depleted or endangered species and to supplement or increase the yield of wild populations. Hatchery releases could also be used to speed recovery of populations damaged by calamities such as drought, storms, floods, toxic chemical releases or other phenomena. It has been suggested that hatchery breeding programmes could be used to improve the genetics of wild populations for production, although little evidence exists to support that view. Examples of breeding success within aquaculture operations are not necessarily relevant to wild populations.

Hatchery studies have clearly increased our understanding of the basic biology and life history of many marine organisms (Richards and Edwards, 1986). For example, descriptions of species' early life history and basic biology of many organisms have been accomplished directly as the result of raising wild-caught eggs or larvae in captivity.

Potential obstacles to hatchery releases

Beyond the many technical difficulties of culturing marine organisms, any programme to supplement wild stocks must address major concerns. Problems include low survival of hatchery releases in the wild and damage to wild populations. For instance, hatchery-reared organisms may:

- fail to survive on release;
- be too expensive to produce;
- be too few to make a quantitative difference in the wild;
- not enter the breeding population;
- delay the employment of more effective management measures to protect remaining wild stocks;
- introduce or facilitate the spread of diseases and parasites in wild populations;
- lead to detrimental exotic invasions of local fauna;

- only benefit selected estuarine, anadromous or freshwater species (Maccall, 1989; Rutledge, 1989);
- alter and damage the genetics of wild stocks;
- damage interest in conserving renewable and native wild stocks;
- not be able to have potential benefits scientifically evaluated from a practical point of view (Maccall, 1989).

Low survival

Reasons for low survival of released organisms include the following.

1. *Production of genetically inferior individuals.* Low survival may be caused by production and release of genetically inferior individuals. These individuals usually arise from inbreeding or from using too few parents. It is also possible that hatchery conditions may artificially select for genetic characteristics that have little survival value in the wild. Individuals from one subpopulation introduced into a new area may not be genetically 'tuned' to the new environment. Presumably, genetic characteristics of different subpopulations have been selected for local survival value out of the general population.
2. *Production of physiologically inferior individuals.* These can result from nutritional deficiencies and environmental effects associated with the hatchery process, such as exposure to disease or acclimation to inappropriate temperature or chemical conditions. Some effects may be subtle, such as reported hearing loss or damage to the lateral line system caused by exposure to normal hatchery noises associated with aerators and pumping equipment (Ha, 1985).
3. *Production of behaviourally inferior individuals.* Hatchery-raised organisms also may have learned inappropriate behaviour for wild conditions. Released individuals may not have learned effectively to avoid predators, identify appropriate habitat or locate food (Fig. 11.1). Roberts *et al.* (1995) report on attempts to condition grouper before release.
4. *Release at inappropriate times or in inappropriate habitats.* These problems arise because of insufficient knowledge about the life history and biology of the released species (Fig. 11.2). Low survival results if hatchery-reared organisms are not released in appropriate habitat. In many cases the appropriate habitat for juveniles is not the same as that for adults. Sandt and Stoner (1993) noted the importance of different habitats to queen conch, *Strombus gigas*, during various growth stages. Unfortunately, for many hatchery-reared organisms, knowledge is lacking about habitat requirements during early phases of life history. Leber *et al.* (1993) noted the importance of pilot studies to

Fig. 11.1 Behavioural problems of hatchery releases.

determine the timing and location of release. It is also necessary to have good knowledge of optimum release size, habitat, season and density.

Damage to wild populations

It is critically important that hatchery releases do not damage wild populations. This danger can occur from the following.

1. *Genetic damage to wild stocks.* Hatchery-raised individuals may be genetically incompatible with wild stocks, perhaps producing infertile hybrids or offspring with reduced fitness. Two processes are important to consider. First, chance selection of parents for broodstock can result in low genetic variability of offspring. This possibility is reduced by rotating and increasing the number of individuals used in the broodstock. Second, hatcheries can select for individuals with specific genetic characteristics that may not have survival value in the wild.

Fig. 11.2 Unanticipated beneficiaries of hatchery release programmes.

This occurs because the conditions in hatcheries are so different from the wild. Often hatcheries have high densities of organisms, poor water quality, a surplus of food and a lack of predators. These are factors that do not occur under natural conditions.

2. *Spread of diseases and parasites into wild stocks.* High densities of individuals in hatcheries make them prone to infection by parasites and diseases, and these could spread into wild populations from released animals. Disease has spread to wild salmon in Europe from culture operations (Sinderman, 1993).

Other considerations

1. *Poorly defined programme goals.* Richards and Edwards (1986) reviewed hatchery release programmes and concluded that many failed because of poorly defined goals. Often projects were not directed at solving any particular biological problem or testing a hypothesis. Often success was measured in terms of the number of hatchlings released and not

by other criteria such as numbers surviving, numbers entering the fishery, increased fishery yield or total contribution to the reproductive population.

2. *Release quantity*. Hatcheries must provide enough individuals to make a difference to the wild population. This is often a problem because of low survival. In many cases survival probabilities can be improved by releasing individuals at larger sizes although this usually reduces the total numbers. Even with good survival, the size of hatchery releases may be insignificant relative to the size of wild populations.
3. *Economic justification*. The costs of hatchery releases must be weighed against potential benefits. This is especially difficult during early phases of research in which culture is rarely cost effective. With increased knowledge, hatchery operations can be expected to improve efficiency and reduce costs. Hatcheries are usually faced with the task of optimizing between choices. Although young individuals are less expensive to culture, they tend to have poor survival. Older individuals are much more expensive to raise but may have increased chances of survival.
4. *Untestable hypotheses*. A major problem is demonstrating that released organisms survive and successfully enter the breeding population. In many cases hypotheses about survival of hatchery-reared organisms cannot be scientifically evaluated (Maccall, 1989). One reason is the difficulty of distinguishing hatchery-reared individuals in the wild population. Tagged individuals are often hard to recognize. This problem may become less of a hurdle with the development and use of genetic markers and biochemical fingerprinting.
5. *Inappropriate response*. A common criticism of hatchery release programmes is that they can reduce interest in conserving native wild stocks, they may divert attention from ultimate population problems, and they may delay efforts to protect wild stocks (Maccall, 1989; Martin *et al.*, 1992; Jones, 1994). Unfortunately, hatchery release programmes may be more appealing than taking necessary tough conservation measures, but should not be used as a substitute for them, even though the latter are unpopular.
6. *Unrealistic expectations*. Hatchery release programmes should foster realistic expectations which can be difficult to do when soliciting funding support.

Case histories of hatchery releases

Few attempts have been made to release hatchery-reared organisms to supplement depleted stocks and restore fisheries productivity in tropical environments. Representative case histories are described below. The last three case histories deal with attempts to restore endangered tropical marine species.

Red drum

After significant declines of red drum, *Sciaenops ocellatus*, in much of the US during the 1970s, in 1982 Florida and the Texas Parks and Wildlife Department began a massive release of hatchery-raised juveniles to enlarge populations. By 1989 over 42 million fingerlings (25 mm TL) had been released into Texas bays and estuaries (Rutledge, 1989). The programme was considered a success by some when fisheries landings eventually increased. However, because of tagging problems, it could never be shown that the observed changes were in fact a result of hatchery releases. Increased landings could also be due to improved natural year-class strength or changes in fishery regulations.

Snook

Populations of snook, *Centropomus undecimalis*, declined in Florida in the 1970s partly in response to destroyed, altered, or polluted juvenile habitat. Starting in 1984, snook were stocked in an attempt to rebuild populations (Richards and Edwards, 1986). After several years, it is not clear what impact, if any, the hatchery programme has had on the wild population.

Striped mullet

Hatchery-reared striped mullet, *Mugil cephalus*, were successfully stocked in stream nursery areas of Kaneohe Bay, Hawaii (Leber *et al.*, 1993). Hatchery-raised individuals were distinguished by coded-wire tagging. Six months after release, they constituted 74% of the total mullet collected. By releasing individuals in appropriate habitat, the yield-per-stocked-juvenile was five times greater than initial releases in Kaneohe Bay. These high survival rates were attributed to the use of pilot studies that optimized release size, sites and season.

Nassau grouper

Roberts *et al.* (1995) experimentally released 29 hatchery-raised Nassau grouper, *Epinephelus striatus*, in the Virgin Islands, an area where the natural population has been severely depleted by fishing. Fish were raised in captivity in Florida to a size of 0.5–1.4 kg and 30–38 cm. After being flown to the Virgin Islands, they were run through a 2 week 'training programme' to facilitate their transition from captivity before being released.

After 5 months, at least five fish remained on the reef where they were released while two were spotted on reefs up to 12 km away. It is yet to be determined whether this experiment will lead to successful recruitment into the spawning stock. The advanced age of these fish means that this approach is not practical on a large scale.

Abalone

Tegner (1993) described an experiment where broodstock of pink *Haliotis corrugata* and green *Haliotis fulgens* abalone were transplanted from offshore islands to a coastal area of California. Stocks on the coast had been depleted and, due to poor larval planktonic dispersal abilities, had not recovered after years of protection from fishing. After 4453 sexually mature animals were transplanted to an area in which oceanographic studies suggested there would be a high probability of larval retention, the stocks showed a dramatic increase. Although not a hatchery-release experiment, this study shows that species can be successfully reintroduced under certain conditions. Saito (1984) reported successful releases of hatchery-raised Ezo abalone, *H. discus hannai*, off Hokkaido, Japan, although recapture rates were less for hatchery-raised abalone (5–10%) than for natural seed (20–25%).

Queen conch

Fishing has depleted queen conch, *Strombus gigas*, populations in many areas of the Caribbean. Laboratory-reared juveniles have been released in attempts to replenish depleted stocks. However, studies in the Bahamas, Puerto Rico, Florida and the Virgin Islands have shown poor survival rates of released conch. Although hatchery production methods have improved, release methods have not (Sandt and Stoner, 1993). Factors influencing mortality include growth rates, predation rates, size at release, density-dependent mortality, season and habitat type where released (Appeldoorn and Ballantine, 1983; Appeldoorn, 1985; Iversen *et al.*, 1987; Lipcius *et al.*, 1992; Sandt and Stoner, 1993; Stoner and Davis, 1994). Hatchery-reared conch have shown lower survival and growth rates compared with wild conch, especially during the first few months after release, although survival was improved with release in suitable seagrass habitat containing resident conch populations (Marshall *et al.*, 1992; Stoner and Sandt, 1992; Stoner and Davis, 1994).

No direct evidence shows the successful replenishment of conch populations based on the release of laboratory-reared juveniles. Releases do not appear practical until further techniques are devised to increase survival (Coulston *et al.*, 1989; Stoner and Davis, 1994). Appeldoorn and Ballantine (1983) and Coulston *et al.* (1989) concluded that strong conservation efforts are a prerequisite for restoring conch populations and that management of natural populations is the most immediate concern.

Tridacnid clams

The giant clam, *Tridacna gigas*, is an endangered reef species over much of the Indo–Pacific region due to excessive exploitation (Villanoy *et al.*, 1988). An attempt was made to reintroduce giant clams to Guam where they had become extinct (Munro and Heslinga, 1983). Trial restocking of wild populations by transplanting approximately 3000 clams (weighing over 18 t) from artificial breeding grounds off Orpheus Island, Australia, to the Great Barrier Reef are also under way (Wheal, 1993). It is too early to evaluate the success of these efforts.

Sea turtles

Sea turtles are considered threatened or endangered over much of their range. Two of the best-documented restocking efforts were made in the US where turtles were 'headstarted' by incubating eggs and raising hatchlings in captivity before releasing them into the wild. It was hoped that high predation rates could be avoided during the vulnerable early stages of juvenile life by raising young turtles and releasing them at a larger size. Programmes in the USA were eventually terminated. Florida terminated a 30 year headstart green and loggerhead turtle programme in 1989 because of concerns about possible interference with imprinting mechanisms that guide turtles to nesting beaches, skewed sex ratios from artificial incubation of eggs, nutritional deficiencies in captivity and behavioural modification of confined hatchlings (National Research Council, 1990).

The US National Marine Fishery Service started in 1978 to headstart endangered Kemp's Ridley sea turtles in Texas for release in the Gulf of Mexico. The Kemp's Ridley is the rarest of all sea turtle species; fewer than 500 nesting turtles are left from an estimated 40 000 females in 1947. The species is only known to nest on one beach in Mexico and is endangered from direct exploitation and mortality in shrimp trawls. Wild-caught eggs were incubated, hatched, and juveniles raised in captivity for a year before release. Approximately 2000 turtles were released each year for 12 years. An effort was made to establish a new nesting area by attempting to imprint hatchlings on a beach at Padre Island National Seashore, Texas. This imprinting effort was ended in 1988 because no turtles had returned to nest and it was felt that if imprinting were successful, a sufficient number of turtles had been released to detect nesting over the next 10 years without additional releases.

After review (Eckert *et al.*, 1992), the Kemp's Ridley headstarting programme was terminated in 1993 because it failed to demonstrate successful entry of headstart turtles into the breeding population. Although survival of headstart turtles had been documented in the wild, no nesting headstart turtles had been found. Currently, headstarting is generally not considered a long-term management tool in the recovery of endangered sea turtles. It is considered unlikely that increased recruitment into the adult populations can occur without simultaneous conservation measures to reduce mortality from other human activities (Crouse *et al.*, 1987; National Research Council, 1990).

Conclusions and research recommendations

Despite claims for success, there are few data showing that hatchery-reared organisms can survive, enter the breeding population and make a significant numerical contribution to wild populations. For every success or partial success, there have been many failures. This fact is not fully reflected in the scientific literature because of a natural bias against reporting failures. As in temperate studies, the few demonstrated successes were confined to enclosed or partially enclosed bodies of water. Demonstrating success will continue to be the most critical research problem for hatchery-release programmes. To help in this effort, improved tagging technology must be developed that allows the identification of hatchery-reared organisms and their offspring in the catch (Maccall, 1989). Development of genetic tags and genetic fingerprinting may help (King *et al.*, 1993).

Unless the original causes of overexploitation are treated, hatchery supplements will be of little benefit. As Richards and Edwards (1986) noted, it is important to learn from, and not repeat, past mistakes. Research should continue to evaluate past hatchery-release efforts to determine what was, and was not, successful. Two of the most important roles of hatchery research will be to test specific hypotheses about survival (Martin *et al.*, 1992) and provide basic life history information about different species (Richards and Edwards, 1986). Hatchery releases are most justified where a stock or species is threatened and is unable to recover without assistance. Success is most likely to occur for species with limited dispersal abilities and low natural mortality, when the biology of the organism is well known, and when appropriate conservation measures are in place to prevent collapse.

11.4 HABITAT ALTERATION

Besides manipulating populations, it may be possible to restore fishery productivity by manipulating habitats. This usually involves either the use of artificial habitats, such as artificial reefs, or the restoration of habitats damaged by human or natural events.

Artificial reefs

Artificial reefs are made in a variety of shapes and sizes with many different materials (Grove *et al.*, 1991). Although they have a variety of purposes, they are commonly used in fisheries and for habitat enhancement and damage mitigation (Seaman and Sprague, 1991). Some have been creatively used to passively protect critical habitat areas from illegal trawling in Spain (Ramos-Espla and Bayle-Sempere, 1990) and the Gulf of Thailand (Polovina, 1991b). Despite the many beneficial uses and popularity of artificial reefs (Stone, 1985), their use has been controversial in terms of effectiveness at increasing total fisheries productivity (Bohnsack, 1989; Polovina, 1991b). Their appeal is based, in part, on the fact that they often have higher densities of organisms than natural habitats; they therefore improve fishing, at least in the short term. It is not uncommon to have fishes caught within hours of deployment. However, artificial reefs may not increase total fisheries productivity over the long term, especially if stocks are already overexploited (Bohnsack, 1989). It must be shown that artificial reefs are not just another fishing method in which habitat is used as an attractant instead of, or in addition to, bait (Fig. 11.3).

The controversy is over the extent to which artificial reefs increase new biomass (production) compared with simply attracting and aggregating organisms from surrounding areas without increasing total biomass (attraction). The ability of artificial habitats to increase production depends on the ecology of the species in question, reef design, and deployment location. Increased production is most likely where reef habitat is limiting, when fishing effort is low and when a large stock reservoir exists relative to harvest (Bohnsack, 1989). Polovina (1991b) noted that artificial reefs can increase the exploitable biomass either by increasing total biomass or by increasing the available exploitable biomass through aggregation, as long as recruitment overfishing does not occur. If the latter occurs, however, artificial reefs can make the problem worse by aggregating remaining fishes, making them more vulnerable to harvest. Artificial reefs are unlikely to increase total productivity of species that are highly mobile or heavily fished, are limited by the supply of settlers or settle in non-reef habitats (Doherty and Williams, 1988; Mapstone and Fowler, 1988; Bohnsack, 1989).

This production controversy has dominated much of the artificial reef literature since the mid 1980s. Progress in resolving it has been slow, in part because effects of artificial reef deployment differ in each situation, and in most cases no appropriate data have been collected to answer the question. Often testing the alternatives is difficult or impossible. Some

Fig. 11.3 Problems with artificial reefs that primarily attract fishes. Before (top) and after (bottom).

studies suggest that total production is increased for certain organisms, such as octopus (Polovina and Sakai, 1989), juvenile spiny lobster, *Panulirus argus*, (Eggleston *et al.*, 1990), and for reef species at oil production structures in the northern Gulf of Mexico where little hard substratum exists. More evidence, however, suggests that artificial reefs do little to increase total productivity, especially for species targeted by fishing (Doherty and Williams, 1988; Bohnsack, 1989; Polovina and Sakai, 1989; Doherty and Fowler, 1994a).

While progress in resolving the controversy has been slow, incorporating new information into management action has been even slower. I suggest that there is sufficient information in most regions to evaluate the potential for stocks to benefit from artificial habitats by reviewing life history and habitat requirements of target populations (Bohnsack *et al.*, 1991; Polovina, 1991b). Unfortunately, a lack of conclusive data has become an excuse to continue questionable practices. Even when sufficient data are available, they tend to get ignored when they do not fit the paradigm of supporting increased total productivity. As with any tool, artificial reefs are subject to good use and misuse. In particular, programmes designed primarily to dispose of unwanted materials or that measure success in terms of the amount of material, or number of reefs deployed, should be carefully examined.

I conclude that although artificial reefs are an effective fishing tool, in most cases they are unlikely to restore fisheries productivity significantly. Only a few species are likely to benefit and many will not be of direct economic importance. Costs are prohibitive and the amount of material deployed is likely to be so small, relative to available natural habitat, that benefits will be negligible. The main problem of artificial reef programmes is their widespread popularity. This diverts attention and resources from more important issues and more effective management actions, such as reducing pollution and habitat loss, or treating overfishing.

Habitat alteration and restoration

Efforts have also been made to create, restore, rehabilitate or mitigate damaged or lost natural habitats. These activities are widely practised and an extensive literature exists, beyond that which can be reviewed in detail here (Stroud, 1992; Thayer, 1992). Restoration is used primarily to repair or replace habitat damaged or destroyed by human activities or natural events such as storms. Tropical habitats most important and suitable for restoration include mangroves (Cintron-Molero, 1992), seagrasses (Fonseca, 1990, 1992; Kirkman, 1992), coral reefs (Maragos, 1992) and tidal marshes (Seneca and Broome, 1992), all of which are major primary producers and critical habitat for many important coastal species.

Many of the criticisms applied to hatchery release programmes also apply to habitat restoration and damage-mitigation efforts. They may promote unrealistic expectations, be ineffective, and distract efforts from treating the primary sources of degradation (Fig. 11.4). Effective restoration programmes must be economically sound, provide functional habitat and successfully treat the cause of loss. Efforts to replant seagrasses killed by pollution or turbidity, for example, are unlikely to be successful unless the sources of stress are treated. In fact, death of seagrasses may increase

Fig. 11.4 Problems with habitat restoration.

turbidity problems by increasing sediment resuspension. Thayer (1992) emphasized the need to demonstrate the restoration of functional attributes of habitats to the level of natural habitats. Success must be measured in terms of function, not just area restored.

Unfortunately, few studies have adequately demonstrated successful restoration (Lewis, 1992; Thayer, 1992). Lewis (1992) found only four studies in three US states that indicated rapid (3–5 year) establishment of comparable fish communities in created or restored coastal wetlands as compared with natural wetlands. Gilmore *et al.* (1982) showed that restoring previously impounded marsh–mangrove wetlands provided habitat that was used by a variety of economically important fish species. Woodley and Clark (1989) examined the restoration of coral reefs and concluded that there were some examples of successful reef rehabilitation but that the cost-effectiveness was debatable when large areas were considered. Rehabilitation of coral reefs is likely to be attempted only in areas of particularly high value, such as marine parks.

Habitat restoration, despite a lack of supporting data, is considered one of the most important approaches for restoring long-term fisheries productivity. This evaluation is based on the realization of the extensive habitat loss that has already occurred in many areas of the world. In the contiguous 48 states of the US, for example, an estimated 54% of the original 915 000 km^2 of wetlands had been lost by the mid 1970s (Chambers, 1992). Research priority should be directed at documenting the value of restoration by testing the hypothesis that restored habitats support populations and provide ecological functions comparable to natural habitats. Too often success is declared prematurely when the restoration effort is completed without showing its effectiveness. Despite the potential of habitat restoration, it is better to avoid habitat loss in the first place than to have to restore or mitigate damage later (Fox, 1992).

11.5 EXOTIC INTRODUCTIONS

The introduction of non-native (exotic) species into new areas for fishery purposes is a common and growing occurrence worldwide (Sinderman, 1993). The most frequent species used for fisheries purposes include various salmon, shrimp, tilapia, cichlids and bivalve molluscs. Although exotic introduction has popular appeal, it has often failed or led to disastrous or unforeseen consequences (Courtenay and Robins, 1975; Sinderman, 1993) (Fig. 11.5). Sinderman (1993) noted widespread introductions of diseases and parasites caused by releases of non-indigenous marine animals.

Well, there goes the neighborhood...

Fig. 11.5 Problems with introduced exotic species. Republished, with permission, from *Environmental Biology of Fishes*, 3(1): 5.

Courtenay and Robins (1989) reviewed fish introductions and noted only a few tropical marine examples. Hawaii had the only documented deliberate introductions of marine species for management purposes; seven out of 21 species became established. An additional six introductions were unplanned. Overall, they were failures. Of the 13 that established, four were rated as mistakes, two were known to cause human poisoning, and the remaining seven had no value except possibly for the bluestriped snapper, *Lutjanus kasmira*.

Many introductions are unplanned and often have catastrophic consequences as exemplified by the invasion of zebra mussels in US waters (Ludyanskiy *et al.*, 1993). An unplanned introduction of the comb jellyfish into the Black Sea has been blamed for destroying fisheries and for dominating the biomass of the system (Travis, 1993).

The history of introductions indicates that extreme caution should be used when introducing species into non-native areas (Courtenay and Robins, 1989). Species should be introduced only after careful research and when no suitable native species are available.

11.6 PERMANENT AREA PROTECTION

Marine protected areas are increasingly used to protect natural resources from various sources of human disturbance (White, 1988b). By 1993, over 100 marine protected areas were listed in the Caribbean alone (Sobel, 1993). Increased interest has developed in using areas permanently closed to all harvest activities to protect and restore fisheries productivity (Polacheck, 1990; DeMartini, 1993; Dugan and Davis, 1993a). I give this topic considerable emphasis because it is a relatively new, potentially important, and yet controversial approach to fisheries management.

I use the term marine fishery reserve (MFR) to refer to areas permanently protected from all fishing and other direct extractive exploitation. Despite efforts to standardize terminology by the World Conservation Union (IUCN) (Ray, 1976; Sobel, 1993), MFRs have been called a variety of names with no consistent definition. Terminology includes 'no fishing' (or 'no take') zones, 'conservation districts', 'core areas', 'sanctuaries', 'non-consumptive' or 'replenishment zones', 'reserves', 'preserves', 'parks', 'protection boxes', 'harvest refugia', and 'wildlife refuges', among others. While many areas protect certain species or protect against specific kinds of fishing, such as trawling or commercial fishing, relatively few permanently protect all species from harvest (Foster and Lemay, 1989). California in 1993, for example, had 107 named protected areas, of which only four or five, covering a total of less than 8 ha, had complete protection from harvest (G. Davis, pers. comm.).

Applications

Marine fishery reserves can potentially help maintain fishery productivity and insure against stock collapse outside reserves by protecting the quantity of larval production and their genetic quality. They may:

- increase stock abundance;
- increase target-organism size;
- increase reproductive output and recruitment;
- protect genetic diversity of stocks from fisheries selection;
- enhance fishery yields in adjacent fishing grounds;
- provide a simple, effective, understandable, and least burdensome method for protecting fisheries;
- provide insurance against fishery collapse;
- reduce chances of overfishing;
- eliminate bycatch and release mortality in protected areas;
- protect rare and vulnerable species;
- simplify compliance, awareness and enforcement;
- require few management data compared with conventional approaches;
- maintain trophy fisheries;
- have equitable impact among users;
- provide better fisheries understanding of exploited species;
- provide a basis for ecosystem management.

They also support many non-fishery uses (Bohnsack, 1993). General (non-fishery) benefits of marine fishery reserves are that they:

- allow research on exploited species;
- increase species diversity;
- increase habitat complexity and quality;
- increase community stability;
- provide scientific control sites for understanding the impacts of human activities;
- provide undisturbed monitoring sites;
- enhance current human activities (tourism, diving, research);
- create new uses (ecotourism, education, enhanced fishery research);
- separate incompatible activities (reduce conflicts);
- eliminate fishing damage to critical habitats;
- improve public awareness, education and understanding;
- create areas with intrinsic value.

By protecting biodiversity and habitat from damage due to fishing (see below), MFRs can be used for educational, recreational and scientific uses that are not possible in a fully exploited fishery. In many cases the economic benefits of non-extractive uses alone may justify the lost fishery use of an area (van't Hof, 1985; Dixon, 1993).

Fisheries interest in MFRs has developed because conventional management options are limited (Munro and Williams, 1985), have failed to be effective (Butler *et al.*, 1993; Pauly, 1993), or are expensive to employ and difficult to enforce (Bohnsack, 1993; Roberts and Polunin, 1993). Quotas and bag limits, for example, are inappropriate for many tropical reef fisheries because they require extensive data, analyses, and information on the biology of individual stocks that are unobtainable (Appeldoorn, Chapter 9). Size limits can be ineffective because of release mortality, especially with increased capture depth. Also, as minimum sizes increase, fewer fishery-dependent data become available because smaller size and age classes are no longer landed. MFRs may be especially well suited for many tropical fisheries where species are considered particularly vulnerable to fishing (Jennings and Lock, Chapter 8).

MFRs have popular appeal because they are based on a simple concept that natural systems can maintain themselves if protected from human interference. From a biological perspective, the theory supporting MFRs is simple and based on the biology and ecology of most marine organisms that have relatively sedentary adult or juvenile stages and that disperse primarily by eggs and larvae (Plan Development Team, 1990; Bohnsack, 1993). From a management perspective, MFRs have appeal because they separate incompatible activities while allowing them to continue. For example, spear fishing can occur outside MFRs while activities incompatible with it, such as ecotourism, behaviour studies and fish photography, can occur inside MFRs. Almost all countries and cultures accept the idea of protecting critical areas and using zones to separate incompatible activities, at least in terrestrial systems. With MFRs, data collection needs are reduced and enforcement may be less expensive compared with conventional fishery management measures. Because it is difficult or impossible to control total fishing effort in practice (Ludwig *et al.*, 1993), protecting a small area may be more acceptable to fishing interests than other actions. Also, protecting small areas does not directly discriminate between fishing interests because all fishing is eliminated in contrast with allowing some types of fishing and not others. MFRs may also provide some insurance against management and recruitment failures (Plan Development Team, 1990).

History of use

MFRs re-establish a process that once occurred naturally in many fisheries (Dugan and Davis, 1993b). Natural refuges from fishing existed because some areas were too deep, too remote, too hard to locate, or unfishable by available gears. Over time, natural refuges have become less effective or ceased to exist with increased fishing effort and better technology.

Although described as a modern measure, the concept of MFRs is not new (Johannes, 1984). In the 19th century, suggestions were made to close portions of the North Sea to give stocks a chance to recover, but were thwarted because nations could not agree on which areas to close or for how long (Hardy, 1959). Unfished areas were more likely to occur when countries were in conflict than when at peace. The first major unintentional experiment with area closure came when the entire North Sea was subjected to military operations during WWI. During the 4.5 years of closure, stocks recovered enormously. A second unintentional large-scale 'experiment' with similar results occurred during WWII.

A major change occurred in the 1970s when coastal nations began to declare their adjacent seas as exclusive economic zones (EEZs). For the first time, large areas could be protected under one legal authority. Australia, New Zealand and a host of other countries established various marine protected areas, although not all were enforced or had good compliance. Unfortunately, few of these areas had data collection programmes to measure biological effects of protection.

Pauly (1993) credited the modern scientific development of fishery interest in protected areas to the classic work of Beverton and Holt (1957). They described and modelled the North Sea plaice fishery, in which *de facto* refugia existed because minefields from WWII made trawling too dangerous. Pauly (1993) predicted that 'refugia or sanctuaries' were 'likely to grow in importance throughout the 1990s and beyond'.

Tests of hypotheses

The growing interest in the use of permanently protected areas to restore and protect fisheries is demonstrated by the number of reviews produced between 1991 and 1993 (Roberts and Polunin, 1991, 1993; Rowley, 1992; Dugan and Davis, 1993b; Towns and Ballantine, 1993). These reviews agreed that MFRs were a potentially important fisheries management tool, that considerable evidence supported some hypotheses, but that insufficient data were available to test others. In the 1990s, the amount of relevant material has continued to grow. Below I briefly discuss some of the major hypotheses.

1. *MFRs increase the abundance (density) and average size of exploited species inside them*

Reports of increased abundance and average size of exploited species in MFRs are common. Out of 31 studies reviewed by Dugan and Davis

(1993b), 24 showed increased population abundance in reserves, four showed decreased abundance, and in three abundance was not examined. Average size of target species increased in 19 studies, decreased in four, and was not investigated in nine. Towns and Ballantine (1993) reported that spiny lobster, *Jasus edwardsi*, abundance in a New Zealand reserve was 10 times higher and the average size larger than in surrounding areas. Previously targeted reef fishes (e.g. *Cheilodactylus spectabilis*) and molluscs (*Haliotis iris*) showed similar changes. Increased abundance and size are well documented for some species in studies from the US in Florida (Bohnsack, 1982; Clark *et al.*, 1989) and Hawaii (Grigg, 1994); from Australia (Harding, 1990), New Zealand (Cole *et al.*, 1990; Towns and Ballantine, 1993), Kenya (McClanahan and Muthiga, 1988), South Africa (Bennett and Attwood, 1991; Buxton, 1993), the Red Sea (Roberts and Polunin, 1992), the Philippines (Russ, 1985; White, 1988a) and from countries in the Mediterranean (Bell, 1983; Garcia-Rubies and Zabala, 1990), the Caribbean (Rowley, 1992; Polunin and Roberts, 1993; Roberts and Polunin, 1993, 1994) and elsewhere (Dugan and Davis, 1993b). Samoilys (1988), however, reported that effects of protection on abundance on some reefs in Kenya were confounded by siltation and dynamiting.

2. *MFRs protect biodiversity*

Reviewers found that MFRs protect biodiversity in terms of species richness, especially for species vulnerable to fishing. White (1988a) reported a higher mean and total number of species from three reserves in the Philippines than from fished areas. Grigg (1994) reported that areas protected from fishing in Hawaii had higher fish diversity (on average four more fish species) than control areas. Higher average species richness has been reported from protected areas in Kenya (Samoilys, 1988; McClanahan, 1989), the US (Bohnsack, 1982), the Mediterranean (Garcia-Rubies and Zabala, 1990), the Caribbean (Roberts and Polunin, 1993, 1994) and elsewhere. Bell (1983), however, found that mean species richness and diversity did not differ significantly between some protected and non-protected sites in the Mediterranean.

Towns and Ballantine (1993) reported that behaviour patterns, such as harem formation, were observed for some species in reserves that were unlike those found along the adjacent coast.

3. *MFRs maintain fishery sustainability*

MFRs are intended to support fisheries by protecting core populations from harvest (Plan Development Team, 1990). This provides some insurance against stock collapse from natural environmental variability or fishing activities. In the event of a fishery collapse, MFRs could rebuild fisheries much faster than would otherwise be possible if the stock collapsed in all areas.

Beverton and Holt (1957) noted that some of the high sustained production from the North Sea could be because about 10% of the habitat was not trawlable because of snags, ship wrecks, uncleared minefields, and other obstructions. The most-cited evidence supporting this is research in the Philippines which indicated that increased catches more than made up for the reduction in fishable area caused by the Sumilon Island reserve (Alcala, 1988; White, 1988a; Alcala and Russ, 1990). In some cases, MFRs may be able to double or triple local catches in a sustainable fashion (Pauly, 1993). Buxton (1993) concluded that South African reserves were a hedge against recruitment failure. Lozano-Alvarez *et al.* (1993) found that the productivity of Mexican lobster fisheries was maintained in part because of natural reserve areas in deep water that could not be fished.

Towns and Ballantine (1993) noted that despite the many obvious benefits of MFRs, the evidence showing that they can augment or restore stocks of exploited species outside reserves was inadequate. Roberts and Polunin (1993) also found that limited evidence existed for enhanced catches in areas adjacent to protected zones, and that evidence was lacking to show that reserves restock fishing grounds. Despite a lack of sound data, they concluded, however, that MFRs are a promising method for providing low-cost management of reef fisheries.

Despite a lack of direct evidence for maintaining fisheries, Davis (1989) noted considerable indirect evidence which has tended to be overlooked. While few MFRs exist that protect all species, many areas have been established to protect particular species or have been closed to specific fisheries. These areas provide partial tests of MFRs, particularly with regard to their value in sustaining fisheries. In southern Florida, for example, the three most important commercial fisheries (pink shrimp, stone crab and spiny lobster) have all had large areas closed to fishing. This circumstance may partially account for their sustainability, in spite of there being more commercial fishing effort than is necessary to catch the same harvest. Spiny lobster refuges, for example, were established in Biscayne Bay, the Dry Tortugas and Everglades National Park (Davis and Dodrill, 1980, 1989) and landings have been stable since the late 1960s despite substantially increased fishing effort (Harper, 1993). Stone crab landings increased although important habitat in Everglades National Park was closed to harvest (Bolden, 1993). Shrimp sanctuaries have been considered successful off Texas (Griffin *et al.*, 1993) and in Florida despite poor compliance (Klima *et al.*, 1986). A model of the tiger prawn, *Penaeus esculentus*, fishery in Australia showed that yield per recruit would decrease with spatial closures but that value per recruit could increase up to 10% with the appropriate closure (Die and Watson, 1992). In contrast, many Florida fisheries that have not been afforded area protection, including those for king mackerel, grouper, jewfish, snook and queen conch, have collapsed or have shown wide population gyrations (Bohnsack *et al.*, 1994).

Next I discuss two mechanisms by which MFRs may improve fishery yields in adjacent areas: migration of fishes out of protected areas into fishing grounds and export of larvae to surrounding areas.

4. *MFRs provide migrants to surrounding areas*

Roberts and Polunin (1991) found almost no information on rates of fish emigration or immigration although some evidence indicated that increased catches occurred near reserves. Rowley (1992) concluded that 'spillover', movement of fish from reserves to surrounding areas, was likely to occur but that there was little direct evidence for it and the magnitude of such effects was impossible to predict. Gitschlag (1986) demonstrated that tagged shrimp moved from the Tortugas Shrimp Sanctuary into fishing grounds. In the same area, Caillouet and Koi (1981) found evidence of increased shrimp size in the catch following sanctuary closure although other studies did not, possibly due to variability in recruitment and illegal trawling inside the sanctuary (T.W. Roberts, 1986). Rigney (1990) noted anecdotal information from Queensland, Australia, that more prawns were caught 'fishing the line' adjacent to MFRs. Funicelli *et al.* (1988) documented that marked red drum, *Sciaenops ocellatus*, spotted seatrout, *Cynoscion nebulosus*, black drum, *Pogonius chromis*, snook, *Centropomus undecimalis*, and striped mullet, *Mugil cephalus*, moved from a small estuarine sanctuary in the Kennedy Space Center, Florida to surrounding fished areas. Holland *et al.* (1993) found little emigration of adult goatfish from a reserve to surrounding areas for most of the year.

One measure of MFR effectiveness is to examine attitudes of fishers with direct MFR experience. Rigney (1990) noted increased, but not universal, acceptance of MFRs by many Australian and New Zealand fishermen. Shorthouse (1990) reported that a survey of 16 prawn vessel operators in northern Australia showed a startling change in attitude 2 years after a closure was imposed. Compliance had increased, a number of fishermen reported increased catches, and some expressed a wish for more permanent closures.

5. *MFRs enhance fecundity*

MFRs are predicted to increase larval supply and spawning potential by maintaining more and larger breeding individuals within reserves than would otherwise occur if all areas were open to fishing. Often fecundity drops dramatically with rarity (Allee effect). This occurs at low densities because males and females of mobile species have difficulty locating each other or poor fertilization occurs, especially for broadcast spawners. MFRs could increase the fecundity of stocks and presumably export larvae by increasing the density and size of spawners. Fertilization success has been demonstrated to be a direct function of spawning adult density for sea urchins (Levitan, 1991; Levitan *et al.*, 1992). Shepherd and Brown (1993) also showed that reproductive potential declined due to rarity for abalone in Australia. Protecting density may be important for other broadcast-spawning species (Dugan and Davis, 1993b).

DeMartini (1993) modelled the effects of MFRs for three different types of fish and concluded that spawning biomass per recruit would increase for fish with moderate vagility but not for highly mobile species. A model of the Torres Strait tiger prawn fishery suggested that eggs per recruit would always increase in the presence of a closure (Die and Watson, 1992).

6. MFRs export larvae to surrounding areas

The capability of MFRs to increase reproductive output is based on larger numbers and increased size of individuals in reserves (Roberts and Polunin, 1991). Few studies have examined their ability to increase reproductive output or to enhance recruitment of larvae to surrounding harvested areas (Roberts and Polunin, 1991; Carr and Reed, 1993; Dugan and Davis, 1993b). Rowley (1992) concluded that there were good reasons to believe that both these effects occurred, although there was little proof. Although difficult to test, it is hard to imagine how this prediction could be falsified considering the fact that eggs and larvae are dispersed widely by ocean currents. Porch (1993) modelled larval dispersal in the Florida Keys and concluded that there was a high probability that larvae and eggs spawned along the Florida Keys could be retained by gyres and returned within 20 km of their birth site. Shepherd and Brown (1993) found limited potential for export of Australian abalone from MFRs because of poor dispersal ability. For the same reason, Tegner (1993) noted the failure of pink *Haliotis corrugata* and green *Haliotis fulgens* abalone populations on offshore islands to successfully recolonize California mainland coastal areas until some adults were transplanted to coastal areas. Populations of red abalone, *Haliotis rufescens*, however, were able to export larvae and did successfully repopulate coastal areas (Tegner, 1989). Another idea of major importance is that MFRs genetically protect stocks.

7. MFRs genetically protect stocks

Fishing tends to selectively remove larger, faster-growing and more aggressive individuals from a population and, over time, could change the genetic characteristics of a population in ways that are undesirable from a human perspective (Plan Development Team, 1990). Individuals would tend to grow more slowly, and to smaller sizes, and to become wary. Reserves could help maintain desirable genetic characteristics and reduce the loss of genetic diversity by protecting a portion of the population from fisheries selection. Within reserves, the wild-type characteristics could maintain their natural selective advantages. This hypothesis has not been tested (Roberts and Polunin, 1991; Dugan and Davis, 1993b) although Shepherd and Brown (1993) concluded that MFRs could play an important role in protecting genetic diversity of abalone.

Finally, I touch briefly on two other hypotheses that relate to better understanding and managing fisheries productivity.

8. MFRs provide biological reference areas

There is a growing need to be able to distinguish the effects of fishing from other human impacts. In this regard, MFRs can serve as critical, minimally disturbed reference areas. Also, studies can be conducted on biological interactions of harvested species that would be impossible in a fully exploited fishery. Measurements of age, growth and natural mortality can be especially important for fisheries purposes. Because of such areas, Polovina (1994) was able to determine that a decline in the Hawaiian lobster fishery was due to climatic changes and not fishing effects. Protected areas in California were useful in determining that disease had helped reduce abalone populations (Richards and Davis, 1993). In Kenya, protected areas were used to document indirect effects of fishing on sea urchins and corals (McClanahan and Muthiga, 1988; McClanahan and Shafir, 1990; Jennings and Lock, Chapter 8).

9. MFRs do not protect highly migratory species

MFRs are not predicted to provide much protection to highly mobile or migratory species because they can be readily caught outside MFRs (Plan Development Team, 1990). This prediction is not controversial and has tended to be supported by modelling studies (Beverton and Holt, 1957; DeMartini, 1993). However, two different species of spiny lobster, *Panulirus argus* in Florida (Davis and Dodrill, 1989) and *Jasus edwardsi* in New Zealand (Cole *et al.*, 1990; Ballantine, 1991), that were thought to be highly migratory, showed unexpectedly that at least some individuals would become resident in protected areas.

Practical design criteria

Optimum size, number, total area and location of reserves are important considerations (Shafer, 1990; Roberts and Polunin, 1991; Ray and McCormick-Ray, 1992). To be effective, reserves must be of sufficient size, include appropriate habitat, and protect critical life history stages of target organisms (Dugan and Davis, 1993b).

Optimum size and shape

Determining optimum size and shape is an important problem that will need to consider a suite of species and social as well as biological impacts. Tisdell and Broadus (1989), using economic analysis, found that optimum size differed when based on social versus ecological criteria. Effective size will depend on the home range and habitat requirements of the target species (Armstrong *et al.*, 1993). Even small reserves have been shown to protect some species with restricted movements (Cole *et al.*, 1990; Holland *et al.*, 1993; Polunin and Roberts, 1993). However, large areas may be needed to protect more mobile species (Beverton and Holt, 1957; DeMartini, 1993).

Optimum number

It is important to know for a given amount of protected area whether it would be more effective to have a few large or a network of small reserves. Several recent reviewers have shown preference for reserve networks (e.g. Dyer and Holland, 1991; Towns and Ballantine, 1993). Shepherd and Brown (1993) noted that many small reserves would be preferable to a few widely dispersed reserves for Australian abalone because of their limited dispersal abilities. Multiple independent reserves would confer a longer persistence time to a population than a single reserve with the same total carrying capacity, if colonization rates were high (Goodman, 1987). This seems to be a reasonable assumption for most marine species with high dispersal abilities. It may also be more socially acceptable to have multiple small reserves.

Location

Practical design considerations for locating marine reserves include the distribution of habitat, patterns of dispersal (Carr and Reed, 1993), locations of population sources and sinks (Pulliam, 1988), socio-economics

(Tisdell and Broadus, 1989), proximity of enforcement (Plan Development Team, 1990), and boundary conditions (Plan Development Team, 1990).

Total area

The World Conservation Union (IUCN) has set a goal of protecting at least one-third of all coastal seascapes (Batisse, 1990). Ballantine (1991) listed general reasons for protecting at least 10% of the New Zealand shelf: (1) a traditional cultural view that 10% signifies importance without serious impact on users (90% is left for exploitation); (2) 10% was a conservative figure compared with 20% to 30% protected in land reserves; (3) 10% provided a reasonable safety factor (the precautionary principle); (4) the importance of setting a clear future goal to ease concerns by resource users that 'it will all be locked up'; and (5) enough to provide worthwhile and widespread benefits. The US South Atlantic Snapper–Grouper Plan Development Team (Plan Development Team, 1990) recommended protecting 20% of the shelf based on theoretical arguments (Goodyear, 1993), empirical evidence from stock collapses (Goodyear, 1993), and the fact that fishing on Gulf of Mexico red snapper had driven the spawning potential to less than 1% (Goodyear and Phares, 1990). The spawning potential is the ratio of the reproductive output under fishing as compared with what would exist without fishing.

In practice, the amount of reserve protection has tended to be very small but in some cases has reached the 30% range. The amount of area protected from all fishing in the Great Barrier Reef was estimated at only 2% (Rigney, 1990). A reserve in the Philippines that protected approximately 25% of the reef was considered effective (Alcala and Russ, 1990). Bermuda recently protected approximately 20% of its shelf (Butler *et al.*, 1993). Proposed reserves in the Florida Keys National Marine Sanctuary included approximately 6–8% of the shelf. In addition to total area protected, it is also important to consider population density and distribution. For example, a Florida Tortugas shrimp sanctuary which represented only 6% of the Tortugas fishing ground contained 36% of the total estimated pink shrimp population (T.W. Roberts, 1986). Few data exist from which to evaluate density and distribution for particular species.

Biology

Clearly the biology of targeted organisms must be considered in the design of MFRs. Important factors to consider include behaviour, home range size, migrations, fecundity, reproduction, interspecific and intraspecific interactions, and dispersal abilities (DeMartini, 1993).

Permanent or periodical rotation

Caddy (1993) suggested that rotating closures were potentially useful for managing a fishery for Mediterranean red coral, *Corallium rubrum*, which is a slow-growing species with poor dispersal abilities. Sluczanowski (1984) concluded that rotating pulse fishing substocks of abalone would be beneficial to egg production with only marginal decreases in yield. Bohnsack (1994), in contrast, concluded that pulse fishing using periodic or rotating area closures would be ineffective for most reef species because closures would have to last for years and the benefits of closure could be rapidly dissipated when fishing resumed. Beinssen (1988), for example, showed that when a closed area at Boult Reef, Australia was opened to recreational fishing, 25% of the grouper were removed within 2 weeks. Caddy (1993) argued that costs of enforcing closed areas would increase over time because the incentive for illegal fishing would increase with stock recovery. Ballantine (1991), however, noted that the public pressure and willingness to enforce closures also increased with stock recovery. Rotating closures can also cause confusion, may create economic hardships for businesses that have adjusted to specific closed areas, and may create administrative and logistical problems for enforcement (Plan Development Team, 1990).

Social factors

Social and economic factors are important considerations for successful design and operation of marine reserves (White, 1986; Batisse, 1990). MFRs are ineffective without public and political support (Tisdell and Broadus, 1989) (Fig. 11.6). Resistance to using reserves is likely to come from (1) local interests in the immediate vicinity, (2) conventional managers who rely on single-species approaches, and (3) claims that there is a lack of sufficient biological data to demonstrate MFR effectiveness and warrant action. White (1986, 1988a) emphasized the importance of education, participation of local communities in reserve planning and implementation, and legal support from national and local officials. As with terrestrial reserves, acceptance of MFRs is likely to increase over time as people become more familiar with them.

Conclusions and recommendations

While general agreement exists that reserves are a good idea, testing of certain predictions has been limited. There is a great need to further test specific hypotheses and evaluate the overall effectiveness of MFRs on larger and more ecologically significant scales (Towns and Ballantine, 1993). Regional efforts are needed to determine the most effective size,

Fig. 11.6 Problems with marine 'protected' areas.

number, location and total area. Sociological and economic research is needed to determine how reserves can best be used on a sustainable basis. Biological research should examine the movements, habitat requirements, and life history of specific target species. While the importance of spillover can be tested by tagging, evaluating the importance of larval transport from reserves to surrounding waters will be the most difficult and expensive research problem. Perhaps the best evidence of dispersal would be sustainability of fisheries in regions with reserves and stock collapses in areas without. The appropriate null model of larval export is that reserves contribute to the surrounding population in proportion to the spawning stock protected in reserves.

Critics and even many proponents of MFRs were very cautious in recommending their use until proof of their effectiveness had been obtained. Of course proof is impossible to obtain unless MFRs are empirically tested at an appropriately large scale. Such standards for proof are far beyond what is normally done in fisheries management; rarely are actions proven before being implemented. There is always considerable doubt about the ultimate effectiveness of any action, whether a size limit, gear restriction or something else. If complete proof of effectiveness were needed, almost no fishery management actions would be taken. I suggest that in most cases the risk of maintaining the *status quo* or taking no action, is greater than the risk of developing an MFR system. Marine fishery reserves are likely to become increasingly important for managing tropical fisheries, especially in coastal areas. Ideally, reserves should be created before stocks collapse and should be used in combination with other management actions.

ACKNOWLEDGEMENTS

I thank Nick Polunin, Callum Roberts, Robert Rowley and an anonymous reviewer for constructive criticisms. Jack Javech and Jeanene McCoy assisted in making figures.

Chapter twelve

Traditional management of reef fishing

Kenneth Ruddle

SUMMARY

The Asia–Pacific region is especially rich in traditional community-based arrangements for reef fisheries management. Such systems are also known from reefs in the Caribbean and the Middle East. Literature on the subject is diverse, but scant and scattered, and many technical terms are used imprecisely. The exact origins of this management are unknown, but most debate surrounds its derivation from conservation practices as against its role in conflict resolution and avoidance. In some areas connections have been established with commercial use and defence of resources, and with resource-related taxation and tribute.

Traditional management systems in the Asia–Pacific region are based on property rights and associated regimes which reflect local structures of power and social organization. The systems are backed by authority which varies from area to area and may include secular leaders, religious leaders or specialists. Sets of rules governing the systems also vary among areas, but common characteristics are exclusivity, transferability, and enforcement. The rules are monitored and sanctions applied where necessary.

These traditional management systems have already disintegrated quite widely, and many factors have contributed to the decline, including processes such as colonialism, replacement of traditional local authority, education, commercialization and economic development. Yet there are strengths of such management, for instance their focus on allocation

Reef Fisheries. Edited by Nicholas V.C. Polunin and Callum M. Roberts.
Published in 1996 by Chapman & Hall, London. ISBN 0 412 60110 9.

problems, their implementation by controlled access to defined areas, and their enforcement by local authority. In many cases such systems may no longer be relevant to management objectives. They must be a viable alternative, however, in the rural hinterlands of far-flung archipelagic states such as Indonesia or Kiribati, where local authority remains and management by central government is cost-ineffective and ineffectual.

12.1 INTRODUCTION

Traditional systems of fishing rights and tenurial relationships of small-scale fishers to resources and areas have been documented throughout the world. Many were common property regimes in which access to a territory was limited to a defined group. Operational rules were specified and control resided in traditional local authorities. Routine decision-making, the implementation and enforcement of decisions, monitoring of the fishery and other basic aspects of management were undertaken exclusively by members of the local community. Mostly, such community-based systems were deliberately or inadvertently weakened or destroyed by colonial administrations. They tended to be replaced by centralized fisheries institutions nominally responsible for all aspects of management, from policy to enforcement.

Alternative systems of management, such as traditional, community-based systems, have rarely been considered in fisheries development programmes. This is largely because they remain relatively little known to outsiders. Such systems are a practical antithesis to conventional models, and embody fundamental concepts relevant to fisheries management worldwide.

Common property management systems have survived in so many widely scattered parts of the world for several reasons. They are a proven means of ensuring community survival through meeting basic human needs. Inclusion rights spread the risks inherent in resource uncertainty and scarcity. They are adapted to fluctuating resource availability and so contribute to community stability. They institutionalize area management, access limitations and regulations for overcoming incompatible gear usage, and problems of allocating time and space to fishers. Gear incompatibilities mainly include siphoning-off fish and direct physical interference. Allocation problems arise through competition for prime spots, or when there are too few spots to sustain the number of fishing units present. This shortage has both social and economic costs. Such problems arise because of the uneven spatial and temporal distribution of fish; all sea areas do not contain productive fishing spots, and not all spots are of equal productivity.

All the statements in the preceding paragraph are in striking contrast to the assumption that fishing strategies evolve as a consequence of competition. Transaction costs (a major cost under other management regimes) are low because resources are managed and controlled by the community, the regime is locally devised and upheld, and individuals are taught to comply with it early in life.

Common property regimes are often tightly bound within and inseparable from the shared cultural matrix of traditional communities. They constitute an element of a lifestyle rather than of just an economic sector or occupation. Compared with their scientific counterparts, they are easily, inexpensively and locally enforceable, based on local authority (Ruddle *et al.*, 1992). Organizational concepts range from the quasi-ownership of specific sites by individuals, families, clans or other small social groups, to the complex statutory legal system of Japan. However, the local community is often the sole owner that controls the local spectrum of marine habitats.

The spatial relationships of one typical such system, that of Ulithi Atoll, Yap State, Federated States of Micronesia (FSM), are illustrated in Fig. 12.1. The atoll is divided into eight ranked districts, of which Mogmog is the main one, each composed of several villages and one or more small islands. Land and sea ownership is vested in matrilineages or clans. These are ranked and the highest is Mogmog, which supplies the hereditary paramount chief who has jurisdiction over the entire atoll. District chiefs control the marine resources of their district on behalf of the paramount chief. Ulithi Atoll is divided into 14 lagoon and 18 reef sections, each controlled by local clans, the members of which have fishing rights in clan waters (Ushijima, 1982; Sudo, 1984). Spatial and administrative characteristics of Ulithi are typical of those in many similar systems worldwide, especially in the Asia–Pacific region.

In the following account I first describe the geographical distribution of traditional coral reef fisheries management systems and suggest some likely reasons for their evolution. I then summarize the rights, rules, types of traditional authority, enforcement and sanctions that characterize traditional systems. Next I examine briefly the factors that lead to the breakdown of systems. A modern role for traditional systems of fisheries management is suggested as a conclusion to the chapter. Although geographically widespread, most information about traditional community-based systems derives from the central and western Pacific.

It is important to keep in mind that what is often labelled 'traditional' may not be especially old-established. Following Nietschmann (1989), in this chapter I do not use the term traditional to connote something necessarily deeply embedded in any local history. Tradition or custom is a practice rather than a principle (Crocombe, 1989).

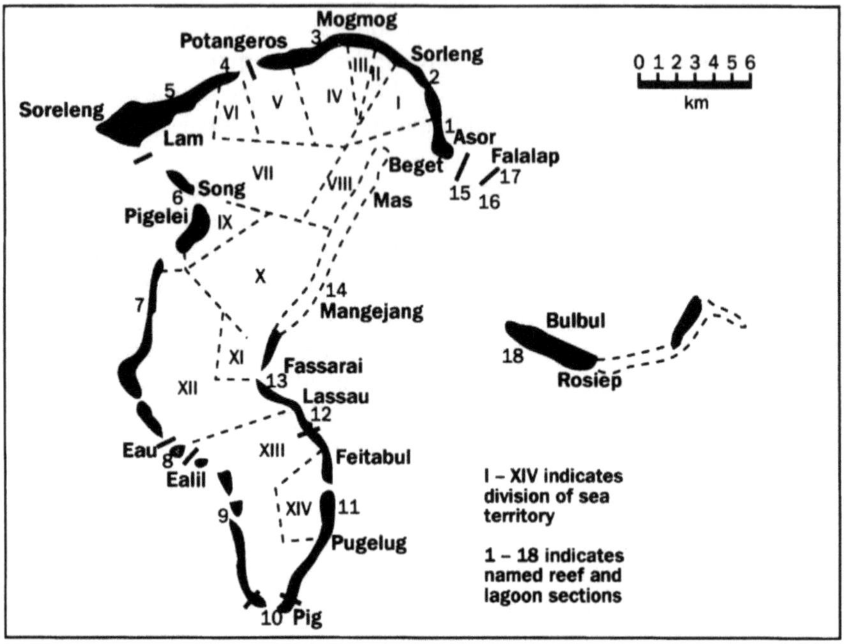

Fig. 12.1 Lagoonal and reef tenure at Ulithi Atoll, Caroline Islands, Micronesia. The clans involved are: Rigipa (reef section 15), Falchugoi (16), Falkel (17), Bogatlaplap (18), Efan (1, lagoon section VIII), Lugalap (2 and 5), Maifan (I), Maiyor (II), Fashilith and Numurui (3, IV and V), Falmay (III), Fashilith (Paramount Chief's clan) (4, 6, VI, VII and IX), Muroch (7, 9, 11, 14, X), Lebogat (8, 13, XII), Tauefan (XI), Fachal (10, 12, XIII) and Ligafaly (XIV). Straight broken lines denote sea territory divisions; broken outlines at 14, 18 denote included continuous sections of named reef. After Ushijima (1982).

12.2 GEOGRAPHICAL DISTRIBUTION OF SYSTEMS

The Asia–Pacific region is especially rich in traditional, community-based reef fishery management systems (Ruddle and Johannes, 1985, 1990; Ruddle, 1994a), but they also occur in the Caribbean (Berkes, 1987; Cordell, 1989; Smith and Berkes, 1991; Espeut, 1992), in Mozambique, East Africa (Wanter, 1990), and in the Middle East (Barth, 1983; R.B. Pollnac, pers. comm.).

Traditional management has been widely, but usually incompletely, described for the Pacific Basin. Such systems have either completely or largely disappeared from the Commonwealth of the Northern Mariana Islands (Freycinet, 1824; Thompson, 1945; Dugan, 1956; Souder, 1987; Amesbury *et al.*, 1989) and parts of Micronesia and Polynesia, whereas they remain in many other parts of those same regions.

The continued importance of traditional systems varies considerably within the FSM. It remains important in Yap State, Chuuk State and the outer islands of Pohnpei State, whereas it has largely disappeared from the main island of Pohnpei and from Kosrae (Foster and Poggie, 1992; Ruddle, 1994a,b). There is little information on the current status of systems in the Marshall Islands. On Nauru it is likely to have disappeared entirely, although in former times there were property rights to inshore waters (Petit-Skinner, 1983). In Palau, traditional resource management has slowly eroded since the Japanese administration formally appropriated the area below the high water mark. Although these rights were re-instated and protected under the *U.S. Trust Territory Code*, old family tenure rights are now ignored (McCutcheon, 1981), traditional rules flouted, the traditional conservation ethic largely abandoned. Traditional authority is ignored and poaching widespread (Johannes, 1981, 1991; McCutcheon, 1981).

In Polynesia, traditional systems have all but disappeared in the State of Hawaii, USA, and have been severely eroded in American Samoa (Wass, 1982; Johannes, 1988b). Elsewhere they remain largely unstudied. In the Kingdom of Tonga, since 1887, by Royal Proclamation of King George Tupou I, ownership of all territorial waters has been vested in the Crown. This was recapitulated in the *Land Act of 1927*. However, Gifford (1929) noted that both inshore and offshore fishing areas were claimed by the adjacent community, and that trespassers were punished. Nowadays fishing is under open access and all traditional rights have lapsed except for the recognition of an owner's exclusive rights, on payment of an annual licence fee, to reef areas surrounded by fish fences and for a distance of 1.6 km around them. In the outer islands, residual notions of village-based rights persist, but have no practical implication these days (Fairbairn, 1992). On Wallis, although in principle family rights extend to the seaward slope of the reef, the lagoon is regarded as being open access. The only exclusive rights are for the removal of sand, which can be done only in waters fronting the collector's own village. Districts control their own sections of reef, but marine boundaries are weakly defined compared with those on land (Pollock, 1992). Little is known of systems in French Polynesia (Tetiarahi, 1987) or Tuvalu.

Traditional systems remain strong in many parts of Micronesia and Polynesia, and especially so in Melanesia. In Kiribati (Micronesia) there is a rich inshore fisheries tradition and lore that includes detailed local rights and regulations. Many traditional management practices have been codified in island by-laws, and so incorporated into contemporary management (Ruddle, 1994a,b). Similarly, in the Cook Islands, Niue, Tokelau, and Western Samoa, elements of traditional systems have been incorporated in contemporary fisheries management (Ruddle, 1994a,b). In Yap State, FSM, traditional fishing rights are among the most complex in the Pacific

Basin (Anon., 1987a; Johannes, 1988b; Smith, 1991). There the supremacy of traditional rights is enshrined in the state Constitution for 12 miles seaward from an island baseline, '...a line following the seaward edge of the reef system...' (*Yap State Code* 18:27). Traditional leaders theoretically have total control over inshore waters, and the government is limited to intervening only for conservation and protection in the State Fishery Zone (Smith, 1991).

Traditional systems remain extensive and diverse in large parts of Melanesia, although relatively little studied. Some have been described for reef fisheries in Ponam (Carrier, 1981; Carrier and Carrier, 1989) and the Trobriand Islands (Williamson, 1989; Tom'tavala, 1990) of Papua New Guinea; for Solomon Islands in general (Ruttley, 1987) and for Lau and Langalanga lagoons of Malaita Province (Akimichi, 1978) and Marovo Lagoon, Western Province (Hviding, 1990) in particular; and for the Torres Strait Islands (Johannes and MacFarlane, 1984, 1990, 1991; Nietschmann, 1989). Briefer accounts exist for the Nenema zone of northwestern New Caledonia (Teulières, 1990, 1991), Fiji (Kunatuba, n.d.; Zann, 1983; Veitayaki, 1990) and Vanuatu (Taurakoto, 1984; Fairbairn, 1990).

In both continental and insular South East Asia, only vestiges remain of what were probably more widespread systems. In the Philippines such systems appear to have been commonplace historically (Blair and Robertson, 1903–1909; Lopez, 1985). Today the coastal or 'municipal' fisheries there operate under open access. However, tradition did not disappear entirely. For example, at Quinlogan Village, a recently settled area on Palawan Island, local management rules were introduced in the 1980s to regulate beach seine operations, allowing equitable access to the prime site for catching shrimp fry (Veloro, 1992). In Indonesia there are few documented examples of traditional management (Polunin, 1984, 1986), although recent research shows that they are still used to manage reef fisheries in parts of the country (Bailey *et al.*, 1990; Zerner, 1991; Bailey and Zerner, 1992a,b,c).

Information from South Asia remains fragmentary (Ruddle, 1994a,b). In Sri Lanka, under statutory law coastal fisheries are managed under open access, but in many localities they operate under local systems of limited entry, based on sociocultural barriers. Traditional rights are limited to beach seines (*madel*), *kattudela* or *bandudela* (stake or pound nets) and *jakotuwa* or *akulwetiya* (fish weirs) (Atapattu, 1987).

12.3 ORIGINS OF SYSTEMS

Origins of traditional fisheries management systems are obscure but must have been diverse. Debate has tended to polarize into origins as conservation devices vs. origins as institutions to prevent or resolve conflict. Some systems might have originated to ensure that taxes were paid. Others might have arisen to prevent resources being pre-empted by other communities, perhaps in response to changes occurring in the larger society. In many areas, systems arose as a response to commercialization, whereas elsewhere they declined under the same external pressure (Ruddle, 1993). In most places, such systems probably evolved for a combination of these and other locally important reasons.

Conservation

Some authors give pre-eminence to considering systems as marine conservation practices based on research in the resource-poor Pacific islands (e.g. Johannes, 1982). Others suggest that ecological conservation, where it can be empirically documented, is essentially an unintended effect of a system of customary law based mainly on political concerns. That is Polunin's (1984) perspective, from Indonesian and New Guinean evidence. He contends that 'sea tenure' is sometimes ecologically dysfunctional. Conservation *per se*, if it was a distinct motive for the origin of systems, is better seen in terms of resource scarcity, which can have both biological and sociocultural causes.

Conflict resolution and avoidance

Evidence from Indonesia and Melanesia indicates an origin in conflict management (Polunin, 1984, 1987). Polunin (1986) shows that conflicts over fisheries access and rights were noted in Irian Jaya (van der Sande, 1907), the Trobriand Islands (Malinowski, 1918), Tanimbar, the Aru Islands and Salayar (Kolff, 1840; van Hoëvell, 1890; Kriebel, 1919), eastern Sumatra (Schot, 1883), and off Aceh, Sumatra, where the shore is partitioned among fishing guilds and an individual was appointed to resolve conflicts (Snouck Hurgronje, 1906). In that region, boundaries of marine territories often seem to have been established arbitrarily (Polunin, 1984). Further, they are subject to quite radical change through time. Patterns have often been determined by opportunism, as when commercialization changed the value of resources, and particularly of *bêche-de-mer*, the edible sea-cucumber. This caused conflict in Tanimbar (Kolff, 1840; van Hoëvell, 1890), and inter-village reserves were established for *bêche-de-mer* on islands off northern Irian Jaya (Feuilleteau de Bruyn, 1920).

Resource scarcity

Resource scarcity was probably a widespread reason for the creation of a management system. A resource was not necessarily scarce for biological reasons, however, thus implying that harvesting had to be regulated to ensure sustained availability. Resource conservation was not necessarily a main motive for the creation of a management system. Examples from Pacific islands demonstrate that scarcity also had cultural roots, and that a management system was designed to conserve first the political structure and not directly the resource involved. In other words, marine resources were deliberately made scarce, in Micronesia for example, by controlling either access to them or the distribution of their benefits, to maintain political order or for other reasons.

Taxation and tribute

Users of marine resources in Indonesia have long been quite heavily taxed, and annual payment to local rulers was often made with part of the turtle-shell or *bêche-de-mer* harvests (Polunin, 1986). The requirement to pay such taxes may be one reason why regulations were established over the resources, to ensure that tax payments could always be made. This was likely also the case in parts of Japan, where resources such as abalone (*Haliotis* sp.) and various seaweeds were for centuries regularly sent as tribute to the court of China (Ruddle, 1985).

Prior appropriation

Systems could sometimes have been instituted to prevent a neighbouring community from claiming an area, prior to any conflict having arisen, in response say, to a resource having acquired a new commercial value. Examples have occurred among the Tigak people of New Ireland Province, Papua New Guinea (Otto, n.d.). Elsewhere the intention could have been to expand the areal extent of a system and institutionalize it in terms of customary law, based on the rumour of impending codification of sea territories within statutory law. This has occurred in Solomon Islands (Ruttley, 1987).

12.4 ORGANIZATIONAL CHARACTERISTICS OF TRADITIONAL REEF FISHERIES MANAGEMENT SYSTEMS

Only in the last two decades has it been realized by scholars that 'sea tenure', or the way in which fishers perceive, define, delimit, 'own' and defend their rights to inshore fishing grounds, exists at all (Emmerson, 1980; Acheson, 1981; Ruddle and Akimichi, 1984; Durrenberger and Pàlsson, 1987). Ironically, some colonial administrators were aware of sea tenure centuries ago. Thus there are few comprehensive accounts of management systems and their institutional arrangements. The problem is compounded by the anecdotal nature of existing literature, lack of recent field work in most places, and rapid decay and disappearance of such systems since Western contact (Ruddle, 1988a). Remaining management systems are commonly hybrids of traditional and modern components, with the former decaying rapidly (Ruddle and Johannes, 1985, 1990).

The difficulties of synthesizing the literature on traditional systems should not be underestimated. A great many problems stem from the lack of precision and definition in the literature. Ruttley (1987) succinctly summarizes the problems in a survey of systems in Solomon Islands, but they are likely to be valid worldwide. The principal difficulties are as follows.

1. Great cultural variety exists, especially within Melanesian nations and South East Asia, making valid generalizations difficult.
2. Available information is scant. Whereas kinship is often stated as the basis for fishing and other resource rights, details are rarely given. Similarly, details of rights are not fully spelled out. For example, it is almost never stated exactly who are outsiders.
3. The terms tribe, clan, reef-owner and leaders are often referred to but almost never defined properly.
4. The 'owners' of reefs as a category are rarely clarified. It is almost never made clear if they are individuals, heads of lineages, clans, tribes, or a village as a whole, or whether the chief acts as a custodian.
5. The role of kinship in the acquisition of right may often be exaggerated, and other mechanisms not considered.
6. Traditional management systems tend to be described in normative, immutable terms, whereas in reality they are known generally to be in a constant state of evolution.

Management systems in the aquatic domain often, but not always, mirror those on land. In traditional systems, an individual's sea rights depend on his or her social status within a corporate community, which ranges from village through clans, subclans, and the like, to the family. Resource territories and user groups are defined. Resource use is governed by rules and controlled by traditional authorities who mete out sanctions and punishments for infringement of regulations.

Authority

In traditional systems, resource control and management is usually vested in traditional authority. Four principal types of authority are recognized: traditional secular leaders, traditional religious leaders, specialists, and rights-owners (Table 12.1). These categories frequently overlap, and responsibility is divided and shared. In Vanuatu, for example, marine

Table 12.1 Principal categories of local authority governing reef fisheries in the Asia–Pacific region

Authority types	Examples and references
Traditional secular leaders	(1) Leaders or a 'village council'. In parts of Papua New Guinea (Williamson, 1989; Tom'tavala, 1990), Solomon Islands (Akimichi, 1978; Ruttley, 1987; Hviding, 1990), Vanuatu (Fairbairn, 1990), Palau (Johannes, 1981), Tokelau (Hooper, 1985, 1990) and Samoa (Buck, 1930). (2) Chiefs. In Micronesia this occurred in the Federated States of Micronesia (Goodenough, 1951; Fischer, 1958; Alkire, 1968; Nason, 1971; Falanruw, 1982, 1992; Ushijiama, 1982; Sudo, 1984; Foster and Poggie, 1992), Kiribati (Zann, 1985) and the Marshall Islands (Tobin, 1958; Sudo, 1976, 1984), and in Polynesia, in the Cook Islands (Beaglehole and Beaglehole, 1938; Crocombe, 1961, 1964, 1967; Utanga, 1988), Niue (Anon., n.d.) and Wallis Island (Pollock, 1992).
Religious specialists	In Maluku, eastern Indonesia (Zerner, 1991), the Marquesas, of French Polynesia (Handy, 1923), Raroia Island, of the Tuamotu Islands (Danielsson, 1956), and Tahiti, in the Society Islands (Tetiarahi, 1987).
Fisheries specialists (under higher authority)	This occurred in Fiji (Thompson, 1940, 1949), Federated States of Micronesia (Emory, 1965; Falanruw, 1982, 1992; Sudo, 1984; Foster and Poggie, 1992) and Kiribati (Teiwaki, 1988) in Micronesia, and in Polynesia in Mangaia, Cook Islands (Buck, 1932), Niue (Anon, n.d.), Tokelau (Hooper, 1985, 1990) and Samoa (Buck, 1930).
Rights-holders (under higher authority)	This occurred on Chuuk (Goodenough, 1951) and Yap (Falanruw, 1992), Federated States of Micronesia, and in Palau (McCutcheon, 1981).

resources are usually controlled by the Village Council, composed of chiefs and elders, or sometimes by an Area Council, made up of leaders from several villages, and by the landowners (Fairbairn, 1990). There is much blurring of authority but usually the Village Council is paramount, although the main chief is often dominant, especially if he is also a major landowner (Fairbairn, 1990). In some places authority is not clearly defined.

Rights

Under traditional systems, marine resource exploitation is governed by use rights to a property. A property right is a claim, consciously protected by customary law and practice, to a resource and/or the services or benefits that derive from it. Such a grant of authority defines uses legitimately viewed as exclusive, as well as penalties for violating those rights. Characteristics of property rights may vary from place to place. They commonly include exclusivity, the right to determine who can use a fishing ground, transferability, the right to sell, lease, or bequeath the rights, enforcement and the right to apprehend and penalize violators (Table 12.2).

The right of enforcement, and in particular that to exclude the free-riding outsider, is a key characteristic, for without it all other rights are diminished. The completeness of a fisher's set of rights provides incentive to invest in the fishery and to act to achieve sustainable benefits.

Almost universal throughout the Asia–Pacific region is the principle that members of fishing communities have primary resource rights by virtue of their status as members of a social group. Rights to exploit fisheries are subject to various degrees of exclusivity, which depend on social organization and local culture. Most commonly, traditional fisheries rights apply to areas, but superimposed on these may be claims held by individuals or groups to a particular species or a specific fishing technology.

Traditional rights to marine resources may be exclusive, primary, or secondary, and may be further classified into rights of occupation and use. Traditional rights are better defined as those to use rather than to own. Rights to use can be exclusive if primary rights holders have a subsidiary right to prevent others from using certain resources within the area over which traditional control is exerted (Pulea, 1985). The relationships between primary and secondary rights are an important and complex characteristic of many traditional management systems, in which overlapping and detailed regulations on the use of technologies and particular species are widespread. Individual rights as subdivisions nested within corporate marine holdings occur widely in Melanesia (Malinowski, 1918; Akimichi, 1978; Carrier, 1981; Johannes and MacFarlane, 1984), Micronesia (Johannes, 1977; McCutcheon, 1981) and Japan (Ruddle, 1985, 1987a,b; Ruddle and Akimichi, 1989). In the Asia–Pacific region, rights of transfer and loan and shared property rights also occur.

Rules

Rules define how a property right is to be exercised by specifying required, permitted and forbidden acts. Thus whereas a right authorizes a fisher to work a specific fishing ground, rules govern options in exercising it and may, for example, specify gear type used or seasonal restrictions, among other limitations. The more complete a set of rights, the less fishers are exposed to the actions of others (Ruddle, 1994a).

Table 12.2 Examples of the principal types of rights governing reef fisheries in the Asia–Pacific region. After Ruddle (1994a)

Types of rights	Examples and references
Exclusive	Exclusive rights have been transmitted through ancestral families, spirits or gods, and are validated by historical–mythological associations. Subsequently, rights in specified territories are defined by customary law (Pulea, 1985).
Primary	Primary rights are usually inherited as a birthright by direct descent from the core of a descent-based corporate group (Ruddle, 1994a). They generally confer access to all resources within a defined territory. For example, in Marovo, Solomon Islands, individuals obtain the primary right to both use and control territories by inheriting group membership from both parents (Hviding, 1990).
Secondary	Secondary rights are more limited and are acquired through affiliation with a corporate group, by marriage, traditional purchase, exchange, as a gift, in exchange for services, and sometimes inheritance. They are often given to residents of historically associated inland villages, as in Western Samoa (Fairbairn, 1991).
'Nested' rights	Areal fisheries rights are sometimes overlain by species and gear rights. Most are simple 'nested' rights, such as to use stone fish traps, as in Palau (McCutcheon, 1981). Complex systems occur, as at Ponam Island, Manus Province, PNG, where overlapping and countervailing sets of 'nested' rights to reef and inshore marine waters, species and fishing techniques occur (Carrier, 1981; Carrier and Carrier, 1989).
Gear	Gear rights are widespread, especially for fixed gear. On Chuuk, Federated States of Micronesia, persons who construct traps have provisional exclusive rights to the area, unless they transfer them to another (Goodenough, 1951). Complex countervailing gear rights sometimes occur, as on Ponam Island, PNG (Carrier, 1981; Carrier and Carrier, 1989).
Species	Access rights and control over species can be an alternative to or complementary to the control of territory. The harvesting and distribution of valuable species was often traditionally controlled by persons in authority. In the Pacific islands, chiefs often have rights to particular species. This is commonly applied to turtles, as at Lifou, New Caledonia (Teulières, 1990, 1991), and Kapingamaringi (Emory, 1965) and Ifaluk (Burrows and Spiro, 1953), Federated Sates of Micronesia. On Butaritari and Makin islands, Kiribati, chiefs had rights to stranded dolphins, large fish and probably also to turtles (Lambert, 1966). On Kapingamaringi (Emory, 1965) and Ifaluk (Burrows and Spiro, 1953), atolls, only older men were permitted to fish for tuna, which then gave them a say in the distribution of the catch (Lieber, 1968).
Habitats	See Ruddle, Chapter 6, this volume.
Of transfer and loan	In some systems the permanent, temporary, or occasional transfer of rights to other social units was permitted. In others, however, it was forbidden. Compensation was often required.

Basic rules define the geographical areas to which rights are applied, persons eligible to fish within a community's sea space, and access by outsiders. Operational rules govern fishing behaviour, gear usage and allocation issues, as well as specifying unacceptable fishing behaviour, conservation practices and distribution of the catch within a community (Ruddle, 1994a) (Table 12.3).

Monitoring, enforcement and sanctions

If fishing rights are to be meaningful, there must be capability for monitoring compliance and imposing sanctions on violators. In the Asia–Pacific region, monitoring and enforcement are generally undertaken within the local community; resource users police themselves and are observed by all others as they do so.

For a variety of reasons, traditional authorities frequently imposed temporary or permanent bans, as well as spatial, temporal, gear or species restrictions on the exploitation of marine resources. These commonly took the form of taboos.

Sanctions were widely invoked for the infringement of fisheries rights and the breaking or ignoring of locally formulated rules. Four principal types of sanctions were widely invoked: social, economic, physical punishment and supernatural (Table 12.4).

12.5 BREAKDOWN OF TRADITIONAL COMMUNITY-BASED MANAGEMENT SYSTEMS

Traditional management systems are increasingly stressed by external factors that cause radical change, including their demise. This is not new, but the intensity and diversity of stresses has increased in recent decades. Thus contemporary systems exist under environmental, social, economic, political and demographic circumstances that often differ greatly from those of even the recent past. Nowadays such systems are swept up in the overall process of national development.

Among the main, all-pervasive external forces are the legacy of colonialism, the impact of Christianity (especially in the Pacific islands), contemporary government policy and legal change, the replacement of traditional local authority, demographic change, urbanization, changes in education systems, modernization and economic development, commercialization and commoditization of living aquatic resources, technological change, the policies of external assistance agencies and national policies for economic sectors other than fisheries. Such external forces usually act

Table 12.3 Examples of rules governing reef fisheries in the Asia–Pacific region. After Ruddle (1994a)

Territorial definition	Territories are usually near or adjacent to a community's settlement. Lateral boundaries are generally the seaward projection of terrestrial boundaries, except where a distinctive physical feature occurs. Most seaward boundaries are marked by the seaward slope of the reef.
Eligibility	In addition to rights, fishing is limited by community-based, national or cultural rules. In most Asia–Pacific societies, membership of a corporate descent group and/or residence are the only rules. Other preconditions also occur, as in Okinawa, Japan, where training, sometimes formalized through apprenticeships (Akimichi, 1984; Ohtsuka and Kuchikura, 1984), was also required.
Inter-community access	Rights of outsiders are usually closely specified by access rules. Prior permission and rules specifying payment were widespread. Neighbouring communities are usually granted access rights, and subsistence fishers generally receive easy access whereas commercial fishers must usually pay.
Gear	Gear rules are widespread. The principal categories are (1) proscribed gear or size regulation, (2) gear regulation for social equity (often combined with temporal allocation rules), (3) regulation of placement to minimize conflicts, and (4) reservation of gear types and fishing techniques to particular social classes or to fishers who fulfil eligibility rules.
Temporal allocation	Rotation systems for allocating space-time among fishing groups are widespread, to promote orderly and equitable fishing, as in reef fisheries in Okinawa Prefecture, Japan (Ruddle, 1987a; Ruddle and Akimichi, 1989).
Fishing behaviour	Commonly, rules aim to promote orderly fishing and protect fish schools. For example, on Niue, behavioural rules used in the round scad fishery ban individual fishing, require fishers to follow strictly the 'fleet leader's' instructions regarding positioning and formation of the fleet, bait type, timing of bait release, noise levels and general behaviour (Anon., n.d.). (These were codified in the *Niue Fish Protection Ordinance*, 1965.)
Species rules	Some societies regulate fishing for particular species, as in villages of the Maluku Islands, eastern Indonesia. These rules applied to different species and are based on whether they are regarded as resident, schooling or non-schooling or migratory fish. In general, all villages have rights to fish in their village sea area, but schooling fish, like tuna, can be harvested only under the direction of a traditional religious specialist (Zerner, 1991; Bailey and Zerner, 1992c).
Conservation rules	In many Pacific island societies, a wide range of conservation rules are applied. These include areal closures, temporal closures (particularly during spawning), live storage or freeing of surplus fish caught during spawning migrations, reservation of areas for fishing during bad weather, size restrictions (although this was rare), and, in recent times, gear restrictions (Johannes, 1978b, 1981, 1982).

Table 12.4 Examples of traditional sanctions applied in reef fishing, by subregion and nation in the Asia–Pacific region. After Ruddle (1994a)

Subregion/nation	Example location	Type of sanction	References
South East Asia			
Indonesia	Porto Village, Saparua Island, Maluku Province	Social: public shaming Physical: corporal punishment (caning) Economic: confiscation of boat and gear, and monetary fines Supernatural: disappearance of gear, sickness, death	Zerner (1991)
Melanesia			
Fiji	Nation-wide	Physical: corporate punishment (beating) Economic: confiscation of catch and destruction of boat and gear	Kunatuba (n.d.); Zann (1983)
New Caledonia	Nenema Zone	Supernatural: sickness	Teulières (1990, 1991)
Papua New Guinea	Labai Village, Trobriand Islands	Physical: capital punishment	Williamson (1989)
Solomon Islands	Nation-wide	Social: reprimands and shaming Economic: compensation (traditional commodities, cash, or return of fish stolen)	Ruttley (1987), Hviding (1990)
	Marovo Lagoon, Western Province	Social: reprimands, shaming, and ostracism in severe cases	Hviding (1990)
Vanuatu	Efate Island	Economic: compensation (traditional commodities)	Fairbairn (1990)
Micronesia			
Federated States of Micronesia	Yap State	Economic: confiscation of gear, boats, land and/or fishing rights Physical: corporal punishment (beating); capital punishment	Anon. (1987a), Falanruw (1982)

Table 12.4 continued

Subregion/nation	Example location	Type of sanction	References
	Pohnpei State	Physical: war	Fischer (1958)
	Pohnpei State Outer Islands Nukuoro Atoll	Economic: Temporary loss of fishing rights	Foster and Poggie (1992)
Kiribati	Nation-wide	Social: violators shamed into committing suicide Economic: monetary fines Physical: capital punishment	Lambert (1966), Zann (1985)
Marshall Islands	Nation-wide	Economic: loss of land rights Physical: capital punishment	Tobin (1952, 1958)
Nauru	Nation-wide	Physical: capital punishment	Petit-Skinner (1983)
Palau	Nation-wide	Economic: monetary fines Physical: capital punishment	Johannes (1981)
Polynesia			
Cook Islands	(Unspecified)	Economic: in-kind compensation Supernatural: (unspecified in source [Utanga])	Mokoroa (1981), Utanga (1988)
Niue	Nation-wide	Social: shaming and humiliation	Anon. (n.d.)
Tokelau	Nation-wide	Social: shaming and humiliation	Hooper (1985, 1990)
Samoa and American Samoa	Nation-wide	Economic: monetary or in-kind fines; progressive loss of fishing rights	Buck (1930)

as a mutually reinforcing and potentially destructive complex (Ruddle, 1993).

A somewhat more recent pressure – but not universally so – is the commercialization of formerly local and mainly subsistence or barter economies, which now links them with external markets. This, in turn, leads to changed perceptions in fishing communities regarding the value of marine products, and often to external factors being internalized by village elites, and so to the breakdown of traditional management systems through the weakening or total collapse of traditional moral authority. Small communities are not immune from the pressures that drive larger polities and commercial elites, and which undermine the moral imperative of local management systems from within. Regional and global markets also have a direct impact on them: external incentives introduce temptations for individual profit at the expense of local social equity and thus undermine systems from within.

Community institutions and management systems are not immutable. They adapt dynamically to external and local experiences and pressures, many of which are not directly related to fisheries. Participants cannot be assumed a priori to act benignly for resource conservation and socially equity, nor should these effects of existing local institutions be romanticized. Hence policy and programme decisions about the present and future usefulness of local management systems must be based on a clearheaded and realistic evaluation of the moral authority, motives, interests and cultural conceptions that underpin them (Ruddle, 1993).

Traditional management systems decline under pressures exerted by internal and external sources, and the latter can trigger the former, such that local phenomena may mask deeper-seated problems afflicting social institutions. Such systems are dynamic, historically conditioned and deeply embedded in larger political, economic and social realms. Traditional management systems vary enormously, so discussion of the external factors that impinge on them must be generalized with many local exceptions.

12.6 A MODERN ROLE: EXPLOITING THE ADVANTAGES OF TRADITIONAL COMMUNITY-BASED SYSTEMS

Unlike conventional fisheries management, traditional systems focus on resolving gear usage and allocation problems, are based on defined areas and controlled access, are self-monitored by local fishers and are enforced by local moral and political authority. These are the great strengths of such systems, and these are what they have to contribute in modern management designs (Ruddle, 1994a).

Problems of gear usage and assignment are overcome in traditional systems at the first level by (1) control of a fishing area as a property, and

(2) defining exactly, by rights, who has access to that area. At the second level, rules of operational behaviour then specify assignments of time and place within that area and group having access. The first level is sustained by rights of exclusion, or limited access, that maintain the private area of a community of local fishers against outsiders. The second level, intra-group operational rules, is sustained by local authority that has the power to invoke sanctions on offenders (Ruddle, 1994a).

Under well-functioning traditional systems, rent dissipation is overcome because access is controlled. Further, in a great many systems, sustainable harvesting practices are enforced, thereby leading to resource conservation.

Although eroded or even broken down in most regions, community-based systems are still used to manage reef fisheries in a wide range of societies. Thus it is increasingly asserted, although usually with scant proof, that traditional management can play a potentially major role in the modern world by ensuring equitable access to fisheries, as well as in managing and enforcing conservation measures to ensure sustainability. The thesis generally is that the more responsibility for control of local resources can be left to local, traditional users, the fewer the social, political, legal, conservation-related and management problems that must be addressed by governments (Ruddle, 1988a). However, not all systems will be of equal usefulness. Cautionary points regarding the modern role of traditional reef fisheries management systems include the following.

1. Most positive assertions regarding the modern role of traditional systems still need verification.
2. The wholesale transfer of concepts is hazardous because, by definition, such systems arise from the deeper cultural patterns of the societies in which they are enmeshed (Ruddle and Akimichi, 1984; Durrenberger and Pàlsson, 1987). So, much more than an understanding of just the local, traditional fishery is required; entire national systems of fishery production, and particularly the relationship between household (traditional) and capitalistic (modern) production, require understanding (Ruddle, 1988b).
3. Traditional systems could be 'fossilized' through explicit, detailed legal definition in the terms of statutory law. This may weaken the adaptive flexibility of a traditional system (Ruddle and Johannes, 1985) unless flexibility is explicitly legislated for.
4. The application of traditional knowledge and practices to contemporary resource management problems is also a relatively new approach, but one that is now the focus of considerable academic and applied interest, partly because of the inadequacy of the biological and economic models usually applied (Ruddle, 1994a). Largely as a consequence of this newness, the relevant concepts and methodologies are not yet well defined.

5. Perhaps most important is that traditional systems are not an automatic godsend to fisheries managers. They create difficulties. Not uncommonly, therefore, governments and entrepreneurs attempt to either weaken or invalidate them.

At first sight the adaptation of traditional systems to a modern purpose may appear to invite strong local resistance, because they are often so much a part of the way of life. But traditional systems already incorporate important elements of conventional fisheries management. For example, parallel management strategies include limited entry, seasonal, spatial, gear, size or species restrictions, prior appropriation rights and the concept of sole ownership, among others (Johannes, 1978b). In fact, the use of many such strategies in the Pacific Basin (Johannes, 1978b) and Japan (Ruddle, 1985, 1987a) antedated their adoption in the West. In conventional economics terms, sole ownership, limited entry, individual transferable quotas and other such management schemes are based on the theory of the firm. On the other hand, in many societies in the Asia–Pacific region the community is the sole owner, and traditions of resource use and management are enforced by community norms that control the behaviour of the membership (Ruddle, 1988a). But this, too, has its parallels in Western fishing communities, as among the lobster fishers of Maine, USA, for example, where socially binding yet unwritten and informal rules carry more weight than official regulations (Acheson, 1987).

Most developing nations face an array of dilemmas in determining rights and responsibilities in marine resources management and development. These include what institutions should manage and enforce regulations for subsistence fisheries, legal support for traditional regulation and enforcement, the managerial and developmental role of the central government in small-scale commercial fisheries, the feasibility of centralized management plans versus local decision making, and the nature of the consultative and collaborative process among fishers, local governments and national authorities. Initially these look like local versus central jurisdictional matters, but the underlying issue is of the policy and means of managing marine resources, and of adapting traditional concepts to modern needs such that the range from subsistence fishery to highly commercialized industrial fishing is served properly.

Alternative arrangements can help overcome the weaknesses of conventional fisheries management. The most appropriate form of fisheries governance is one in which management authority is decentralized, within a broad policy framework, to enable local governments to fundamentally control local fishing via community-based systems. In co-management systems, decision making is shared between central and/or provincial governments and community-based authorities. Such arrangements have long existed *de facto* in many parts of the developing world. Thus, especially in the archipelagic nations, co-management has, in effect, long been practised.

Invalidation of systems

It makes little sense to prolong the existence of community-based management systems that have outlived their historical usefulness. Such a situation arises most clearly near urban-industrial centres, as noted above, where depending on the density of onshore developments, the invalidation of systems could also be justified by the potential health hazard of fish taken from polluted waters. Weakening or invalidating traditional systems is a course of action that can be justified where such systems impede alternative and more important uses of coastal marine space.

Negative consequences of invalidation

Whereas in many cases community-based systems ought to be invalidated or weakened in the national or regional interest, when such a policy is implemented nation-wide it carries with it enormous costs. This is particularly obvious in far-flung archipelagic nations, such as Indonesia, the Philippines, Kiribati, Tuvalu, Solomon Islands or Vanuatu, but no less of any developing nation that lacks the financial, physical and personnel capacity to police its inshore waters. Solving this major problem of costs provides one of the most valid reasons for retaining well-functioning community-based management systems. In dispensing with traditional systems a 'government would thus be disposing of services it got for free and assuming expensive new responsibilities it was ill-equipped to handle' (Johannes, 1988b). As Bailey and Zerner (1992a) observe of Indonesia:

> The Indonesian government is incapable of designing effective fishery management systems due to limited understanding of the complex and highly variable nature of fisheries resources. Government management policies which fail to recognise local institutions and economic needs may be creating more problems than they solve. Moreover, the Indonesian government has limited ability to enforce what regulations are in place due to staff and budgetary constraints.

But the management ability of local community systems, based on a depth of local knowledge, is quite the opposite. However, apart from the exceptions noted in the previous section, to be effective these local rules require recognition, acceptance and protection under statutory law.

Practical considerations, such as the ability of a poor and/or large nation to police its fisheries, often lead logically to the alternative of community-based management. Under certain conditions, traditional systems solve inadequacies of centralized, or single-sector, decision-making. Co-management is a logical approach (Bailey and Zerner, 1992a).

There is a great future role for community-based systems in providing regular information on activities, particularly on coasts along the unvisited rural hinterlands of far-flung archipelagic states. In those mostly subsistence-level societies, where traditional authority remains strong, enforcement and punishment are often largely traditional but can also be used to serve a modern purpose. Traditional punishment can be severe and feared more than that meted out by government, as in Okinawa (Ruddle and Akimichi, 1989), Palau (Johannes, 1981) or American Samoa (Wass, 1982). As Wass (1982) observes of American Samoa, 'Management regulations instituted on the village level are much more effective than those of the territorial or federal governments because they are promulgated within the cultural context by traditional leaders and, consequently, are more likely to receive the approval and fealty of the villagers.' Thus where traditional authority remains strong, a community-based system can still provide a solid foundation for modern fisheries management.

Replacing a traditional system with open access would entail the risk of the discouraging results of conventional management (McManus, Chapter 10). But in tropical regions, those problems would be compounded by:

1. the multispecies nature of fisheries, which would require more cumbersome regulations and correspondingly more enforcement than systems in temperate waters;
2. the scantiness of biological data for use in management and the large percentage of the small-scale catch that is used for subsistence which create immense logistical problems with developing essential data sets from very widely scattered fishing communities;
3. the vast number of geographically scattered fishing units which create almost insuperable financial and logistical problems for regulation and monitoring compared with Western commercial fisheries;
4. the zeal with which data are collected and analysed, together with poor official enforcement of regulations and lack of professionalism among officials, which leave much to be desired; and
5. most governments, which are too poor, or give fisheries too low a priority, to implement conventional regulatory systems that are required by open access regimes or to handle the resultant problems (Appeldoorn, Chapter 9).

ACKNOWLEDGEMENT

I thank Nicholas Polunin for arranging the illustration.

Chapter thirteen

Modern institutional framework for reef fisheries management

Timothy J.H. Adams

SUMMARY

Alternative goals in tropical reef fisheries management include maintenance of similar total catch levels, but with a changed, or changing, balance of species; maintenance of a long-term cumulative catch via a series of short-term fluctuations; prevention of serious damage to stocks; avoidance of conflicts over resource use; or simply prevention of overfishing. Management options include a continuum of possibilities from dialogue between individuals to international conventions. All reef fisheries are subject to some convention or agreement on exploitation, although in remote areas there may be no effective control on use. There are two contrasting poles of practical management for reef fisheries. Pre-emptive management requires detailed knowledge of the target organisms and fishing community. By contrast, retrospective management is a form of adaptive management in which rules are developed on the basis of experience of their effects on stocks. Continued feedback theoretically provides an increasingly precise system of management. Pre-emptive management requires much expensive research whereas retrospective management is risky; the ideal management plan combines elements of both. Much information is needed for developing management models, coming

Reef Fisheries. Edited by Nicholas V.C. Polunin and Callum M. Roberts.
Published in 1996 by Chapman & Hall, London. ISBN 0 412 60110 9.

from a broad range of sources from resource users and exploited stocks to societal, national and international levels. Good management plans involve feedback, cycling information from one level to another and particularly involving return of information to resource users.

Most fisheries require some form of regulation. Regulatory approaches available include investment guidelines, fishing licence issue and renewal criteria, trade licensing guidelines, community or government regulations and regional or international access agreements. Community-level regulation is probably the most cost-effective and practical for countries with a high proportion of subsistence and small-scale fisheries. However, national legislation can empower and support community management without necessarily giving up sovereignty over resources. In many developing countries the national level is also the lowest at which sufficient resources can be dedicated to supporting fishery specialists and conservation measures. Export fisheries are most cost-effectively managed at the national level. Regulations are useless if they cannot be enforced but compliance is always more desirable than enforcement. Community-appointed wardens, recognized and empowered by government, can be very useful in enforcing regulations, both traditional and modern. If biologically appropriate management measures are also appropriate to the human element of the fishery, they will be easier to comply with and will require less enforcement. Conflict between the fishing community and government managers can occur because measures are not usually introduced until a resource is demonstrated to be overfished, and catch or effort has to be cut. Avoidance of extreme conflicts can be a powerful argument to convince senior civil servants to promote measures in advance of actual problems in the fishery. Conditions fostering good relations, or at least understanding of the aims of every group involved, should be inherent in every institutional management framework.

13.1 INTRODUCTION

Some management of reef fisheries is based on long-established precedents, and implemented at the community level through traditional means (Ruddle, Chapter 12). In this chapter I shall treat traditional management as just one, but often a very significant, component in a continuum of possibilities ranging from dialogue between individuals to international conventions. This chapter will not go into great detail about the reasons why management might be imposed, or try to summarize the full range of measures that might be used (Munro and Fakahau, 1993a), but will concentrate on the underlying processes and practicalities of managing tropical reef fisheries. Fundamental to this discussion is that the consideration of a fishery is not limited to consideration of the resource only. The human element is crucial (Hilborn, 1985; Ruddle, Chapter 6; McManus, Chapter 10). Human beings are, after all, the reason why fisheries management is necessary in the first place.

13.2 TYPES OF INSTITUTION

Institutions are mechanisms for the formal expression of common principles. They exist to facilitate cooperation among human beings. The institutional framework for reef fisheries management covers a range of increasing levels of agreement.

1. *No institutions*: Reef fisheries not subject to any kind of agreement concerning their exploitation are non-existent, although there are many remote reefs where no effective control exists. Since the United Nations Convention on the International Law of the Sea legitimized the claims of coastal States to 200-mile exclusive economic zones (EEZ), even the remotest reefs have become significant as potential EEZ boundary base-points.
2. *Agreements among resource users*: Many traditional fishery management measures have evolved out of agreements among individuals to maintain exploitation of a shared resource. Such agreements usually occur where there is access to limited areas, or where fishing is carried out by individuals within the same clan or village. Informal management of this kind is much less effective where the management unit covers a large area, or where the resource users themselves have little in common.
3. *Agreements between resource users and owners*: At the local community level, this is difficult to separate from the previous class because many resource owners are also resource users. A community often controls fishing by outsiders on a certain reef whilst continuing to fish there themselves. However, in some societies this is more formalized, with a community leader or committee taking overall responsibility for controlling exploitation. Such management, where control is at the community level, with the management unit on the scale of an individual reef, and with resource owners in day-to-day contact with resource users, can be extremely effective (Ruddle, Chapter 12). Feedback is implicit and the fine-grained nature of the system is well adapted to patchily distributed resources.
4. *Agreements at the provincial, state or national level*: Here contact among resource users and owners is indirect, mediated by specialists, and formal, codified procedures are followed. Many modern fisheries are managed at this level, particularly those of the industrialized nations,

although this is perhaps more due to the nature of fisheries in temperate waters than to the effect of industrialization on social systems. Temperate and continental shelf fisheries tend to be less patchy than reef fisheries, with a higher biomass concentrated in relatively few major species. As a consequence, management at this broader scale tends to be most effective in temperate waters. However, the national level is often the lowest level at which sufficient resources can be pooled to support fisheries research and fundamental conservation measures in developing countries, and is usually the optimum level for controlling commodity export fisheries.

5. *Agreements at the regional level*: Whilst the geographical region is probably the optimum level for management of highly migratory species such as tuna, and high-seas fisheries, reef fisheries require a much finer-grained approach to management than can be provided by regional institutions, particularly when sovereignty is purely vested in nations. However, regional-level agreements can be a useful tool in the management of export fisheries and, for small nations, are the most effective level for pooling research resources.
6. *Agreements at the global level*: There is probably little place for global institutions in direct fisheries management in the foreseeable future. Most fisheries are most effectively managed, or monitored, at the spatial scale of the unit stock. Yet there are few marine species where the unit stock can be said to be global in distribution. Industrial fisheries management regimes, which the United Nations has actively promoted, are usually implemented at the regional or subregional level. International agreements that affect trade, such as the Convention on International Trade in Endangered Species (CITES) and the General Agreement on Tariffs and Trade (GATT) can, however, have considerable indirect consequences for reef fishery management (Daly, 1993).

Global consideration of fisheries management issues often means that 'economic minority' fisheries, including most reef fisheries, receive a disproportionately small share of attention, and dialogue is monopolized by industrial fisheries. Total global marine fishery production was around 80 million tonnes in 1990 (FAO Fisheries Department, 1992), of which approximately 1% was derived from reefs. The importance of reef fisheries depends on how they are evaluated. If weighted instead by the number of economically significant species involved, reef fisheries might be considered as important as many others. Fiji, for example, records commercial fishery statistics for over 100 finfish species and nearly 50 reef invertebrates (Fiji Fisheries Division, 1992), whilst the western seaboard of the United States and Canada reports on a total of 31 commercial fish and invertebrate species (FAO Fisheries Department, 1992). If weighted by the number of people currently dependent on them for their livelihood, reef fisheries would be amongst the most important in the world, and if weighted by the number of countries in which reef fisheries are important (many of which are small island countries), consideration of these fisheries would be paramount.

It is difficult to judge the level in the institutional continuum at which management of a fishery would be most effective. It depends not only on the biology of the target species, particularly the extent of the unit stock (Appeldoorn, Chapter 9), but also on the social organization of the fishing community targeting that stock (Ruddle, Chapter 6; McManus, Chapter 10). Reef fisheries are probably most effectively managed on a day-to-day basis at the local community level, but without governmental support local communities sometimes have problems in dealing with outsiders. Conversely, some invertebrate export fisheries are most cost-effectively managed at the national level, through control of trade outlets, but may require local community cooperation to be effective. For example, a total national allowable catch (TAC) for export will not take into account the possibility that this TAC is taken from limited reef areas and not spread across the whole area assumed in the original calculation, unless there is some method of locally certifying the origin of each shipment.

13.3 TYPES OF MANAGEMENT

Fisheries management is not an exact science. There are too many unsecured variables to make accurate predictions. One way of trying to cope with this is to discount the human element in fishery management and to concentrate research effort on resource management instead. One school of thought says that the human element can be treated simply as a single controllable parameter of the resource–behaviour model. However, experience of the resultant 'fishery' management regimes in action tends to suggest otherwise. Such regimes cannot cope with market booms caused by the installation of a new airline connection, or a sudden increase in the use of explosives for fishing due to a change in mining methods. Only a very broad information base, a modicum of intuition and a great deal of flexibility can accomplish that. It is occasionally useful to recall that there is no such thing as a single-species fishery. All fisheries involve at least two species and one of them is *Homo sapiens*.

Much has been written about management of temperate, shelf and fluvial (and lately, tuna) fisheries, and most theories are designed around single- or few-species fisheries targeted by industrial methods. The diversity of edible fishes and invertebrates is much higher in the tropics, and is one of the reasons why tropical near-shore (including reef) fisheries are considered to present the most complex management problems in the world (Longhurst and Pauly, 1987; Hatcher *et al.*, 1989). This chapter is based mainly on my own practical experience of overseeing the management of the fisheries of a Pacific small-island nation.

To apply temperate fishery management methods effectively to a reef fishery often costs more than the possible commercial value of the species targeted. This rules out any possibility of resource users eventually bearing the full cost. Most legal measures in place to control reef fisheries in small island countries are low-cost, passive measures. Minimum size limits, enforced mainly at the point of sale, are common, and many countries do not even have a licensing system for domestic commercial fisheries. The recent intensification of research into traditional and community fisheries is welcome because community-level management is potentially far more cost-effective than day-to-day governmental control (Ruddle, Chapter 12). Being relatively geographically fine-grained, it also has much greater potential to be biologically effective. The study of community fisheries should not be dismissed as the mere recording, for posterity, of obsolete traditions 'before they are lost for ever'. Local systems are evolving all the time, and the mosaic of different management measures that are introduced by local groups to cope with widespread change constitute an extensive series of experiments (McManus, Chapter 10). Such patchworks are most noticeable with the prosecution of a new fishery (for example, the periodic demand for certain types of invertebrate for export markets).

Adaptive fisheries management assumes that information can be gained about a fishery by observing the result of applying contrasting management measures. Altering the value of one of the variables (such as massively increasing fishing effort for one season) and then observing what result this has on the fishery is a type of management by experiment. For such experiments to be effective, the results must be observed (Hilborn and Walters, 1992). Local communities observe the results of their management changes and adapt accordingly, whilst researchers observe the results of their own management experiments and amend theory accordingly. Scientists, however, rarely observe the results of local community adaptations. Whilst these adaptations are not scientific, they can provide a very cost-effective source of information for the formulation of hypotheses to be tested scientifically.

The fundamental basis of effective institutional fisheries management is feedback, particularly if management involves an element of experiment. Whilst feedback between harvester and manager is implicit in most local communities, more formal monitoring systems are needed if the management institution is broader than the community. Regulation without feedback is useless unless the fishery is completely steady-state and invariable, which is probably never the case.

For reef fisheries there are two contrasting poles of practical management.

1. *Pre-emptive, or strategic, management*: This requires much knowledge about the biology of the target organism(s), about the fishing community, and about the effect it will have on the resource. It formulates a model of a fishery based on this prior knowledge and applies management measures to the fishery designed (theoretically) to produce a particular result (usually a certain level of catch) at a future period.
2. *Retrospective, or evolutionary, management*: This assumes no detailed initial knowledge. Fishing occurs under a certain set of rules and the result is observed. If catch rates decline, the rules are made more restrictive, and if they do not, the rules can be relaxed. Continued feedback provides (theoretically) an increasingly precise system of management. Adaptive management, in the sense of Walters (1986), is a type of evolutionary management where the set of rules is manipulated to produce an unambiguous and testable 'signal' in the fishery.

Both extremes have their limitations. Pre-emptive management requires much expensive research whilst retrospective management is risky. There is a danger, albeit remote, of an extreme set of rules adversely affecting a fishery for years. The more likely danger is that feedback may not come through in time, or that the management system is too inflexible to adapt quickly, leading to suboptimal catches. Perhaps the most insidious drawback of retrospective management is that it rarely takes account of multi-year climate fluctuations such as the El Niño southern oscillation. Polovina *et al.* (1994) point out the need for resource managers to heed interdecadal climate events. Recent restrictive amendments to the management plan for the Hawaiian spiny lobster fishery (WPFMC, 1991) illustrate how a management plan, based on cycle upswing years, may need drastic revision to account for reduced recruitment during a downturn.

The ideal management plan will, of course, combine elements of both approaches.

13.4 REASONS FOR MANAGEMENT

The most fundamental question of fisheries management is why manage? The introductory paragraphs on types of management systems assume that the purpose of management is to maintain a steady-state fishery with a constant, economically optimal, catch level for each component species into the indefinite future. However, other goals are often more achievable and these include the following.

1. Maintenance of similar total catch levels, but with a changed, or changing, balance of species. Depending on the sequence of fishing methods, this may be due to selective removal of top predators allowing an increased biomass of prey species, with the ultimate climax assemblage consisting of a preponderance of high-mortality-rate species.
2. Maintenance of a cumulative catch over the long term, but via a series of short-term fluctuations. As long as the resource concerned is not too fragile (it is not recommended, for example, for turtles, deep-water snapper, or giant clams), this is a viable low-cost management aim for an export commodity fishery such as sea cucumber (Adams, 1993). However, long-term cumulative catches are likely to be lower than under a steady-state regime, and it is unlikely to encourage stable investment. Such management is definitely not recommended for a subsistence, or staple-food fishery.
3. Prevention of major damage if the aim of effectively influencing, or fine-tuning, the biological aspects of a fishery is either not possible or hopelessly expensive. Measures such as minimum sizes for capture and gear regulations can be designed to provide effective protection to a portion of the stock, and provision can be made for introducing selective bans on threatened species as necessary.
4. Avoidance of conflicts over resource use or resource ownership by concentrating on managing the human element of the fishery. This could be to protect the rights of certain elements of the population (such as subsistence coastal dwellers), to attract foreign investment, or even to maximize the take-home emoluments of fisheries officials. Such factors are often of far more significance to reef fisheries management than biology.
5. Prevention of overfishing is possibly the simplest aim of management. The definition of overfishing, however, is difficult (Pauly *et al.*, 1989; Jennings and Lock, Chapter 8).

We can now consider mechanisms of management. To do this I divide the management regime into fundamental components of information, regulation and enforcement. I will concentrate on measures used at the national government and local community levels.

13.5 INFORMATION SOURCES

If a fishery is to be managed for a particular purpose, information is essential. Surprisingly this is often overlooked when human and financial resources are allocated by institutions involved in fisheries management.

Some reef fisheries are, in fact, managed by inertia, with little reference to changing circumstance.

Information is needed for developing management models, using general systems or primary data. Detail and secondary information about changes with time are also needed for putting management into practice – for fine-tuning the pre-emptive components of the management model, and for adapting to changes in the fishery itself, these changes resulting from the imposition of the management model or from exogenous influences. Information about all of the factors that affect a fishery must be included. The biology of the resource is only one component of the information needs of effective fishery management, which cover the following range.

1. *Feedback from resource users*: This is the most important information linkage to maintain. At the local community level this comes from day-to-day conversations and semi-formal community meetings, but is not usually documented and is difficult for external researchers and national-level managers to access. More formal types of feedback are catch returns (not usually feasible for small-scale and subsistence fishing activities) or occasional surveys of landing points and fishing grounds. Such numerical channels of information sometimes become routine to the point of inaccuracy, as harvesters' responses drift gradually from the standards initially calibrated. For example, a reporting form that measures the landings of reef fish catches by 'string' may not take account of changes in the number of fish on the string as marketing methods change. Fisheries officers should also be encouraged to provide intelligence; they can report anecdotally on events in the fishery. The most relaxed and gossip-prone offices often have a better idea of how a fishery is performing than those with a slavish devotion to numerical data. They are less able to justify their conclusions, however, and a balance must be struck.
2. *Feedback from direct resource status monitoring*: Most finfish fisheries are not amenable to direct resource status monitoring because the records that are most easily collected, such as catch returns and creel surveys, monitor catch status, not resource status. No fishing gear is completely efficient. Catch rates vary (Dalzell, Chapter 7) and catch per unit effort is often difficult to relate to abundance (Hilborn, 1985; Jennings and Lock, Chapter 8). Invertebrates, being relatively sedentary, tend to be more amenable to direct observation. Estimates of resource status derived from area surveys are usually more reliable, given equal sampling effort, than those derived from catch monitoring. There are of course exceptions. Some invertebrates are cryptic and some fishes are sedentary. Most fisheries departments collect data from fishing only, and should occasionally calibrate the resultant

abundance estimates against fishery-independent estimates. Where it is not cost effective to use fishery-independent methods for resource status monitoring (where the attainment of a useful degree of precision would require an unrealistically large survey effort) and only fisheries data are available, the estimation of effort should not be ignored. Recently, the assessment of tropical fisheries through length–frequency sampling has become common. Whilst this reduces the need for fishing effort quantification for stock assessment, it introduces its own set of assumptions. It does not give the fishery manager much information about the human component of the fishery.

3. *Feedback from communities*: This is important to local government and broader levels of management. In societies where subsistence fishing is significant, or where reef tenure prevails, government must take account of communities. In many societies, regular meetings occur between community leaders and government, but attendance can become extremely time consuming for little real return. In such cases the fisheries management institution should maintain people within the community with whom it can liaise regularly.
4. *Feedback from theoreticians*: Specialists who can take the available information, analyse it, and then predict how a fish stock, a coral reef community, a village society, or even an entire fishery will react, are rare at the community or local government level. Because most reef fisheries are in developing countries, fishery modellers and synthesists are not even common at the national level. This is unfortunate, as such people need to maintain close contact with fisheries and the actual problems of managers if their work is to remain relevant. At the same time, feedback from theoreticians is essential to the fisheries management institution. A reasonable model of how target resources behave and interact is necessary to speed up the process of designing and fine-tuning management plans. Management can be completely retrospective, moving slowly towards the optimum by a series of experiments, but this process may take generations. Experience in several regions has shown that the linkage between academic theoretician and reef fishery manager often requires considerable mediation and interpretation. International scientific journals are not accessible to everyone and are not the quickest way of getting essential information into the public domain. Fisheries-aware information specialists are an extremely desirable commodity to any institution.

Academic research in reef fisheries management often fails to adequately consider invertebrate fisheries. Although only 15% of the total world catch by weight, invertebrates form nearly 58% of the total value of the top eight most valuable species groups in the global catch (calculated from FAO Fisheries Department, 1992). Invertebrates form over 50% of the live weight of commercially marketed catches from Pacific island coastal (mainly reef) fisheries (Dalzell and Adams, 1994) and they are usually by far the major portion of reef fishery products for export. They are also a major component of subsistence fisheries, particularly the reef-gleaning subsistence fisheries that are carried out mainly by women and children (Ruddle, Chapter 6).

5. *Feedback from commerce and the economy*. This can help to predict possible trends in the fishery. In the Pacific islands, 80% of the total catch is consumed directly by communities (Dalzell and Adams, 1994). Most reef fisheries are subsistence, and demand is fairly predictable and slow-changing. However, commercial market information is important because it is the commercial fisheries that cause most of the rapidly evolving crises in management. Even community-level managers need information on export market trends for certain products.

   Another aspect of feedback from trade, particularly from customs and quarantine services, concerns the need to manage the introduction of exotic species. Species introductions for aquaculture are increasing, and the trade in live reef fishes for aquaria is diversifying rapidly. Such introductions have the potential not only to disrupt local reef ecosystems, but also to introduce exotic parasites and diseases (Bohnsack, Chapter 11). Unfortunately, facilities for marine quarantine, and even for making basic assessments of the risks of certain introductions, are rudimentary in most developing countries. Discussion of this issue is outside the scope of this chapter, and the reader is referred to Eldredge (1994) and Humphrey (1994) for further information.

6. *Feedback from society*: Because most fishery products are eaten, the nutritional preferences of a society are significant and nutritional surveys can often provide supplementary information useful for quantification of subsistence fisheries. A knowledge of human population distributions is sometimes as important as fish population distribution in reef fishery management, and exploitation is almost always higher closer to urban centres. Politics is one of the most significant factors of all in the institutional management of fisheries, although a description of the role of politics would require a volume in itself, and will not be attempted here.

7. *Feedback from other coastal resource users*: Reef fisheries are downstream of almost all human-influenced coastal processes. A major task of management institutions is to make other coastal users aware of the vulnerabilities of reef fisheries. One notable exception is commercial spear-fishing which can rapidly deplete the most visible reef fishes and make the remainder wary of divers, with obvious impacts on other

resource users, particularly recreational divers. Many governments are now looking at broader levels of integration in the management of coastal zones, with opportunities for greatly improved flow of information between different resource users. Integrated coastal zone management committees are likely to be involved in prioritizing the rights of different resource users, including reef fisheries (McManus, Chapter 10).

8. *Feedback from similar fisheries in other areas*: The knowledge gained in one area can be very useful in the management of similar fisheries in others (Munro and Fakahau, 1993b). Also, fast-developing export fisheries driven by foreign capital tend to move gradually across regions, and advance warning of the type of problems encountered can be invaluable. Communication among different national fisheries administrations is already relatively good, with several regional or global meetings available annually to national fisheries staff. Most such meetings are driven by the need for international management of highly migratory species, but because most small-nation tuna fishery managers are also reef fisheries managers, the benefits are more widespread. Regional institutions and newsletters also play a major role in seeking out and sharing relevant experiences (e.g. Adams, 1993). Countries where national reef fishery management responsibility has been devolved to the state or provincial level sometimes lose access to this information, because only the national coordinators may attend the meetings, and their priorities may not include reef fisheries.
9. *Feedback from the broader natural sciences*: Fisheries science is generally considered to be a branch of zoology, but most hard fisheries science is actually population ecology, and certain aspects of stock assessment resemble demography. Fisheries management is broader still (McManus, Chapter 10). It needs to draw on information and models developed for psychology, meteorology, ethnography, wildlife management, forestry and many other disciplines to function well. It is a rare institution indeed that can maintain such diverse linkages effectively.

The word feedback has been used in this section instead of the more usual 'information' because feedback implies a two-way, or circular, process. Good information systems will return processed, or broader, information to the people who supply the primary information. Fisheries institutions should supply useful information to the community that fills in the catch returns and completes the questionnaires, as well as using the information internally. Most survey planning meetings spend all of their time designing a form or stratifying a sample area, with little thought to how the derived information is going to benefit the primary sources. For a regular monitoring activity such feedback is vital, and may even preclude the need for data sources to be forced into cooperation by legislation and expensive enforcement.

A few examples of such feedback may be useful. Individual catch returns may be compiled into a cartographical summary of fishing effort for that month and sent back to each fishing boat with the next month's catch-return form. An institution may be able to maintain a newsletter or radio broadcast passing useful news, and not just institutional directives, back to the community. A national institution can provide a central point for obtaining export market information. Institutional staff can address community meetings. Unfortunately, information necessary for reef fishery management is sometimes collected only by scientists and hoarded against future publications (Mathews, 1993). All too rarely is the time taken to synthesize, summarize, and turn this information around to the benefit of the primary resource user, yet it would pay considerable dividends in the long term.

All such sources of information should be part of the practical management of reef fisheries. They are necessary both for the models of pre-emptive management and for the evolution of retrospective management systems. They are necessary both for hard-science and for seat-of-the-pants management. Information acquisition, processing and feedback is probably the most important function of reef fishery management institutions.

13.6 REGULATION

To most people, regulation means legislation. Unfortunately, it also has a similar connotation to many government-employed reef fishery managers. However, regulation is literally the application of a set of rules to a potentially chaotic situation to achieve a desirable result. Quite often it does not require either formal legislation, or intervention by that penultimate arbiter of human dispute, the courts. The rules that make up a fisheries management regime can range widely. They can be informal guidelines on whether or not to issue a licence to a particular fishing vessel. They might be unwritten community rules that prevent the fishing of spawning aggregations of certain groupers in certain areas. They could be international agreements on the use, or not, of certain types of fishing gear to take a particular species on the high seas.

Available regulatory measures include the following.

1. *Investment guidelines.* Developing-country governments can offer considerable incentive to investment, particularly of foreign capital. Sustainable long-term productivity of a fishery may, however, compete with the need for short-term economic growth. Foreign fishing companies may be offered tax concessions and accelerated depreciation to set

up business by the ministry responsible for economic planning, whilst the ministry responsible for fisheries is desperately trying to stop over-capitalization in certain high-interest fisheries, often reef fisheries. If there is adequate coordination within government, the fisheries department may be able to take part in the economic planning process and set some of the criteria for encouraging investment related to marine resources, or to issue its own guidelines for investors (Lewis, 1985). Environmental impact assessments (EIA) of new investment in fisheries may provide some safeguards, but are usually activated only for terrestrial developments and usually then only for very large projects. Investment guidelines and EIA can only be used as fisheries management tools before fishing has started, however. Once a company has set up business, the permission to continue operation usually becomes much less stringent.

2. *Fishing licence issue criteria.* Licences to take fish are usually annual or seasonal, and can thus be used to help moderate the continued operation of a fishery. Whilst the power to issue licences is a legislated power, it is usually exercised at the discretion of the licensing officer. Within this discretion, licences can be issued according to guidelines mandated by the existing fishery management regime. Non-issuance under certain conditions can restrict the number of people, or companies, that have access. Limited entry will not conserve a resource unless other measures restrict the total catch, but it can help those who are licensed to operate at a profit, and mitigate the economic problems associated with gold-rush fisheries. Limited-entry licensing is not a regulatory method that should be used on subsistence, or diffuse, fisheries.

   A major reason for setting up a licensing system for a reef fishery is to quantify it. It is useful to census how many people are fishing and what vessels and gear they are using. Such licences are normally issued on request and licence fees tend to be minimal. Fishing licences are less useful as regulatory measures than as identity cards. But, as they become accepted and widespread, they can assume broader roles, as tickets of admission to certain traditional fishing grounds or for catching certain species. To avoid losing the census value of the original licence, additional licensing systems can be set up to mandate these broader roles.

3. *Fishing licence renewal conditions.* Guidelines governing the issue, or not, of fishing licences are measures that can usually be applied only once per year. Once the licence is issued, the owners of that licence can do whatever they want without fear of it being taken away, except on conviction by a court. In general, only courts can rescind ownership of anything. However, if conditions are attached to the

licence, the licensing authority does at least have grounds for refusing to renew it if these conditions are contravened. In countries where legal systems are overburdened, and where prosecution of fisheries laws is a low priority, licence conditions can be a powerful incentive for good behaviour.

4. *Trade licensing guidelines.* Many of the products of reef fisheries do not pass through commercial channels, so trade restrictions are of limited utility. Export fisheries are particularly amenable to such measures. If export permits are issuable on a per-shipment basis, and at the discretion of the licensing officer, a fairly fine level of regulation is possible. Export commodities are channelled through processing points and at ports of entry, and this permits efficient enforcement. In fact, they can be so effectively enforced that institutional staff may sometimes concentrate on these almost to the exclusion of other fisheries.
5. *Community regulations.* Local and traditional regulation of reef fisheries have been discussed already (Ruddle, Chapter 12). If national or local government institutions fail to acknowledge such systems, they have little hope of effectively managing a reef fishery. Even if complete government ownership of reefs and their resources is assumed, the continuing activities and expectations of the communities that actually fish those reefs must be taken into account. Some exercise of authority at the community level is almost always beneficial. Although the entrepreneurialism of a high-ranking individual may dominate the decisions of a small community for long enough to decimate a key resource, such conditions may occur in any institution. This is particularly so in those which are not guided by centuries of precedent.

   Community management is usually beneficial to fisheries because it operates at the scale of the individual reef. In small-island countries, where almost all members of the local community make full and regular use of reef resources, anomalies and trends are quickly spotted and corrective action taken. The role of government institutions should be supportive, and there should be some designated contact point within the fishing community. Outside support is also useful in arbitration because community decision-making is not necessarily harmonious.

   In Fiji, for example, the linkage between community and government is assisted by several factors:

   - Government formally recognizes the traditional right to exploit reef fisheries by designated communities, although it retains sovereign ownership of the seabed and, by implication, absolute ownership of resources.

- Government formally recognizes the right of these communities to recommend restrictions on fishing gear, area, or target species on any licence.
- Government formally records the boundaries of traditional fishing rights areas, and mediates the settlement of such boundaries by agreement between neighbouring communities, or any changes that might ensue.
- The community can formally nominate one of its number to be an honorary fish warden, with powers to investigate any fishing within that area and to escort suspected offenders to the nearest authority.
- Government fisheries officers cannot issue a fishing licence to any person who has not already obtained the written permission of the representative of the customary fishing rights area where he or she intends to fish. A separate permit must be obtained for each such target area (there are several hundred), each of which may apply a different set of access conditions.
- Government informally recognizes the traditional right of the community to exact a levy on outsiders wishing to have access to the resources of the customary fishing rights area. Cash is often required today, particularly for non-Fijians, but the fulfilment of customary courtesies is important.
- There are regular general meetings between the community administration and government officers, at which fisheries management issues can be aired.

6. *Fishing associations and lobbies.* These are rare in developing countries, where commercial and recreational fishing is not predominant, but in some cases they can be the non-traditional equivalent of community management systems. Generally, they exist to advance the interests of groups of individuals within the commercial or recreational fishing community. Such associations often have a longer-term view than individuals, however. If sufficiently forward-thinking, they can be entrusted with many of the regulatory aspects of management. Japan has developed the regulatory role of the commercial fishing association further than most.
7. *Government regulations.* These are the most absolute form of control. The national legislative system has two main roles in fisheries management. It defines the framework within which other levels of management can operate, and acts as the ultimate resort in cases of dispute. It is unrealistic to expect legislation to define every detail of the management of a fishery. Legislative systems must be refined over a period of decades, by trial and error, and by a process of adapting principles from other areas. Legislation as framework, for example,

empowers licensing officers to set certain conditions for the issue of fishing licences. It defines the powers of the body which arbitrates customary fishing boundary disputes, or it specifies the type of information that commercial fishermen must supply about their catches. It should not define, for example, the maximum allowable catch of a certain species from a certain reef, or the number of vessels that will be allowed to fish in a certain lagoon, because these management parameters need to be flexible and to change from season to season.

Legislation defines punishments for infringement of rules and guides legislative authority in deciding guilt. One aspect of this dispute settlement is that legislation should also provide some measure of protection for the individual against ill-informed, arbitrary or corrupt decisions by those in charge of fisheries management. For example, it is safer that committees decide on the allocation of licences than individuals and it is advisable to make financial procedures explicit. If no fee is to be charged for a certain service, then the legislation should make this clear.

8. *International access agreements.* Agreements between governments to allow access of fishing vessels from one country to the resources of another are rare for reef fisheries. Despite the suspicion with which foreign tuna vessels under international agreements are still often regarded, most foreign reef poaching today is by boats that are either unlicensed or individually licensed. Few reef fisheries are large scale, thus few governments consider it worthwhile to regularize the activities of any of its reef fishing vessels that range too widely. Remote reefs subject to a high degree of extranational exploitation thus currently have little hope of ever being effectively managed, short of posting a gunboat on permanent duty.
9. *Regional agreements.* These are usually agreements between nations for a common purpose, such as improving trade. If they involve reef fishery products, they can also be used in management. In theory, groups of small nations can thereby gain sufficient economic clout to control the supply of a rare resource and obtain the highest price whilst maximizing long-term yields. In practice, the effort involved is only worthwhile for a very high unit-value resource. In eastern Polynesia, for example, a regional marketing agreement is currently being considered for black pearls, *Pinctada margaritifera*, annual trade of which is currently worth nearly US$50 million.

Reef fisheries management thus has a variety of regulatory options to consider. In general, community-level regulation is considered to be both the most cost-effective and the most immediately practical for countries that maintain a high proportion of subsistence and small-scale fisheries.

National fisheries legislation can empower and support community management without necessarily giving up sovereignty over resources. The main need for primary government regulation is in previously unexploited fisheries, in industrial fisheries where the fate of entire groups of resources may be in the hands of a few company managers, and in fisheries where unwise exploitation in one reef area (for example, of nesting or spawning aggregations) can severely affect future fisheries in another community's area.

Formal management plans are becoming increasingly popular for reef fisheries. Even if formalization is not possible, or warranted, every fisheries department should informally maintain a set of written management policies, and update them occasionally in the light of experience, even if only to harmonize the decisions of different officers, or successive heads of department. For occasional reef fisheries these plans might consist of the single phrase 'Leave decisions to the community', where it is the community that benefits from fishing and which would suffer if a resource is imprudently managed. For other fisheries, particularly export fisheries, these plans may be more codified and complex, or even formally legislated, but they should all have one thing in common: they should make allowance for future change.

13.7 ENFORCEMENT

Regulations are almost useless if they cannot be enforced. The most biologically appropriate management plan is the regime that is most likely to maintain the resource at maximum levels of long-term productivity. It will not function if people cannot be induced to comply with it. Compliance is always more desirable than enforcement, and regulations should be designed with this in mind (Fig. 13.1). For example, it is a lot easier for the fishing community to comply with a minimum size limit for taking certain species of fishes if the minimum allowable mesh size for gill nets maximizes the escape of the prohibited size classes. Such a regulatory interaction has been overlooked in at least one country. 'Compliance facilitation' is particularly important in small-island countries where law enforcement staff may be related to a significant proportion of the fishing community, and where family ties may prevent the arrest or censure of a relative (Berg and Olsen, 1989).

Governmental enforcement of regulations is not easy because reef fisheries are usually small scale, with many harvesters landing an extremely heterogeneous mix of species at a multitude of landing points (Munro, Chapter 1). Enforcement has more similarity to temperate-water sport fisheries than to industrial fisheries. As mentioned previously, industrial methods of management are relatively costly, particularly for subsistence

Fig. 13.1 Temperate fishery management methods are not always appropriate in the tropics.

fisheries. As a result, many developing country governments play little part in the active management of reef fisheries. Those regulations which are in place are mainly passive, with occasional deterrent enforcement. Many of these countries also have some measure of community control of fisheries in place, whether acknowledged by government or not.

Many of the enforcement activities of government fisheries officers are directed at fish traders (Fig. 13.2). Traders are usually less numerous than harvesters and are thus a convenient bottleneck at which to maximize the effectiveness of scarce enforcement resources. Traders are vulnerable to inspections for contravention of size limits, or of out-of-season or forbidden species, and will not normally buy such produce from the fishing community if prosecuted too often. The effects of regulation thus eventually filter down to the target community. Where a species is much sought after and subject to heavy commercial competition, the harvester may force the trader to purchase the illegal bycatch in order to acquire the legitimate commodity. Such cases are a signal for enforcement to move out into the fishing community. Either community management is not working, or the human population is outstripping the carrying capacity of its local reef environment.

Fig. 13.2 Many of the enforcement activities of government fisheries officers are directed at fish traders.

Community-appointed fish wardens, recognized and empowered by government, can be very useful in enforcing reef fishery regulations, both traditional and modern. Very often, the authority of the post itself is sufficient reward for part-time activity, particularly as the warden would normally be out fishing for subsistence purposes anyway. If expenses need to be met, these should preferably be administered by the community out of levies charged for access, otherwise government may end up employing more administrative staff than it would have invested in enforcement officers. Such wardens should not be neglected or taken for granted, and should be able to take part in regular consultations with the local fisheries officer, and in annual meetings.

If biologically appropriate management measures are also appropriate to the human element of a fishery, they will be easier to comply with and require less enforcement. Most traditional measures, by their very nature, are appropriate to the community, but governmental fishery managers must not rely blindly on community reef fishery management, especially where communities are under stress. Immigration has obvious destabilizing effects on long-established communities. Emigration and urban drift have more insidious consequences, where expectations of future entry to the cash economy by one's children may undermine any desire to conserve resources for the next generation. It may become more important to acquire cash to pay for schooling than to practise sustainable fishing.

If there is no community management in operation, the appropriateness of proposed regulations should at least be tested on the community before they are committed to legislation. If regulations are issued as a set of guidelines and tried, then discussed, with various fishing communities, it will quickly become apparent which suggestions the harvesters will be more inclined to follow. For those measures where compliance is not easy, but which it is still considered essential to enforce, incentives for compliance can be considered. For example, duty-free fuel for small-scale fishermen who voluntarily complete catch-return forms is an incentive that is used by the Vanuatu government. Some fisheries departments supply subsidized ice as an incentive to taking out a fishing licence, and others may subsidize the price of larger-mesh gill nets in an effort to reduce the catch of immature fish. Whilst these incentives are positive measures to assist enforcement, they should be used with caution. Government subsidies may prevent future private-sector development and government may be forced to supply fishing gear, or ice, for ever after.

Another aspect of enforcement is bribery. Only 3 months after appointment, the Secretary to the Department of Fisheries and Marine Resources in Papua New Guinea told a meeting of law-enforcement and surveillance officers that he had already been offered (and turned down) a total of US$23 000 in bribes (Anon., 1993). The potential for the unequal distribution of favours, in return for personal gain, is inherent in every system of authority. In some societies, 'extra-curricular' fees are the norm, and are so overt that they cannot really be considered corruption. Such fees, like tips or gratuities, may even be taken into account when setting the basic wage for institutional officers. In some societies, a token gift customarily accompanies any request to authority, and a failure to make this gesture is considered ill-mannered. True corruption is when individuals use their positions of privilege, or authority, to favour one applicant over another in return for personal gain. Where a management regime is designed to limit the number of licences, the presence of individuals who issue permits in order to advance their own personal aims can entirely destroy that regime.

Such corruption is difficult to pin-point and even more difficult to prove. Extreme security of tenure is a feature of many government jobs, and a corrupt officer is difficult to remove. One method of countering such problems, where suspected, is to make enforcement and inspection officers work as teams, with rotating membership.

Where staff for enforcement are scarce, there is often the temptation to use researchers to assist in supplying information towards arrests, or even in active enforcement. This should be avoided. Information is as essential to management as enforcement and much information is voluntarily provided. Neither the fishing community nor traders will be cooperative in talking to someone who may arrest them, or who may lead to them paying more tax. Information and enforcement functions of reef fishery management should be exercised by separate institutions, or at least by physically separate branches of the same institution.

13.8 INTERACTIONS AMONG THE FISHING COMMUNITY, MANAGERS AND SCIENTISTS

Distinctions between the fishing community, managers and scientists are often blurred in the context of tropical reef fisheries. Harvesters are often part of local community management or may own some component of the reef or resource themselves. Shortages of skilled people mean that government fisheries scientists often fulfil many of the tasks of fishery managers, short of enforcement. The two main types of interaction are conflict and cooperation.

Conflict

Conflict between the fishing community and government managers can occur because measures are not usually introduced until a resource is demonstrated to be overfished, and catch or effort has to be cut. Management costs money, therefore there is a natural tendency not to exercise it until it is absolutely necessary, and even then there may be political pressures to avoid effective action (Aiken and Haughton, 1990; MacKenzie, 1993). Although commercial fishers, and particularly entrepreneurs, may demur, it is always more desirable to approach the level of optimum yield (or the point beyond which overfishing takes place) asymptotically. The conflicts generated by the loss of entrepreneurial expectations are always less than the conflicts involved in trying to cut back an overcapitalized fishery.

Avoidance of extreme conflicts can be a powerful argument to convince senior civil servants to promote measures in advance of actual problems in the fishery. One of the advantages of retrospective management in this regard is that it evolves. The initial management plan can be a simple framework, concentrating mainly on information feedback and requiring minimal investment in enforcement, and thus be relatively painless to introduce. If sufficient flexibility is incorporated, the plan can become more or less stringent as necessary, by small increments if monitoring is sufficiently comprehensive.

Tension between managers and fisheries scientists tends to occur because managers are obliged to maintain a holistic approach. Scientists need to break the problem down into smaller parts amenable to measurement. At the very least, they will separate the resource aspects from its human elements. The recommendations of fisheries scientists may thus be different from what managers intuitively feel is best, given the many other factors under consideration, but for which they have no other source of hard evidence to justify their intuition. Evolutionary management may provide some reconciliation, by allowing managers to test their feelings whilst scientists can rigorously observe the results of the experiment and fine-tune the theory.

The need for managers often to make sweeping, or rapid, decisions based on minimal evidence also conflicts with the tendency of scientists to hedge all opinions about with provisos. There are many management meetings where scientists are accused of sitting on the fence. Scientists occasionally overcompensate, and present wild guesses as justified conclusions, but this is rare. To mitigate such conflicts, scientists should try to present an estimate of confidence in every presentation of results. Even if an estimate of, say, biomass, is inconclusive, to say that it is 95% likely that the real value is within 70% of, say, 25 000 tonnes (7500–42 500 t) may at least divert accusations of sitting on the fence. This is particularly so if the scientist can then point out that a 10% increase in research funding would improve this confidence interval to less than 20% (20 000–30 000 t). However, some statistical methods currently employed in fisheries science lack any theoretical method of quantifying confidence in the result. Even empirical methods of estimating confidence may take an impossible amount of computer time for the more complex models.

Cooperation

Conditions fostering good relations, or at least understanding the aims of every group involved, should be inherent in every institutional management framework. This principle is automatically taken into account in most modern institutions. Indeed, some new institutions attempt to go so far in fostering cooperation that they lose efficiency.

Meetings, in the form of staff meetings, annual institutional meetings, community meetings, seminars and workshops, summits, industry–government councils, fishing industry advisory boards, integrated coastal zone planning meetings, regional and international fisheries meetings, or whatever, can all take their toll on the work programme if not managed properly. The time spent at meetings is an investment; there must be clear returns on that investment.

Some apparent tension between departments may even be desirable under certain conditions. For example, fishers and traders must see a clear demarcation between enforcers and information gatherers. However, it is easy for modern fisheries management institutions, particularly those with a strong enforcement role, to begin demanding information as a right. Whoever coined the term civil servant for a government officer was optimistic!

Cooperation is a multi-faceted process, and information is one of its main expressions. One of the main shortcomings of government in fisheries management is that there is often poor communication between different departments. There must be mechanisms to discuss and resolve interdepartmental conflicts at the technical level. In many countries, departments only communicate at the apex, through the chain of command, and the apex is not necessarily aware of all that is happening further down in the hierarchy. Apart from general administrative issues, government departments responsible for reef fisheries management interact with others responsible for such functions as economic planning, rural development, environment, customs, police, ports and shipping and legal drafting. Without dialogue, departments will develop dissimilar attitudes towards fisheries if they do not have access to appropriate information. For example, economic planners usually consider fisheries in terms of their potential for short-term economic growth, and their views are extremely influential in setting overall government policy. Conversely, environment departments have a conservative attitude, or occasionally distaste, towards exploitation of natural resources and may try to relieve local communities of all management decision-making capacity.

Fisheries management is not always a popular pretext in itself for calling regular interdepartmental technical meetings, and the recent trend towards interdisciplinary coastal zone and environmental planning provides useful opportunities to open general discussion on fisheries management issues. The documents resulting from the 1992 United Nations Conference on Environment and Development (UNCED), particularly Agenda 21 Chapter 17, provide a framework for considering fisheries management in the broader context (Anon., 1992b). Programme area D is of particular interest. Reef ecosystems are identified as being critical habitat areas where states should accord the highest priority to providing 'necessary limitations on use', whilst taking into account the knowledge and interests of local communities and small-scale artisanal fisheries. As with the United Nations Law of the Sea Conferences, arguments over the exact interpretation of the output of UNCED will occupy institutions for years to come.

ACKNOWLEDGEMENT

I would like to acknowledge the drawings by Jipe Le Bars.

Chapter fourteen

Developments in tropical reef fisheries science and management

Nicholas V.C. Polunin, Callum M. Roberts and Daniel Pauly

SUMMARY

Extinction is conceivable for some species, including aquarium-fish stocks of high value and limited geographical range, but a biological basis exists for management of such small fishes. Immediate data needs are much greater for management of stocks of larger species which most reef fishing targets. The strengths and limitations of underwater visual census are now better recognized, and there is improved capacity for fish ageing, especially through otolith analysis, as a basis for growth and mortality assessments. Information technology has grown rapidly in scope, and modelling approaches offer a better foundation now for incorporating biological data into analytical approaches to sustainability. Reef fishery dynamics may widely be simpler than implied by their 'multispecies' condition, and in spite of their assumptions, surplus production models have proved useful in stock assessment. Empirical and exploratory approaches to sustainability, however, are more desirable than ever, as uncertainty about long-term ecosystem effects of fishing increases. Work of an experimental nature urgently needs also to be directed at rehabilitation techniques, especially marine fishery reserves. Successful management is, however, unlikely without local community involvement. Understanding

Reef Fisheries. Edited by Nicholas V.C. Polunin and Callum M. Roberts.
Published in 1996 by Chapman & Hall, London. ISBN 0 412 60110 9.

the scope of such participation demands a greater social science input to reef fishery studies; here as elsewhere there are encouraging trends in recent studies.

14.1 INTRODUCTION

It is not so long since tropical reef fishery yields were thought, admittedly on the basis of limited data, to range from 0.5 to 5 $t\,km^{-2}\,year^{-1}$ (Stevenson and Marshall, 1974; Marten and Polovina, 1982). Much higher estimates, around 20 $t\,km^{-2}\,year^{-1}$, by Alcala (1981) and Alcala and Luchavez (1981) in the Philippines, and by Wass (1982) in Samoa, were thought at first not to be representative (Marshall, 1980). However, high yields of this order have now been independently estimated for a number of sites in the South Pacific and South East Asia (Dalzell, Chapter 7), and the higher estimates are now close to the maximum levels of fish production predicted by trophic and other models of reef ecosystems (Fig. 14.1). One research problem that remains, however, is the rigorous quantification of the effects of factors such as primary productivity, depth, sampling

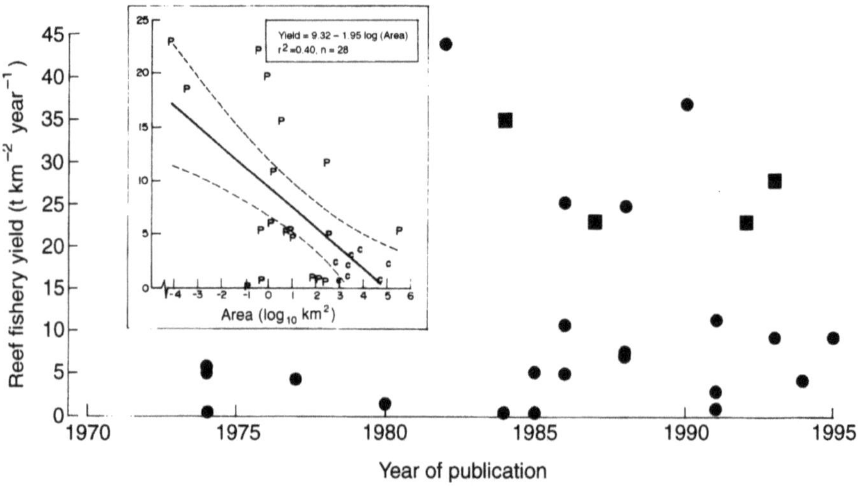

Fig. 14.1 Recent estimates of tropical reef fisheries yield have in some instances reached the high value of fish production indicated by trophic models. Filled circles (from Dalzell, Chapter 7) show observed yields, filled squares show model estimates, both plotted against year of publication. The model data are those of Galzin (1987), Polunin and Klumpp (1992b), Arias-Gonzalez (1993) and Polovina (1984). The insert, adapted from Arias-Gonzalez *et al.* (1994), shows the relationship between observed yields ($t\ km^{-2}\ year^{-1}$) and the area from which yields were estimated, in the Caribbean (c) and the South Pacific (p).

area, or coral cover, on yield estimates (Marten and Polovina, 1982; Munro, 1984a). Figure 14.1 illustrates correlation of one such factor, sampling area, with reef fisheries yields. Beyond the capacity to predict we must also ensure that the predicted yields will be sustainable. Our ability to do so depends on the basis of our predictions, which may be either analytic or empirical. The former implies that a fishery is simulated, species by species, by incorporating all available estimates of rates (growth, mortality, recruitment, predation, etc.) into species-specific sub-models, then integrating these into a comprehensive model (Larkin and Gazey, 1982). This approach is very data-intensive. Estimation of the parameters required for this, enhancing access to, and synthesis of the parameter values which have already been estimated, will therefore be the theme of the first section of this chapter.

Alternatively, in the absence of detailed information upon which to base an analytic approach, experimental and exploratory approaches must be used to obtain information on productivity and sustainability. These will be the focus of the second section of this chapter. Beyond both approaches to estimating yields and their sustainability, a basis for fishery regulation must be created, and this is a more people-orientated activity. Understanding the dynamics of reef fisheries, the perceptions of reef fishers, and their historical and social contexts all have implications for management. The current status of, and needs for, scientific research in these areas will thus be the focus of the third section. While the scientific basis for good advice to managers is still weak, we will show that there are encouraging trends. These represent potential growth areas with respect to the science and management of tropical reef fisheries.

14.2 ASSESSMENT OF TROPICAL REEF FISHERIES

Knowledge of reef fish biology has been greatly influenced by ecological studies of small species (contributions in Sale, 1991). This knowledge is highly relevant to fisheries supplying the aquarium trade, but is beside the point where larger species such as groupers are concerned. Indeed, substantial gaps remain in knowledge of reproduction (Sadovy, Chapter 2), larval life (Boehlert, Chapter 3) and juvenile stages (Roberts, Chapter 4) of reef fishery species, and of the productive base of reef fisheries (Polunin, Chapter 5). There are limited catch and yield data on the fisheries themselves (Dalzell, Chapter 7), and on fishing effects (Jennings and Lock, Chapter 8). Also, much work needs to be done on the geographical, social and economic facets of reef fishing (Ruddle, Chapter 6; McManus, Chapter 10), on the dynamics of fisheries (Appeldoorn, Chapter 9), and on the roles of traditional (Ruddle, Chapter 12) and modern (Adams, Chapter 13) institutions in development and management of reef fisheries. A traditional practice, area closure, is being revived as a modern management tool, but the underlying mechanisms still need to be explored, to derive a sound basis for functional design of reserves (Bohnsack, Chapter 11).

Sampling and reduction of stock assessment data

The visibility of many reef fishery species to scuba divers implies that underwater visual census (UVC) will continue to be a useful stock assessment technique (Harmelin-Vivien *et al.*, 1985). However, there have been only few tests of how well UVC does sample a target fish community (Samoilys, 1992; Acosta, 1993; Watson *et al.*, 1995; Jennings and Polunin, in press b). Questions of observer error and sampling design apart, the value of UVC to stock assessment or to determining effects of exploitation depends on the site and species involved. In Fiji and New Caledonia, emperors (Lethrinidae) are poorly sampled by UVC (Kulbicki *et al.*, 1987; Jennings and Polunin, in press b), and yet these species are very important in the catch (Dalzell *et al.*, 1992). It is clear that where spear fishing is common, the utility of the technique will be limited by the flight behaviour of resident fishes. Some important species, however, including groupers and snappers, appear to be more abundant in UVC than in the catches of bottom-set lines. Species exhibiting equal densities in the two techniques are likely to be few (Kulbicki *et al.*, 1987). Information on size at age is essential for stock assessment, and can be obtained in two ways: direct ageing of hard parts like otoliths (contributions in Summerfeldt and Hall, 1987), or analysis of length–frequency data (contributions in Pauly and Morgan, 1987; Appeldoorn, Chapter 9).

Annual banding is not marked in otoliths of fishes from low latitudes (Munro and Williams, 1985). This forces researchers interested in absolute ages to rely on daily rings (Pannella, 1971), especially in small species (Polunin and Brothers, 1989) and the juveniles of larger species (Lou, 1992). While the fine banding of older members of larger species is widely agreed to be difficult or impossible to discern, seasonal and/or spawning checks do occur under some circumstances (Longhurst and Pauly, 1987), along with annuli that can be validated (Manickchand-Dass, 1987; Lou, 1992). Annual banding is certainly clear enough to provide a basis for ageing at mid latitudes, and recent studies have made use of this for some important fishery species such as groupers (Bullock *et al.*, 1992; Ferreira and Russ, 1992; Sadovy *et al.*, 1992; Bullock and Murphy, 1994).

Difficulties remain, despite these and similar studies. One of the difficulties still to be addressed is the often fragmentary nature of the otolith record in large (old) fish. In those parts of an otolith sequence where daily bands cannot be counted, they have to be assumed to be of similar width to those of areas where they can be counted and measured (Ralston, 1985). Techniques such as scanning electron microscopy can be used to increase resolution, at least to check on features in a few otolith sections, and such results can be further refined using image analysis software, which enables band counts to be based on objective densitometric criteria.

Estimating growth parameters (usually those of the von Bertalanffy growth function, or VBGF) is straightforward when (absolute) otolith ages are available, but should be done using one of the now widely available non-linear fitting routines, rather than through a linear transformation of the VBGF.

The other approach for obtaining (relative) ages and estimating growth parameters, the analysis of length–frequency data, can be implemented using techniques such as ELEFAN I (Pauly, 1987) or SLCA (Shepherd, 1987), both incorporated in the FiSAT software package recently released by FAO (Gayanilo *et al.*, 1994), the MIX package (MacDonald and Pitcher, 1979), or the somewhat costly MULTIFAN software of Fournier *et al.* (1990). Besides estimating the parameters L_∞, t_0 and k of the VBGF, several of these techniques also allow quantification of seasonal and other growth oscillations in reef fishes (Pauly and Ingles, 1981; Longhurst and Pauly, 1987; Appeldoorn, Chapter 9). Whether age- or length-based methods are used for estimation, information is lost, when, in assessments, only a single set of growth parameters is used. The growth of individual fish being variable, incorporating that variability would make the uncertainty inherent in estimates of yield, for example, more explicit. Thus the variance estimates output by the techniques listed above must be considered when performing assessments. There are many ways in which this and other elements of uncertainty and risk can be formally incorporated into stock assessment models (Hilborn and Walters, 1992).

If not accounted for, migrations and variable recruitment, both within and among years, will compromise mortality results obtained using techniques that assume a steady state (Appeldoorn, Chapter 9). On the other hand, there is no reason to believe that coral reef fish recruitment is significantly more variable than recruitment in high-latitude species, to which catch curves and similar techniques have been successfully applied (Ricker, 1975). Particularly, it would be appropriate to estimate mortality in unexploited stocks using such techniques. This is both because their assumptions are more easily met when new recruits form a relatively small fraction of overall biomass, and because mortality in such stocks consists only of natural mortality (M), a parameter otherwise difficult to estimate reliably.

Difficulty in estimating M in tropical fishes is the reason why Pauly (1980b) derived his empirical model (Appeldoorn, Chapter 9, Equations 9.13–14) for prediction of M from VBGF parameters and mean environmental temperature (*T*). Although, at the time, only 17 estimates of *M* were available for reef fishes, out of the 175 used to derive his model, this group did not appear to display any significant pattern of deviation from the values predicted by the model (Pauly, 1980b). With more values of *M* having become available, there is growing interest in re-assessment of Pauly's (1980b) model (e.g. Trenkel, 1993). Given enough data, such analysis could also consider Ursin's (1984) point that specific groups may show distinct trends in effects of temperature on population parameters (Polunin and Brothers, 1989).

Overall, it is our impression that the complexity of reef fisheries may not always be as great as the ecological and fishery literature would have us believe. Thus, for example, although many species may ultimately contribute to the catch, a small number typically constitutes a very large part of the biomass (e.g. McManus *et al.*, 1992; Smith and Dalzell, 1993). It is likely, therefore, that the data requirements of reef fishery stock assessment may often be easier to meet than is supposed.

How can existing data be better accessed?

Information technology provides the means for storage, transfer and dissemination of the large amounts of data required for fishery stock assessment. The technology brings with it a growing potential for computer-based statistics- or graphics-orientated analyses. There is now a wealth of means for manipulating, exploring, visualizing and interpreting complex and voluminous data. There is also much information on reef fishes and fisheries in publications, both formal and 'grey', but much of this, unfortunately, is not always available when and where needed. Especially in developing countries, there is a strong tendency for valuable data sets and data-rich reports to be lost (Polunin, 1983; Janzen, 1986; Mathews, 1993). This problem is being addressed for tropical reef fisheries by comprehensive accounts of the resources of whole regions, such as the books of Munro (1983g) on the Caribbean and of Wright and Hill (1993) on the South Pacific. It is also an objective of the FishBase Project of the International Center for Living Aquatic Resources Management (ICLARM), in Manila, Philippines (Pauly and Froese, 1991). The latter project has recently released its first CD-ROM, providing fully referenced taxonomic, biological and ecological information on 12 000 fish species, of which 1980 are reef-associated (Froese and Pauly, 1994). By way of illustration, FishBase contains 189 sets of VBGF growth parameters, and the parameters of 385 length–weight relationships, for reef fishes, including those of Bohnsack and Harper (1988) and of Kulbicki *et al.* (1993). ICLARM should be contacted on how to obtain this CD-ROM or to collaborate with the FishBase Project; see the last author's address.

Has our capacity for synthesis increased?

Simulation and mass-balance models are two major tools for synthesizing the multifaceted information available on tropical reef fisheries. These models may take a variety of forms, but all have in common that they simplify, sometimes strongly so, all but the process that is of interest to the model's architect (Larkin and Gazey, 1982). Still, they can be used to make sense of complex data, to derive new hypotheses and to quantify the roles of processes or factors allegedly affecting reef fisheries. We present three examples, each emphasizing one (set of) process(es), with simplifying assumptions for what may be called the 'boundary conditions'.

The first example refers to a simulation model emphasizing the recruitment to a fish stock consisting of a metapopulation, located in a number of habitat patches interconnected through larval or other means of dispersal. Man *et al.* (1995) used this to examine the role of marine fishery reserves in enhancing recruitment to fishing grounds. They showed that reserves become highly beneficial when the local extinction rate caused by fishing is large. In addition, sustainable yield was maximized when half of the patches were occupied by a population.

The second example pertains to a simulation model of a temperate population of slow-moving molluscs (abalone), but could easily be adapted to reef species. The model, called AbaSim, incorporates, in a spatially structured context, growth, mortality and a variable recruitment to the stock, which may be exploited at different rates (Prince *et al.*, 1991). The results, presented in the form of vivid graphics, allow assessing the short-, medium- and long-term impacts of different management interventions in a way that is intuitive, yet rigorous, and thus can be used as a communication channel among fishers, managers and scientists (Sluczanowski, 1992). One can only hope that this approach will be emulated by reef researchers.

The third example pertains to the ECOPATH II modelling approach and software of Christensen and Pauly (1992a,b). This mass-balance approach, which builds on Polovina (1984) and emphasizes the trophic relationship within ecosystems, has been widely applied to reef systems (Arias-Gonzalez, 1993; Pauly and Christensen, 1993; contributions in Christensen and Pauly, 1993).

The data required for construction of ECOPATH II models are the same rates and biomass estimates that field researchers tend to publish, and hence this approach can be used to verify the mutual compatibility of such estimates (e.g. is the production rate of herbivore X at least as high as the consumption rate of its predator Y?). Further, once the trophic flows linking the various 'boxes' of an ECOPATH II model are estimated (Fig. 14.2), their topology can be analysed, for example in terms of trophic

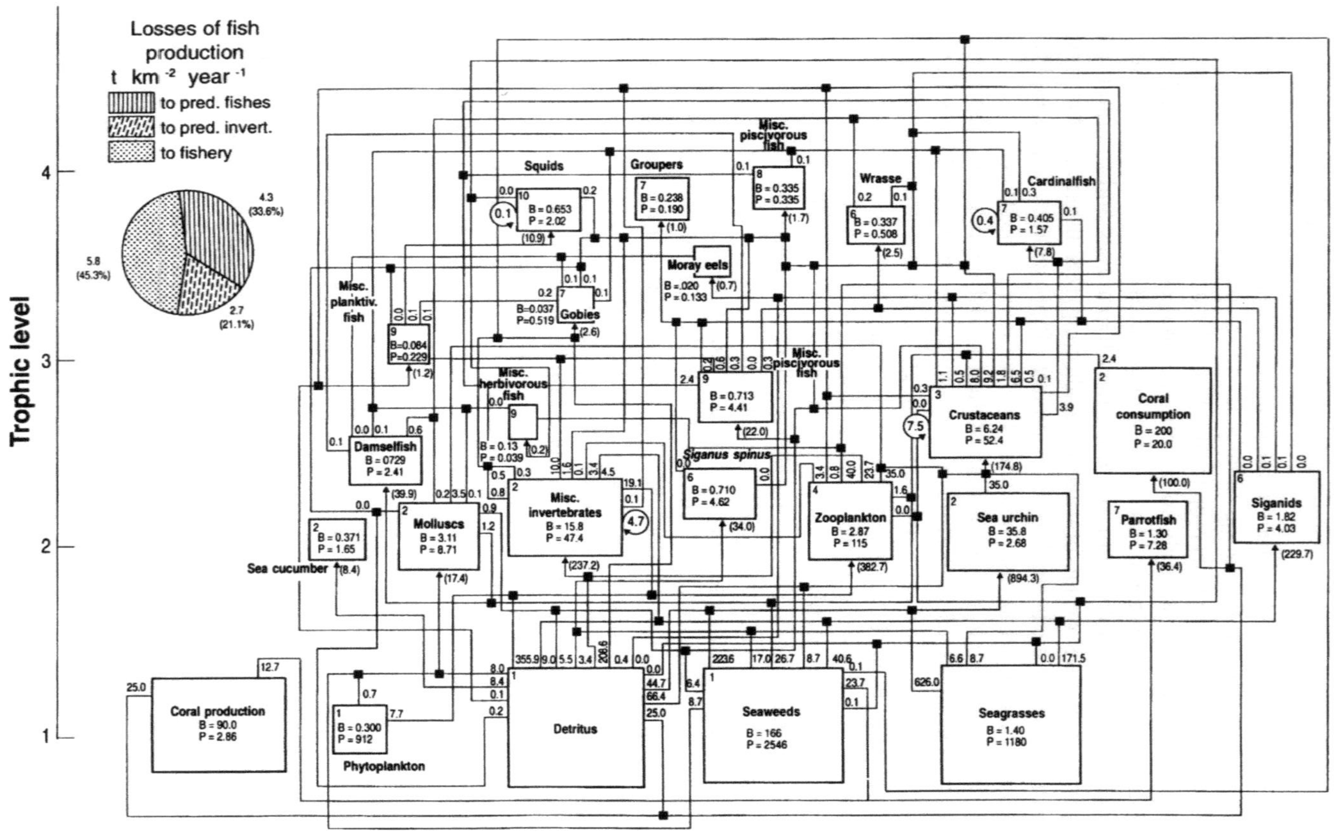

Trophic level
Losses of fish production
t km -2 year -1
to pred. fishes
to pred. invert.
to fishery
4.3 (33.6%)
2.7 (21.1%)
5.8 (45.3%)
Squids
B = 0.653
P = 2.02
Groupers
B = 0.238
P = 0.190
Misc. piscivorous fish
B = 0.335
P = 0.335
Wrasse
B = 0.337
P = 0.508
Cardinalfish
B = 0.405
P = 1.57
Gobies
B=0.037
P=0.519
Moray eels
B =.020
P = 0.133
Misc. planktiv. fish
B=0.084
P=0.229
Misc. herbivorous fish
B = 0.13
P = 0.039
Misc. piscivorous fish
B = 0.713
P = 4.41
Damselfish
B = 0729
P = 2.41
Siganus spinus
B = 0.710
P = 4.62
Crustaceans
B = 6.24
P = 52.4
Coral consumption
B = 200
P = 20.0
Sea cucumber
B = 0.371
P = 1.65
Molluscs
B = 3.11
P = 8.71
Misc. invertebrates
B = 15.8
P = 47.4
Zooplankton
B = 2.87
P = 115
Sea urchin
B = 35.8
P = 2.68
Parrotfish
B = 1.30
P = 7.28
Siganids
B = 1.82
P = 4.03
Coral production
B = 90.0
P = 2.86
Phytoplankton
B = 0.300
P = 912
Detritus
Seaweeds
B = 166
P = 2546
Seagrasses
B = 1.40
P = 1180

impacts (Fig. 14.3) and compared, using the rich theory of flow network recently developed by Ulanowicz (1986) and others (contributions in Christensen and Pauly, 1993). Moreover, the newly released Windows version (3.0) of ECOPATH II includes a routine that allows entry of ranges of rates or biomasses, rather than point estimates, and of distributions (uniform, triangular, normal, log-normal). This feature allows for a Monte Carlo resampling of the distributions, leading to a multitude of model realizations. These realizations are evaluated for their conformity to constraints set by the user, and the selection of a 'best' solution in a least-squares sense. Through this new routine, called EcoRanger, ECOPATH II has acquired the ability, previously lacking, to answer 'what if?' questions.

The three examples above indicate that our capacity for synthesizing data from tropical reef fisheries has much improved in recent years. Therefore, it seems, one important task now should be to use models such as these, to answer practical questions, then use the answers to improve the models and/or develop alternatives. We suggest (along with Pauly and Christensen, 1994) that this approach beats any abstract debate on whether the complexity of tropical reefs precludes them from being modelled. Nevertheless, at the same time, efforts must continue to be made to expand the basic data available for such synthesis.

14.3 RESILIENCE AND SUSTAINABILITY

Lack of information on component populations makes prediction of overall productivity (Polunin, Chapter 5; Appeldoorn, Chapter 9) and fishing effects (Dalzell, Chapter 7; Jennings and Lock, Chapter 8; McManus, Chapter 10) difficult. Habitat is manifestly fragile. Communities may change dramatically as exploitation begins, so that virgin stocks offer a poor basis for predicting later productivity. Stocks with slow growth, low reproduction rates and high susceptibility to fishing, such as sharks, may be inherently prone to depletion (Adams, 1980).

What are maximal and sustainable yields?

Surplus-production models (Schaefer, 1954, 1957; Ricker, 1975) provide, in principle, an approach to estimate the catch that can be taken on a

---

Fig. 14.2 Mass-balance model of trophic interactions in the Bolinao coral reef flat of Bolinao, Philippines, constructed using the ECOPATH II software. Box sizes are proportional to (log) biomasses, expressed in t wet weight km^{-2}; flows are in t ww km^{-2} $year^{-1}$; backflows to the detritus are omitted for clarity. Adapted from Aliño *et al.* (1993).

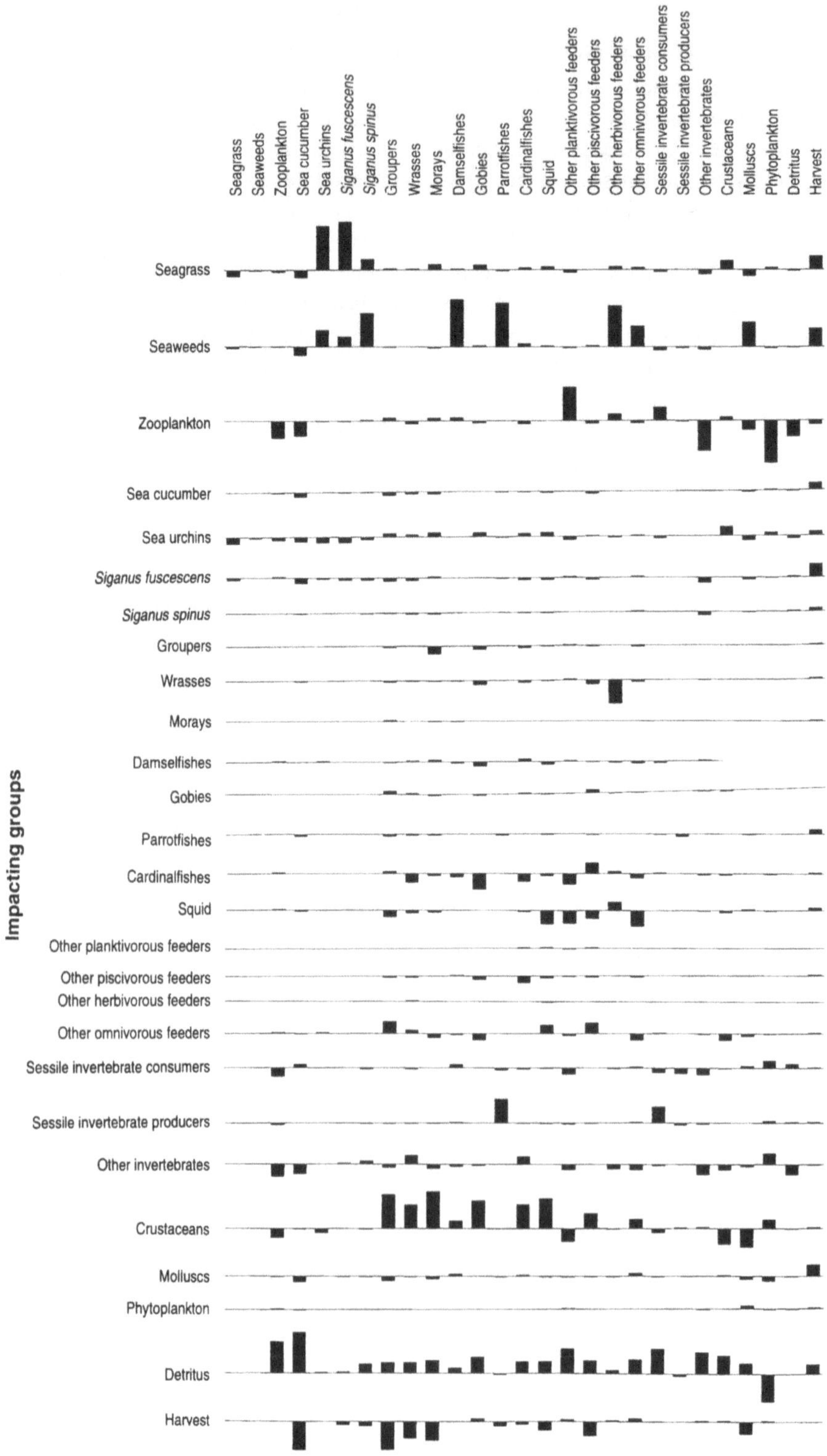
Seagrass
Seaweeds
Zooplankton
Sea cucumber
Sea urchins
Siganus fuscescens
Siganus spinus
Groupers
Wrasses
Morays
Damselfishes
Gobies
Parrotfishes
Cardinalfishes
Squid
Other planktivorous feeders
Other piscivorous feeders
Other herbivorous feeders
Other omnivorous feeders
Sessile invertebrate consumers
Sessile invertebrate producers
Other invertebrates
Crustaceans
Molluscs
Phytoplankton
Detritus
Harvest
Impacting groups
Seagrass
Seaweeds
Zooplankton
Sea cucumber
Sea urchins
Siganus fuscescens
Siganus spinus
Groupers
Wrasses
Morays
Damselfishes
Gobies
Parrotfishes
Cardinalfishes
Squid
Other planktivorous feeders
Other piscivorous feeders
Other herbivorous feeders
Other omnivorous feeders
Sessile invertebrate consumers
Sessile invertebrate producers
Other invertebrates
Crustaceans
Molluscs
Phytoplankton
Detritus
Harvest

sustained basis from a given single-species stock, given a time-series of catch and effort data. The approach has been much criticized (Hilborn and Walters, 1992; Koslow *et al.*, 1994), notably because most of its users fail to account for large and rapid changes in fishing effort (i.e. for non-equilibrium situations). It continues to be widely used, however, as it appears resistant to violation of many of its assumptions and can implicitly account for multispecies interactions (Pope, 1979a).

It is thus not any inherent feature of production models that has prevented their wide application to reef fisheries, but the absence of sufficiently long time series of mutually compatible catch and effort data. The replacement, by Marten (1979a,b), working on Lake Victoria, and by Munro (1980), working on the Jamaican reef fishery, of different times by different sites (assumed to have been similar before exploitation) was therefore a major advance in tropical fisheries research, and in fisheries research generally (Caddy and García, 1983; Pauly, 1994). Further development of this approach by Polovina (1989b) has led to a composite model, incorporating both time and space, and accounting for non-equilibrium effects. With such a powerful tool, tropical reef researchers can now estimate sustainable yield wherever enough replicates can be found in time and/or space – and if fisheries-induced changes can be assumed to be reversible, our next topic.

Reductions of predator or competitor biomass through fishing can be expected to have numerous indirect effects (Jennings and Lock, Chapter 8). One of the best-documented effects to date has been the increased incidence of low-value species in the catch, as shown by Koslow *et al.* (1988) for Jamaica, Butler *et al.* (1993) for Bermuda, and Appeldoorn *et al.* (1992) for Puerto Rico and the US Virgin Islands.

Although Jennings *et al.* (in press) and Jennings and Polunin (in press e), in combined fishery and UVC studies, found no evidence for increased abundance of prey fishes in the Seychelles or Fiji, sea urchins do appear widely to have increased in abundance following removal of predators by fishing (McClanahan, 1992, 1994). Increased sea-urchin density with increase of fishing effort may not appear to be of immediate consequence to fisheries (unless sea urchins are consumed locally or can be exported), but there are potential side-effects for the ecosystem, for example through bioerosion of the reef (McClanahan and Muthiga, 1988). In such a case,

Fig. 14.3 Trophic impact matrix of the reef ecosystem model in Fig. 14.2, estimated by the ECOPATH II software and showing how an increased consumer biomass would affect the biomass of all other elements in the ecosystem, assuming that diet compositions are conserved. The changes (increases above, or decreases below the baseline) are relative, but comparable within rows.

exploitation, through its impact on a keystone species (Paine, 1966) or group, may lead to ecological changes, and the yields predicted by surplus production models, however sophisticated, would not be sustainable.

Changes may also occur because of nutrient removal by fishing, although the evidence is equivocal. After years of contention, it is still unclear whether tropical reefs are widely 'nutrient-limited' or not (Polunin, Chapter 5). Internal stores of nutrients have been little studied (Entsch *et al.*, 1983) and external sources have rarely been quantified (Meyer *et al.*, 1983; Lewis, 1987; Hamner *et al.*, 1988), but it seems possible in any case that space is a major determinant of algal biomass and productivity (Grigg *et al.*, 1984). Intense fishing evidently would remove a large part of the store of nitrogen and phosphorus in reef biomass. If external supply is slow, and internal cycling fast, the effects could be large and long term, and conversely otherwise.

There is indication that nutrient export through fishing may be important for oceanic atolls. Thus, Salvat *et al.* (1985) calculated that the lagoon of Takapoto could export only 3.5% of the biomass of bivalve molluscs per year without nitrogen deficit. Here again, strong exploitation would not be sustainable.

We can perhaps generalize these observations by suggesting that extensive harvesting should be limited to areas with high nutrient inputs, such as continental margins, compared with nutrient-poor regions such as oceanic gyres. This would match the situation occurring in non-reef fisheries, which have high catches in areas with high value of new production and low catches in areas where regenerated production predominates (Longhurst and Pauly, 1987).

How readily do stocks and systems recover?

Two key aspects of sustainability are the recovery rate of depleted species, once fishing pressure is relaxed, and extinctions, which preclude recovery.

Intensive episodic exploitation of small reef areas by spear fishing and leaf-sweeps in Yap (Smith and Dalzell, 1993) rapidly depletes fish biomass, but there are indications that this grows back to pre-exploitation levels in 1 or 2 years. In the Philippines, abundances of several reef fishes have increased in small reserves within a few years of their establishment (White, 1988a; Russ and Alcala, 1994), although recovery in numbers of fish is much faster than recovery of biomass, especially in larger species such as groupers. About 50% of the target species on two Caribbean reefs apparently recovered in protected areas which were less than 2 km across, within 4 years of the fishery being closed (Polunin and Roberts, 1993). Roberts (in press) recorded a 60% increase in overall fish biomass within the Saba Marine Park in the Caribbean, within 2 years. These findings indicate that local populations have the potential rapidly to recover from depletion. Tagging and tracking work (Holland *et al.*, 1993) indicates that most reef fishes are strongly site attached. However, many larger species, such as jacks (Carangidae) and some snappers (Lutjanidae) are mobile, and movement of large fishes is another mechanism of local recovery following fishing (DeMartini, 1993).

Estimating the rate of recovery from, and reversibility of, fishing effects over large reef areas appears more difficult. Where growth overfishing predominates, recovery following effort reduction may be rapid if the fish in question are fast-growing, as in the case of mullids (Garcia and Demetropoulos, 1986). Recovery may be slower if the biomass reduction was due to recruitment overfishing, as it takes time to rebuild adult spawning biomass and high fecundities (Polunin and Morton, 1992). Regarding extinctions, McAllister *et al.* (1994) found that 17% of the 800 reef fish species which they reviewed had been recorded from areas of less than 50 000 km^2. Many species occur in only a single island or cluster of islands. Widespread heavy fishing could cause global extinctions of some such species, particularly where there is also habitat damage. A complicating factor is that the majority of species with restricted ranges are of high value to the aquarium trade. Species-by-species restrictions on capture and trade (through the CITES lists) may be appropriate to prevent overexploitation, but at present there are few controls on the industry (Wood, 1992). Issues relating to the aquarium trade can be expected to increase in importance for reef fishery managers. Fortunately, much of the basic information to devise management strategies for these fisheries already exists. The same cannot be said for species such as sharks, which grow and reproduce slowly.

Real closure to fishing of areas at the scale of hundreds of km will be difficult in most cases. A mix of protective and rehabilitative measures may be more desirable, particularly where valuable species are depleted, while the fishery as a whole is still healthy (Munro, Chapter 1; Ruddle, Chapter 6). The rearing and wild release of giant clams may be considered in this context, along with the careful use of artificial reefs (Polovina, 1991b; Bohnsack, Chapter 11). Another possibility is the provision of shelter habitat to reduce mortality in species such as spiny lobsters (Eggleston *et al.*, 1992). Important management questions remain also as to the rates and mechanisms of recovery of communities and ecosystems. The extent and nature of these effects is only now being appreciated; few case studies or time series data exist which could be used to draw inferences on these processes. Data sets relating the state of reef fishery resources to episodes of pulse fishing (Smith and Dalzell, 1993), changes in fishing effort over time (Koslow *et al.*, 1988; Alcala and Russ, 1990; Russ and Alcala, 1994) or spatial differences in effort (Jennings and Polunin, 1995) are very few. Assembling such data sets is an urgent task.

More on the multispecies question

In tropical multispecies fisheries, catch composition usually changes as fishing pressure increases. Vulnerable species decline, but overall catches may be maintained through species replacements, or by shifts in gear use (Pauly, 1979; Pope, 1979a; Munro, 1980). Such maintenance of overall yield may prevent important actions from being taken until it is too late, for example when grouper stocks are heavily depleted (Butler *et al.*, 1993).

Low gear selectivity is often cited as a problem in that it renders application of single-species approaches to stock assessment and management of questionable value. However, even if more selective gears could be devised, they would be of limited value as yields of single species would be small. Besides, it would be difficult, given highly variable growth and mortality schedules (Adams, 1980; Buesa, 1987; Kirkwood *et al.*, 1994), to devise implementable management strategies that would simultaneously optimize overall yield, consisting mainly of small species, while protecting stocks of larger, vulnerable target species. Marine fishery reserves, in combination with restrictions on gear and take, offer a way out of this dilemma, and this is the reason why they are given so much emphasis in this contribution and in this book.

14.4 MANAGEMENT IN SOCIETAL, ECONOMIC AND GLOBAL CONTEXTS

Beyond the natural science base of tropical reef fisheries lie their little-explored, but crucial, social science aspects. In addition, the global context of tropical reef fisheries needs consideration.

What management measures can be implemented?

To be successful, management measures must appear reasonable and advantageous to the fishing communities concerned, i.e. to the major users of fisheries resources (Ruddle, Chapter 12). This perception will obviously depend on many factors, including the condition of a fishery (Appeldoorn, Chapter 9), which may dictate interventions ranging from mild to harsh, and objectives which, as we have seen, may be determined at different levels in the community (Adams, Chapter 13). Feasibility will be governed by many factors, including political considerations and costs and benefits of action, which may be assessed in a variety of ways. In countries lacking strong administrative and scientific infrastructures, simplicity and cost-effectiveness will be major determinants of success.

Such criteria mitigate against techniques such as individual transferable quotas, fashionable as they may be among fisheries scientists (Pauly, in press), and favour blanket measures such as area closure (Roberts and Polunin, 1991; Medley *et al.*, 1993; Bohnsack, Chapter 11).

Another important aspect of proposed reef fisheries management schemes is whether they be made to relate to traditional management systems. The strengths of these systems lie in their focus on spatial and temporal structures of gear deployment (Beeching, 1993) and resource allocation, and on the fact that decisions are enforced by local moral and political authority (Ruddle, Chapter 12). While the origin and nature of traditional, community-based systems for managing reef fisheries is debated, the key issue is whether or not these systems can meet modern pressures.

Cooke (1994) quantified fishing pressure and management aptitude in eight Fijian customary fishing rights areas, or *qoliqoli*. Of the *qoliqoli* identified as being subject to high subsistence and commercial fishing pressure, half showed low aptitude and half high aptitude to management (Fig. 14.4). Thus, even within one relatively small country, large differences occur in the way those responsible for tenured areas meet the challenges of management.

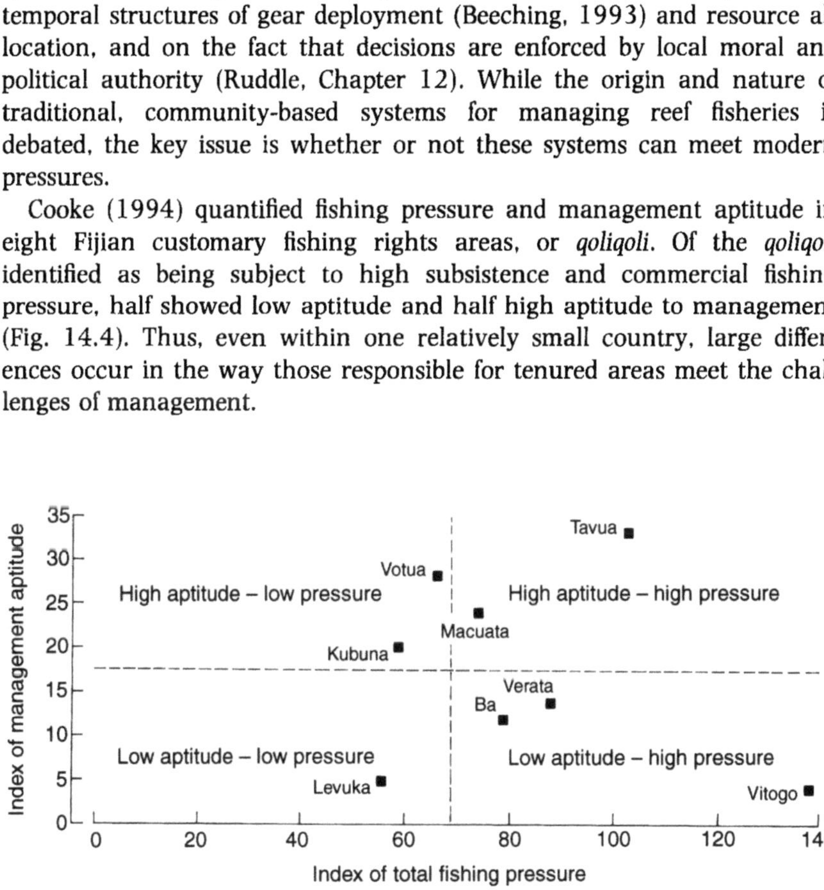

Fig. 14.4 The management aptitude of Fijian customary fishing rights areas (*qoliqoli*) varies with overall pressure on fishery resources within the area (Cooke, 1994). The index of management aptitude was derived from the sum of indices of: ability to marshall relevant information + approach to goodwill payments + management measures taken + consultation/cooperation with the Fisheries Department + patrolling and enforcement. Total resource pressure was measured as the sum of commercial fishing pressure + poaching supplement + subsistence fishing pressure. Figure redrawn by permission of Andrew Cooke Esq.

The better-managed grounds included be those which collaborated actively with the Fisheries Division, although at least one rights-owning group showed high management aptitude without state support (Cooke, 1994). In some cases, therefore, such an approach may help identify those rights owners in need of state assistance with management.

Modern governance of reef fisheries

When central governments can offer expertise and facilities, while the fishing community can provide knowledge and a monitoring function, co-management (Pinkerton, 1989) may be considered. However, such conditions do not necessarily occur in all developing countries; it is not by accident that the co-management concept emerged in Canada. In many of the developing countries where reefs occur, government cannot offer scientific expertise for local fisheries management, and neither are all fishing communities committed to sustainable resource utilization, or able effectively to protect 'their' resources. Thus, fisheries reserves in the Philippines stand or fall at the whim of local political forces (Alcala and Russ, 1990; Russ and Alcala, 1994). One reason for this may be the basic, if implied, tenet of co-management, emanating from its very definition (contributions in Pinkerton, 1989). This is that central governments and fishing communities are, with regards to fisheries resources, the only legitimate users.

For reef resources, which can be extracted, but may also form the basis of ecotourism (e.g. support scuba diving resorts and their staff) or be conserved as sources of germplasm, there are clearly more users than the fishing communities and governments. Modern concepts of governance (contributions in Kooiman, 1993) offer a framework for arrangements enabling different groups of users to interact, and governments to limit their role to providing a level field for these interactions.

Whether such governance arrangements would be conducive to adaptive management schemes (Russ, 1991; Hilborn and Walters, 1992; Medley *et al.*, 1993) is unclear. This, and related studies, dealing with the appropriateness of adaptive management schemes for subsistence fisheries, seem worthwhile areas for future socio-economic and policy research on tropical reef fisheries.

Tropical reef fisheries in a global context

The data now available (e.g. Fig. 14.1) suggest that (potential) tropical reef fisheries catches are, on a global basis, much higher than estimated by Smith (1978). Within specific regions, such as the South Pacific, present catches already are significant, although to a modest population scattered over a vast area (Dalzell and Adams, 1994).

Escalating human populations are increasing pressure on reef resources throughout most of the intertropical belt. At the same time, human activities are causing widespread degradation, especially on coastlines and to coral reefs. To this must be added the possible consequences of global climate change for reef fisheries, which are likely because the reefs themselves stand to be affected (Glantz and Feingold, 1990).

In many places, simply maintaining present fisheries production in the face of these pressures, as opposed to increasing yields, may be the primary challenge to managers. However, as mentioned above, fisheries management extends far beyond fisheries biology. Where the pressures on fishery resources are great, regulation will demand broadly based and imaginative approaches, encompassing both human and natural resources. Successful management schemes will increasingly have to look beyond a narrow fishery focus. In high-pressure areas, they will have to deal with the causes of massive entry into fishing by displaced farmers (Pauly, 1994) and of inshore habitat degradation, through comprehensive coastal zone management strategies (see McManus *et al.*, 1992, for a model study). In addition, these schemes will have to incorporate means of providing alternative livelihoods to displaced fishers, such as in cases where fishery reserves are designated. Only through use of such integrative approaches are reef fishery yields likely to be sustained over the long term.

ACKNOWLEDGEMENTS

We are grateful to the other authors in this book and to many colleagues for the ideas which they have shared with us. We thank Drew Wright, John Caddy, John Tarbit and Chris Mees for their comments on drafts. The senior author thanks the Overseas Development Administration for funding through its Fish Management Science Programme. This is ICLARM Contribution No.1129.

References

Abdul Nabi, A.H. (1980) Taxonomy, biometry, length–weight relationship and age and growth studies of family Mullidae, goatfishes, in the Gulf of Aqaba (Jordan). *Proc. Symp. Coastal Marine Env. Red Sea, Gulf of Aden and Trop. W. Indian Ocean, Khartoum*, **2**, 193–226.

Abou-Seedo, F., Wright, J.M. and Clayton, D.A. (1990) Aspects of the biology of *Diplodus sargus kotschyi* (Sparidae) from Kuwait Bay. *Cybium*, **14**, 217–23.

Abu-Hakima, R. (1984) Comparison of aspects of the reproductive biology of *Pomadasys, Otolithes* and *Pampus* spp. in Kuwaiti waters. *Fish. Res.*, **2**, 177–200.

Abu-Hakima, R. (1987) Aspects of the reproductive biology of the grouper, *Epinephelus tauvina* (Forskål), in Kuwaiti waters. *J. Fish Biol.*, **30**, 213–22.

Acheson, J.M. (1981) Anthropology of fishing. *A. Rev. Anthropol.*, **10**, 275–316.

Acheson, J.M. (1987) The lobster fiefs revisited: economic and ecological effects of territoriality in the Maine lobster fishery, in *The Question of the Commons: the Culture and Ecology of Communal Resources* (eds B.M. McCay and J.M. Acheson), Univ. Arizona Press, Tucson, pp. 37–65.

Acosta, A.R. (1993) Factors influencing gillnet and trammel net selectivity on a Caribbean coral reef. PhD dissertation, University of Puerto Rico, Mayagüez, Puerto Rico, 174 pp.

Acosta, A.R. (1994) Soak time and net length effects on catch rate of entangling nets in coral reef areas. *Fish. Res.*, **19**, 105–19.

Acosta, A.R. and Appeldoorn, R.S. (1992) Estimation of growth, mortality and yield per recruit for *Lutjanus synagris* (Linnaeus) in Puerto Rico. *Bull. mar. Sci.*, **50**, 282–91.

Acosta, A.R. and Appeldoorn, R.S. (1995) Catching efficiency and selectivity of gillnets and trammel nets in coral reefs from Southwestern Puerto Rico. *Fish. Res.*, **22**, 175–96.

Acosta, A.R. and Recksiek, C.W. (1989) Coral reef fisheries at Cape Bolinao, Philippines: an assessment of catch, effort and yield. *Asian mar. Biol.*, **6**, 101–14.

Acosta, A.R., Turingan, R.G., Appeldoorn, R.S. and Recksiek, C.W. (1994) Reproducibility of estimates of effective area fished by Antillean fish traps in coral reef environments. *Proc. Gulf Carib. Fish. Inst.*, **43**(B), 346–54.

Adams, P.B. (1980) Life history patterns in marine fishes and their consequences for management. *Fish. Bull. U.S.*, **78**, 1–12.

Adams, T.J.H. (1993) Management of bêche-de-mer (sea cucumber) fisheries. South Pacific Commission, Nouméa, New Caledonia: *Bêche-de-mer Special Interest Group Infor. Bull.*, **5**, 15–21.

Aiken, K. (1983a) The biology, ecology and bionomics of the butterfly and angelfishes, Chaetodontidae, in *Caribbean Coral Reef Fishery Resources*, (*ICLARM Stud. Rev.* 7) (ed. J.L. Munro), ICLARM, Manila, Philippines, pp. 155–65.

Aiken, K. (1983b) The biology, ecology and bionomics of the triggerfishes, Balistidae, in *Caribbean Coral Reef Fishery Resources* (*ICLARM Stud. Rev.* 7), (ed. J.L. Munro), ICLARM, Manila, Philippines, pp. 191–205.

Aiken, K.A. and Haughton, M.O. (1990) Regulating fishing effort: the Jamaican experience. *Proc. 40th Gulf Carib. Fish. Inst. Meeting*, 139–50.

Aitken, R.T. (1930) *Ethnology of Tubuai*. Bernice P. Bishop Museum, Honolulu, HI, *Bulletin* **70**.

Akimichi, T. (1978) The ecological aspect of Lau (Solomon Islands) ethnoichthyology. *J. Polynesian Soc.*, **87**, 301–326.

Akimichi, T. (1984) Territorial regulations in the small-scale fisheries of Itoman, Okinawa, in *Maritime Institutions in the Western Pacific* (*Senri Ethnological Studies* **17**) (eds K. Ruddle and T. Akimichi), National Museum of Ethnology, Osaka, pp. 89–120.

Al-Ghais, S.M. (1993) Some aspects of the biology of *Siganus canaliculatus* in the southern Arabian Gulf. *Bull. mar. Sci.*, **52**, 886–97.

Al-Ogaily, S.M., Hanafi, N. el D. and Hussain, A. (1992) Some aspects of the biology of red snapper *Lutjanus campechanus* (Rivas, 1966) from the Red Sea. *Asian Fish. Sci.*, **5**, 327–39.

Alcala, A.C. (1981) Fish yields of coral reefs of Sumilon Island, central Philippines. *Bull. nat. Res. Council Philippines*, **36**, 1–7.

Alcala, A.C. (1988) Effects of marine reserves on coral fish abundance and yields of Philippine coral reefs. *Ambio*, **17**, 194–9.

Alcala, A.C. and Gomez, E.D. (1985) Fish yields of coral reefs in the central Philippines. *Proc. 5th Int. Coral Reef Symp.*, **5**, 521–4.

Alcala, A.C. and Gomez, E.D. (1987) Dynamiting coral reefs: a resource destructive fishing method, in *Human Impacts on Coral Reefs: Facts and Recommendations* (ed. B. Salvat), Antènne du Muséum EPHE, French Polynesia, pp. 51–60.

Alcala, A.C. and Luchavez, T. (1981) Fish yield of a coral reef surrounding Apo Island, central Visayas. *Proc. 4th Int. Coral Reef Symp.*, **1**, 69–73.

Alcala, A.C. and Russ, G.R. (1990) A direct test of the effects of protective management on the abundance and yield of tropical marine resources. *J. Conseil. Cons. Int. Explor. Mer*, **46**, 40–47.

Aldenhoven, J.M. (1986) Local variation in mortality rates and life-expectancy estimates of the coral-reef fish *Centropyge bicolor* (Pisces: Pomacanthidae). *Mar. Biol.*, **92**, 237–44.

Alder, J. and Wicaksono, A. (1992) Community and outside fishers' use of marine resources at Taka Bone Rate Atoll. Environmental Management Development in Indonesia, Jakarta.

Alexander, A.B. (1902) Notes on boats, apparatus and fishing methods employed by the natives of the South Sea Islands, and the results of fishing trials by the Albatross. U.S. Commissioner of Fish and Fisheries, Doc. 509. Govt Printing Office, Washington, DC, pp. 741–829.

Aliño, P.M., McManus, L.T., McManus, J.W., Nañola, C.L., jun., Fortes, M.D., Trono, M.D., jun. and Jacinto, G.S. (1993) Initial parameter estimation of a coral reef flat ecosystem in Bolinao, Pangasinan, Northwestern Philippines, in *Trophic Models of Aquatic Ecosystems* (*ICLARM Conf. Proc.* **26**) (eds V. Christensen and D. Pauly), ICLARM, Manila, ICES, Copenhagen, and DANIDA, Copenhagen, pp. 252–8.

Alkire, W.H. (1965) *Lamotrek Atoll and Inter-island Socio-economic Ties* (*Illinois Stud. Anthropol.*, **5**), Univ. Illinois Press, 180 pp.

Alkire, W.H. (1968) An atoll environment and ethnography. *Geographia*, **4**, 54–9.

Alkire, W.H. (1977) *An Introduction to the Peoples and Cultures of Micronesia*, Cummings Publishing Co., Menlo Park, CA.

Alldredge, A.L. and King J.M. (1985) The distance demersal zooplankton migrate above the benthos: implications for predation. *Mar. Biol.*, **84**, 253–60.

Allen, G.R. (1972) *The Anemonefishes: their Classification and Biology*. T.F.H. Publications, Neptune City, NJ.

Allen, G.R. (1985) *FAO Species Catalogue*. Vol. 6. *Snappers of the World. An Annotated and Illustrated Catalogue of Lutjanid Species Known to Date. FAO Fish. Synopses*, **125**, 208 pp.

Allen, K.R. (1971) Relation between production and biomass. *J. Fish. Res. Bd Can.*, **28**, 1573–81.

Alongi, D.M. (1988) Detritus in coral reef ecosystems: fluxes and fates. *Proc. 6th Int. Coral Reef Symp.*, **1**, 29–36.

Amesbury, J.R., Hunter-Anderson, R.L. and Wells, E.F. (1989) *Native Fishing Rights and Limited Entry in the CNMI*, Micronesian Archaeological Research Services, Guam.

Amesbury, S.S., Cushing, F.A. and Sakamoto, R.K. (1986) *Fishing on Guam* (Guide to the Coastal Resources of Guam, Vol. 3), University of Guam.

Andersen, K.P. and Ursin, E. (1977) A multispecies extension to the Beverton and Holt theory of fishing, with accounts of phosphorous circulation and primary production. *Meddelelser fra Danmarks Fiskeri-og Havundersoegelser*, N.S. 7, 319–435.

Anderson, G.R.V., Ehrlich, A.H., Ehrlich, P.R., Roughgarden, J.D., Russell, B.C. and Talbot, F.H. (1981) The community structure of coral reef fishes. *Am. Nat.*, **117**, 476–95.

Anderson, L.G. (1980) An economic analysis of joint recreational and commercial fisheries, in *Allocation of Fishery Resources: Proc. Tech. Consultation on Allocation of Fishery Resources, Vichy, France, 20–23 April 1980*, (ed. J.H. Grover), United Nations Food and Agriculture Organization, Rome, pp. 16–26.

Anderson, R.C. (1992) A second look at the reef fish resources of the Maldives. *Bay of Bengal News*, **45**, 6–9.

Anderson, R.C., Naheed, Z., Rasheed, M. and Arif, A. (1992) Reef fish resources survey in the Maldives – Phase II. Bay of Bengal Programme, Madras, BOBP/WP/80,MDV/88/007, 51 pp.

Anell, B. (1955) *Contribution to the History of Fishing in the Southern Seas*, Almquist and Wiksell, Uppsala.

Anon. (1982) Further fishing trials with bottom-set longlines in Sri Lanka. Bay of Bengal Programme, Madras. *Working Paper* **16**, 24 pp.

Anon. (A.D. Lewis) (1983a) The fishery resources of Rabi. Fisheries Divn, Min. Agric. Fish., Fiji. 30 pp.

Anon. (A.D. Lewis) (1983b) The fishery resources of Rotuma. Fisheries Divn, Min. Agric. Fish., Fiji. 31 pp.

Anon. (1984) Annual Report – 1983. Research and Surveys Branch, Fisheries Divn, Dept Primary Industry, Port Moresby, Papua New Guinea. 76 pp.

Anon. (1985) Traditional muro-ami, an effective but destructive coral reef fishing gear. *ICLARM Newsl.*, **1985**, 12–13.

Anon. (1987a) Survey of reef ownership, ownership enforcement, and fishing rights in Yap Proper: 1986. Mar. Resour. Manage. Divn, Yap State Dept Resour. Devel., Colonia.

Anon. (1987b) A survey of village fishermen: 1987. Mar Res. Manage. Divn, Yap State Dept Resour. Devel., Colonia. 66 pp.

Anon. (1988) The marine resources of Palmerston Island. South Pacific Commission Inshore Fish. Res. Proj., unpubl. doc. 61 pp.

Anon. (1989) Annual Report – 1989. Fisheries Divn, Min. Nat. Resour. Devel., Kiribati. 38 pp.

Anon. (1990) Annual Report – 1990. Divn Mar. Resour., Min. Resour. Devel., Palau. 40 pp.

Anon. (1991) Annual Report – 1991. Divn Mar. Resour., Min. Resour. Devel., Palau. 66 pp.

Anon. (1992a) Annual Report – 1992. Divn Mar. Resour., Min. Resour. Devel., Palau. 98 pp.

Anon. (1992b) *Agenda 21. Proc. UN Conf. Environment and Development, Rio de Janeiro, 3–14 June 1992*, United Nations, New York, USA.

Anon. (1993) Fisheries "bribes" in PNG. *South Seas Digest* **13**(7) (18th June).

Anon. (n.d.) Traditional Fisheries Management. Unpubl. ms, Min. Agric., Forestry Fish., Gov of Niue, Alofi.

Appeldoorn, R.S. (1985) Growth, mortality and dispersion of juvenile laboratory-reared conchs, *Strombus gigas* and *S. costatus*, released at an offshore site. *Bull. mar. Sci.*, **37**, 785–93.

Appeldoorn, R.S. (1987a) Modification of a seasonally oscillating growth function for use with mark–recapture data. *J. Conseil, Cons. int. Explor. Mer*, **43**, 194–98.

Appeldoorn, R.S. (1987b) Development and testing of a new methodology for assessing lumped-species unit stocks in tropical fisheries. University of Puerto Rico Sea Grant Program, Proposal R/LR-06-87-RAP3. 18 pp.

Appeldoorn, R.S. (1988) Ontogenetic changes in natural mortality rate of queen conch, *Strombus gigas* (Mollusca: Mesogastropoda). *Bull. mar. Sci.*, **42**, 159–65.

Appeldoorn, R.S. (1993) Interspecific relationships between growth parameters, with application to haemulid fishes. *Proc. 7th Int. Coral Reef Symp.*, **2**, 899–904.

Appeldoorn, R.S. and Ballantine, D.L. (1983) Field release of cultured queen conch in Puerto Rico: implications for stock restoration. *Proc. Gulf Carib. Fish. Inst.*, **36**, 89–98.

Appeldoorn, R.S. and Lindeman, K.C. (1985) Multispecies assessment in coral reef fisheries using higher taxonomic categories as unit stocks, with an analysis of an artisanal haemulid fishery. *Proc. 5th Int. Coral Reef Congr.*, **5**, 507–14.

Appeldoorn, R.S. and Meyers, S. (1993) Puerto Rico and Hispaniola. *FAO Fish. tech. Pap.*, **326**, 99–158.

Appeldoorn, R.S., Beets, J., Bohnsack, J., Bolden, S., Matos, D., Meyers, S., Rosario, A., Sadovy, Y. and Tobias, W. (1992) Shallow water reef fish stock assessment for the U.S. Caribbean. *NOAA tech. Memo.* NMFS-SEFSC-304, 70 pp.

Appeldoorn, R.S., Hensley, D.A., Shapiro, D.Y., Kioroglou, S. and Sanderson, B.G. (1994) Egg dispersal in a Caribbean coral reef fish, *Thalassoma bifasciatum*. II. Dispersal off the reef platform. *Bull. mar. Sci.*, **54**, 271–80.

Arias-Gonzalez, E. (1993) Fonctionnement trophique d'un ecosystème récifal: secteur de Tiahura, Ile de Moorea, Polynésie Française, PhD thesis, École Pratique des Hautes Etudes, Perpignan, xix + 358 pp.

Arias-Gonzalez, J.E., Galzin, R., Neilson, J., Mahon, R. and Aiken, K. (1994) Reference area as a factor affecting potential yield estimates of coral reef fishes. *NAGA, The ICLARM Quarterly*, **17**(4), 37–40.

Armstrong, D.A., Wainwright, T.C., Jensen, G.C., Dinnel, P.A. and Andersen, H.B. (1993) Taking refuge from bycatch issues: red king crab (*Paralithodes camtschaticus*) and trawl fisheries in the Eastern Bering Sea. *Can. J. Fish. aquat. Sci.*, **50**, 1993–2000.

Arnold, E.L., jun. and Thompson, J.R. (1958) Offshore spawning of the striped mullet, *Mugil cephalus* in the Gulf of Mexico. *Copeia*, **1958**, 130–32.

Arreguín-Sánchez, F. (1987) Present status of the red grouper fishery of the Campeche Bank. *Proc. Gulf Carib. Fish. Inst.*, **38**, 498–509.

Asch, S.E. (1972) Group pressure, in *Social Psychology: Experimentation, Theory, Research* (ed. W.S. Sahakian), Intext Education Publishers, San Francisco, CA, pp. 309–19.

Atapattu, A.R. (1987) Territorial use rights in fisheries (TURFs) in Sri Lanka: case studies on Jakottu fisheries in the Madu Ganga estuary and Kattudel fishery in the Negombo lagoon, in *Papers Presented at the Symposium on the Exploitation and Management of Marine Fishery Resources in Southeast Asia held in conjunction with the 22nd session of the IPFC, Darwin, 16–26 February, 1987* (ed. Anon.), FAO, Bangkok, Regional Office for Asia and the Pacific, *RAPA Report* 1987/10, pp. 379–401.

Atz, J.W. (1964) Intersexuality in fishes, in *Intersexuality in Vertebrates including Man* (eds C.N. Armstrong and A.J. Marshall), Academic Press, London, pp. 145–232.

Austin, H.M. (1971) A survey of the ichthyofauna of the mangroves of western Puerto Rico during December, 1967 – August, 1968. *Carib. J. Sci.*, **11**, 27–39.

Axelrod, H.R. (1971) *The Aquarium Fish Industry – 1971*, T.F.H. Publications, Neptune City, NJ.

Ayling, A.M. and Ayling, A.L. (1986a) A biological survey of selected reefs in the Capricorn Section of the Great Barrier Reef Marine Park. Great Barrier Reef Marine Park Authority, Townsville, Qld, 61 pp.

Ayling, A.M. and Ayling, A.L. (1986b) Coral trout survey data: raw data sheets and abundance summaries from all surveys of coral trout species (*Plectropomus* spp.) carried out by Sea Research for the Great Barrier Reef Marine Park Authority between February 1983 and July 1986. Unpubl. rep., GBRMPA, Townsville, Qld, 212 pp.

Baelde, P. (1990) Differences in the structure of fish assemblages in *Thalassia testudinum* beds in Guadeloupe, French West Indies, and their significance. *Mar. Biol.*, **105**, 163–173.

Bagenal, T.B. (1978) Aspects of fish fecundity, in *Methods for Assessment of Fish Production in Fresh Waters*, (ed. S.D. Gerking), Blackwell, Oxford, pp. 75–101.

Bailey, C. and Zerner, C. (1992a) Role of traditional fisheries resource management systems for sustainable resource utilization, in *Prosiding Forum Perikanan II. Sukabumi, 18–21 June 1991*. Central Res. Inst. Fisheries, Agency Agric. Res. Devel., Min. Agric., Jakarta, pp. 307–15.

Bailey, C. and Zerner, C. (1992b) Community-based fisheries management institutions in Indonesia. Unpubl. paper presented at the Society for Applied Anthropology, Memphis, 25–29 March, 1992.

Bailey, C. and Zerner, C. (1992c) Local management of fisheries resources in Indonesia: opportunities and constraints, in *Contributions to Fishery Development Policy in Indonesia* (eds R.B. Pollnac, C. Bailey and A. Poernomo), Central Res. Inst. Fisheries, Agency Agric. Res. Devel., Min. Agric., Jakarta, pp. 38–56.

Bailey, C., Pollnac, R.B. and Malvestuto, S. (1990) The Kapuas River fishery: problems and opportunities for local resource management. Paper presented at

1st Meeting, Int. Ass. Study of Common Property, Duke University, Durham, NC, September 27–30.
Bailey, K.M. and Houde, E.D. (1989) Predation on eggs and larvae of marine fishes and the recruitment problem. *Adv. mar. Biol.*, **25**, 1–83.
Bak, R.P.M. (1990) Patterns of echinoid bio-erosion in two Pacific coral reef lagoons. *Mar. Ecol. – Progr. Ser.*, **66**, 267–72.
Bak, R.P.M. (1994) Sea-urchin bioerosion on coral reefs – place in the carbonate budget and relevant variables. *Coral Reefs*, **13**, 99–103.
Bak, R.P.M. and Engel, M.S. (1979) Distribution, abundance and survival of juvenile hermatypic corals (scleractinia) and the importance of life history strategies in the parent coral community. *Mar. Biol.*, **54**, 341–52.
Bak, R.P.M., Carpay, M.J.E. and de Ruyter van Steveninck, E.D. (1984) Densities of the sea urchin *Diadema antillarum* (Philippi) before and after mass mortalities on the coral reefs of Curaçao. *Mar. Ecol. – Progr. Ser.*, **17**, 105–108.
Bakun, A. (1988) Local retention of planktonic early life stages in tropical demersal reef bank systems: the role of vertically-structured hydrodynamic processes, in *IOC/FAO Workshop on Recruitment in Tropical Coastal Demersal Communities*, International Oceanographic Commission, Paris, *IOC Workshop Rep.*, no. 44, Suppl., pp. 15–32.
Ballantine, B. (1991) *Marine Reserves for New Zealand*. University of Auckland, *Leigh Lab. Bull.* **25**, 196 pp.
Bannerot, S.P., Fox, W.W., jun. and Powers, J.E. (1987) Reproductive strategies and the management of snappers and groupers in the Gulf of Mexico and the Caribbean, in *Tropical Snappers and Groupers: Biology and Fisheries Management* (eds J.J. Polovina and S. Ralston), Westview Press, Boulder, CO, pp. 561–603.
Baquie, B. (1977) Fishing in Rarotonga, MA thesis, University of Auckland, New Zealand, 163 pp.
Baranov, F.I. (1948) *Theory and Assessment of Fishing Gear*, Pishchepromizdat, Moscow (trans. Ontario Dept Lands Forests, Maple, Ontario).
Bardach, J.E. (1958) On the movements of certain Bermuda reef fishes. *Ecology*, **39**, 139–46.
Bardach, J.E. (1959) The summer standing crop of fish on a shallow Bermuda reef. *Limnol. Oceanogr.*, **4**, 77–85.
Bardach, J.E., Smith, C.L. and Menzel, D.W. (1958) Bermuda Fisheries Research Program Final Report. Bermuda Trade Devel. Bd, Hamilton, 59 pp.
Barlow, G.W. (1981) Patterns of parental investment, dispersal and size among coral-reef fishes. *Env. Biol. Fishes*, **6**, 65–85.
Barss, P.G. (1982) Injuries caused by garfish in Papua New Guinea. *Br. med. J.*, **284**, 77–9.
Barth, F. (1983) *Sohar: Culture and Society in an Omani Town*, The Johns Hopkins University Press, Baltimore, MD.
Bartsch, J. (1993) Application of a circulation and transport model system to the dispersal of herring larvae in the North Sea. *Continental Shelf Res.*, **13**, 1335–61.
Batisse, M. (1990) Development and implementation of the biosphere reserve concept and its applicability to coastal regions. *Env. Conserv.*, **17**, 111–16.
Bauer, J.A., jun. and Bauer, S.E. (1981) Reproductive biology of pigmy angelfishes of the genus *Centropyge* (Pomacanthidae). *Bull. mar. Sci.*, **31**, 495–513.
Bax, N.J. (1991) A comparison of the biomass flow to fish, fisheries and mammals in six marine ecosystems. *ICES mar. Sci. Symp.*, **193**, 217–24.
Bayliss-Smith, T. (MS) The price of protein: marine fisheries in Pacific subsistence. Unpubl. MS, Dept Geography, Univ. Cambridge, Cambridge CB2 3EN, UK, 29 pp.

Bayliss-Smith, T.P. (1977) Energy use and economic development in Pacific communities, in *Subsistence and Survival – Rural Ecology in the Pacific* (eds T.P. Bayliss-Smith and R.G. Feacham), Academic Press, London, pp. 317–59.

Beaglehole, A. and Beaglehole, P. (1938) *Ethnology of Pukapuka*. Bernice P. Bishop Museum, Honolulu, HI, *Bull* **150**.

Beasley, H.G. (1928) *Pacific Island Records: Fish Hooks*, Seeley, Service and Co., London.

Beddington, J.R. (1984) The response of multispecies systems to perturbations, in *Exploitation of Marine Communities* (ed. R.M. May), Springer-Verlag, Berlin, pp. 209–25.

Beddington, J.R. and Cooke, J.G. (1983) Potential yield of fish stocks. *FAO Fish. tech. Pap.*, **242**, 47 pp.

Beeching, A.J. (1993) A description of temporal and spatial fishing patterns in Suva, Fiji. MSc Thesis, University of Newcastle, Centre for Trop. Coastal Manage. Stud., Newcastle NE1 7RU, UK, 153 pp.

Beets, J. and Friedlander, A. (1992) Stock analysis and management strategies for red hind, *Epinephelus guttatus*, in the U.S. Virgin Islands. *Proc. Gulf Carib. Fish. Inst.*, **42**, 66–80.

Beets, J.P. and Hixon, M.A. (1994) Distribution, persistence, and growth of groupers on natural and artificial patch reefs in the Virgin Islands. *Bull. mar. Sci.*, **55**, 470–83.

Begon, M. (1979) *Investigating Animal Abundance: Capture–Recapture for Biologists*, Edward Arnold, London.

Behrents, K.C. (1987) The influence of shelter availability on recruitment and early juvenile survivorship of *Lythrypnus dalli* Gilbert (Pisces: Gobiidae). *J. exp. mar. Biol. Ecol.*, **107**, 45–59.

Beinssen, K. (1988) Boult Reef revisited. Great Barrier Reef Authority, Townsville, Australia, *Reeflections*, March 1988, 8–9.

Beinssen, K. (1989) Results of the Boult Reef replenishment area study. Unpubl. rep., National Parks and Wildlife Service, Brisbane, 40 pp.

Bell, J.D. (1983) Effects of depth and marine reserve fishing restrictions on the structure of a rocky reef fish assemblage in the north-western Mediterranean Sea. *J. appl. Ecol.*, **20**, 357–69.

Bell, J.D. and Galzin, R. (1984) Influence of live coral cover on coral-reef fish communities. *Mar. Ecol. – Progr. Ser.*, **15**, 265–74.

Bell, J.D., Harmelin-Vivien, M. and Galzin, R. (1985a) Large scale variation in abundance of butterflyfishes (Chaetodontidae) on Polynesian reefs. *Proc. 5th Int. Coral Reef Symp.*, **5**, 421–6.

Bell, J.D., Craik, G.J.S., Pollard, D.A. and Russell, B.C. (1985b) Estimating length frequency distributions of large reef fish underwater. *Coral Reefs*, **4**, 41–4.

Bell, J.D., Lyle, J.M., Bulman, C., Graham, K.J., Newton, G.M. and Smith, D.C. (1992) Spatial variation in reproduction, and occurrence of non-reproductive adults, in orange roughy, *Hoplostethus atlanticus* Collet (Trachichthyidae), from south-eastern Australia. *J. Fish Biol.*, **40**, 107–22.

Bell, L.J. and Colin, P.L. (1986) Mass spawning of *Caesio teres* (Pisces: Caesionidae) at Enewetak Atoll, Marshall Islands. *Env. Biol. Fishes*, **15**, 69–74.

Bellwood, D.R. (1988) Seasonal changes in the size and composition of the fish yield from reefs around Apo Island, central Philippines, with notes on methods of yield estimation. *J. Fish Biol.*, **32**, 881–93.

Bellwood, D.R. (1994) A phylogenetic study of the parrotfishes, family Scaridae (Pisces: Labroidei), with a revision of genera. *Rec. Aust. Mus.*, **20**, (Suppl.) 1–86.

Bellwood, D.R. (1995) Direct estimate of bioerosion by two parrotfish species, *Chlorurus gibbus* and *C. sordidus*, on the Great Barrier Reef, Australia. *Mar. Biol.* **121**, 419–29.

Bellwood, D.R. and Alcala, A.C. (1988) The effect of a minimum length specification on visual estimates of density and biomass of coral reef fishes. *Coral Reefs*, **7**, 23–28.

Bennett, B.A. and Attwood, C.G. (1991) Evidence for recovery of a surf-zone fish assemblage following the establishment of a marine reserve on the southern coast of South Africa. *Mar. Ecol. – Progr. Ser.*, **75**, 173–81.

Berg, C.J. jun. and Olsen, D.A. (1989) Conservation and management of queen conch (*Strombus gigas*) fisheries in the Caribbean, in *Marine Invertebrate Fisheries: Their Assessment and Management* (ed. J.F. Caddy), Wiley, New York, pp. 421–42.

Berger, W.H. (1989) Global maps of ocean productivity, in *Productivity of the Ocean: Present and Past* (eds W.H. Berger, V.S. Smetacek and G. Wefer), Wiley, New York, pp. 429–55.

Berkes, F. (1985) Fishermen and 'the tragedy of the commons'. *Env. Conserv.*, **12**, 199–205.

Berkes, F. (1987) The common property resource problem and the fisheries of Barbados and Jamaica. *Env. Manage.*, **11**, 225–35.

Berntsen, J.D., Skagen, W. and Svendsen, E. (1994) Modelling the transport of particles in the North Sea with reference to sandeel larvae. *Fish. Oceanogr.*, **3**, 81–91.

Bertrand, J. (1986) Donneés concernant la reproduction de *Lethrinus mahsena* (Forsskål 1775) sur les bancs de Saya de Malha (Ocean Indien). *Cybium*, **10**, 15–29.

Bertrand, J. (1988) Selectivity of hooks in the hand-line fishery of the Saya de Malha Banks (Indian Ocean). *Fish. Res.*, **6**, 249–55.

Beverton, R.J.H. (1984) Dynamics of single species, in *Exploitation of Marine Communities* (ed. R.M. May), Springer-Verlag, Berlin, pp. 13–58.

Beverton, R.J.H. (1987) Longevity in fish: some ecological and evolutionary perspectives in *Ageing Processes in Animals*, (eds A.D. Woodhead, M. Witten and K. Thompson), Plenum Press, New York, pp. 161–86.

Beverton, R.J.H. and Holt, S.J. (1956) A review of methods for estimating mortality rates in exploited fish populations, with special reference to sources of bias in catch sampling. *Rapp. P.-V. Réun., Cons. int. Explor. Mer*, **140**, 67–83.

Beverton, R.J.H and Holt, S.J. (1957) On the dynamics of exploited fish populations. UK Min. Agric. Fish., *Fisheries Invest.*, Ser. 2, **19**. [Also available in facsimilie reprint, 1993, Chapman & Hall, London.]

Beverton, R.J.H. and Holt, S.J. (1959) A review of the lifespan and mortality rates of fish in nature and their relation to growth and other physiological characteristics. *Ciba Foundation Colloq. on Ageing*, **5**, 142–80.

Beverton, R.J.H. and Holt, S.J. (1964) Manual of methods for fish stock assessment. Part 2. Tables of yield functions. *FAO Fish. tech. Paper*, **38** (Rev. 1), 67 pp.

BFAR, (1988). (1987) *Fisheries Statistics of the Philippines*, Vol. 37. Bureau of Fisheries and Aquatic Resources, Quezon City, Philippines.

Billard, R. (1987) Testis growth and spermatogenesis in teleost fish; the problem of the large interspecies variability in testis size, in *Reproductive Physiology of Fish* (ed. D.R. Idler). *Publ. Memorial Univ. Nfld. Mar. Sci. Res. Lab.*, pp.183–6.

Birkeland, C. (1977) The importance of rate of biomass accumulation in early successional stages of benthic communities to the survival of coral recruits. *Proc. 3rd Int. Coral Reef Symp.*, 1, 15–21.

Birkeland, C. (1982) Terrestrial runoff as a cause of outbreaks of *Acanthaster planci* (Echinodermata: Asteroidea). *Mar. Biol.*, **69**, 175–85.

Birkeland, C. (1989) The influence of echinoderms on coral-reef communities, in *Echinoderm Studies*, Vol. 3 (eds M. Jangoux and J.M. Lawrence), Balkema, Rotterdam, pp. 1–79.

Birkeland, C. and Amesbury, S.S. (1987) Fish-transect surveys to determine the influence of neighboring habitats on fish community structure in the tropical Pacific, in *Regional Co-operation on Environmental Protection of the Marine and Coastal Areas of the Pacific* (ed. A.L. Dahl), *UNEP Regional Seas Reports and Studies* **97**. United Nations Environment Programme, Nairobi, pp. 195–202.

Blaber, S.J.M. and Copeland J.W. (eds) (1990) *Tuna Baitfish in the Indo–Pacific Region*. Australian Council Int. Agric. Res., Canberra, *Proceedings* no. 30.

Blaber, S.J.M., Milton, D.A., Rawlinson, N.J.F., Tiroba, G. and Nichols, P.V. (1990) Subsistence fishing in the Solomon Islands and the possible conflict with commercial baitfishing, in *Tuna Baitfish in the Indo–Pacific Region* (eds S.J.M. Blaber and J.W. Copeland), Australian Council Int. Agric. Res., Canberra, *Proceedings* no. 30, pp 159–68.

Blaber, S.J.M., Brewer, D.T., Salini, J.P., Kerr, J.D. and Conacher, C. (1992) Species composition and biomasses of fishes in tropical seagrasses at Groote Eylandt, Northern Australia. *Estuarine, Coastal Shelf Sci.*, **35**, 605–20.

Black, K.P. (1993) The relative importance of local retention and inter-reef dispersal of neutrally buoyant material on coral reefs. *Coral Reefs*, **12**, 43–53.

Black, K.P. and Moran, P.J. (1991) Influence of hydrodynamics on the passive dispersal and initial recruitment of larvae of *Acanthaster planci* (Echinodermata, Asteroidea) on the Great Barrier Reef. *Mar. Ecol. – Progr. Ser.*, **69**, 55–65.

Black, K.P., Moran, P.J. and Hammond, L.S. (1991) Numerical models show coral reefs can be self-seeding. *Mar. Ecol. – Progr. Ser.*, **74**, 1–11.

Blair, E.H. and Robertson, J.A. (1903–1909) The Philippine Islands: 1493–1898, A.H. Clark Co., Cleveland, OH (55 vols).

Blaxter, J.H.S. and Hunter, J.R. (1982) The biology of the clupeoid fishes, in *Advances in Marine Biology*, Vol. 20 (eds J.H.S. Blaxter, F.S. Russell and M. Yonge), Academic Press, London, pp. 1–223.

Boehlert, G.W. (1988) An approach to recruitment research in insular ecosystems, in *IREP (OSLR) Workshop on Recruitment in Tropical Coastal Demersal Communities*, International Oceanographic Commission, Paris, *IOC Workshop Report* no. 44, Suppl., pp. 33–44.

Boehlert, G.W. and Mundy, B.C. (1993) Ichthyoplankton assemblages at seamounts and oceanic islands. *Bull. mar. Sci.*, **53**, 336–61.

Boehlert, G.W. and Mundy, B.C. (1994) Vertical and onshore–offshore distributional patterns of tuna larvae in relation to physical habitat features. *Mar. Ecol. – Progr. Ser.*, **107**, 1–13.

Boehlert, G.W., Watson, W. and Sun, L.C. (1992) Horizontal and vertical distributions of larval fishes around an isolated oceanic island in the tropical Pacific. *Deep-Sea Res.*, 39A, 439–66.

Bohnsack, J.A. (1982) Effects of piscivorous predator removal on coral reef fish community structure, in *Gutshop '81: Fish Food Habits and Studies* (eds G.M. Caillet and C.A. Simenstad), Washington Sea Grant Publ., University of Washington, Seattle, pp. 258–67.

Bohnsack, J.A. (1989) Are high densities of fishes at artificial reefs the result of habitat limitation or behavioral preference? *Bull. mar. Sci.*, **44**, 631–45.

Bohnsack, J.A. (1990) The potential of marine fishery reserves for reef fish management in the U.S. southern Atlantic. *NOAA tech. memo.* NMFS-SEFC 261, 40 pp.

Bohnsack, J.A. (1993) Marine reserves: they enhance fisheries, reduce conflicts, and protect resources. *Oceanus*, **36**, 63–71.

Bohnsack, J.A. (1994) How marine fishery reserves can improve fisheries. *Proc. Gulf Carib. Fish. Inst.*, **43**, 217–41.

Bohnsack, J.A. and Bannerot, S.P. (1986) A stationary visual census technique for quantitatively assessing community structure of coral reef fishes. *NOAA tech. Rep.* NMFS-41, 15 pp.

Bohnsack, J.A. and Harper, D.E. (1988) Length–weight relationships of selected marine reef fishes from the southeastern United States and the Caribbean. *NOAA tech. Rep.* NMFS-SEFC, **215**, iii + 31 pp.

Bohnsack, J.A. and Talbot, F.H. (1980) Species packing by reef fishes on Australian and Caribbean reefs: an experimental approach. *Bull. mar. Sci.*, **30**, 710–23.

Bohnsack, J.A., Sutherland, D.L., Harper, D.E., McClellan, D.B., Hulsbeck, M.W. and Holt, C.M. (1989) The effects of fish trap mesh size on reef fish catch off southeastern Florida. *Mar. Fish. Rev.*, **51**, 36–46.

Bohnsack, J.A., Johnson, D.L. and Ambrose, R.F. (1991) The ecology of artificial reef habitats and fishes, in *Artificial Habitats for Marine and Freshwater Fisheries*, (eds W. Seaman, jun. and L.M. Sprague), Academic Press, San Diego, pp. 61–107.

Bohnsack, J.A., Harper, D.E. and McClellan, D.B. (1994) Fisheries trends from Monroe County, Florida. *Bull. mar. Sci.*, 54, 982–1018.

Bolden, S.K. (1993) Summary of the Florida Gulf Coast commercial stone crab fishery 1962–1992. Southeast Fisheries Science Center, National Marine Fisheries Service, NOAA, Miami Laboratory Contribution: MIA-92/93–84, 32 pp.

Boo, E. (1990) *Ecotourism: The Potentials and Pitfalls*, Vol. 1. World Wildlife Fund, Baltimore, MD, 73 pp.

Bouain, A. (1980) Sexualité et cycle sexuel des merous (Poissons, Téléostéens, Serranidés) des côtes sud tunisiens. *Bulletin d'Office Nationale des Pêches, Tunisie*, **4**, 215–29.

Bouain, A. and Siau, Y. (1983) Observations on the female reproductive cycle and fecundity of three species of groupers (*Epinephelus*) from the southeast Tunisian seashores. *Mar. Biol.*, **73**, 211–20.

Bouchon-Navaro, Y. and Bouchon, C. (1989) Correlations between chaetodontid fishes and coral communities of the Gulf of Aqaba (Red Sea). *Env. Biol. Fishes*, **25**, 1–3.

Bouchon-Navaro, Y., Bouchon, C. and Harmelin-Vivien, M.L. (1985) Impact of coral degradation on a chaetodontid fish assemblage (Moorea, French Polynesia). *Proc. 5th Int. Coral Reef Congr.*, **5**, 427–32.

Bouchon-Navaro, Y., Bouchon, C. and Louis, M. (1992) L'ichtyofaune des herbiers de phanerogames marines de la Baie de Fort-de-France (Martinique, Antilles Françaises). *Cybium*, **16**, 307–30.

Boudreau, B., Bourget, E. and Simard, Y. (1993) Behavioural responses of competent lobster postlarvae to odor plumes. *Mar. Biol.*, **117**, 63–70.

Boulon, R. (1990) Mangroves as nursery grounds for recreational fisheries. Final report U.S. Virgin Islands Dingell–Johnson Expansion Project F–7, Study V. Division of Fish and Wildlife, St Thomas. 23 pp.

Braley, R.D. (1987) Spatial distribution and population parameters of *Tridacna gigas* and *T. derasa*. *Micronesica*, **20**, 225–46.

Braley, R.D. (1988) Recruitment of the giant clams *Tridacna gigas* and *T. derasa* at four sites on the Great Barrier Reef, in *Giant Clams in Asia and the Pacific*, (eds J.W. Copeland and J.S. Lucas), Australian Centre Int. Agric. Res. Canberra, pp. 73–7.

Breder, C.M. jun. and Rosen, D.E. (1966) *Modes of Reproduction in Fishes*. T.F.H. Publications, Neptune City, NJ.

Brock, R.E. (1979) An experimental study on the effects of grazing by parrotfishes and the role of refuges in benthic community structure. *Mar. Biol.*, **51**, 381–8.

Brock, R.E. (1982) A critique of the visual census method for assessing coral reef fish populations. *Bull. mar. Sci.*, **32**, 269–76.

Brock, R.E., Lewis, C. and Wass, R.C. (1979) Stability and structure of a fish community on a coral patch reef in Hawaii. *Mar. Biol.*, **54**, 281–92.

Brock, V.E. (1954) A preliminary report on a method of estimating reef fish populations. *J. Wildl. Manage.*, **18**, 297–308.

Brogan, M.W. (1994) Two methods of sampling fish larvae over reefs – a comparison from the Gulf of California. *Mar. Biol.*, **118**, 33–44.

Brothers, E.B. (1980) Age and growth studies in tropical fishes, in *Stock Assessment for Tropical Small-scale Fisheries*, (eds S.B. Saila and P.M. Roedel), Int. Center Mar. Resour. Devel., University of Rhode Island, Kingston, RI, pp. 119–36.

Brothers, E.B. (1984) Otolith studies, in *Ontogeny and Systematics of Fishes* (ed. H.G. Moser), Allen Press, KS, pp 50–57.

Brothers, E.B. (1987) Methodological approaches to the examination of otoliths in aging studies, in *Age and Growth of Fish*, (eds R.C. Summerfelt and G.E. Hall), Iowa State University Press, Ames, IA, pp. 319–30.

Brothers, E.B., Williams, D.M. and Sale, P.F. (1983) Length of larval life in twelve families of fishes of "One Tree Lagoon", Great Barrier Reef, Australia. *Mar. Biol.*, **76**, 319–24.

Brouns, J.J.W.M. and Heijs, F.M.L. (1985) Tropical seagrass systems in Papua New Guinea. A general account of the marine flora and fauna. *Proc. K. ned. Akad. Wet. Ser. C Biol. Med. Sci.*, **88**, 145–82.

Brown, B., Shepherd, A.D., Weir, I. and Edwards, A. (1989) Effects of degradation of the environment on local reef fisheries in the Maldives. *Rastain Newsl.*, **2**, 1–12.

Brown, B.E., Brennan, J.A., Grosslein, M.D., Heyerdahl, E.G. and Hennemuth, R.C. (1976) The effect of fishing on the marine finfish biomass in the Northwest Atlantic from the Gulf of Main to Cape Hatteras. *ICNAF res. Bull.*, **12**, 49–68.

Brown, D.M., and Cheng, L. (1981) New net for sampling the ocean surface. *Mar. Ecol. – Progr. Ser.*, **5**, 225–7.

Brown, I.W., Doherty, P., Ferreira, B. *et al.* (eds) (1994) Growth, reproduction and recruitment of Great Barrier Reef food fish stocks. Southern Fisheries Centre, Queensland Dept of Primary Ind., Final Project Report to the Fish. Res., and Development Corporation, Project No. 90/18. 154 pp.

Bruggeman, J.H., van Kessel, A., van Rooij, J. and Breeman, A. (1994a) A parrotfish bioerosion model: implications of fish size, feeding mode and habitat use for the destruction of reef substrates, in *Parrotfish Grazing on Coral Reefs: a Trophic Novelty*, Ponsen and Looijen, Wageningen, pp. 131–52.

Bruggemann, J.H., Begeman, J., Bosma, E.M., Verburg, P. and Breeman, A.M. (1994) Foraging by the stoplight parrotfish *Sparisoma viride* 2. Intake and assimilation of food, protein and energy. *Mar. Ecol. – Progr. Ser.*, 106, 57–71.

Brulé, T., Maldonado Montiel, T., Rodriguez Canche, L. Gpe. and Mexicano Cintora, G. (in press) Aspectos sobre la biologia reproductiva y trofica del mero *Epinephe-*

lus morio (Valenciennes, 1828) del Banco de Campeche, Yucatan, Mexico. *Proc. Gulf Carib. Fish. Inst.*, **44**.

Bruslé, J. and Bruslé, S. (1976) Contribution à l'étude de la reproduction de deux espèces de merous *E. aeneus* G. Saint-Hilaire, 1809 (Linné, 1758) et *E. guaza* des côtes de Tunisie. *Rev. Trav. Inst. Pêches Marit.*, **39**, 313–20.

Buchheim, J.R. and Hixon, M.A. (1992) Competition for shelter holes in the coral-reef fish *Acanthemblemaria spinosa* Metzelaar. *J. exp. mar. Biol. Ecol.*, **164**, 45–54.

Buck, P.H. (1930) *Samoan Material Culture*. Bernice P. Bishop Museum, Honolulu, HI, *Bulletin* **75**.

Buck, P.H. (1932) *Ethnology of Tongareva*. Bernice P. Bishop Museum, Honolulu, HI, *Bulletin* **92**.

Buck, P.H. (1934) *Mangaian Society*. Bernice P. Bishop Museum, Honolulu, HI, *Bulletin* **122**.

Buck, P.H. (1938) *Ethnology of Mangareva*. Bernice P. Bishop Museum, Honolulu, HI, *Bulletin* **157**.

Buck, P.H. (1950) *Material Culture of Kapingamarangi*. Bernice P. Bishop Museum, Honolulu, HI, *Bulletin* **200**.

Buesa, R.J. (1987) Growth rate of tropical demersal fishes. *Mar. Ecol. – Progr. Ser.*, **36**, 191–9.

Buesa Mas, R.J. (1960) Pesca exploratoria de la langosta con nasas, al sur de Camaguey, Cuba. *Contribucion Centro Investigacion Pesquera, Habana*, **11**, 30 pp.

Buesa Mas, R.J. (1961) Segunda pesca exploratoria y datos biologicos de la langosta, *Panulirus argus*, en Cuba. *Contribucion Centro Investigacion Pesquera, Habana*, **12**, 69 pp.

Buhat, D. (1994) Community-based coral reef and fisheries management, San Salvador Island, Philippines, in *Collaborative and Community-based Management of Coral Reefs: Lessons From Experience* (eds A.T. White, L.Z. Hale, Y. Renard and L. Cortesi), Kumarian Press, West Hartford, CT, pp. 33–50.

Bullock, L.H. and Murphy, M.D. (1994) Aspects of the life history of the yellow-mouth grouper, *Mycteroperca interstitialis*, in the eastern Gulf of Mexico. *Bull. mar. Sci.*, **55**, 30–45.

Bullock, L.H. and Smith, G.B. (1991) Seabasses (Pisces: Serranidae). *Mem. Hourglass Cruises, Florida Mar. Res. Inst.*, pp. 1–243.

Bullock, L.H., Murphy, M.D., Godcharles, M.F. and Mitchell, M.E. (1992) Age, growth, and reproduction of jewfish *Epinephelus itajara* in the eastern Gulf of Mexico. *Fishery Bull. U.S.*, **90**, 243–9.

Burnett-Herkes, J. (1975) Contribution to the biology of the red hind, *Epinephelus guttatus*, a commercially important serranid fish from the tropical western Atlantic, PhD dissertation, University of Miami, Coral Gables, FL, 154 pp.

Burnett-Herkes, J., Luckhurst, B. and Ward, J. (1989) Management of Antillean trap fisheries – Bermuda's experience. *Proc. Gulf Carib. Fish. Inst.*, **39**, 5–11.

Burnham, K.P., Anderson, D.R., White, G.C., Brownie, C. and Pollock, K.H. (1987) *Design and Analysis Methods for Fish Survival Experiments Based on Release–Recapture. American Fisheries Society Monograph* **5**. Bethesda, MD.

Burrows, E. and Spiro, M. (1953) *An Atoll Culture: Ethnography of Ifaluk in the Central Carolines*, Human Relations Area Files, New Haven, CT.

Burrows, E.G. (1937) *Ethnology of Uvea (Wallis Island)*. Bernice P. Bishop Museum, Honolulu, HI, *Bulletin* **145**.

Butler, J.N., Burnett-Herkes J., Barnes, J.A. and Ward, J. (1993) The Bermuda fisheries: a tragedy of the commons averted? *Environment*, **35**, 7–33.

Butler, V. (1988) Lapita fishing strategies: the faunal evidence, in *Archaeology of the Lapita Cultural Complex: a Critical Review* (eds P. Kirch and T. Hunt). Washington State Museum, Seattle, *Res. Rep.*, **5**, pp. 99–115.

Buxton, C.D. (1990) The reproductive biology of *Chrysoblephus laticeps* (Teleostei: Sparidae). *J. Zool., Lond.*, **220**, 497–511.

Buxton, C.D. (1993) Life-history changes in exploited reef fishes on the east coast of South Africa. *Env. Biol. Fishes*, **36**, 47–63.

Buxton, C.D., and Clarke, J.R. (1992) The biology of the bronze bream, *Pachymetopon grande* (Teleostei: Sparidae) from the south-east Cape coast, South Africa. *S. Afr. J. Zool.*, **27**, 21–32.

Buxton, C.D. and Smale, M.J. (1989) Abundance and distribution patterns of three temperate marine reef fish (Teleostei: Sparidae) in exploited and unexploited areas off the Southern Cape Coast. *J. appl. Ecol.*, **26**, 441–51.

BWAM (1986) Employment and working conditions of muroami fishermen: a report submitted to the International Labor Organization. Unpubl. rep., ILO, Geneva, 72 pp.

Cabanban, A.S. (1984) Some aspects of the biology of *Pterocaesio pisang* in the Central Visayas, MSc thesis, College of Science, University of the Philippines, Quezon City. 69 pp.

Cabanban, A.S. and White, A.T. (1981) Marine conservation program using non–formal education at Apo Island, Negros Oriental, Philippines. *Proc. 4th Int. Coral Reef Symp.*, **1**, 317–21.

Caces-Borja, P. (1975) On the ability of otter trawls to catch pelagic fish in Manila Bay. *Philippine J. Fish.*, **10**, 39–56.

Caddy, J.F. (1984) Displaying multispecies information in relation to fishing intensity: a sampling and production modelling approach. *FAO Fish. Rep.*, **327** (Suppl.), 286–90.

Caddy, J.F. (1991) Death rates and time intervals: is there an alternative to the constant natural mortality axiom? *Rev. Fish Biol. Fish.*, **1**, 109–38.

Caddy, J.F. (1993) Background concepts for a rotating harvesting strategy with particular reference to the Mediterranean red coral, *Corallium rubrum*. *Mar. Fish. Rev.*, **55**, 10–18.

Caddy, J.F. and Csirke, J. (1983) Approximations to sustainable yield for exploited and unexploited stocks. *Océanographie Trop.*, **18**, 3–15.

Caddy, J.F. and García, S. (1983) Production modelling without long data series. *FAO Fish. Rep.* **268**, (Suppl.), 309–31.

Caillart, B. and Morize, E. (1985) La production de la pêcherie de l'atoll de Tikehau en 1985. ORSTOM, Tahiti, *Notes et Documents Oceanographiques*, **30**, 45–71.

Caillouet, C.W. jun. and Koi, D.B. (1981) Trends in ex-vessel value and size composition of reported annual catches of pink shrimp from the Tortugas fishery, 1960–1978. *Gulf Res. Rep.*, **7**, 71–8.

Caldwell, D.K. (1957) The biology and systematics of the pinfish, *Lagodon rhomboides* (Linnaeus). *Bull. Fla State Mus., Biol. Sci.*, **2**, 77–173.

Caley, M.J. (1993) Predation, recruitment and the dynamics of communities of coral-reef fishes. *Mar. Biol.*, **117**, 33–44.

Calow, P. (1979) The cost of reproduction – a physiological approach. *Biol. Rev.*, **54**, 23–40.

Cammen, M.L. (1980) Ingestion rate: an empirical model for aquatic deposit-feeders and detritivores. *Oecologia*, **44**, 303–10.

Campana, S.E. and Jones, C.M. (1992) Analysis of otolith microstructure data. *Can. spec. Publ. Fish. aqua. Sci.*, **117**, 73–100.

Campos, A.G. and Bashirullah, A.K.M. (1975) Biologia del pargo *Lutjanus griseus* (Linn.) de la Isla de Cubagua, Venezuela. II. Maduracion sexual y fecundidad. *Bol. Inst. Oceanogr. Univ. Oriente*, **14**, 109–16.

Carpenter, K.E. (1977) Philippine coral reef fishery resources. *Philippine J. Fish.*, **17**, 95–125.

Carpenter, K.E. and Alcala, A.C. (1977) Philippine coral reef fisheries resources. Part II. Muro-ami and kayakas reef fisheries, benefit or bane? *Philippine J. Fish.*, **15**, 217–35.

Carpenter, K.E. and Allen, G.R. (1989) Emperor fishes and large-eye breams of the world (Family Lethrinidae). FAO species catalogue. *FAO Fish. Synop.*, **9**, 1–118.

Carpenter, K.E., Miclat, R.I., Albaladejo, V.D. and Corpuz, V.T. (1981) The influence of substrate structure on the local abundance and diversity of Philippine reef fishes. *Proc. 4th Int. Coral Reef Symp.*, **2**, 497–502.

Carpenter, R.C. (1984) Predation and population density control of homing behaviour in the Caribbean echinoid *Diadema antillarum*. *Mar. Biol.*, **82**, 101–8.

Carpenter, R.C. (1985) Sea urchin mass-mortality: effects on reef algal abundance, species composition and metabolism and other coral reef herbivores. *Proc. 5th Int. Coral Reef Cong.*, **4**, 53–60.

Carpenter, R.C. (1986) Partitioning herbivory and its effects on coral reef algal communities. *Ecol. Monogr.*, **56**, 345–63.

Carpenter, R.C. (1988) Mass mortality of a Caribbean sea urchin: immediate effects on community metabolism and other herbivores. *Proc. US Nat. Acad. Sci.*, **85**, 511–14.

Carpenter, R.C. (1990) Mass mortality of *Diadema antillarum*. II. Effects on population densities and grazing intensity of parrotfishes and surgeonfishes. *Mar. Biol.*, **104**, 79–86.

Carr, M.H. (1991) Habitat selection and recruitment of an assemblage of temperate zone reef fishes. *J. exp. mar. Biol. Ecol.*, **146**, 113–37.

Carr, M.H. and Reed, D.C. (1993) Conceptual issues relevant to marine harvest refuges: examples from temperate reef fishes. *Can. J. Fish. aquat. Sci.*, **50**, 2019–28.

Carrier, J.G. (1981) Ownership of productive resources on Ponam Island, Manus Province. *J. Société Océanistes*, **72–3**, 206–17.

Carrier, J.G. (1982) Fishing practices on Ponam Island (Manus Province, Papua New Guinea). *Anthropos*, **77**, 904–15.

Carrier, J.G. and Carrier, A.H. (1989) Marine tenure and economic reward on Ponam Island, Manus Province, in *A Sea of Small Boats: Customary Law and Territoriality in the World of Inshore Fishing* (ed. J.C. Cordell), Cultural Survival, Cambridge, MA, *Report* no. 62, pp. 94–120.

Carroll, V. (ed.) (1975) *Pacific Atoll Populations*. (Assoc. Social Anthropol. S. Pacific Monogr. no. 3), East–West Centre, Honolulu, HI.

Carter, J. (1989) Grouper sex in Belize. *Nat. Hist.*, **98**, 61–8.

Carter, J., Marrow, G.J. and Pryor, V. (1994) Aspects of the ecology and reproduction of Nassau grouper, *Epinephelus striatus*, off the coast of Belize, Central America. *Proc. Gulf Carib. Fish. Inst.*, **43**, 64–110.

Castañeda, P. and Sy, J.C. (1983) A modified muro-ami scareline. *Fish. Newsl., Philippine Bur. Fish. Aquat. Resour.*, **12**(1), 14–18.

Castañeda, P.G. and Miclat, R.I. (1981). The municipal coral reef park in the Philippines. *Proc. 4th Int. Coral Reef Symp.*, **1**, 285.

Catala, R. (1957) Report on Gilbert Islands: some aspects of human ecology. *Atoll Res. Bull.*, **59**, 1–187.

Cavarivière, A. (1982) Les balistes des côtes africaines (*Balistes carolinensis*). Biologie, prolifération et possibilités d'exploitation. *Oceanologica Acta*, **5**, 453–60.

Chakroun-Marzouk, N. and Kartas, F. (1987) Reproduction de *Pagrus caeruleostictus* (Valenciennes, 1830) (Pisces, Sparidae) des côtes tunisiennes. *Bull. Inst. nat Sci. tech. Océanogr. Pêche, Salammbo*, **14**, 33–45.

Chambers, J.R. (1992) Coastal degradation and fish population losses, in *Stemming the Tide of Coastal Fish Habitat Loss* (ed. Richard H. Stroud), National Coalition for Marine Conservation, Savannah, GA, pp. 45–51.

Chansang, H. (1984) Coral reef survey methods in the Andaman Sea. *UNESCO Rep. mar. Sci.*, **21**, 27–35.

Chao, A. (1989) Estimating population size for sparse data in capture–recapture experiments. *Biometrics* **45**, 427–38.

Chapau, M.R. and Lokani, P. (1986) Manus west coast fisheries resources survey. Unpubl. rep., Fish. Divn, Dept Primary Industry, Port Moresby, Papua New Guinea, 88 pp.

Chapman, D.G. (1970) Reanalysis of Antarctic fin whale population data. *Rep. Int. Whaling Commission*, **20**, 54–9.

Chapman, L.B. and Cusack, P. (1988) Deep Sea Fisheries Development Project. Report on second visit to Tuvalu (30 August – 7 December 1983). Unpubl. rep, S. Pacific Commission, New Caledonia, 51 pp.

Chapman, L.B. and Cusack, P. (1989) Deep Sea Fisheries Development Project. Report on fourth visit to the Territory of New Caledonia at the Belep Islands (18 August – 15 September 1986). Unpubl. rep., S. Pacific Commission, New Caledonia, 30 pp.

Chapman, L.B. and Lewis, A.D. (1982) UNDP/MAF survey of Walu and other large coastal pelagics in Fiji waters. Unpubl. rep., Min. Agric. Fish., Fisheries Divn, Suva, Fiji, 36 pp.

Chapman, M.D. (1987) Women's fishing in Oceania. *Human Ecol.*, **15**, 267–88.

Charnov, E.L. (1982) *The Theory of Sex Allocation* (Monographs in Population Biology **18**), Princeton University Press, Princeton, NJ, 335 pp.

Chartock, M.A. (1983) The role of *Acanthurus guttatus* (Bloch and Schneider 1801) in cycling algal production to detritus. *Biotropica*, **15**, 117–21.

Chatwin, B.M. (1958) Mortality rates and estimates of theoretical yield in relation to minimum commercial size of lingcod (*Ophiodon elongatus*) from the Strait of Georgia, British Columbia. *J. Fish. Res. Bd Can.*, **15**, 831–49.

Chen, C.-P., Hsieh, H.-L. and Chang, K.-H. (1980) Some aspects of the sex change and reproductive biology of the grouper, *Epinephelus diacanthus* (Cuvier et Valenciensis). *Bull. Inst. Zool., Acad. Sin.*, **19**, 11–17.

Chittleborough, R.G. (1970) Studies on recruitment in the western Australian rock lobster, *Panulirus longipes cygnus* George: density and natural mortality of juveniles. *Aust. J. mar. Freshwat. Res.*, **21**, 131–48.

Choat, J.H. (1991) The biology of herbivorous fishes on coral reefs, in *The Ecology of Fishes on Coral Reefs* (ed. P.F. Sale), Academic Press, San Diego, pp. 120–55.

Choat, J.H. and Bellwood, D.R. (1991) Reef fishes: their history and evolution, in *The Ecology of Fishes on Coral Reefs* (ed. P.F. Sale), Academic Press, San Diego, pp. 39–66.

Choat, J.H. and Randall, J.E. (1986) A review of the parrotfishes (Family Scaridae) of the Great Barrier Reef of Australia with description of a new species. *Rec. Aust. Mus.*, **38**, 175–228.

Choat, J.H. and Robertson, D.R. (1975) Protogynous hermaphroditism in fishes of the family Scaridae, in *Intersexuality in the Animal Kingdom* (ed. R. Reinboth), Springer Verlag, Berlin, pp. 263–83.

Choat, J.H., Ayling, A.M. and Schiel, D.R. (1988) Temporal and spatial variation in an island fish fauna. *J. exp. mar. Biol. Ecol.*, **121**, 91–111.

Choat, J.H., Doherty, P.J., Kerrigan, B.A. and Leis, J.M. (1993) A comparison of towed nets, purse seine, and light-aggregation devices for sampling larvae and pelagic juveniles of coral reef fishes. *Fishery Bull. U.S.*, **91**, 195–209.

Christensen, V. and Pauly, D. (1992a) A guide to the ECOPATH II software system (version 2.1). *ICLARM Software*, 6, 72 pp.

Christensen, V. and Pauly, D. (1992b) ECOPATH II – a software for balancing steady-state ecosystem models and calculating network characteristics. *Ecol. Modeling*, **61**, 169–85.

Christensen, V. and Pauly, D. (1993) *Trophic Models of Aquatic Ecosystems* (*ICLARM Conf. Proc.* **26**), ICLARM, Manila, 390 pp.

Christie, P., White, A.T. and Buhat, D. (1990) San Salvador Island marine conservation project: some lessons for community-based resource management, in *Proc. First National Symp. Mar. Sci.: Network of Institutions in Marine Science, 16–18 May 1990, Bolinao, Pangasinan, Philippines* (eds A.C. Alcala and L.T. McManus), Mar. Sci. Inst., Univ. Philippines, Quezon City, Philippines, pp. 193–207.

Chullasorn, S. and Martosubroto, P. (1986) Distribution and important biological features of coastal fish resources in Southeast Asia. *FAO Fish. tech. Pap.*, **278**, 84 pp.

Cintron-Molero, G. (1992) Restoring mangrove systems, in *Restoring the Nation's Marine Environment* (ed. G. Thayer), Maryland Sea Grant, College Park, MD, pp. 223–77.

Clark, C.W. (1989) Bioeconomics, in *Perspectives in Ecological Theory* (eds J. Roughgarden, R.M. May and S.A. Levin), Princeton University Press, Princeton, NJ, pp. 275–86.

Clark, C.W. (1990) *Mathematical Bioeconomics: The Optimal Management of Renewable Resources*, 2nd edn. John Wiley and Sons, New York, 386 pp.

Clark, J.R., Causey, B. and Bohnsack, J.A. (1989) Benefits from coral reef protection: Looe Key Reef, Florida, in *Coastal Zone '89: Proc. 6th Symp. Coastal and Ocean Management*, **4**, (eds O.T. Magoon, H. Converse, D. Miner, L.T. Tobin and D. Clark), Am. Soc. Civil Engineers, Charleston, SC, 3076–86.

Clarke, R.D. (1977) Habitat distribution and species diversity of pomacentrid and chaetodontid fishes near Bimini, Bahamas. *Mar. Biol.*, **40**, 277–89.

Clarke, R.D. (1989) Population fluctuation, competition and microhabitat distribution of two species of tube blennies, *Acanthemblemaria* (Teleostei: Chaenopsidae). *Bull. mar. Sci.*, **44**, 1174–85.

Clarke, T.A. (1983) Comparison of abundance estimates of small fishes by three towed nets and preliminary results of the use of small purse seines and sampling devices. *Biol. Oceanogr.*, **2**, 311–40.

Clarke, T.A. (1991) Larvae of nearshore fishes in oceanic waters near Oahu, Hawaii. *NOAA tech. Rep. NMFS*, **101**, 1–19.

Claro, R. (1983) Dinamica estacional de algunos indicadores morfofisiologicos del pargo criollo, *Lutjanus analis* (Cuvier), en la plataforma cubana. *Reporte de Investigacion, Instituto de Oceanologia*, **22**, 1–14.

Clausade, M., Gravier, N., Picard, J. Pichon, M., Roman, M.L., Thomassin, B., Vasseur, P., Vivien, M. and Weydert, P. (1971) Coral reef morphology in the vicinity of Tuléar (Madagascar): contribution to a coral reef terminology. *Téthys* (Suppl.), **2**, 74 pp.

Coates, D., Crane, P., Miller, D. and Theisen, D. (1984) The fish and prawn resource survey of Milne Bay by F.R.V. Melisa, June/July/August 1994. Fish. Divn, Dept Primary Industry, Port Moresby, *Research and Surveys Branch Res. Rep.* 84–11, 25 pp.

Cobb, J.S., Wang, D., Campbell, D.B. and Rooney, P. (1989) Speed and direction of swimming by postlarvae of the American lobster. *Trans. Am. Fish. Soc.*, **118**, 82–6.

Cochran, W.G. (1977) *Sampling Techniques*, 3rd edn, Wiley, New York.

Cole, R.G., Ayling, T.M. and Creese, R.G. (1990) Effects of marine reserve protection at Goat Island, northern New Zealand. *N.Z. J. mar. freshwat. Res.*, **24**, 197–210.

Colin, P.L. (1977) Daily and summer–winter variation in mass spawning of the striped parrotfish, *Scarus croicensis*. *Fishery Bull., U.S.*, **76**, 117–24.

Colin, P.L. (1992) Reproduction of the Nassau grouper, *Epinephelus striatus* (Pisces: Serranidae) and its relationship to environmental conditions. *Env. Biol. Fishes*, **34**, 357–77.

Colin, P.L. and Bell, L.J. (1991) Aspects of the spawning of labrid and scarid fishes (Pisces: Labroidei) at Enewetak Atoll, Marshall Islands with notes on other families. *Env. Biol. Fishes*, **31**, 229–60.

Colin, P.L. and Clavijo, I E. (1978) Mass spawning by the spotted goatfish, *Pseudupeneus maculatus* (Black) (Pisces: Mullidae). *Bull. mar. Sci.*, **28**, 780–82.

Colin, P.L. and Clavijo, I.E. (1988) Spawning activity of fishes producing pelagic eggs on a shelf edge coral reef, southwestern Puerto Rico. *Bull. mar. Sci.*, **43**, 249–79.

Colin, P.L., Shapiro, D.Y. and Weiler, D. (1987) Aspects of the reproduction of two groupers, *Epinephelus guttatus* and *E. striatus* in the West Indies. *Bull. mar. Sci.*, **40**, 220–30.

Collins, L.A., Johnson, A.G. and Keim, C.P. (1993) Spawning and fecundity in the red snapper, *Lutjanus campechanus* from the northeastern Gulf of Mexico. *Proc. EPOMEX Int. Snapper–Grouper Symp., Campeche, Mexico, October 1993* (Abstract).

Collins, M.R., Waltz, C.W., Roumillat, W.A. and Stubbs, D.L. (1987) Contribution to the life history and reproductive biology of gag, *Mycteroperca microlepis* (Serranidae), in the South Atlantic Bight. *Fishery Bull., U.S.*, **85**, 648–53.

Conand, F. (1988) *Biologie et Écologie des Poissons Pélagiques du Lagon de Nouvelle Calédonie Utilisables comme Appat Thonier* (ORSTOM Études et Thèses), Institut Français de Recherche Scientifique pour le Développment en Coopération, Paris.

Connell, J. (1983) Migration, employment and development in the South Pacific. Country Report No. 3, Federated States of Micronesia. International Labour Organisation and South Pacific Commission, Nouméa, New Caledonia. 65 pp.

Connell, S.D. and Jones, D.P. (1991) The influence of habitat composition on post recruitment processes in a temperate reef fish population. *J. exp. mar. Biol. Ecol.*, **151**, 271–94.

Conroy, D.A. (1975) An evaluation of the present state of the world trade in ornamental fish. *FAO Fish. tech. Pap.*, **146**. 128 pp.

Cook, S.B. (1980) Fish predation on pulmonate limpets. *Veliger*, **22**, 380–81.

Cooke, A.J. (1994) The qoliqoli of Fiji – management of resources in traditional fishing grounds, MSc thesis, University of Newcastle, Centre Trop. Coastal Manage. Stud., Newcastle NE1 7RU, UK, 165 pp.

Cordell, J.C. (ed.) (1989) *A Sea of Small Boats: Customary Law and Territoriality in the World of Inshore Fishing*. Cultural Survival, Cambridge, MA, Rep. no. 62.

Corpuz, V.T., Castañeda, P. and Sy, J.C. (1983a) Muro-ami. *Fish. Newsl., Philippine Bur. Fish. aquat. Resour.*, **12**(1), 2–13.

Corpuz, V.T., Castañeda, P. and Sy, J.C. (1983b) A modified muro-ami scareline. *Fish. Newsl., Philippine Bur. Fish. aquat. Resour.*, **12**(1), 14–17.

Corpuz, V.T., Castañeda, P. and Sy, J.C. (1985) Traditional muro–ami, an effective but destructive coral reef fishing gear. *ICLARM Newsl.*, **8**, 12–13.

Cortes, E. and Gruber, H.S. (1990) Diet, feeding habits and estimates of daily ration of young lemon sharks, *Negaprion brevirostris* (Poey). *Copeia*, **1990**, 204–18.

Cortez-Zaragosa, E., Dalzell, P. and Pauly, D. (1989) Hook selectivity of yellowfin tuna (*Thunnus albacares*) caught off la Union, Philippines. *J. appl. Ichthyol.*, **1**, 12–17.

Costanza, R. (ed.) (1991) *Ecological Economics: The Science and Management of Sustainability*, Columbia University Press, New York, NY, 525 pp.

Coulston, M.L., Berey, R.W., Dempsey, A.C. and Odum, P. (1989) Assessment of the queen conch (*Strombus gigas*) population and predation studies of hatchery reared juveniles in Salt River Canyon, St. Croix, U.S. Virgin Islands. *Proc. Gulf Carib. Fish. Inst.*, **38**, 294–305.

Courtenay, W.R., jun. and Robins, C.R. (1975) Exotic organisms: an unsolved, complex problem. *BioScience*, **25**, 306–12.

Courtenay, W.R., jun. and Robins, C.R. (1989) Fish introductions: good management, mismanagement, or no management? *Rev. aquat. Sci.*, **1**, 159–72.

Cowen, R.K. (1990) Sex change and life history patterns of the labrid, *Semicossyphus pulcher*, across an environmental gradient. *Copeia*, **1990**, 787–95.

Cowen, R.K. and Castro, L.R. (1994) Relation of coral reef fish larval distributions to island scale circulation around Barbados, West Indies. *Bull. mar. Sci.*, **54**, 228–44.

Cox, E.F. (1994) Resource use by corallivorous butterflyfishes (family Chaetodontidae) in Hawaii. *Bull. mar. Sci.*, **54**, 535–45.

Craig, J.F. (1985) Ageing in fish. *Can. J. Zool.*, **63**, 1–8.

Craik, W.J.S. (1981a) Recreational fishing on the Great Barrier Reef. *Proc. 4th Int. Coral Reef Symp.*, **1**, 47–52.

Craik, W.J.S. (1981b) Underwater survey of coral trout *Plectropomus leopardus* (Serranidae) populations in the Capricornia section of the Great Barrier Reef Marine Park. *Proc. 4th Int. Coral Reef Symp.*, **1**, 53–8.

Crecco, V., Savoy, T. and Whitworth, W. (1986) Effects of density-dependent and climatic factors on American shad, *Alosa sapidissima*: a predictive approach. *Can. J. Fish. aquat. Sci.*, **43**, 457–63.

Crocombe, R.G. (1961) Land tenure in the Cook Islands. PhD dissertation, Australian National University, Canberra.

Crocombe, R.G. (1964) *Land Tenure in the Cook Islands*, Melbourne University Press, Melbourne.

Crocombe, R.G. (1967) From ascendancy to dependency: the politics of Atiu. *J. Pac. Hist.*, **2**, 97–112.

Crocombe, R.G. (1989) Tomorrow's customary tenure in relation to conservation, in *Report on the Workshop on Customary Tenure, Traditional Resource Management and Nature Conservation, Noumea, New Caledonia, 28 March – 1 April, 1988* (ed. P.E.J. Thomas), S. Pacific Regional Env. Prog., South Pacific Commission, Nouméa, New Caledonia, pp. 21–7.

Crouse, D.T., Crowder, L.B. and Caswell, H. (1987) A stage-based population model for loggerhead sea turtles and implications for conservation. *Ecology*, **68**, 1412–23.

Cruz, L.P. de la (ed.) (1993) *Our Sea: Our Life*. Voluntary Services Overseas (VSO), Quezon City, Philippines, 98 pp.

Cury, P. and Roy, C. (1989) Optimal environmental window and pelagic fish recruitment success in upwelling areas. *Can. J. Fish. aquat. Sci.*, **46**, 670–80.

Cushing, D.H. (1975a) The natural mortality of the plaice. *J. Conseil, Cons. int. Explor. Mer*, **36**, 150–57.

Cushing, D.H. (1975b) *Marine Ecology and Fisheries*, Cambridge University Press, New York, 278 pp.

Cushing, D.H. (1988) The study of stock and recruitment, in *Fish Population Dynamics*, 2nd edn, (ed. J.A. Gulland), Wiley, Chichester, pp. 105–28.

Dahl, A.L. (1982) *Coral Reef Monitoring Handbook*. South Pacific Commission, Nouméa, New Caledonia.

Daly, H.E. (1993) The perils of free trade. *Scient. Am.*, (November) 24–9.

Dalzell, P. (1989) The biology of surgeonfishes (Family: Acanthuridae), with particular emphasis on *Acanthurus nigricauda* and *A. xanthopterus* from northern Papua New Guinea, M.Phil. thesis, University of Newcastle upon Tyne, 285 pp.

Dalzell, P. (1993a) Developments in pelagic fisheries in Papua New Guinea. *S. Pac. Commission Fish. Newsl.*, **65**, 37–42.

Dalzell, P. (1993b) The fisheries biology of flying fishes (Families: Exocoetidae and Hemiramphidae) from the Camotes Sea, Central Philippines. *J. Fish Biol.*, **43**, 19–32.

Dalzell, P. and Adams, T.J.H. (1994) The present status of coastal fisheries production in the South Pacific Islands. South Pacific Commission, Nouméa, New Caledonia, *S. Pac. Commission 25th Reg. Tech. Meeting on Fisheries, Working Paper*, **8**.

Dalzell, P. and Aimi, J. (1989) Catch rates and catch composition of Antillean style fish traps deployed on coral reefs in northern Papua New Guinea. Dept Primary Ind., Port Moresby, Papua New Guinea, *Tech. Rep.* **89/1**, 13 pp.

Dalzell, P. and Aini, J. (1992) The performance of Antillean wire mesh fish traps set on coral reefs in northern Papua New Guinea. *Asian Fish. Sci.*, **5**, 89–102.

Dalzell, P. and Debao, A. (1994) Coastal fisheries production on Nauru. South Pacific Commission, Nouméa, New Caledonia, *Inshore Fish. Res., Project Country Assignment Rep.*, 19 pp.

Dalzell, P. and Lewis, A.D. (1989) A review of the South Pacific tuna baitfisheries: small pelagic fisheries associated with coral reefs. *Mar. Fish. Rev.*, **51**, 1–10.

Dalzell, P. and Pauly, D. (1990) Assessment of the fish resources of southeast Asia, with emphasis on the Banda and Arafura seas. *Neth. J. Sea Res.*, **25**, 641–50.

Dalzell, P. and Preston, G.L. (1992) Deep slope fishery resources of the South Pacific. South Pacific Commission, Nouméa, New Caledonia, *Inshore Fish. Res., Proj. tech. Doc.* **2**. 299 pp.

Dalzell, P.J. and Wright, A. (1986) An assessment of the exploitation of coral reef fishery resources in Papua New Guinea, in *The First Asian Fisheries Forum*, Vol. I (eds J.L. Maclean, L.B. Dizon and L.V. Hosillos), Asian Fisheries Society, Manila, Philippines, pp. 477–81.

Dalzell, P.J. and Wright, A. (1990) Analysis of catch data from an artisanal coral reef fishery in the Tigak Islands, Papua New Guinea. *Papua New Guinea J. Agric., Forestry Fish.*, **35**, 23–36.

Dalzell, P., Corpuz, P., Ganaden, R. and Arce, F. (1990) Philippine small pelagic fisheries and their management. *Aquacult. Fish. Manage.*, **21**, 77–94.

Dalzell, P., Sharma, S. and Nath, G. (1992) Estimation of exploitation rates in a multi-species emperor (Pisces: Lethrinidae) fishery in Fiji based on length–frequency data. South Pacific Commission, Nouméa, New Caledonia, *Pap. Fish. Sci. Pacific Islands*, **1**, 43–50.

Dalzell, P., Lindsay, S.R. and Patiale, H. (1993) Fisheries resources survey of the island of Niue. South Pacific Commission, Nouméa, New Caledonia, *Inshore Fish. Res., Proj. tech. Doc.*, **3**, 68 pp.

Dan, S.S. (1977) Intraovarian studies and fecundity in *Nemipterus japonicus* (Bloch). *Indian J. Fish.*, **24**, 48–55.

Dandonneau, Y. and Charpy, L. (1985) An empirical approach to the island mass effect in the south tropical Pacific based on sea surface chlorophyll concentrations. *Deep-Sea Res.*, **32**, 707–21.

Danielsson, B. (1956) *Work and Life on Raroia*. George Allen and Unwin, London.

Darcy, G.H. (1983a) Synopsis of biological data on the grunts *Haemulon aurolineatum* and *H. plumieri* (Pisces: Haemulidae). *NOAA tech. Rep. NMFS Circ.*, **448**, 37 pp.

Darcy, G.H. (1983b) Synopsis of biological data on the pigfish, *Orthopristis chrysoptera* (Pisces: Haemulidae). *NOAA tech. Rep. NMFS Circ.*, **449**, 23 pp.

Davies, C.R. (1989) The effectiveness of non-destructive sampling of coral reef fish populations, BSc honours thesis, James Cook University of North Queensland, Townsville, Australia, 87 pp.

Davis, G.E. (1989) Designated harvest refugia: the next stage of marine fishery management in California. *Calif. Coop. Oceanic Fish. Invest. Rep.*, **30**, 53–8.

Davis, G.E. and Dodrill, J.W. (1980) Marine parks and sanctuaries for spiny lobster fisheries management. *Fishery Bull., U.S.*, **78**, 979–84.

Davis, G.E. and Dodrill, J.W. (1989) Recreational fishery and population dynamics of spiny lobster, *Panulirus argus*, in Florida Bay, Everglades National Park, 1977–1980. *Bull. mar. Sci.*, **44**, 78–88.

Davis, T.L.O. and West, G.J. (1993) Maturation, reproductive seasonality, fecundity, and spawning frequency in *Lutjanus vittus* (Quoy and Gaimard) from the north west shelf of Australia. *Fishery Bull., U.S.*, **91**, 224–36.

Davis, W.D. and Birdsong, R.S. (1973) Coral reef fishes which forage in the water column. *Helgoländer wiss. Meeresunter.*, **24**, 292–306.

Dawson Shepherd, A.R., Warwick, R.M., Clarke, K.R. and Brown, B.E. (1992) An analysis of fish community responses to coral mining in the Maldives. *Env. Biol. Fishes*, **33**, 367–80.

Dayaratne, P. (1988) Gill-net selectivity for *Amblygaster* (= *Sardinella*) *sirm*. *Asian Fish. Sci.*, **2**, 71–82.

de Boer, B.A. (1978) Factors influencing the distribution of the damselfish *Chromis cyanea* (Poey), Pomacentridae, on a reef at Curacao, Netherlands Antilles. *Bull. mar. Sci.*, **28**, 550–65.

de Moussac, G. (1986a) Mise en evidence de l'hermaphrodisme protogyne *d'Epinephelus chlorostigma* (Valenciennes, 1828) aux Seychelles (Pisces, Serranidae). *Cybium*, **10**, 249–62.

de Moussac, G. (1986b) Basket traps in the region. *South West Indian Ocean Fish. Bull.* no. 18. 5 pp.

de Moussac, G. (1987) Seychelles artisanal fisheries statistics for 1985. Seychelles Fishing Authority, Port Victoria, *Tech. Rep.* **4**. 79 pp.

De Silva, M.W.R.N. (1984) Coral reef assessment and management methodologies currently used in Malaysia. *UNESCO Rep. mar. Sci.*, **21**, 47–56.

De Silva, M.W.R.N. (1985) Status of the coral reefs of Sri Lanka. *Proc. 5th Int. Coral Reef Congr.*, **6**, 515–18.

De Silva, M.W.R.N. and Rahaman, R.A. (1982) Management plan for the coral reefs of Palau Paya/Segantang group of islands. Unpubl. rep., Faculty of Fisheries and Marine Science, University Pertanian Malaysia, and World Wildlife Fund, Malaysia, Serdang, Selangor and Kuala Lumpur.

De Vantier, L.M. and Deacon, G. (1990) Distribution of *Acanthaster planci* at Lord Howe Island, the southernmost Indo–Pacific reef. *Coral Reefs*, **9**, 145–8.

DeCrosta, M.A. (1984) Age determination, growth and energetics of three species of carcharhinid sharks in Hawaii. MS thesis, Dept Oceanography, University of Hawaii, 66 pp.

DeMartini, E.E. (1993) Modeling the potential of fishery reserves for managing Pacific coral reef fisheries. *Fishery Bull., U.S.*, **91**, 414–27.

DeMartini, E.E., Ellis, D.M. and Honda, V.A. (1992) Comparisons of spiny lobster *Panulirus marginatus* fecundity, egg size and spawning frequency before and after exploitation. *Fishery Bull., U.S.*, **91**, 1–7.

Dennis, G.D. (1988) Commercial catch length–frequency data as a tool for fisheries management with an application to the Puerto Rico trap fishery. *Memoria de la Sociedad de Ciencias Naturales La Salle*, **48** (Suplemento 3), 289–310.

Dennis, G.D. (1991) The validity of length–frequency derived growth parameters from commercial catch data and their application to stock assessment of the yellowtail snapper (*Ocyurus chrysurus*). *Proc. Gulf Carib. Fish. Inst.*, **40**, 126–38.

Dennis, G.D. (1992) Resource utilization by members of a guild of benthic feeding coral reef fish. PhD dissertation, University of Puerto Rico, Mayagüez, Puerto Rico, 224 pp.

Deshmukh, V.M. (1973) Fishery and biology of *Pomadasys hasta* (Bloch). *Indian J. Fish.*, **20**, 497–522.

Desurmont, A. (1989) Essais de pêche aux casiers profonds en Nouvelle-Calédonie. South Pacific Commission, Nouméa, New Caledonia, *S. Pac. Commission 21st Reg. tech. Meeting on Fisheries, Info. Paper*. 18 pp.

DeVlaming, V.L., Grossman, G. and Chapman, F. (1982) On the use of the gonosomatic index. *Comp. Biochem. Physiol.*, **73A**, 31–9.

Die, D.J. and Watson, R.A. (1992) A per-recruit simulation model for evaluating spatial closures in an Australian penaeid fishery. *Aquat. Living Resour.*, **5**, 145–53.

Diplock, J.H. and Dalzell, P. (1991) Summary of the results from the NFCF–OFCF survey of the deep slope fishery resources of the outer banks and seamounts in the Federated States of Micronesia, September 1989 to February 1991. South Pacific Commission, Inshore Fish. Res., Project, Nouméa, New Caledonia. 7 pp.

Dixon, J.A. (1993) Economic benefits of marine protected areas. *Oceanus*, **36**, 35–40.

Doherty, P.J. (1981) Coral reef fishes: recruitment-limited assemblages? *Proc. 4th Int. Coral Reef Symp.*, **2**, 465–70.

Doherty, P.J. (1982) Some effects of density on the juveniles of two species of tropical, territorial damselfishes. *J. exp. mar. Biol. Ecol.*, **65**, 249–61.

Doherty, P.J. (1983) Tropical territorial damselfishes: is density limited by aggression or recruitment? *Ecology*, **64**, 176–90.

Doherty, P.J. (1987) The replenishment of populations of coral reef fishes, recruitment surveys, and the problems of variability manifest on multiple scales. *Bull. mar. Sci.*, **41**, 411–22.

Doherty, P.J. (1991) Spatial and temporal patterns in recruitment, in *The Ecology of Fishes on Coral Reefs* (ed. P.F. Sale), Academic Press, San Diego, pp. 261–93.

Doherty, P. and Fowler, T. (1994a) An empirical test of recruitment limitation in a coral reef fish. *Science*, **263**, 935–9.

Doherty, P.J. and Fowler, A. (1994b) Demographic consequences of variable recruitment to coral reef fish populations: a congeneric comparison of two damselfishes. *Bull. mar. Sci.*, **54**, 297–313.

Doherty, P.J. and Sale, P.F. (1985) Predation on juvenile coral reef fishes: an exclusion experiment. *Coral Reefs*, **4**, 225–34.

Doherty, P.J. and Williams, D.McB. (1988) The replenishment of coral reef fish populations. *Oceanogr. mar. Biol. A. Rev.*, **26**, 487–551.

Doherty, P.J., Williams, D.M. and Sale, P.F. (1985) The adaptive significance of larval dispersal in coral reef fishes. *Env. Biol. Fishes*, **12**, 81–90.

Domeier, M.L., Koenig, C.C. and Coleman, F.C. (in press) Reproductive biology of the gray snapper (*Lutjanus griseus*), with notes on spawning for other western Atlantic snappers (Lutjanidae). *Proc. EPOMEX Snapper–Grouper Symp., Campeche, Mexico, October 1993.*

Donaldson, T.J. (1990) Lek-like courtship by males, and multiple spawnings by females of *Synodus dermatogenys* (Synodontidae). *Jap. J. Ichthyol.*, **37**, 292–302.

Done, T.J. (1982) Patterns in the distribution of coral communities across the central Great Barrier Reef. *Coral Reefs*, 1, 95–107.

Done, T.J. (1987) Simulation of the effects of *Acanthaster planci* on the population structure of massive corals of the genus *Porites*: evidence of population resilience. *Coral Reefs*, **6**, 75–90.

Done, T.J. (1988) Simulation of recovery of predisturbance size structure in populations of *Porites* corals damaged by crown of thorns starfish *Acanthaster planci* L. *Mar. Biol.*, **100**, 51–61.

Done, T.J. (1992) Phase-shifts in coral-reef communities and their ecological significance. *Hydrobiologia*, **247**, 121–32

Done, T.J., Osborne, K. and Navin, K.F. (1988) Recovery of corals post-*Acanthaster*: progress and prospects. *Proc. 6th Int. Coral Reef Symp.*, **2**, 137–42.

Doty, M.S. and Oguri, M. (1956) The island mass effect. *J. Conseil, Cons. int. Explor. Mer*, **22**, 33–7.

Dow, R.L. (1977) Effects of climatic cycles on the relative abundance and availability of commercial marine and estuarine species. *J. Conseil, Cons. int. Explor. Mer*, **37**, 274–80.

Dragovich, A. and Potthoff, T. (1972) Comparative study of food of skipjack and yellowfin tunas off the coast of west Africa. *Fishery Bull., U.S.*, **70**, 1087–1101.

Dredge, M. (1988) Queensland near reef trawl fisheries. SPC, Nouméa, New Caledonia. *South Pacific Commission Workshop on Pacific Inshore Fishery Resources, March 14–25th, 1988, Background Paper*, **80**, 21 pp.

Druzhinin, A.D. (1970) The range and biology of snappers (Family Lutjanidae). *J. Ichthyol.*, **10**, 717–36.

Ducklow, H.W. (1990) The biomass production and fate of bacteria in coral reefs, in *Coral Reefs*, (Ecosystems of the World 25), (ed. Z. Dubinsky), Elsevier, Amsterdam, pp. 265–89.

Dufour, V. (1991) Variations of fish larvae abundance in reefs – effect of light on the colonization of the reefs by fish larvae. *C. r. Acad. Sci. Paris, Serie III – Sciences de La Vie*, **313**, 187–194.

Dufour, V. and Galzin, R. (1993) Colonization patterns of reef fish larvae to the lagoon at Moorea Island, French Polynesia. *Mar. Ecol. – Progr. Ser.*, **102**, 143–52.

Dugan, J.E. and Davis, G.E. (1993a) Introduction to the international symposium on marine harvest refugia. *Can. J. Fish. aquat. Sci.*, **50**, 1991–2.

Dugan, J.E. and Davis, G.E. (1993b) Applications of marine refugia to coastal fisheries management. *Can. J. Fish. aquat. Sci.*, **50**, 2029–42.

Dugan, P.F. (1956) *The Early History of Guam, 1521–1698*, San California State College Press, San Diego.

Durrenberger, E.P. and Pàlsson, G. (1987) The grass roots and the state: resource management in Icelandic fishing, in *The Question of the Commons: the Culture and Ecology of Communal Resources* (eds B.J. McCay and J.M. Acheson), University of Arizona Press, Tucson, pp. 370–92.

Dye, T. (1983) Fish and fishing on Niuatoputapu. *Oceania*, **53**(3), 242–71.

Dyer, M.I. and Holland, M.M. (1991) The biosphere-reserve concept: needs for a network design. *BioScience*, **41**, 319–24.

Eberhardt, L.L. (1978) Transect methods for population studies. *J. Wildl. Manage.*, **42**, 1–31.

Ebert, T.A., and Russell, M.P. (1988) Latitudinal variation in size structure of the west coast purple sea urchin: a correlation with headlands. *Limnol. Oceanogr.*, **33**, 286–94.

Ebisawa, A. (1990) Reproductive biology of *Lethrinus nebulosus* (Pisces: Lethrinidae) around the Okinawan waters. *Nippon Suisan Gakkaishi*, **56**, 1941–54.

Eckersley, R. (1992) *Environmentalism and Political Theory*, State University of New York Press, Albany, NY, 274 pp.

Eckert, G.J. (1984) Annual and spatial variation in recruitment of labroid fishes among seven reefs in the Capricorn/Bunker Group, Great Barrier Reef. *Mar. Biol.*, **78**, 123–7.

Eckert, G.J. (1987) Estimates of adult and juvenile mortality for labrid fishes at One Tree Reef, Great Barrier Reef. *Mar. Biol.*, **95**, 167–71.

Eckert, S.A., Crouse, D., Crowder, L.B., Maceina, M. and Shah, A. (1992) Review of the Kemp's Ridley sea turtle headstart experiment, 22–23 September 1992, Galveston, Texas. National Marine Fisheries Service, Galveston, Texas, 9 pp.

Edwards, R.C. (1983) The Taiwanese pair trawler fishery in tropical Australian waters. *Fish. Res.*, **2**, 47–60.

Eggers, D.M., Bartoo, N.W., Rickard, N.A., Nelson, R.E., Wissmar, R.C., Burgner, R.L. and Devol, A.H. (1978) The Lake Washington ecosystem: the perspective from the fish community and forage base. *J. Fish. Res. Bd Can.*, **35**, 1553–71.

Eggers, D.M., Rickard, N.A., Chapman, D.G. and Whitney, R.R. (1982) A methodology for estimating area fished for baited hooks and traps along a groundline. *Can. J. Fish. aquat. Sci.*, **39**, 448–53.

Eggleston, D. (1972) Patterns of biology in the Nemipteridae. *J. mar. biol. Ass. India*, **14**, 357–64.

Eggleston, D.B., Lipcius, R.N., Miller, D.L. and Coba-Cetina, L. (1990) Shelter scaling regulates survival of juvenile Caribbean spiny lobster *Panulirus argus*. *Mar. Ecol. – Progr. Ser.*, **63**, 79–88.

Eggleston, D.B., Lipcius, R.N. and Miller, D.L. (1992) Artificial shelters and survival of juvenile Caribbean spiny lobsters *Panulirus argus*: spatial, habitat and lobster size effects. *Fishery Bull., U.S.*, **90**, 691–702.

Egretaud, C. (1992) Étude de la biologie générale, et plus particulièrement du régime alimentaire de *Lethrinus nebulosus* du lagon d'Ouvea (Nouvelle Calédonie), thesis – Mémoires de Stage, ENSA, Rennes, France, 102 pp.
Ehrhardt, N.M. and Ault, J.S. (1992) Analysis of two length-based mortality models applied to bounded catch length frequencies. *Trans. Am. Fish. Soc.*, **121**, 115–22.
Ehrhardt, N.M. and Die, D.J. (1988) Selectivity of gill nets used in the commercial Spanish Mackerel fishery of Florida. *Trans. Am. Fish. Soc.*, **117**, 574–80.
Eldredge, L.G. (1987) Poisons for fishing on coral reefs, in *Human Impacts on Coral Reefs: Facts and Recommendations* (ed B. Salvat), Antènne du Muséum École Pratique des Hautes Études, French Polynesia, pp. 61–6.
Eldredge, L.G. (1994) Introductions of commercially significant aquatic organisms to the Pacific Islands. South Pacific Commission, Nouméa, New Caledonia, *Inshore Fish. Res., Proj. tech. Rep.* **6**, 120 pp.
Ellis, R.W. (1969) The rational exploitation of the shrimp of El Salvador. *Proc. Gulf Carib. Fish. Inst.*, **21**, 126–34.
Emery, A.R. and Winterbottom, R. (1991) Lagoonal fish and fisheries of Dravuni, Great Astrolabe Reef, Fiji, South Pacific. University of the South Pacific, Marine Studies Programme, Suva, Fiji, *Tech. Rep.* **3**, 22 pp.
Emmerson, D.K. (1980) Rethinking artisanal fisheries development: western concepts, Asian experiences. World Bank, Washington, DC, *Staff Working Paper* **423**, x + 97 pp.
Emory, K.P. (1965) *Kapingamarangi: Social and Religious Life of a Polynesian Atoll.* Bernice P. Bishop Museum, Honolulu, HI, *Bulletin* **228**.
Endean, R. and Stablum, W. (1973) A study of some aspects of the crown-of-thorns starfish (*Acanthaster planci*) infestations of reefs of Australia's Great Barrier Reef. *Atoll Res. Bull.*, **167**, 1–60.
Endean, R., Cameron, A.M. and de Vantier, L.M. (1988) *Acanthaster planci* predation on massive corals: the myth of rapid recovery of devastated reefs. *Proc. 6th Int. Coral Reef Symp.*, **2**, 143–55.
Entsch, B., Boto, K.G., Sim, R.G. & Wellington, J.T. (1983) Phosphorus and nitrogen in coral reef sediments. *Limnol. Oceanogr.*, **28**, 465–76.
Erdman, D.S. (1976) Spawning patterns of fishes from the northeastern Caribbean. Dept Agric. Commonwealth of Puerto Rico, *Agric. Fish. Contrib.*, **8**(2), 1–37.
Erickson, D.L., Hightower, J.E. and Grossman, G.D. (1985) The relative gonadal index: an alternative index for quantification of reproductive condition. *Comp. Biochem. Physiol.*, **81**A, 117–20.
Espeut, P. (1992) Fishing for finfish in Belize and the South Coast of Jamaica: a socioeconomic analysis. Center Mar. Sci. Res., University of the West Indies, Mona, Jamaica, *Res. Rep.* **3**.
Everson, A.R., Williams, H.A. and Ito, B.M. (1989) Maturation and reproduction in two Hawaiian eteline snappers, uku, *Aprion virescens*, and onaga, *Etelis coruscans*. *Fishery Bull., U.S.*, **87**, 877–88.
Fairbairn, T.I.J. (1990) Reef and Lagoon Tenure in the Republic of Vanuatu and prospects for mariculture development. Dept Economics, University of Queensland, St Lucia. *Res. Rep. Pap. Economics Giant Clam Mariculture* no. 13.
Fairbairn, T.I.J. (1991) Traditional reef and lagoon tenure in Western Samoa and its implications for giant clam mariculture. Dept Economics, University of Queensland, St Lucia. *Res. Rep. Pap. Economics Giant Clam Mariculture* no. 17.
Fairbairn, T.I.J. (1992) Marine property rights in relation to giant clam mar-

iculture in the Kingdom of Tonga, in *Giant Clams in the Sustainable Development of the South Pacific* (ed. C. Tisdell), Australian Centre Int. Agric. Res., Canberra, pp. 119–33.

Falanruw, M.V.C. (1982) People pressure and management of limited resources on Yap, in *National Parks, Conservation and Development: the Role of Protected Areas in Sustaining Society* (eds J.A. McNeely and K.R. Miller), Smithsonian Institution Press, Washington, DC, pp. 348–54.

Falanruw, M.V.C. (1992) Traditional use of the marine environment on Yap. Paper presented at the Science of Pacific Peoples Conference, University of the South Pacific, Suva, July 5–10. (MS in the library of the Yap Institute of Science, Colonia, Yap, FSM.).

FAO Fisheries Department (1992) Review of the state of world fishery resources. Part 1. The marine resources. FAO, Rome, *FAO Fish. Cir.*, **710**, Rev. 8, Part 1, 114 pp.

Faure, G. (1989) Degradation of coral reefs at Moorea Island (French Polynesia) by *Acanthaster planci. J. Coastal Res.*, **5**, 295–305.

Felfoldy-Fergusson, K. (1988) The collection and uses of inshore reef fisheries information to assess and monitor the shelf fisheries of the Kingdom of Tonga using the ICLARM approach. Summary of the first year's activities and results. Presented at the South Pacific Commission Inshore Fisheries Resources Workshop, Nouméa, March 1988. SPC, Nouméa, New Caledonia. *SPC Inshore Fish. Res./BP* 41.

Ferreira, B.P. (1993) Reproduction of the inshore coral trout *Plectropomus maculatus* (Perciformes: Serranidae) from the Central Great Barrier Reef, Australia. *J. Fish Biol.*, **42**, 831–44.

Ferreira, B.P. & Russ, G.R. (1992) Age, growth and mortality of the inshore coral trout *Plectropomus maculatus* (Pisces: Serranidae) from the Central Great Barrier Reef, Australia. *Aust. J. mar. Freshwat. Res.*, **43**, 1301–12.

Ferrer, E.M. (1989) People's participation in coastal area management, in *Coastal Area Management in Southeast Asia: Policies, Management Strategies and Case Studies* (*ICLARM Conf. Proc.* **19**) (eds T.-E. Chua and D. Pauly), ICLARM, Manila, pp. 117–27.

Ferrer, E.M. (1991) Territorial use rights in fisheries and the management of artificial reefs in the Philippines, in *Towards an Integrated Management of Tropical Coastal Resources* (*ICLARM Conf. Proc.* **22**) (eds L.M. Chou, T.-E. Chua, H.W. Khoo, P.E. Lim, J.N. Paw, G.T. Silvestre, M.J. Valencia, A.T. White and P.K. Wong), ICLARM, Manila, pp. 299–302.

Ferry, R.E. and Kohler, C.C. (1987) Effects of trap fishing on fish populations inhabiting a fringing coral reef. *N. Am. J. Fish. Manage.*, **7**, 580–88.

Festinger, L. (1972) Theory of cognitive dissonance, in *Social Psychology: Experimentation, Theory, Research* (ed. W.S. Sahakian), Intext Education Publishers, San Francisco, CA, pp. 254–7.

Feuilleteau de Bruyn, W.K.H. (1920) *Schouten-en Padaido-Eilanden*. Mededeelingen van het Bureau voor de Bestuurszaken, Encyclopedisch Bureau, **21**, 193 pp.

Fiedler, P.C. (1986) Offshore entrainment of anchovy spawning habitat, eggs, and larvae by a displaced eddy in 1985. *Calif. Coop. Oceanic Fish. Invest. Rep.*, **27**, 144–52.

Fiji Fisheries Division (1992) Annual Report 1991. Ministry of Primary Industries, Suva, Fiji. 48 pp.

Fine, J.C. (1990) Groupers in love. *Sea Frontiers*, 36, 42–5.

Firth, R. (1965) *Primitive Polynesian Economy*, Routledge and Kegan Paul, London.

Fischer, J.L. (1958) Contemporary Ponape Island land tenure, in *Land Tenure Patterns in Trust Territory of the Pacific Islands* (ed. J.E. de Young), Trust Territory Government, Agana, Guam, pp. 76–159.

Fishelson, L. (1990) *Rhinomuraena* spp. (Pisces: Muraenidae): the first vertebrate genus with post-anally situated urogenital organs. *Mar. Biol.*, **105**, 253–7.

Fishelson, L., Montgomery, L.W. and Myrberg, A A., jun. (1985) A new fat body associated with the gonad of surgeonfishes (Acanthuridae, Teleostei). *Mar. Biol.*, **86**, 109–12.

Fishelson, L., Montgomery, L.W. and Myrberg, A.A., jun. (1987) Biology of surgeonfish *Acanthurus nigrofuscus* with emphasis on changeover in diet and annual gonadal cycles. *Mar. Ecol. – Progr. Ser.*, **39**, 37–47.

Flowers, J.M. and Saila, S.B. (1972) An analysis of temperature effects on the inshore lobster fishery. *J. Fish. Res. Bd Can.*, **29**, 1221–5.

Fonseca, M.S. (1990) Regional analysis of the creation and restoration of seagrass systems, in *Wetland Creation and Restoration: The Status of the Science* (eds J.A. Kusler and M.E. Kentula), Island Press, Washington, DC, pp. 171–94.

Fonseca, M.S. (1992) Restoring seagrass systems in the United States, in *Restoring the Nation's Marine Environment* (ed. G. Thayer), Maryland Sea Grant, College Park, MD, pp. 79–110.

Foster, K.B. and Poggie, J.J. (1992) Customary marine tenure practises for mariculture management in outlying communities of Pohnpei, in *Coastal Aquaculture in Developing Countries: Problems and Perspectives* (eds R.B. Pollnac and P. Weeks), Int. Center Mar. Resour. Devel., University of Rhode Island, Kingston, RI, pp. 33–53.

Foster, N. and Lemay, M.H. (eds) (1989) Managing marine protected areas: an action plan. US Man and the Biosphere Program, US Dept of State, Washington, DC, 63 pp.

Foster, S.A. (1987) The relative impacts of grazing by Caribbean coral reef fishes and *Diadema*: effects of habitat and surge. *J. exp. mar. Biol. Ecol.*, **105**, 1–20.

Foucher, R.P., and Beamish, R.J. (1980) Production of nonviable oocytes by the Pacific hake (*Merluccius productus*). *Can. J. Fish. aquat. Sci.*, **37**, 41–8.

Fournier, D.A., Sibert, J.R., Majkowski, J. and Hampton, J. (1990) MULTIFAN, a likelihood-based method for estimating growth parameters and age composition from multiple length frequency data sets illustrated using data for southern bluefin tuna (*Thunnus maccoyii*). *Can. J. Fish. aquat. Sci.*, **47**, 301–13.

Fowler, A.J. (1987) The development of sampling strategies for population studies of coral reef fishes. A case study. *Coral Reefs*, **6**, 49–58.

Fowler, A.J. (1990) Spatial and temporal patterns of distribution and abundance of chaetodontid fishes at One Tree Reef, southern GBR. *Mar. Ecol. – Progr. Ser.*, **64**, 39–53.

Fowler, A.J., Doherty, P.J. and Williams, D.M. (1992) Multi-scale analysis of recruitment of a coral reef fish on the Great Barrier Reef. *Mar. Ecol. – Progr. Ser.*, **82**, 131–41.

Fox, P.J. (1986) A manual of rapid appraisal techniques for Philippine coastal fisheries: problem solving and project identification. Bureau of Fisheries and Aquatic Resources, Research Division, Quezon City, Philippines, 40 pp.

Fox, W.W., jun. (1992) Stemming the tide: challenges for conserving the nation's coastal fish habitats, in *Stemming the Tide of Coastal Fish Habitat Loss*, National Coalition for Marine Conservation, Savannah, GA, pp. 9–13.

Francis, R.C. (1974) Relationship of fishing mortality to natural mortality at the level of maximum sustainable yield under the logistic stock production model. *J.*

Fish. Res. Bd Can., **31**, 1539–42.

Frank, K.T. and Leggett, W.C. (1994). Fisheries ecology in the context of ecological and evolutionary theory. *A. Rev. Ecol. Syst.*, **25**, 401–22.

Freycinet, L. de (1824) *Voyage Autour du Monde*, Pilet Ainé, Paris.

Fricke, H.W. (1980) Control of different mating systems in a coral reef fish by one environmental factor. *Anim. Behav.*, **28**, 561–9.

Frielink, A.B. (1983) Coastal fisheries in Papua New Guinea: the current situation. Dept Primary Industry, Port Moresby, *Fish. Res. Rep.* 83/10.

Froese, R. (1994) ReefBase: a global database on coral reefs, in *The Management of Coral Reef Resource Systems* (*ICLARM Conf. Proc.*) (eds J.L. Munro and P.E. Munro), ICLARM, Manila, pp. 52–4.

Froese, R. and Pauly, D. (eds) (1994) *FishBase User's Manual: a Biological Database on Fish. ICLARM Software* 7, pp. var. (with CD-ROM).

Frydl, P. and Stearn, C.W. (1978) Rate of bioerosion by parrotfishes in Barbados reef environments. *J. sedim. Petrol.*, **48**, 1149–57.

Funicelli, N.A., Johnson, D.R. and Meineke, D.A. (1988) Assessment of the effectiveness of an existing fish sanctuary within the Kennedy Space Center. A special purpose report to the Florida Marine Fisheries Commission, Tallahassee, FL, USA, 54 pp.

Furnas, M.J. and Mitchell, A.W. (1988) Shelf-scale estimates of phytoplankton primary production in the Great Barrier Reef. *Proc. 6th Int. Coral Reef Symp.*, **2**, 557–62.

Fusimalohi, T. and Preston, G.L. (1983) Report on the South Pacific Commission Deep Sea Fisheries Development Project's visit to the Republic of Vanuatu (12 August 1980–14 June 1981). Unpubl. rep., South Pacific Commission, Nouméa, New Caledonia, 41 pp.

Gabriel, W.L., Sissenwine, M.P. and Overholtz, W.J. (1989) Analysis of spawning stock biomass per recruit: an example for Georges Bank haddock. *N. Am. J. Fish. Manage.*, **9**, 383–91.

Galef, B.G. jun. (1976) Social transmission of acquired behavior: a discussion of tradition and social learning in vertebrates, in *Advances in the Study of Behavior*, Vol. 6 (eds J.S. Rosenblatt, R.A. Hinde, E. Shaw and C. Beer), Academic Press, New York, pp. 77–100.

Galvez, R. and Sadorra, M.S.M. (1988) Blast fishing: a Philippine case study. *Trop. Coastal Area Manage.*, **3**, 9–10.

Galzin, R. (1987) Potential fisheries yield of a Moorea fringing reef (French Polynesia) by the analysis of three dominant fishes. *Atoll Res. Bull.*, **305**, 21 pp.

Garcia, S. and Demetropoulos, A. (1986) Management of Cyprus fisheries. *FAO Fish. tech. Pap.* **250**, v + 40 pp.

García, S., Sparre, P. and Csirke, J. (1989) Estimating surplus production and maximum sustainable yield from biomass data when catch and effort time series are not available. *Fish. Res.*, **8**, 13–23.

Garcia-Cagide, A. (1986) Caracteristicas de la reproduccion del ronco amarillo *Haemulon sciurus*, en la region oriental del Golfo de Batabano, Cuba. *Reporte de Investigacion, Instituto de Oceanologia*, **48**, 1–28.

Garcia-Cagide, A. (1987) Caracteristicas de la reproduccion del ronco arara, *Haemulon plumieri* (Lacépède), en la region oriental del Golfo de Batabano, Cuba. *Revta Invest. Marinas*, **8**, 39–55.

Garcia-Cagide, A. and Claro, R. (1983) Datos sobre la reproduccion de algunos peces comerciales del Golfo de Batabano. *Reporte de Investigacion, Instituto de Oceanologia*, **12**, 1–20.

Garcia-Rubies, A. and Zabala, M. (1990) Effects of total fishing prohibition on the rocky fish assemblages of Medes Islands marine reserve (NW Mediterranean). *Scient. Marina*, **54**, 317–28.

Garratt, P.A., Govender, A. and Punt, A.E. (1993) Growth acceleration at sex change in the protogynous hermaphrodite *Chrysoblephus puniceus* (Pisces: Sparidae). *S. Afr. J. mar. Sci.*, **13**, 187–93.

Gates, C.E. (1979) Line transects and related issues, in *Sampling Biological Populations* (eds R.M. Cormack, G.P. Patil and D.S. Robson), International Co-operative Publishing House, Fairland, MD, pp. 71–154.

Gauldie, R.W. (1991) Taking stock of genetic concepts in fisheries management. *Can. J. Fish. aquat. Sci.*, **48**, 722–31.

Gaut, V.C. and Munro, J.L. (1983) The biology, ecology and bionomics of the grunts, Pomadasyidae, in *Caribbean Coral Reef Fishery Resources, (ICLARM Stud. Rev.* 7) (ed. J.L. Munro), ICLARM, Manila, Philippines, pp. 110–41.

Gawel, M. (1981) Marine resources development planning for tropical Pacific islands. *Proc. 4th Int. Coral Reef Symp.*, **1**, 247–57.

Gayanilo, F.C., jun., Soriano, M. and Pauly, D. (1989) *A Draft Guide to the Complete ELEFAN. ICLARM Software*, **2**, 70 pp.

Gayanilo, F.C., Sparre, P. and Pauly, D. (1994) The FAO–ICLARM stock assessment tools (FiSAT) user's guide. *FAO Computerized Information Series/Fisheries (FiSAT)*, **6**, pp. var.

George, C.J. (1972) Notes on the breeding and movements of the rabbitfishes, *Siganus rivulatus* (Forsskål) and *S. luridus* Ruppell, in the coastal waters of the Lebanon. *Annali Mus. civ. Stor. nat. Giacomo Doria*, **79**, 32–44.

Ghorab, H.M., Bayoumi, A.R., Bebars, M.I. and Hassan, A.A. (1986) The reproductive biology of the grouper *Epinephelus chlorostigma* (Pisces, Serranidae) from the Red Sea. *Bull. Inst. Oceanogr. Fish., Arab Rep. Egypt*, **12**, 13–33.

Gifford, E.W. (1929) *Tongan Society*. Bernice P. Bishop Museum, Honolulu, HI, *Bulletin* **61**.

Gilmore, R.G. and Jones, R.S. (1992) Color variation and associated behavior in the epinepheline groupers, *Mycteroperca microlepis* (Goode and Bean) and *M. phenax* Jordan and Swain. *Bull. mar. Sci.*, **51**, 83–103.

Gilmore, R.G., Cooke, D.W. and Donohoe, C.J. (1982) A comparison of the fish populations and habitats in open and closed salt marsh impoundments in east-central Florida. *Northeast Gulf Sci.*, **5**, 25–7.

Gitschlag, G.R. (1986) Movement of pink shrimp in relation to the Tortugas Sanctuary. *N. Am. J. Fish. Manage.*, **6**, 328–38.

Gladfelter, W.B. and Gladfelter, E.H. (1978) Fish community structure as a function of habitat structure on West Indian patch reefs. *Revista Biol. Trop.*, (Suppl. 1), 65–84.

Gladfelter, W.B. and Johnson, W.S. (1983) Feeding niche separation in a guild of tropical reef fishes (Holocentridae). *Ecology*, **64**, 552–63.

Gladstone, W. (1986) Spawning behavior of the bumphead parrotfish *Bolbometopon muricatum* at Yonge Reef, Great Barrier Reef. *Jap. J. Ichthyol.*, **33**, 326–8.

Gladstone, W. (1987) The eggs and larvae of the sharpnose pufferfish *Canthigaster valentini* (Pisces: Tetraodontidae) are unpalatable to other reef fishes. *Copeia*, **1987**, 227–30.

Gladstone, W. and Westoby, M. (1988) Growth and reproduction in *Canthigaster valentini* (Pisces: Tetraodontidae): a comparison of a toxic reef fish with other reef fishes. *Env. Biol. Fishes*, **21**, 207–21.

Glantz, M.H and Feingold, L.E. (eds) (1990) *Climate Variability, Climate Change and Fisheries*. ESIG, National Center for Atmospheric Research, Boulder, CO, 139 pp.

Glynn, P.W. (1973) *Acanthaster*: effect on coral reef growth in Panama. *Science*, **180**, 504–6.

Glynn, P.W. (1990) Feeding ecology of selected coral-reef macroconsumers: patterns and effects on coral community structure, in *Coral Reefs* (ed. Z. Dubinsky) (Ecosystems of the World, 25). Elsevier, Amsterdam, pp. 365–400.

Glynn, P.W. and Wellington, G.M. (1983) *Corals and Coral Reefs of the Galápagos Islands* (with an annotated list of the scleractinian corals of the Galápagos by J.W. Wells). University of California Press, Berkeley, 330 pp.

Glynn, P.W., Wellington, G.M. and Birkeland, C. (1979) Coral reef growth in Galápagos: limitation by sea urchins. *Science*, **203**, 47–9.

Gobert, B. (1989) Effort de pêche et production des pêcheries artisanales Martiniquaises. Institut Français de Recherche Scientifique pour le Développement en Coopération, (ORSTOM) Martinique, *Doc. scient.* **22**, 98 pp.

Gobert, B. (1992) Impact of the use of trammel nets on a tropical reef resource. *Fish. Res.*, **13**, 353–67.

Goeden, G.B. (1978) A monograph of the coral trout, *Plectropomus leopardus* (Lacepede). Queensland Fisheries Service, *Res. Bull.*, **1**, 1–42.

Goldman, B. and Talbot, F.H. (1976) Aspects of the ecology of coral reef fishes, in *Biology and Geology of Coral Reefs*, Vol. 3 (eds O.A. Jones and R. Endean). Academic Press, New York, pp. 125–54.

Gomez, E.D. (1980) Status report on research and degradation problems of the coral reefs of the East Asian seas. Paper presented at the Meeting of Experts to review the Draft Action Plan for the East Asian Seas, Baguio, Philippines, 17–21 June, 1980. South China Sea Fisheries Development and Coordination Programme, Manila. *UNEP/WG 41/INF*, **15**.

Gomez, E.D., Alcala, A.C. and San Diego, A.C. (1981) Status of Philippine coral reefs. *Proc. 4th Int. Coral Reef Symp.*, **1**, 275–82.

Gomez, E.D., Alcala, A.C. and Yap, H.T. (1987) Other fishing methods destructive to coral, in *Human Impacts on Coral Reefs: Facts and Recommendations*, (ed. B. Salvat), Antènne du Muséum École Pratique des Hautes Études, French Polynesia, pp. 67–75.

Gonzalez, J.A., Lozano, I.J. and Hernandez-Cruz, C.M. (1993) Fecundidad de *Sparisoma* (*Euscarus*) *cretense* (L.) (Osteichthyes, Scaridae) en Canarias. *Bol. Inst. Espanol Oceanogr.*, **9**, 123–31.

Goodenough, W.H. (1951) *Property, Kin and Community on Truk*. Yale University Press, New Haven, CT.

Goodman, D. (1987) Consideration of stochastic demography in the design and management of biological reserves. *Natural Resource Modeling*, **1**, 205–35.

Goodyear, C.P. (1989) Spawning stock biomass per recruit: the biological basis for a fisheries management tool. *ICCAT Working Doc.* SCRS/89/82, 10 pp.

Goodyear, C.P. (1993) Spawning stock biomass per recruit in fisheries management: Foundation and current use, in *Risk Evaluation and Biological Reference Points for Fisheries Management* (eds S.J. Smith, J.H. Hunt and D. Rivard), *Can. spec. Publ. Fish. aquat. Sci.*, **120**, 67–81.

Goodyear, C.P. and Phares, P. (1990) Status of red snapper stocks of the Gulf of Mexico. Unpubl. rep., NOAA NMFS Southeast Fisheries Science Center, Miami Laboratory, Coastal Resource Division CRD 89/90-05, 72 pp.

Gordon, H.S. (1954) The economic theory of a common property resource: the fishery. *J. polit. Econ.*, **62**, 124–42.

Goreau, T.F. and Wells, J.W. (1967) The shallow-water Scleractinia of Jamaica: revised list of species and their vertical distribution range. *Bull. mar. Sci.*, **17**, 442–53.

Goreau, T.F., Lang, J.C., Graham, E.H. and Goreau, P.D. (1972) Structure and ecology of the Saipan reefs in relation to predation by *Acanthaster planci* (L). *Bull. mar. Sci.*, **22**, 113–52.

Govoni, J.J. (1993) Flux of larval fishes across frontal boundaries: examples from the Mississippi River plume front and the western Gulf Stream front in winter. *Bull. mar. Sci.*, **53**, 538–66.

Graham, W.M., Field, J.G. and Potts, D.C. (1992) Persistent 'upwelling shadows' and their influence on zooplankton distributions. *Mar. Biol.*, **114**, 561–70.

Grand, S. (1985) The importance of the reef lagoon fishery in French Polynesia. *Proc. 5th Coral Reef Congr.*, **2**, 495–500.

Grant, C. (1981) High catch rates in Norfolk Island dropline fishery. *Aust. Fish.*, **40**(3), 10–13.

Green, A.L. (1993) Damselfish territories: focus sites for studies of the early post-settlement biology and ecology of labrid fishes (Family Labridae). *Proc. 7th Int. Coral Reef Symp.*, **1**, 601–5.

Griffin, W., Hendrickson, H., Oliver, C., Matlock, G., Bryan, C.E., Riechers, R. and Clark, J. (1993) An economic analysis of Texas shrimp season closures. *Mar. Fish. Rev.*, **54**, 21–8.

Grigg, R.W. (1994) Effects of sewage discharge, fishing pressure and habitat complexity on coral ecosystems and reef fishes in Hawaii. *Mar. Ecol. – Progr. Ser.*, **103**, 25–34.

Grigg, R.W., Polovina, J.J. & Atkinson, M.J. (1984) Model of a coral ecosystem III. Resource limitation, community regulation, fisheries yield and resource management. *Coral Reefs*, **3**, 23–7.

Grimes, C.B. (1987) Reproductive biology of the Lutjanidae: a review, in *Tropical Snappers and Groupers: Biology and Fisheries management* (eds J.J. Polovina and S. Ralston), Westview Press, Boulder, CO, pp. 239–94.

Grimes, C.B., and Huntsman, G.R. (1979) Reproductive biology of the vermilion snapper, *Rhomboplites aurorubens*, from North Carolina and South Carolina. *Fishery Bull., U.S.*, **78**, 137–46.

Grimes, C.B., Idelberger, C.F., Able, K.W. and Turner, S.C. (1988) The reproductive biology of tilefish, *Lopholatilus chamaeleonticeps* Goode and Bean, from the United States mid-Atlantic Bight, and the effects of fishing on the breeding system. *Fishery Bull., U.S.*, **86**, 745–62.

Grove, R.S., Sonu, C.J. and Nakamura, M. (1991) Fisheries applications and biological impacts of artificial reefs, in *Artificial Habitats for Marine and Freshwater Fisheries* (eds W. Seaman, jun. and L.M. Sprague), Academic Press, San Diego, pp. 109–52.

Grover, J.J., Olla, B.L. and Wicklund, R.I. (1992) Food habits of Nassau grouper (*Epinephelus striatus*) juveniles in three habitats in the Bahamas. *Proc. Gulf Carib. Fish. Inst.*, **42**, 247.

Guerin, J.M. and Cillauren, E. (1989) Pêche profonde aux casiers à Vanuatu: résultats des campagnes expérimentales. Institut Français de Recherche Scientifique pour le Développment en Coopération, Port Vila, Vanuatu. *Notes Doc.* **21**, 48 pp.

Guitart-Manday, D. and Juarez-Fernandez, F. (1966) Desarrollo embrionario y primeros estudios larvales de la cherna criolla, *Epinephelus striatus* (Bloch) (Perciformes: Serranidae). *Academia de Ciencias de Cuba, Instituto de Oceanologia, La Habana*, **1**, 35–45.

Gulland, J.A. (1955) Estimation of growth and mortality in commercial fish populations. Min. Agric. Fish., London, *Fish. Invest. Ser. 2*, **18**(9), 46 pp.

Gulland, J.A. (1971) *The Fish Resources of the Ocean*, Fishing News (Books) Ltd, Surrey, England.

Gulland, J.A. (1983a) Can a study of stock and recruitment aid management decisions? *Rapp. P.-V. Réun. Cons. perm. Int. Explor. Mer*, **164**, 368–72.

Gulland, J.A. (1983b) *Fish Stock Assessment: A Manual of Basic Methods*, Wiley, New York, 223 pp.

Gulland, J.A. (1984) Best estimates. *Fishbyte*, **2**(3), 3–5.

Gulland, J.A. and Garcia, S. (1983) Are catch and effort statistics essential? *Fishbyte*, **1**(2), 2–4.

Gundermann, N., Popper, D.M. and Lichatowich, T. (1983) Biology and life cycle of *Siganus vermiculatus* (Siganidae, Pisces). *Pac. Sci.*, **37**, 165–80.

Gygi, R.A. (1969) An estimate of the erosional effect of *Sparisoma viride* (Bonaterre), the green parrotfish, on some Bermuda reefs. *Spec. Publ. Bermuda Biol. Stn*, **2**, 137–43.

Gygi, R.A. (1975) *Sparisoma viride* (Bonnaterre) the stoplight parrotfish, a major sediment producer on coral reefs of Bermuda? *Eclogae Geol. Helv.*, **68**, 327–59.

Ha, S.J. (1985) Evidence of temporary hearing loss (temporary threshold shift) in fish subjected to laboratory ambient noise. *Proc. PA Acad. Sci.*, **59**, 78. (abstract)

Haight, W.R., Parrish, J.D. and Hayes, T.A. (1993) Feeding ecology of deepwater lutjanid snappers at Penguin Bank, Hawaii. *Trans. Am. Fish. Soc.*, **122**, 328–47.

Haimovici, M. and Cousin, J.C.B. (1988) Reproductive biology of the castanha *Umbrina canosai* (Pisces, Sciaenidae) in southern Brazil. *Revista brasil. Biol.*, **49**, 523–37.

Halapua, S. (1982) *Fishermen of Tonga: their Means of Survival*, University of the South Pacific, Suva, Fiji, 100 pp.

Hall, C.A.S. (1988) An assessment of several of the historically most influential theoretical models used in ecology and of the data provided in their support. *Ecol. Modeling*, **43**, 5–31.

Hamley, J.M. (1975) Review of gill net selectivity. *J. Fish. Res. Bd Can.*, **32**, 1943–69.

Hamley, J.M. and Skud, B.E. (1978) Factors affecting longline catch and effort: II. Hook-spacing. *Int. Pac. Halibut Commission Sci. Rep.*, **65**, 15–24.

Hammerman, T. (1986) Fish trap trials in Sri Lanka. Bay of Bengal Programme, *Working Paper* **42**.

Hamner, W.M., Jones, M.S., Carleton, J.H., Hauri, I.R. and Williams, D.M. (1988) Zooplankton, planktivorous fish, and water currents on a windward reef face: Great Barrier Reef, Australia. *Bull. mar. Sci.*, **42**, 459–79.

Handy, E.C.S. (1923) *The Native Culture of the Marquesas*. Bernice P. Bishop Museum, Honolulu, HI, *Bulletin* **9**.

Handy, E.C.S. (1932) *Housing, Boats and Fishing in the Society Islands*. Bernice P. Bishop Museum, Honolulu, HI, *Bulletin* **90**.

Hanski, I. (1991) Single-species metapopulation dynamics: concepts, models and observations. *Biol. J. Linn. Soc.*, **42**, 17–38.

Hanson, J.M. and Leggett, W.C. (1982) Empirical production of fish biomass and yield. *Can. J. Fish. aquat. Sci.*, **39**, 257–63.

Hardin, G. (1968) The tragedy of the commons. *Science*, **162**, 1243–8.

Hardin, G. (1991) Paramount positions in ecological economics, in *Ecological Economics: The Science and Management of Sustainability* (ed. R. Costanza), Columbia University Press, New York, NY, pp. 47–57.

Harding, J.H. (1990) Shark: return of the gray nurse. *Sea Frontiers*, **34**, 30–33.

Hardy, A. (1959) The over-fishing problem, Chapter 13 in *The Open Sea: its Natural History*, Part II, *Fish and Fisheries*, Houghton Mifflin Company, Boston, MA, 322 pp.

Hare, J.A., Cowen, R.K., Zehr, J.P., Juanes, F. and Day, K.H. (1994) Biological and oceanographic insights from larval labrid (Pisces, Labridae) identification using mtDNA sequences. *Mar. Biol.*, **118**, 17–24.

Harmelin-Vivien, M.L. (1981) Trophic relationships of reef fishes in Tulear (Madagascar). *Oceanologica Acta*, **4**, 365–74.

Harmelin-Vivien, M.L. (1983) Étude comparative de l'ichthyofaune des herbiers de phanérogames marines en milieu tropical et temperé. *Revue d'Ecologie Terre et Vie*, **38**, 179–210.

Harmelin-Vivien, M.L. (1989a) Implications of feeding specialization on the recruitment processes and community structure of butterflyfishes. *Env. Biol. Fishes*, **25**, 101–10.

Harmelin-Vivien, M.L. (1989b) Reef fish community structure: an Indo–Pacific comparison. *Ecol. Stud.*, **69**, 21–60.

Harmelin-Vivien, M.L., Harmelin, J.G., Chauvet, C., Duval, C., Galzin, R., Lejeune, P., Barnabe, G., Blanc, F., Chevalier, R., Duclerc, J. and Lasserre, G. (1985) Évaluation visuelle des peuplements et populations de poissons: méthodes et problèmes. *Revue d'Ecologie Terre et Vie*, **40**, 467–539.

Harper, D.E. (1993) The 1993 spiny lobster monitoring report on trends in landings, CPUE, and size of harvested lobster. NOAA, Southeast Fisheries Science Center, Miami Laboratory Contribution MIA-92/93-92, 36 pp. Chapter 11

Hart, P.J.B. (1986) Foraging in teleost fishes, in *The Behaviour of Teleost Fishes*, (ed. T.J. Pitcher), Croom Helm, London, pp. 211–35.

Hartsuijker, L. and Nicholson, W.E. (1981) Results of potfishing survey on Pedro Bank (Jamaica): the relations between catch rates, catch composition, the size of fish and their recruitment to the fishery. Technical Report of Potfisheries Survey of Pedro Bank, 2. Project FAO/TCP/JAM 8902, Fisheries Division, Jamaica.

Hasse, J.J., Madraisau, B.B. and McVey, J.P. (1977) Some aspects of the life history of *Siganus canaliculatus* (Park) (Pisces: Siganidae) in Palau. *Micronesica*, **13**, 297–312.

Hatcher, B.G. (1981) The interaction between grazing organisms and the epilithic algal community of a coral reef: a quantitative assessment. *Proc. 4th Int. Coral Reef Symp.*, **2**, 515–24.

Hatcher, B.G., Johannes, R.E. and Robertson, A.I. (1989) Review of research relevant to the conservation of shallow tropical marine ecosystems. *Oceanogr. mar. Biol. A. Rev.*, **27**, 337–414.

Hateley, J.G. and Sleeter, T.D. (1993) A biochemical genetic investigation of spiny lobster (*Panulirus argus*) stock replenishment in Bermuda. *Bull. mar. Sci.*, **52**, 993–1006.

Hawkins, C.M. (1981) Efficiency of organic matter absorption by the tropical echinoid *Diadema antillarum* Philippi fed non-macrophytic algae. *J. exp. mar. Biol. Ecol.*, **49**, 245–53.

Hawkins, C.M. and Lewis, J.B. (1982) Ecological energetics of the tropical sea urchin *Diadema antillarum* Philippi in Barbados, West Indies. *Estuar. Coastal Shelf Sci.*, **15**, 645–69.

Hay, M.E. (1981) Herbivory, algal distribution and the maintenance of between-habitat diversity on a tropical fringing reef. *Am. Nat.*, **118**, 520–40.

Hay, M.E. (1984) Patterns of fish and urchin grazing on Caribbean coral reefs: are previous results typical? *Ecology*, **65**, 446–54.

Hay, M.E. (1985) Spatial patterns of herbivore impact and their importance in maintaining algal species richness. *Proc. 5th Int. Coral Reef Symp.*, **4**, 29–34.

Hay, M.E. (1991) Fish–seaweed interactions on coral reefs: effects of herbivorous fishes and adaptations of their prey, in *The Ecology of Fishes on Coral Reefs* (ed. P.F. Sale), Academic Press, San Diego, pp. 96–119.

Hay, M.E. and Taylor, P.R. (1985) Competition between herbivorous fishes and urchins on Caribbean coral reefs. *Oecologia*, **65**, 591–8.

He, X. and Wright, R.A. (1992) An experimental study of piscivore–planktivore interactions – population and community responses to predation. *Can. J. Fish. aquat. Sci.*, **49**, 1176–83.

Heath, M.R. (1992) Field investigations of the early-life stages of marine fish. *Adv. mar. Biol.*, **28**, 1–174.

Heemstra, P.C. and Randall, J.E. (1993) *Groupers of the World (Family Serranidae, Subfamily Epinephelinae). An Annotated and Illustrated Catalogue of the Grouper, Rockcod, Hind, Coral Grouper and Lyretail Species Known to Date.* (FAO species catalogue.) *FAO Fish. Synop.* **16**, 382 pp.

Helfman, G.S. and Schultz, E.T. (1984) Social transmission of behavioural traditions in a coral reef fish. *Anim. Behav.*, **32**, 379–84.

Helfrich, P. and Allen, P.M. (1975) Observations on the spawning of mullet, *Crenimugil crenilabis* (Forskøl), at Enewetak, Marshall Islands. *Micronesica*, **11**, 219–25.

Helgason, T. and Gislason, H. (1979) VPA-analysis with species interactions due to predation. *ICES CM* 1979/G:52, 10 pp.

Helm, N. (1992) A report on the market survey of reef and lagoon fish catches in Western Samoa. South Pacific Commission, Nouméa, New Caledonia. Papers on Fisheries Science from the Pacific Islands. *Inshore Fish. Res. Proj., tech. Doc.*, **1**, 1–5.

Hensley, D.A., Appeldoorn, R.S., Shapiro, D.Y., Ray, M. and Turingan, R.G. (1994) Egg dispersal in a Caribbean coral reef fish, *Thalassoma bifasciatum*. I. Dispersal over the reef platform. *Bull. mar. Sci.*, **54**, 256–70.

Hernandez-Leon, S. (1991) Accumulation of mesozooplankton in a wake area as a causative mechanism of the island-mass effect. *Mar. Biol.*, **109**, 141–7.

Hiatt, R.W. and Strasburg, D.W. (1960) Ecological relationships of the fish fauna on coral reefs of the Marshall Islands. *Ecol. Monogr.*, **30**, 65–127.

Hilborn, R. (1985) Fleet dynamics and individual variation: why some people catch more fish than others. *Can. J. Fish. aquat. Sci.*, **42**, 2–13.

Hilborn, R. and Walters, C.J. (1992) *Quantitative Fisheries Stock Assessment: Choice, Dynamics and Uncertainty*, Chapman & Hall, New York, 570 pp.

Hill, H.B. (1978) The use of nearshore marine life as a food resource by American Samoans. University of Hawaii, Honolulu, HI. Pacific Islands Program, *Misc. Work Papers* **1**.

Hixon, M.A. (1982) Differential fish grazing and benthic community structure on Hawaiin reefs, in *Gutshop '81: Fish Food Habits and Studies* (eds G.M. Caillet and C.A. Simenstad), Washington Sea Grant Publication, University of Washington, Seattle, pp. 249–57.

Hixon, M.A. (1991) Predation as a process structuring coral reef fish communities, in *The Ecology of Fishes on Coral Reefs* (ed. P.F. Sale), Academic Press, San Diego, pp. 475–508.

Hixon, M.A. and Beets, J.P. (1989) Shelter characteristics and Caribbean fish assemblages: experiments with artificial reefs. *Bull. mar. Sci.*, **44**, 666–80.

Hixon, M.A. and Beets, J.P. (1993) Predation, prey refuges, and the structure of coral-reef fish assemblages. *Ecol. Monogr.*, **63**, 77–101.

Hobson, E.S. (1968) Predatory behavior of some shore fishes in the Gulf of California. *US Fish Wild. Serv., Bur. Sport Fish. Wildlife, Res. Rep.* **73**, vi + 92 pp.

Hobson, E.S. (1974) Feeding relationships of teleostean fishes on coral reefs in Kona, Hawaii. *Fishery Bull., U.S.*, **72**, 915–1031.

Hobson, E.S. and Chess, J.R. (1978) Trophic relationships among fishes and plankton in the lagoon at Enewetak Atoll, Marshall Islands. *Fishery Bull., U.S.*, **76**, 133–53.

Hocart, A.M. (1929) *Lau Islands, Fiji*. Bernice P. Bishop Museum, Honolulu, HI, *Bulletin* **62**.

Hodgson, G. and Dixon, J.A. (1988) *Logging Versus Fisheries and Tourism in Palawan: An Environmental and Economic Analysis*. East–West Environment and Policy Institute, University of Hawaii, Honolulu, HI, *Occ. Pap.* **7**, 95 pp.

Hoenig, J.M. (1983) Empirical use of longevity data to estimate mortality rates. *Fishery Bull., U.S.*, **82**, 898–903.

Hoenig, J.M. (1987) Estimation of growth and mortality parameters for use in length-structured stock production models, in *Length-based Methods in Fisheries Research* (*ICLARM Conf. Proc.* **13**) (eds D. Pauly and G.R. Morgan), ICLARM, Manila, Philippines and Kuwait Institute of Scientific Research, Safat, Kuwait, pp. 121–8.

Hoenig, J.M., Lawing, W.D. and Hoenig, N.A. (1983) Using mean age, mean length and median length data to estimate the total mortality rate. *ICES CM* 1983/D:23, 11 pp.

Hoenig, J.M., Csirke, J., Sanders, M.J., Abella, A., Andreoli, M.G., Levi, D., Ragonese, S., Al-Shoushani, M. and El-Musa, M.M. (1987) Data acquisition for length-based stock assessment: report of Working Group I, in *Length-based Methods in Fisheries Research* (*ICLARM Conf. Proc.* **13**) (eds D. Pauly and G.R. Morgan), ICLARM, Manila, Philippines and Kuwait Institute of Scientific Research, Safat, Kuwait, pp. 343–52.

Hoffman, D.S., and Grau, E.G. (1989) Daytime changes in oocyte development with relation to the tide for the Hawaiian saddleback wrasse *Thalassoma duperrey*. *J. Fish Biol.*, **34**, 529–46.

Hoffman, S.G. (1983) Sex-related foraging behavior in sequentially hermaphroditic hogfishes (*Bodianus* spp.). *Ecology*, **64**, 798–808.

Hoffman, S.G. and Robertson, D.R. (1983) Foraging and reproduction of two Caribbean reef toadfishes (Batrachoididae). *Bull. mar. Sci.*, **33**, 919–26.

Holland, K.N., Peterson, J.D., Lowe, C.G. and Wetherbee, B.M. (1993) Movements, distribution and growth rates of the white goatfish *Mulloides flavolineatus* in a fisheries conservation zone. *Bull. mar. Sci.*, **52**, 982–92.

Holt, S.J. (1963) A method of determining gear selectivity and its application. *Spec. Pub. Int. Comm. NW Atl. Fish.*, **5**, 106–15.

Hood, P.B. and Schleider, R.A. (1992) Age, growth, and reproduction of gag, *Mycteroperca microlepis* (Pisces: Serranidae), in the eastern Gulf of Mexico. *Bull. mar. Sci.*, 51, 337–52.

Hooper, A. (1985) Tokelau fishing in traditional and modern contexts, in *The Tra-*

ditional Knowledge and Management of Coastal Systems in Asia and the Pacific (eds K. Ruddle and R.E. Johannes), UNESCO–ROSTSEA, Jakarta, pp. 7–38.

Hooper, A. (1990). Tokelau fishing in traditional and modern contexts, in *Traditional Marine Resource Management in the Pacific Basin: an Anthology* (eds K. Ruddle and R.E. Johannes), UNESCO–ROSTSEA, Jakarta, pp. 213–240.

Horwood, J.W., Bannister, R.C.A. and Howlett, G.J. (1986) Comparative fecundity of North Sea plaice (*Pleuronectes platessa* L.). *Proc. R. Soc.*, **228B**, 401–31.

Hosaka, E.Y. (1973) *Shore Fishing in Hawaii*. Petroglyph Press, Hilo, HI.

Houde, E.D., and Lovdal, J.A. (1984) Seasonality of occurrence, foods, and food preferences of ichthyoplankton in Biscayne Bay, Florida. *Estuar. Coastal Shelf Sci.*, **18**, 403–20.

Houde, E.D. and Lovdal, J.A. (1985) Patterns of variability in ichthyoplankton occurrence and abundance in Biscayne Bay, Florida. *Estuar. Coastal Shelf Sci.*, **20**, 79–104.

Houde, E.D. and Zastrow, C.E. (1993) Ecosystem- and taxon-specific dynamic and energetics properties of larval fish assemblages. *Bull. mar. Sci.*, **53**, 290–335.

Hourigan, T.F. (1989) Environmental determinants of butterflyfish social systems. *Env. Biol. Fishes*, **25**, 61–78.

Hourigan, T.F., Tricas, T.C. and Reese, E.S. (1988) Coral reef fishes as indicators of environmental stress in coral reefs, in *Marine Organisms as Indicators* (eds D.F. Soule and G.S. Keppel), Springer-Verlag, New York, pp. 107–36.

Hughes, T.P. (1994) Catastrophes, phase shifts, and large-scale degradation of a Caribbean coral reef. *Science*, **265**, 1547–51.

Hughes, T.P., Keller, B.D., Jackson, J.B.C. and Boyle, M–J. (1987a) Mass mortality of the echinoid *Diadema antillarum* Philippi in Jamaica. *Bull. mar. Sci.*, **36**, 377–84.

Hughes, T.P., Reed, D.C. and Boyle, M–J. (1987b) Herbivory on coral reefs: community structure following mass mortalities of sea urchins. *J. exp. mar. Biol. Ecol.*, **113**, 39–59.

Hulo, J. (1984) Fishing practices in Buka Island, North Solomons Province, in *Subsistence Fishing Practices of Northern Papua New Guinea* (eds N.J. Quinn, B. Kojis and P. Warhepa), Appropriate Technology Development Institute, Lae, Papua New Guinea, *Trad. Technol. Ser.* no 2, pp. 28–33.

Humphrey, J.D. (1994) Risks associated with movements of marine animals in the South Pacific. South Pacific Commission, Nouméa, New Caledonia, *Inshore Fish. Res. Proj. techn. Rep.* **7**, 50 pp.

Hunte, W., Cote, I. and Tomascik, T. (1986) On the dynamics of the mass mortality of *Diadema antillarum* in Barbados. *Coral Reefs*, **4**, 135–9.

Hunte von Herbing, I. and Hunte, W. (1991) Spawning and recruitment in the bluehead wrasse *Thalassoma bifasciatum* in Barbados, West Indies. *Mar. Ecol. – Progr. Ser.*, **72**, 49–58.

Hunter, J.R. (1981) Feeding ecology and predation of marine fish larvae, in *Marine Fish Larvae* (ed. R. Lasker), University of Washington Press, Seattle, pp. 33–79.

Hunter, J.R. and Goldberg, S.R. (1980) Spawning incidence and batch fecundity in northern anchovy, *Engraulis mordax*. *Fishery Bull., U.S.*, **77**, 641–52.

Hunter, J.R. and Macewicz, B.J. (1985) Measurement of spawning frequency in multiple spawning fishes, in *An egg production method for estimating spawning biomass of pelagic fish: application to the northern anchovy*, Engraulis mordax (ed. R. Lasker), *NOAA NMFS tech. Rep.* **36**, 67–78.

Hunter, J.R., Macevicz, B.J., Lo, N.C. and Kimbrell, C.A. (1992) Fecundity, spawning, and maturity of female Dover sole *Microstomus pacificus*, with an evaluation of assumptions and precision. *Fishery Bull., U.S.*, **90**, 101–28.

Huntsman, G.R., Nicholson, W.R. and Fox, W.W., jun. (eds) (1982) The biological bases for reef fishery management. *NOAA Tech. Memo.* NMFS-SEFC-80, 216 pp.

Huntsman, G.R., Manooch, C.S. III and Grimes, C.B. (1983) Yield per recruit models of some reef fishes of the U.S. South Atlantic Bight. *Fishery Bull., U.S.*, **81**, 679–95.

Hussain, N.A. and Abdullah, M.A.S. (1977) The length–weight relationship, spawning season and food habits of six commercial fishes in Kuwaiti waters. *Indian J. Fish.*, **24**, 181–94.

Hviding, E. (1990) Keeping the sea: aspects of marine tenure in Marovo lagoon, Solomon Islands, in *Traditional Marine Resource Management in the Pacific Basin: an Anthology* (eds K. Ruddle and R.E. Johannes), UNESCO–ROSTSEA, Jakarta, pp. 1–43.

Hviding, E. and Ruddle, K. (1991) A regional assessment of the potential role of customary marine tenure (CMT) systems in contemporary fisheries management in the South Pacific. FFA Report, South Pacific Forum Fisheries Agency, Honiara, Solomon Islands. 20 pp.

Ikehara, I.I., Kami, H.T. and Sakomoto, R.K. (1970) Exploratory fishing survey of the inshore fisheries resources of Guam. *Proc. Second Coop. Study Kuroshio Symp., Tokyo, Japan*, pp. 425–36.

Ivens, W.G. (1930) *The Island Builders of the Pacific*, Seeley, Service and Co., London.

Iversen, E.S., Rutherford, E.S., Bannerot, S.P. and Jory, D.E. (1987) Biological data on Berry Island (Bahamas) queen conchs, *Strombus gigas*, with mariculture and fisheries management implications. *Fishery Bull., U.S.*, **85**, 299–310.

Jahn, A.E., and Lavenberg, R.J. (1986) Fine-scale distribution of nearshore, suprabenthic fish larvae. *Mar. Ecol. – Progr. Ser.*, **31**, 223–31.

Janzen, D. (1986) Science is forever. *Oikos*, **46**, 281–83.

Jeffries, M.J. and Lawton, J.H. (1984) Enemy-free space and the structure of ecological communities. *Biol. J. Linn. Soc.*, **23**, 269–86.

Jennings, S. and Beverton, R.J.H. (1991) Intraspecific variation in the life history tactics of Atlantic herring (*Clupea harengus* L.) stocks. *ICES J. mar. Sci.*, **48**, 117–25.

Jennings, S. and Polunin, N.V.C. (1995) Comparative size and composition of yield from six Fijian reef fisheries. *J. Fish Biol.*, **46**, 28–46.

Jennings, S. and Polunin, N.V.C. (in press a) Relationships between catch and effort in multispecies reef fisheries subject to different levels of exploitation. *Fish. Manage. Ecol.* **2**, 89–101.

Jennings, S. and Polunin, N.V.C. (in press b) Biased underwater visual census biomass estimates for target-species in tropical reef fisheries. *J. Fish Biol.*

Jennings, S. and Polunin, N.V.C. (in press c) Impacts of fishing on tropical reef ecosystems. *Ambio.*

Jennings, S. and Polunin, N.V.C. (in press d) Fishing strategies, fishery development and socioeconomics in traditionally managed Fijian reef fisheries. *Fish. Manage. Ecol.*

Jennings, S. and Polunin, N.V.C. (in press e) Effects of fishing effort and catch rate upon the structure and biomass of Fijian reef fish communities. *J. appl. Ecol.*

Jennings, S., Brierley, A.S. and Walker, J.W. (1994) The inshore fish assemblages of the Galápagos archipelago. *Biol. Conserv.*, **70**, 49–57.

Jennings, S., Boullé, D. and Pounin, N.V.C. (in press a) Habitat correlates of the distribution and biomass of Seychelles' reef fishes. *Env. Biol. Fishes.*

Jennings, S., Grandcourt, E.M. and Polunin, N.V.C. (in press b) Effects of fishing on the diversity, biomass and trophic structure of Seychelles' reef fish communities. *Coral Reefs*.

Jennings, S., Marshall, S.S. and Polunin, N.V.C. (in press c) Seychelles marine protected areas: comparative structure and status of reef fish communities. *Biol. Conserv.*

Johannes, R.E. (1974) Sources of nutritional energy for reef coral. *Proc. 2nd Int. Coral Reef Symp.*, **1**, 133–7.

Johannes, R.E. (1977) Traditional law of the sea in Micronesia. *Micronesica*, **13**, 121–7.

Johannes, R.E. (1978a) Reproductive strategies of coastal marine fishes in the tropics. *Env. Biol. Fishes*, **3**, 65–84.

Johannes, R.E. (1978b) Traditional marine conservation methods in Oceania and their demise. *A. Rev. Ecol. Syst.*, **9**, 349–64.

Johannes, R.E. (1980) Using knowledge of the reproductive behavior of reef and lagoon fishes to improve fishing yields, in *Fish Behavior and its Use in the Capture and Culture of Fishes* (*ICLARM Conf. Proc.* **5**) (eds J.E. Bardach, J.J. Magnuson, R.C. May and J.M. Reinhart), International Center for Living Aquatic Resources Management, Manila, Philippines, 247–70.

Johannes, R.E. (1981) *Words of the Lagoon: Fishing and Marine Lore in the Palau District of Micronesia*, University of California Press, Berkeley and Los Angeles, 320 pp.

Johannes, R.E. (1982) Traditional conservation methods and protected marine areas in Oceania. *Ambio*, **11**, 258–61.

Johannes, R.E. (1984) Traditional conservation methods and protected marine areas in Oceania, in *National Parks, Conservation, and Development: The Role of Protected Areas in Sustaining Society* (eds J.A. McNeely and K.R. Miller), Smithsonian Institution Press, Washington, DC, pp. 344–7.

Johannes, R.E. (1988a) Spawning aggregation of the grouper, *Plectropomus areolatus* (Rüppell) in the Solomon Islands. *Proc. 6th Int. Coral Reef Symp.*, **2**, 751–5.

Johannes, R.E. (1988b) The role of marine resource tenure systems (TURFs) in sustainable nearshore marine resource development and management in U.S.-affiliated tropical Pacific islands, in *Topic Reviews in Insular Resource Development and Management in the Pacific U.S.-Affiliated Islands*, University of Guam Marine Laboratory, Agana, Guam. *Tech. Rep.* **88**.

Johannes, R.E. (1989) A spawning aggregation of the grouper *Plectropomus aureolatus* (Rüppell) in the Solomon Islands. *Proc. 6th Int. Coral Reef Symp.*, **2**, 751–5.

Johannes, R.E. (1991) Some suggested management initiatives in palau's nearshore fisheries, and the relevance of traditional management. Unpubl. rep., Marine Resources Division, Palau and South Pacific Commission, Noumea, New Caledonia. 39 pp.

Johannes, R.E. (1993) The plight of the osfish, or why quantitative sophistication is no substitute for asking the right questions. *Naga, the ICLARM Quarterly*, **16**(1), 4–5.

Johannes, R.E. and Gerber, R. (1975) Import and export of net plankton by an Eniwetok coral reef community. *Proc. 2nd Int. Coral Reef Symp.*, 97–104.

Johannes, R.E. and Hviding, E. (1987) *Traditional Knowledge of Marine Resources of the People of Marovo Lagoon, Solomon Islands, with Comments on Marine Conservation*, Commonwealth Science Council, London.

Johannes, R.E. and MacFarlane, J.W. (1984) Traditional sea rights in the Torres Strait Islands, with emphasis on Murray Island, in *Maritime Institutions in the*

Western Pacific, (*Senri Ethnol. Stud.*, **17**) (eds K. Ruddle and T. Akimichi), National Museum of Ethnology, Osaka, 253–66.

Johannes, R.E. and MacFarlane, J.W. (1990) Assessing traditional fishing rights systems in the context of marine resource management: a Torres Strait example, in *Traditional Marine Resource Management in the Pacific Basin: an Anthology* (eds K. Ruddle and R.E. Johannes), UNESCO–ROSTSEA, Jakarta, pp. 241–61.

Johannes, R.E. and MacFarlane, J.W. (1991) *Traditional Fishing in the Torres Strait Islands*, CSIRO, Division of Fisheries, Hobart.

Johnson, G.D. (1984) Percoidei: development and relationships, in *Ontogeny and Systematics of Fishes* (*Am. Soc. Ichthyol. Herpetol. Spec. Publ.* 1), (eds H.G. Moser, W.J. Richards, D.M. Cohen, M.P. Fahay, A.W. Kendall, jun. and S.L. Richardson), Allen Press, Lawrence, KS, pp. 464–98.

Johnson, J.C., Griffith, D.C. and Murray, J.D. (1989) Recreational fishermen's perceptions and preferences for marine fish: some methodological considerations, in *Proc. 39th Ann. Mg Gulf Carib. Fish. Inst., Nov. 1986*, (eds G.T. Waugh and M.H. Goodwin), Hamilton, Bermuda and Charleston, SC, pp. 146–67.

Jokiel, P.L. (1989) Rafting of reef corals and other organisms at Kwajalein Atoll. *Mar. Biol.*, **101**, 483–93.

Jones, E.C. (1962) Evidence of an island effect upon the standing crop of zooplankton near the Marquesas Islands, Central Pacific. *J. Conseil, Cons. int. Explor. Mer*, **27**, 223–31.

Jones, G.P. (1988) Experimental evaluation of the effects of habitat structure and competitive interactions on the juveniles of two coral reef fishes. *J. exp. mar. Biol. Ecol.*, **123**, 115–26.

Jones, G.P. (1991) Postrecruitment processes in the ecology of coral reef fish populations: a multifactorial perspective, in *The Ecology of Fishes on Coral Reefs* (ed. P.F. Sale), Academic Press, San Diego, pp. 294–328.

Jones, G.P., Sale, P.F. and Ferrell, D.J. (1988) Do large carnivorous fishes affect the ecology of macrofauna in shallow lagoonal sediments?: a pilot experiment. *Proc. 6th Int. Coral Reef Symp.*, **2**, 77–82.

Jones, R. (1982) Ecosystems, food chains and fish yields, in *Theory and Management of Tropical Fisheries* (*ICLARM Conf. Proc.* **9**) (eds D. Pauly and G.I. Murphy), ICLARM, Manila, Philippines, pp. 195–239.

Jones, R.S. (1968) Ecological relationships in Hawaiian and Johnston Island Acanthuridae (surgeonfishes). *Micronesica*, **4**, 309–61.

Jones, R.S. and Chase, J.A. (1975) Community structure and distribution of fishes in an enclosed high island lagoon in Guam. *Micronesica*, **11**, 127–48.

Jones, S. (1994) Endangered Species Act Battles. *Fisheries*, **19**, 22–5.

Josephides, L. (1982) The socio-economic condition of women in some fisherfolk communities of Papua New Guinea. Unpubl. rep. on an ESCAP/FAO survey.

Kailola, P.J., Williams, M.J., Steward, P.C., Reichelt, R.E., McNee, A. and Grieve, C. (1993) *Australian Fisheries Resources*, Bureau of Resource Sciences, Commonwealth of Australia.

Kanda, K.A., Koike, A., Takeuchi, S. and Ogura, M. (1978) Selectivity of the hook for mackerel, *Scomber japonicus* Houttuyn, pole fishing. *J. Tokyo Univ. Fish.*, **64**, 109–14.

Kartas, F. and Quignard, J.-P. (1984) *La Fecondité des Poissons Teleostéens*, Collection de Biologie des Milieux Marins, Masson, Paris.

Katnik, S.E. (1982) Effects of fishing pressure on the reef flat fisheries of Guam. MSc thesis, University of Guam, 62 pp.

Kaufman, L.S. and Ebersole, J.P. (1984) Microtopography and the organisation of two assemblages of coral reef fishes in the West Indies. *J. exp. mar. Biol. Ecol.*, **78**, 253–68.

Kaufman, L.[S.], Ebersole, J., Beets, J. and McIvor, C.C. (1992) A key phase in the recruitment dynamics of coral reef fishes: post-settlement transition. *Env. Biol. Fishes*, **34**, 109–18.

Kawaguchi, K. (1974) Handline and longline fishing explorations for snapper and related species in the Caribbean and adjacent waters. *Mar. Fish. Rev.*, **36**, 8–31.

Kedidi, S.M. (1984a) The Red Sea reef associated fishery of the Sudan, catches, efforts and catches per fishing effort survey conducted during 1982–1984. Unpubl. rep., UNDP/FAO, Cairo. 17 pp.

Kedidi, S.M. (1984b) Description of the artisanal fishery at Tuwwal, Saudi Arabia: catches, efforts and catches per unit effort. Survey conducted during 1981–1982. Unpubl. rep., UNDP/FAO, Cairo. Project for Development of Fisheries in the Areas of the Red Sea and Gulf of Aden, FAO/UNDP RAB/81/002/16, 17 pp.

Kenchington, R.A. and Hudson, B.E.T. (eds) (1984) *Coral Reef Handbook*. UNESCO Reg. Off. Sci. Tech. SE Asia, Jakarta, Indonesia, 281 pp.

Kennedy, T.F. (1962) *Fishermen of the Pacific Islands*, A.H. and A.W. Reid, Wellington and Sydney.

Kimmel, J.J. (1985) A characterization of Puerto Rican fish assemblages, PhD dissertation, University of Puerto Rico, Mayagüez, Puerto Rico, 206 pp.

King, T.L., Ward, R. and Blandon, I.R. (1993) Gene marking: a viable assessment method. *Fisheries*, **18**, 4–5.

Kingsford, M.J. (1990) Linear oceanographic features – a focus for research on recruitment processes. *Aust. J. Ecol.*, **15**, 391–401.

Kingsford, M.J. (1992) Spatial and temporal variation in predation on reef fishes by coral trout (*Plectropomus leopardus*, Serranidae). *Coral Reefs*, **11**, 193–8.

Kingsford, M.J. (1993) Biotic and abiotic structure in the pelagic environment – importance to small fishes. *Bull. mar. Sci.*, **53**, 393–415.

Kingsford, M.J. and Choat, J.H. (1989) Horizontal distribution patterns of presettlement reef fish: are they influenced by the proximity of reefs? *Mar. Biol.*, **101**, 285–97.

Kingsford, M.J., Wolanski, E. and Choat, J.H. (1991) Influence of tidally induced fronts and langmuir circulations on distribution and movements of presettlement fishes around a coral reef. *Mar. Biol.*, **109**, 167–80.

Kinsey, D.W. (1985) Metabolism, calcification and carbon production I – Systems level studies. *Proc. 5th Int. Coral Reef Congr.*, **4**, 505–26.

Kirch, P.V., Hunt, T.L., Weisler, M., Butler, V. and Allen, M.S. (1991) Mussau islands prehistory: results of the 1985–1986 excavations. Dept Prehistory, Aust. National Univ., Canberra, Australia, *Occ. Pap. Prehist.*, **20**, 144–63.

Kirkman, H. (1992) Large-scale restoration of seagrass meadows, in *Restoring the Nation's Marine Environment*, (ed. G. Thayer), Maryland Sea Grant, College Park, MD, pp. 111–40.

Kirkwood, G.P. (1982) Simple models for multispecies fisheries, in *Theory and Management of Tropical Fisheries*, (*ICLARM Conf. Proc.* **9**) (eds D. Pauly and G.I. Murphy), ICLARM, Manila, Philippines and Division of Fisheries Research, CSIRO, Cronulla, Australia, pp. 83–98.

Kirkwood, G.P., Beddington, J.R. and Rossouw, J.A. (1994) Harvesting species of different lifespans, in *Large-scale Ecology and Conservation Biology* (*Symp. Brit. Ecol. Soc.* **35**), (eds P.J. Edwards, R. May and N.R. Webb), Blackwell Scientific Pubs, Oxford, pp. 199–227.

Kitalong, A.H. and Dalzell, P. (1994) A preliminary assessment of the status of inshore coral reef fish stocks in Palau. South Pacific Commission, Nouméa, New Caledonia, *Inshore Fish. Res. Proj. tech. Doc.* 6, 37 pp.

Kitchell, J.F. (ed.) (1992) *Food Web Management: a Case Study of Lake Mendota*, Springer Verlag, Berlin, 553 pp.

Klumpp, D.W. and McKinnon, A.D. (1989) Temporal and spatial patterns in primary production of a coral-reef epilithic algal community. *J. exp. mar. Biol. Ecol.*, **131**, 1–22.

Klumpp, D.W. and McKinnon, A.D. (1992) Community structure, biomass and productivity of epilithic algal communities on the Great Barrier Reef: dynamics at different spatial scales. *Mar. Ecol. – Progr. Ser.*, **86**, 77–89.

Klumpp, D.W. and Polunin, N.V.C. (1989) Partitioning among grazers of food resources within damselfish territories on a coral reef. *J. exp. mar. Biol. Ecol.*, **125**, 145–69.

Klumpp, D.W. and Polunin, N.V.C. (1990) Algal production, grazers and habitat partitioning on a coral reef: positive correlation between grazing rate and food availability, in *Trophic Relationships in the Marine Environment Proc. 24th Eur. Mar. Biol. Symp.*, (eds M. Barnes and R.N. Gibson), Aberdeen University, Aberdeen, pp. 372–88.

Knowlton, N. (1992) Thresholds and multiple stable states in coral-reef community dynamics. *Am. Zool.*, **32**, 674–82.

Kobayashi, D.R. (1989) Fine-scale distribution of larval fishes: patterns and processes adjacent to coral reefs in Kaneohe Bay, Hawaii. *Mar. Biol.*, **100**, 285–93.

Koenig, C.C., Coleman, F.C., Collins, L.A., Sadovy, Y. and Colin, P.L. (in press) Reproduction in gag, *Mycteroperca microlepis*, (Pisces: Serranidae) in the eastern Gulf of Mexico and the consequences of fishing spawning aggregations, in *Proc. EPOMEX Int. Snapper–Grouper Symp., Campeche, Mexico, October 1993.*

Koike, A., Takeuchi, S., Ogura, M., Kanda, K. and Arihara, C. 1968 Selection curve of the hook of the longline. *J. Tokyo Univ. Fish.*, **55**, 77–82.

Kolff, D.H. (1840) *Voyage of the Dutch Brig of War Dourga*. James Madden, London.

Kooiman, J. (ed.) (1993) *Modern Governance–Society Interactions*. Sage Publications, London, 280 pp.

Koslow, J.A., Hanley, F. and Wicklund, R. (1988) Effects of fishing on reef fish communities at Pedro Bank and Port Royal cays, Jamaica. *Mar. Ecol. – Progr. Ser.*, **43**, 201–12.

Koslow, J.A., Aiken, K., Auil, S. and Clementson, A. (1994) Catch and effort analysis of the reef fisheries of Jamaica and Belize. *Fishery Bull., U.S.*, **92**, 737–47.

Kriebel, D.J.C. (1919) Grond en Watersrechten in der onderafdeeling Saleijer. *Koloniaal Tijdschriften*, **8**, 1086–1109.

Krishnamoorthi, B. (1972) Biology of the threadfin bream *Nemipterus japonicus* (Block). *Indian J. Fish.*, **18**, 1–21.

Kulbicki, M. (1988a) Patterns in the trophic structure of fish populations across the SW lagoon of New Caledonia. *Proc. 6th Int. Coral Reef Symp.*, **2**, 89–94.

Kulbicki, M. (1988b) Correlation between catch data from bottom longlines and fish censuses in the SW lagoon of New Caledonia. *Proc. 6th Int. Coral Reef Symp.*, **2**, 305–12.

Kulbicki, M. and Grandperrin, R. (1988) Survey of the soft bottom carnivorous fish population using bottom longline in the south-west lagoon of New Caledonia. Presented at the South Pacific Commission Inshore Fisheries Resources Workshop, Nouméa, March 1988, *SPC Inshore Fish. Res./BP* 41.

Kulbicki, M. and Mou-Tham, G. (1987) Essais de pêche au casier à poissons dans le lagon de Nouvelle-Calédonie. Institut Français de Recherche Scientifique pour le Développement en Coopération, Noumea, New Caledonia. *ORSTOM Rapp. scient. tech., Sci. Mer, Biol. mar.*, **47**. 22 pp.

Kulbicki, M. and Wantiez, L. (1990) Comparison between fish bycatch from shrimp trawl net and visual census in St Vincent Bay, New Caledonia. *Fish. Bull., U.S.*, **88**, 667–75.

Kulbicki, M., Mou-Tham, G., Bargibant, G., Menou, J.-L. and Tirard, P. (1987) Résultats préliminaires des pêches expérimentales à la palangre dans le lagon sud–ouest de Nouvelle Calédonie. Institut Français de Recherche Scientifique pour le Développement en Coopération, Nouméa, New Caledonia. *ORSTOM Rapp. scient. tech., Sci. Mer, Biol. mar.*, **49**, pp. 104.

Kulbicki, M., Baillon, N., Morize, E. and Thollot, P. (1990) Campagne Corail 1 de chalutage exploratoire aux îles Chesterfield et à la Lansdowne (N.O. Alis - 15 août au 4 septembre 1988). Institut Français de Recherche Scientifique pour le Développment en Coopération, Nouméa, New Caledonia. *Rapp. scient. tech., Sci. Mer, Biol. mar.*, **56**, 448 pp.

Kulbicki, M., Mou Tham, G., Thollot, P. and Wantiez, L. (1993) Length–weight relationships of fish from the lagoon of New Caledonia. *Naga, the ICLARM Quarterly*, 2(3), 26–30.

Kunatuba, P. (n.d.) Traditional sea tenure: conservation in the South Pacific - Fiji. Unpubl. MS, Fisheries Dept, Govt of Fiji, Suva.

Künzel, T., Löwenberg, U. and Weber, W. (1983) Demersal fish resources of the Mahé Plateau/Sechelles. *Arch. Fisch. Wiss.*, **34**, 1–22.

Kunzmann, A. (1988) Commercial trials with a coral reef longline. *Jurnal Penelitian Perikanan Laut*, **45**, 33–9.

Kuo, C.L. (1988) The study of fishery biology on porgies, *Lethrinus nebulosus* (Forsskål), in waters of Australia. *Acta oceanogr. Taiwanica*, **19**, 125–31.

Lacson, J.M. (1994) Fixed allele frequency differences among Palauan and Okinawan populations of the damselfishes *Chrysipterus cyanea* and *Pomacentrus coelestis*. *Mar. Biol.*, **118**, 359–65.

Lacson, J.M. and Morizot, D.C. (1991) Temporal genetic variation in subpopulations of bicolor damselfish (*Stegastes partitus*) inhabiting coral reefs in the Florida Keys. *Mar. Biol.*, **110**, 353–57.

Lal, P., Swamy, K. and Singh, P. (1984) "Mangrove ecosystem" fisheries associated with mangroves and their management. Mangrove fishes in Wairiki Creek and their implications on the management of resources in Fiji, in *Productivity and Processes in Island Marine Ecosystems. UNESCO Rep. mar. Sci.*, **27**, 93–108.

Lal, P.N. and Slatter, C. (1982) The integration of women in fisheries development in Fiji: report of an ESCAP/FAO initiated project on improving the socio-economic condition of women in fisherfolk communities. Unpubl. rep., Fish. Divn, Min. Agric. Fish. and Centre for App. Stud. Devel., Univ. S. Pacific, Suva.

Lam, T. J. (1974) Siganids: their biology and mariculture potential. *Aquaculture*, **3**, 325–54.

Lambert, B. (1966) The economic activities of a Gilbertese chief, in *Political Anthropology* (eds M. Swartz, A. Tuden, and V.W. Turner), University of Chicago Press, Chicago, pp. 155–72.

Langham, N.P.E. and Mathias, J.A. (1977) The problems of conservation of coral reefs in Northwest Sabah. *Mar. Res. Indonesia*, **17**, 53–8.

Larkin, P.[A.] and Gazey, W. (1982) Application of ecological simulation models to management of tropical multispecies fisheries, in *Theory and Management of*

Tropical Fisheries, (*ICLARM Conf. Proc.* **9**) (eds D. Pauly and G.I. Murphy), ICLARM, Manila, Philippines, pp. 123–40.

Lasker, R. (1975) Field criteria for survival of anchovy larvae: the relation between inshore chlorophyll maximum layers and successful first feeding. *Fishery Bull., U.S.*, **73**, 453–62.

Lassig, B.R. (1982) The minor role of large transient fishes in structuring small-scale coral patch reef assemblages, PhD thesis, Macquarie University, Australia, 206 pp.

Lassuy, D.R. (1984) Diet, intestinal morphology and nitrogen assimilation efficiency in the damselfish, *Stegastes lividus*, in Guam. *Env. Biol. Fishes*, **10**, 183–93.

Law, R. (1991) On the quantitative genetics of correlated characters under directional selection in age structured populations. *Phil. Trans. R. Soc.*, **331B**, 213–23.

Law, R. and Grey, D.R. (1989) Evolution of yields from populations with age-specific cropping. *Evol. Ecol.*, **3**, 343–59.

Lawton, J.H. (1983) Plant architecture and the diversity of phytophagous insects. *A. Rev. Entomol.*, **28**, 23–39.

Laxton, J.H. (1974) Aspects of the ecology of the coral eating starfish *Acanthaster planci*. *Biol. J. Linn. Soc.*, **6**, 19–45.

Leak, J.C. and Houde, E.D. (1987) Cohort growth and survival of bay anchovy *Anchoa mitchilli* larvae in Biscayne Bay, Florida. *Mar. Ecol. – Progr. Ser.*, **37**, 109–22.

Leber, K.M., Sterrit, D.A., Arce, S.M. and Brennan, N.P. (1993) A test of marine stock enhancement concept: importance of pilot experiments to establish release protocol, in *American Fisheries Society, 123rd Annual Meeting*, Abstracts, p. 70.

LeCren, E.D., Kipling, C. and McCormack, J. (1977) A study of the numbers, biomass and yearclass strengths of perch (*Perca fluviatilis* L.) in Windermere from 1941 to 1966. *J. Anim. Ecol.*, **46**, 281–307.

Lee, C.K.C. (1974) The reproduction, growth and survival of *Upeneus moluccensis* (Bleeker) in relation to the commercial fishery in Hong Kong. *Hong Kong Fish. Bull.*, **4**, 17–32.

Lee, J.U. and Al-Baz, A.F. (1989) Assessment of fish stocks exploited by fish traps in the Arabian Gulf area. *Asian Fish. Sci.*, **2**, 213–31.

Lee, T.N., Rooth, C., Williams, E., McGowan, M., Szmant, A.F. and Clarke, M.E. (1992) Influence of Florida Current, gyres and wind-driven circulation on transport of larvae and recruitment in the Florida Keys Coral Reefs. *Cont. Shelf Res.*, **12**, 971–1002.

Leis, J.M. (1982) Nearshore distributional gradients of larval fish (15 taxa) and planktonic crustaceans (6 taxa) in Hawaii. *Mar. Biol.*, **72**, 89–97.

Leis, J.M. (1983) Coral reef fish larvae (Labridae) in the east Pacific barrier. *Copeia*, **1983**, 826–8.

Leis, J.M. (1986) Vertical and horizontal distribution of fish larvae near coral reefs at Lizard Island, Great Barrier Reef. *Mar. Biol.*, **90**, 505–16.

Leis, J.[M.] (1991a) The pelagic stage of reef fishes: the larval biology of coral reef fishes, in *The Ecology of Fishes on Coral Reefs* (ed. P.F. Sale), Academic Press, San Diego, pp. 183–230.

Leis, J. M. (1991b) Vertical distribution of fish larvae in the Great Barrier Reef Lagoon, Australia. *Mar. Biol.*, **109**, 157–66.

Leis, J.M. (1993) Larval fish assemblages near Indo–Pacific coral reefs. *Bull. mar. Sci.*, **53**, 362–92.

Leis, J.M. (1994) Coral Sea atoll lagoons: closed nurseries for the larvae of a few coral reef fishes. *Bull. mar. Sci.*, **54**, 206–27.

Leis, J.M. and Miller, J.M. (1976) Offshore distributional patterns of Hawaiian fish larvae. *Mar. Biol.*, **36**, 359–67.

Leis, J.M. and Rennis, D.S. (1983) *The Larvae of Indo–Pacific Coral Reef Fishes*, University of Hawaii Press, Honolulu, HI, 269 pp.

Leis, J.M. and Trnski, T. (1989) *The Larvae of Indo–Pacific Shore Fishes*. University of Hawaii Press, Honolulu, HI, 371 pp.

Leis, J.M., Goldman, B. and Reader, S.E. (1989) Epibenthic fish larvae in the Great Barrier Reef Lagoon near Lizard Island, Australia. *Jap. J. Ichthyol.*, **35**, 428–33.

Lemay, M.H., Ansavajitanon, S. and Hale, L.Z. (1991) A national coral reef management strategy for Thailand, in *Coastal Zone '91, Proc. Seventh Symp. Coastal and Ocean Manage., Long Beach, CA, July 8–12, 1991* (eds O.T. Magoon, H. Converse, V. Tippie, L.T. Tobin and D. Clark), American Society of Civil Engineers, New York, Vol 2. pp. 1698–1712.

Lessa, W. (1966) *Ulithi – A Micronesian Design for Living*, Holt, Rinehart and Winston, New York and London.

Lessios, H.A. (1988) Mass mortality of *Diadema antillarum* in the Caribbean: what have we learned? *A. Rev. Ecol. Syst.*, **19**, 371–93.

Lessios, H.A., Cubit, J.D., Robertson, D.R., Shulman, M.J., Parker, M.R., Garrity, S.D. and Levings, S.C. (1984a) Mass mortality of *Diadema antillarum* on the Caribbean coast of Panama. *Coral Reefs*, **3**, 173–82.

Lessios, H.A., Robertson, D.R. and Cubit, J.D. (1984b) Spread of *Diadema* mass mortality through the Caribbean. *Science*, **226**, 335–8.

Levin, L.A. (1990) A review of methods for labeling and tracking marine invertebrate larvae. *Ophelia*, **32**, 115–44.

Levitan, D.R. (1991) Influence of body size and population density on fertilization success and reproductive output in a free-spawning invertebrate. *Biol. Bull. (Woods Hole)*, **181**, 261–68.

Levitan, D.R., Sewell, M.A. and Chia, F. (1992) How distribution and abundance influence fertilization success in the sea urchin *Strongylocentrotus franciscanus*. *Ecology*, **73**, 248–54.

Lewin, R. (1986) Supply-side ecology. *Science*, **234**, 25–7.

Lewis, A.D. (1985) *Fishery Resource Profiles: Information for Development Planning*, Fish. Divn, Min. Primary Ind., Govt. Fiji, Suva, 90 pp.

Lewis, A.D., Chapman, L.B. and Sesewa, A. (1983) Biological notes on coastal pelagic fishes in Fiji. Min. Agric. Fish., Suva, Fiji, *Fish. Divn tech. Rep.* **4**, 68 pp.

Lewis, J.B. (1987) Measurements of groundwater seepage flux onto a coral reef: spatial and temporal variations. *Limnol. Oceanogr.*, **32**, 1165–9.

Lewis, R.R. III (1992) Coastal habitat restoration as a fishery management tool, in *Stemming the Tide of Coastal Fish Habitat Loss*, (ed. G. Thayer). National Coalition for Marine Conservation, Savannah, GA, pp. 169–73.

Lewis, S.M. (1986) The role of herbivorous fishes in the organization of a Caribbean reef community. *Ecol. Monogr.*, **56**, 183–200.

Lieber, M. (1968) The nature and relationships between land tenure and kinship on Kapingamaringi atoll, PhD dissertation, Dept Anthropol., University of Pittsburgh, Pittsburgh, PA.

Ligtvoet, W. and Witte, F. (1991) Perturbation through predator introduction: effects on the food web and fish yields in Lake Victoria (East Africa), in *Terrestrial and Aquatic Ecosystems: Perturbation and Recovery* (ed. O. Ravera), Ellis Horwood, New York, pp. 263–8.

Lincoln-Smith, M.P. (1989) Improving multispecies rocky fish census by counting different groups of species using different procedures. *Env. Biol. Fishes*, **26**, 29–37.

Lincoln-Smith, M.P., Bell, J.D., Pollard, D.A. and Russell, B.C. (1989) Catch and fishing effort of competition spearfishermen in southeastern Australia. *Fish. Res.*, **8**, 45–61.

Lipcius, R.N., Marshall, L.S., jun. and Cox, C. (1992) Regulation of mortality rates in juvenile queen conch. *Proc. Gulf Carib. Fish. Inst.*, **41**, 444–50.

Littler, M.M. and Littler, D.S. (1984) Models of tropical reef biogenesis, in *Progress in Phycological Research*, Vol. 3 (eds F.E. Round and D.J. Chapman), Biopress, Bristol, pp. 323–63.

Liu, H.-C. and Su, M.-S. (1971) Maturity and fecundity of yellow sea bream (*Dentex tumifrons*) in the southern area of the East China Sea and the northern area of the South China Sea. *Acta Oceanogr. Taiwanica*, **1**, 89–100.

Liu, H.-C. and Su, M.-S. (1992) Maturity and spawning of golden thread (*Nemipterus virgatus*) from the northern area of the South China Sea. *J. Fish. Soc. Taiwan*, **1**, 39–46.

Lobel, P.S. (1978) Diel, lunar, and seasonal periodicity in the reproductive behavior of the pomacanthid *Centropyge potteri*, and some other reef fishes in Hawaii. *Pac. Sci.*, **32**, 193–207.

Lobel, P.S. (1989) Ocean current variability and the spawning season of Hawaiian reef fishes. *Env. Biol. Fishes*, **24**, 161–71.

Lobel, P.S. and Johannes, R.E. (1980) Nesting, eggs, and larvae of triggerfishes (Balistidae). *Env. Biol. Fishes*, **5**, 251–2.

Lobel, P.S. and Robinson, A.R. (1986) Transport and entrapment of fish larvae by ocean mesoscale eddies and currents in Hawaiian waters. *Deep-Sea Res.*, **33A**, 483–500.

Lock, J.M. (1986a) Study of the Port Moresby artisanal reef fishery. Dept Primary Ind., Port Moresby, Papua New Guinea, *Tech. Rep. Fish. Divn*, 86/1, 56 pp.

Lock, J.M. (1986b) Fish yields of the Port Moresby barrier and fringing reefs. Dept Primary Ind., Port Moresby, Papua New Guinea, *Tech. Rep. Fish. Divn*, 86/2, 17 pp.

Lock, J.M. (1986c) Effects of fishing pressure on the fish resources of the Port Moresby barrier and fringing reefs. Dept Primary Ind., Port Moresby, Papua New Guinea, *Tech. Rep. Fish. Divn*, 86/3, 31 pp.

Lock, J.M. (1986d) Economics of the Port Moresby artisanal fishery. Dept Primary Ind., Fish. Divn, Port Moresby, Papua New Guinea. *Tech. Rep. Fish. Divn*, 86/4, 35 pp.

Lockwood, B. (1971) *Samoan Village Economy*, Oxford University Press, Melbourne.

Loeb, E.M. (1926) *History and Traditions of Niue*. Bernice P. Bishop Museum, Honolulu, HI, *Bulletin* **32**.

Lokkeborg, S. and Åsmund, B. (1992) Species and size selectivity in longline fishing: a review. *Fish. Res.*, **13**, 311–22.

Longhurst, A. (1960) Mesh selection factors in the trawl fishery off tropical West Africa. *J. Conseil, Cons. int. Explor. Mer*, **25**, 320–5.

Longhurst, A.R. and Pauly, D. (1987) *Ecology of Tropical Oceans*, Academic Press, Orlando, FL, 407 pp.

Longley, N.A. and Hildebrand, S.F. (1941) *Systematic Catalogue of the Fishes of Tortugas, Florida, with Observations on Colour, Habits and Local Distribution*, Carnegie Institute, Washington, DC, Publ. no. 535, 331 pp.

Lopez, M.D.G. (1985) Notes on traditional fisheries in the Philippines, in *The Tradi-*

tional Knowledge and Management of Coastal Systems in Asia and the Pacific (eds K. Ruddle and R.E. Johannes), UNESCO–ROSTSEA, Jakarta, pp. 190–206.

Lopez, M.D.G. (1986) An invertebrate resource survey of Lingayen Gulf, Philippines, in *North Pacific Workshop on Stock Assessment and Management of Invertebrates* (eds G.S. Jamieson and N. Bourne), *Can. spec. Publ. Fish. aquat. Sci.* **92**, 402–9.

Losanes, L.P., Matuda, K. and Fujimori, Y. (1992a) Outdoor tank experiments on the influence of soaking time on catch efficiency of gill nets and entangling nets. *Fish. Res.*, **15**, 217–28.

Losanes, L.P., Matuda, K. and Fujimori, Y. (1992b) Estimating the entangling effects of trammel and semi-trammel net selectivity on rainbow trout (*Oncorhynchus mykiss*). *Fish. Res.*, **15**, 229–42.

Losanes, L.P., Matuda, K., Machii, T. and Koike, A. (1992c) Catching efficiency and selectivity of entangling nets. *Fish. Res.*, **13**, 9–24.

Lou, D.C. (1992) Validation of annual growth bands in the otolith of tropical parrotfishes (*Scarus schlegeli* Bleeker). *J. Fish Biol.*, **41**, 775–90.

Loubens, G. (1978) Biologie de quelques espèces de poissons du lagon néo-calédonien. I. Détermination de l'age (otolithometrie). *Cah. ORSTOM, Ser. Oceanogr.*, **16**, 263–83.

Loubens, G. (1980a) Biologie de quelques espèces de poissons du lagon néo-calédonien. III. Croissance. *Cah. Indo-pacifique*, **2**, 101–53.

Loubens, G. (1980b) Biologie de quelques espèces de poissons du lagon néo-calédonien. II. Sexualité et reproduction. *Cah. Indo-pacifique*, **2**, 41–72.

Louis, M., Bouchon, C. and Bouchon–Navaro, Y. (1992) L'Ichtyofaune de mangrove dans la Baie de Fort-de-France (Martinique). *Cybium*, **16**, 291–305.

Lowry, K. and Sadacharan, D. (1993) Coastal management in Sri Lanka. *Coastal Management in Tropical Asia*, **1**, 1, 3, 5–7.

Lozano-Alvarez, E., Briones-Fourzan, P. and Negrete-Soto, F. (1993) Occurrence and seasonal variations of spiny lobsters, *Panulirus argus* (Latreille), on the shelf outside Bahia de la Ascension, Mexico. *Fishery Bull., U.S.*, **91**, 808–15.

Luchavez, T.F. and Alcala, A.C. (1988) Effects of fishing pressure on coral reef fishes in the central Philippines. *Proc. 6th Int. Coral Reef Symp.*, **2**, 251–3.

Luckhurst, B. and Ward, J. (1987) Behavioral dynamics of coral reef fishes in Antillean fish traps at Bermuda. *Proc. Gulf Carib. Fish. Inst.*, **38**, 528–48.

Luckhurst, B.E. and Luckhurst, K. (1978) Analysis of the influence of substrate variables on coral reef fish communities. *Mar. Biol.*, **49**, 317–23.

Ludwig, D., Hilborn, R. and Walters, C. (1993) Uncertainty, resource exploitation, and conservation: lessons from history. *Science*, **260**, 17–18.

Ludyanskiy, M.L., McDonald, D. and MacNeill, D. (1993) Impact of the zebra mussel, a bivalve invader. *BioScience*, **43**, 533–44.

Lutjeharms, J.R.E. and Heydorn, A.E.F. (1981a) Recruitment of rock lobster on Vema Seamount from the islands of Tristan da Cunha. *Deep-Sea Res.*, **28A**, 1237.

Lutjeharms, J.R.E. and Heydorn, A.E.F. (1981b) The rock-lobster (*Jasus tristani*) on Vema Seamount: drifting buoys suggest a possible recruiting mechanism. *Deep-Sea Res.*, **28A**, 631–6.

Lutnesky, M.M.F. (1994a) Size-dependent rate of protogynous sex change in the pomacanthid angelfish, *Centropyge potteri*, in *1994 Annual Meeting (Los Angeles), American Society Ichthyologists and Herpetologists*, p. 118 (abstract).

Lutnesky, M.M.F. (1994b) Density-dependent protogynous sex change in territorial-haremic fishes: models and evidence. *Behav. Ecol.*, **5**, 375–83.

McAllister, D.E. (1988) Environmental, economic and social costs of coral reef destruction in the Philippines. *Galaxea*, 7, 161–78.

McAllister, D.E., Schueler, F.W., Roberts, C.M. and Hawkins, J.P. (1994) Mapping and GIS analysis of the global distribution of coral reef fishes on an equal-area grid, in *Mapping the Diversity of Nature* (ed. R. Miller), Chapman & Hall, London, pp 155–75.

McArthur, N. (1981) New Hebrides population 1840–1967: a re-interpretation. South Pacific Commission, Nouméa, New Caledonia, *Occ. Pap.* no. 18.

MacArthur, R.H. and MacArthur, J.W. (1961) On bird species diversity. *Ecology*, **42**, 594–8.

Maccall, A.D. (1989) Against marine fish hatcheries: Ironies of fishery politics in the technological era. *Calif. Coop. Oceanic Fish. Invest. Rep.*, **30**, 46–8.

McClanahan, T.R. (1989) Kenyan coral reef-associated gastropod fauna: a comparison between protected and unprotected reefs. *Mar. Ecol. – Progr. Ser.*, **53**, 11–20.

McClanahan, T.R. (1990) Hierarchical control of coral reef ecosystems, PhD thesis, University of Florida, 218 pp.

McClanahan, T.R. (1992) Resource utilization, competition and predation: a model and example from coral reef grazers. *Ecol. Modelling*, **61**, 195–215.

McClanahan, T.R. (1994) Kenyan coral reef lagoon fish: effects of fishing, substrate complexity, and sea urchins. *Coral Reefs*, **13**, 231–41.

McClanahan, T.R. (in press) Fish predators and scavengers of the sea urchin *Echinometra mathaei* in Kenyan coral-reef marine parks. *Env. Biol. Fishes*, **43**, 187–93.

McClanahan, T.R. and Muthiga, N.A. (1988) Changes in Kenyan coral reef community structure and function due to exploitation. *Hydrobiologia*, **166**, 269–76.

McClanahan, T.R. and Shafir, S.H. (1990) Causes and consequences of sea urchin abundance and diversity in Kenyan coral reef lagoons. *Oecologia*, **83**, 362–70.

McCormick, M.I. and Choat, J.H. (1987) Estimating total abundance of a large temperate-reef fish using visual strip-transects. *Mar. Biol.*, **96**, 469–78.

McCormick, M.I. and Milicich, M.J. (1993) Late pelagic-stage goatfishes – distribution patterns and inferences on schooling behaviour. *J. exp. mar. Biol. Ecol.*, **174**, 15–42.

McCoy, J.L. (1980) Biology, exploitation and management of giant clams (Tridacnidae) in the Kingdom of Tonga. Fish. Divn, Min. Agric., Forestry Fish., Nuku'alofa, *Fish. Bull.*, 1.

McCracken, F.D. (1963) Selection by codend meshes and hooks on cod, haddock, flatfish and redfish, in *The Selectivity of Fishing Gear* (Special Publication 5), Int. Comm. NW Atl. Fish., pp. 131–55.

McCutcheon, M.M. (1981) Resource exploitation and the tenure of land and sea in Palau. PhD dissertation, University Microfilms, Ann Arbor, MI.

MacDonald, P.D.M. and Pitcher, T.J. (1979) Age groups from size-frequency data: a versatile and efficient method for analysing distribution mixtures. *J. Fish. Res. Bd Can.*, **36**, 987–1001.

Mace, P.M. and Sissenwine, M.P. (1993) How much spawning per recruit is enough? *Can. spec. Publ. Fish. aquat. Sci.*, **120**, 101–18.

McFarland, W.N., Ogden, J.C. and Lythgoe, J.N. (1979) The influence of light on the twilight migrations of grunts. *Env. Biol. Fishes*, **4**, 9–22.

McFarland, W.N., Brothers, E.B., Ogden, J.C., Shulman, M.J., Bermingham, E.L. and Kotchian-Prentiss, N.M. (1985) Recruitment patterns in young French grunts, *Haemulon flavolineatum* (family Haemulidae) at St. Croix, U.S.V.I. *Fishery Bull., U.S.*, **83**, 413–26.

MacGregor, G. (1937) *Ethnology of the Tokelau Islands*. Bernice P. Bishop Museum, Honolulu, HI, *Bulletin* **146**.

Macintyre, R.G., Glynn, P.W. and Cortés, J. (1992) Holocene reef history in the eastern Pacific: mainland Costa Rica, Cano Island, Cocos Island and Galapagos Islands. *Proc. 7th Int. Coral Reef Symp.*, **2**, 1174–84.

MacKenzie, D. (1993). Europe's fishing rules slip through the net. *New Scientist*, no. 1883, 7.

McManus, J.W. (1988) Coral reefs of the ASEAN Region: status and management. *Ambio*, **17**, 189–93.

McManus, J.W. (1992) How much harvest should there be? in *Resource Ecology of the Bolinao Coral Reef System* (*ICLARM Stud. Rev.* **22**) (eds J.W. McManus, C. Nañola, R. Reyes and K. Kesner), ICLARM, Manila, Philippines, pp. 52–6.

McManus, J.W. (1995) Future prospects for artificial reefs in the Philippines, in *Artificial Reefs in the Philippines*, (*ICLARM Conf. Proc.* **49**) (eds J.L. Munro and M. Balgas), ICLARM, Manila, Philippines, pp. 33–9.

McManus, J.W., Ferrer, E.M. and Campos, W.L. (1988) A village-level approach to coastal adaptive management and resource assessment (CAMRA). *Proc. 6th Int. Coral Reef Symp.*, **2**, 381–6.

McManus, J.W., Nañola, C.L., Reyes, R.B. and Kesner, K.N. (eds) (1992) *Resource Ecology of the Bolinao Coral Reef System* (*ICLARM Stud. Rev.* **22**), ICLARM, Manila, Philippines, 117 pp.

McManus, L.T. (1989) The gleaners of northwest Lingayen Gulf, Philippines. *Naga, the ICLARM Quarterly*, **12**(2), 13.

McPherson, G.R., Squire, L. and O'Brien, J. (1992) Reproduction of three dominant *Lutjanus* species of the Great Barrier Reef inter-reef fishery. *Asian Fish. Sci.*, **5**, 15–24.

Mahedevan, P.K. and Devadoss, P. (1975) A note on the fecundity and spawning period of *Drepane punctata* (Linneaus). *Indian J. Fish.*, **22**, 262–4.

Malinowski, B. (1918) Fishing in the Trobriand Islands *Man*, **18**, 87–92.

Man, A., Law, R. and Polunin, N.V.C. (1995) Role of marine reserves in recruitment to reef fisheries: a metapopulation model. *Biol. Conserv.*, **71**, 197–204.

Manickchand-Dass, S. (1987) Reproduction, age and growth of the lane snapper, *Lutjanus synagris* (Linnaeus), in Trinidad, West Indies. *Bull. mar. Sci.*, **40**, 22–8.

Manickchand-Heileman, S. and Kenny, J.S. (1990) Reproduction, age, and growth of the whitemouth croaker *Micropogonias furnieri* (Desmarest 1823) in Trinidad waters. *Fishery Bull., U.S.*, **88**, 523–9.

Mann, K.H. (1993) Physical oceanography, food chains, and fish stocks: a review. *ICES J. mar. Sci.*, **50**, 105–19.

Manooch, C.S. III (1976) Reproductive cycle, fecundity, and sex ratios of the red porgy, *Pagrus pagrus* (Pisces: Sparidae) in North Carolina. *Fishery Bull., U.S.*, **74**, 775–81.

Manooch, C.S. III (1982) Aging of reef fishes in the Southeast Fisheries Center, in *The Biological Bases for Reef Fishery Management* (eds G.R. Huntsman, W.R. Nicholson and W.W. Fox), *NOAA tech. Memo.* NMFS-SEFC-80, pp. 24–43.

Manooch, C.S. III (1987) Age and growth of snappers and groupers, in *Tropical Snappers and Groupers: Biology and Fisheries Management* (eds J.J. Polovina and S. Ralston), Westview Press, Boulder, CO, pp. 329–73.

Mapstone, B.D. and Fowler, A.J. (1988) Recruitment and the structure of assemblages of fish on coral reefs. *Trends Ecol. Evol.*, **3**, 72–7.

Maragos, J.E. (1992) Restoring coral reefs with emphasis on Pacific Reefs, in *Restoring the Nation's Marine Environment* (ed. G. Thayer), Maryland Sea Grant, College Park, MD, pp 141–221.

Marchal, E., Stequert, B., Intes, A., Cremoux, J.L. and Piton, B. (1981) Ressources pélagiques et demersals des îles Seychelles: résultats de la deuxième campagne du N/O Coriolis. Unpubl. rep., Institut Français de Recherche Scientifique pour le Développement en Coopération (ORSTOM), Paris and Ministry of Planning and Development, Seychelles.

Marshall, L.S., Lipcius, R.N. and Cox, C. (1992). Comparison of mortality rates of hatchery-reared and wild juvenile queen conch in natural habitats. *Proc. Gulf Carib. Fish. Inst.*, **41**, 445–6.

Marshall, N. (1980) Fishery yields of coral reefs and adjacent shallow water environments, in *Stock Assessment for Tropical Small Scale Fisheries* (eds S.B. Saila and P.M. Roedel). Proceedings of an International Workshop, 19–21 September 1979, University of Rhode Island, Kingston, RI, pp. 103–9.

Marshall, N. (1985) Ecological sustainable yield (fisheries potential) of coral reef areas as related to physiographic features of coral reef environments. *Proc. 5th Int. Coral Reef Cong.*, **5**, 525–30.

Marten, G.G. (1979a) Predator removal: effects on fisheries yields in Lake Victoria (East Africa). *Science*, **203**, 646–8.

Marten, G.G. (1979b) Impact of fishing on the inshore fishery of Lake Victoria (East Africa). *J. Fish. Res. Bd Can.*, **36**, 891–900.

Marten, G.G. and Polovina, J.J. (1982) A comparative study of fish yields from various tropical ecosystems, in *Theory and Management of Tropical Fisheries* (*ICLARM Conf. Proc.* **9**) (eds D. Pauly and G.I. Murphy), ICLARM, Manila, Philippines and Division of Fisheries Research, CSIRO, Cronulla, Australia, pp. 255–89.

Martin, J., Webster, J. and Edwards, G. (1992) Hatcheries and wild stocks: are they compatible? *Fisheries*, **17**, 4.

Matheson, R.H. III and Huntsman, G.R. (1984) Growth, mortality, and yield-per-recruit models for speckled hind and snowy grouper from the United States South Atlantic Bight. *Trans. Am. Fish. Soc.*, **113**, 607–16.

Mathews, C.P. (1991) Spawning stock biomass-per-recruit analysis: a timely substitute for stock recruitment analysis. *Fishbyte*, **9**(1), 7–11.

Mathews, C.P. (1993) On the preservation of data. *NAGA, The ICLARM Quarterly*, April–July, 39–41.

Mathews, C.P. and Samuel, M. (1987) Growth and mortality assessments for groupers, *Epinephelus* spp. from Kuwait. *Kuwait Bull. mar. Sci.*, **9**, 173–92.

Mathews, C.P. and Samuel, M. (1991) Growth, mortality and length–weight parameters for some Kuwaiti fish and shrimp. *Fishbyte*, **9**(2), 30–3.

Mathias, J.A. and Langham, N.P.E. (1978) Coral reefs, in *Coastal Resources of West Sabah* (eds T.E.Chua and J.A. Mathias), Universiti Sains Malaysia Press, Penang, Malaysia, pp. 117–51.

May, R.M., Beddington, J.R., Clark, C.W., Holt, S.J. and Laws, R.M. (1979) Management of multispecies fisheries. *Science*, **205**, 267–77.

Mead, M. (1969) *Social Organization of Manua*. Bernice P. Bishop Museum, Honolulu, HI, *Bulletin* **76**.

Mead, P. (1987) Deep Sea Fisheries Development Project. Report of the third visit to Tonga (6 September 1980 – 7 May 1981). South Pacific Commission, Nouméa, New Caledonia. 44 pp.

Medley, P.A., Gaudian, G. and Wells, S. (1993). Coral reef fisheries stock assessment. *Rev. Fish Biol. Fish.*, **3**, 242–85.

Meekan, M.G. (1988) Settlement and mortality patterns of juvenile reef fishes at Lizard Island northern Great Barrier Reef. *Proc. 6th Int. Coral Reef Symp.*, **2**, 779–84.

Meekan, M.G., Milicich, M.J. and Doherty, P.J. (1993) Larval production drives temporal patterns of larval supply and recruitment of a coral reef damselfish. *Mar. Ecol. – Progr. Ser.*, **93**, 217–25.

Mees, C.C. (1993) Population biology and stock assessment of *Pristipomoides filamentosus* on the Mahé Plateau, Seychelles. *J. Fish Biol.*, **43**, 695–708.

Mees, C.C., Yeeting, B.M. and Taniera, T. (1988) Small scale fisheries in the Gilbert Group of the Republic of Kiribati. Unpubl. report, Fish. Divn, Min. Nat. Resour. Devel., Tarawa, 63 pp.

Mendez Rebolledo, F. (1989) Contribucíon al estudio de la biología y la pesquería del pargo guanapo, *Lutjanus synagris* Linnaeus, 1758 (Pisces Lutjanidae), en el parque nacional Archipielago de Los Roques, Venezuela. Trabajo Especíal de Grado [thesis], Universidad Central de Venezuela, 103 pp.

Menzel, D.W. (1959) Utilization of algae for growth by the angelfish, *Holacanthus bermudensis*. *J. Conseil, Cons. int. Explor. Mer*, **29**, 308–13.

Methot, R.D., jun. (1984) Seasonal variation in survival of larval northern anchovy, *Engraulis mordax*, estimated from the age distribution of juveniles. *Fishery Bull., U.S.*, **81**, 741–50.

Meyer, J.L., Shultz, E.T. and Helfman, G.S. (1983) Fish schools: an asset to corals. *Science*, **220**, 1047–9.

Milicich, M.J., Meekan, M.G. and Doherty, P.J. (1992) Larval supply – a good predictor of recruitment of three species of reef fish (Pomacentridae). *Mar. Ecol. – Progr. Ser.*, **86**, 153–66.

Miller, D.L. (1989) Technology, territoriality and ecology: the evolution of Mexico's spiny lobster fishery, in *Common Property Resources: Ecology and Community-based Sustainable Development* (ed. F. Berkes), Valhaven Press, London, pp. 185–98.

Miller, R.J. (1975) Density of the commercial spider crab, *Chironectes opilio*, and calibration of effective area fished per trap using bottom photography. *J. Fish. Res. Bd Can.*, **32**, 761–8.

Miller, R.J. and Hunte, W. (1987) Effective area fished by Antillean traps. *Bull. mar. Sci.*, **40**, 484–93.

Minns, C.K. and Hurley, D.A. (1988) Effects of net length and set time on fish catches in gill nets. *N. Am. J. Fish. Manage.*, **8**, 216–23.

Moe, M.A., jun. (1966) Tagging fishes in Florida offshore waters. *Tech. Ser. Fla State Bd Conserv.*, **49**, 1–40.

Moe, M.A. (1969) Biology of the red grouper, *Epinephelus morio*, (Valenciennes) from the eastern Gulf of Mexico. Florida Dept. Nat. Resour. Mar. Res. Lab., *Prof. Pap. Ser.*, **10**, 1–95.

Mohan, R.S.L. (1985) A note on the changing catch trend in the traditional trap-fishery of Keelakarai and Rameswaram. *Indian J. Fish.*, **32**, 387–91.

Mokoroa, P. (1981) Traditional Cook Islands' fishing techniques. *J. Soc. Océanistes*, **37**, 267–70.

Montgomery, W.L. and Galzin, R. (1993) Seasonality in gonads, fat deposits and condition of tropical surgeonfishes (Teleostei: Acanthuridae). *Mar. Biol.*, **115**, 529–36.

Moore, C.M. and Labisky, R.F. (1984) Population parameters of a relatively unexploited stock of snowy grouper in the lower Florida Keys. *Trans. Am. Fish. Soc.*, **113**, 322–9.

Morales-Nin, B. and Ralston, S. (1990) Age and growth of *Lutjanus kasmira* (Forskål) in Hawaiian waters. *J. Fish Biol.*, **36**, 191–203.

Moran, P.J. (1990) *Acanthaster planci* (L.): biographical data. *Coral Reefs*, **9**, 95–6.

Morgan, G.R. (1974) Aspects of the population dynamics of the western rock lobster, *Panulirus cygnus* George. I: Estimation of population density: II: Seasonal changes in the catchability coefficient. *Aust. J. mar. Freshwat. Res.*, **25**, 235–59.

Morrison, D. (1988) Comparing fish and urchin grazing in shallow and deeper coral reef algal communities. *Ecology*, **69**, 1367–82.

Moyer, J.T. and Nakazono, A. (1978) Population structure, reproductive behavior, and protogynous hermaphroditism in the angelfish *Centropyge interruptus* at Miyake-jima, Japan. *Jap. J. Ichthyol.*, **25**, 25–39.

Moyer, J.T., Thresher, R.E. and Colin, P.L. (1983) Courtship, spawning and inferred social organization of American angelfishes (genera *Pomacanthus*, *Holacanthus* and *Centropyge*: Pomacanthidae). *Env. Biol. Fishes*, **9**, 25–39.

Moyle, P.B. and Cech, J.J., jun. (1988) *Fishes: an Introduction to Ichthyology*, Prentice-Hall, Englewood Cliffs, NJ.

Muehlstein, L.K. and Beets, J. (1992) Seagrass declines and their impact on fisheries. *Proc. Gulf Carib. Fish. Inst.*, **42**, 55–65.

Munbodh, M., Raymead, T.S. and Kallee, P. (1988) L'importance économique de récifs coralliens et tentatives d'aquaculture à Maurice. *Le Journal de la Nature*, **1**(1), 56–68.

Munro, A.D., Scott, A.P. and Lam, T.J. (eds) (1990) *Reproductive Seasonality in Teleosts: Environmental Influences*, CRC Press, Boca Raton, FL.

Munro, J.L. (1969) The sea fisheries of Jamaica: past, present and future. *Jamaica J.*, **3**, 16–22.

Munro, J.L. (1974) The mode of operation of Antillean fish traps and the relationship between ingress, escapement, catch and soak. *J. Conseil, Cons. int. Explor. Mer*, **35**(3), 337–50.

Munro, J.L. (1975) Assessment of the potential productivity of Jamaican fisheries. Scientific Report to [United Kingdom] Overseas Development Administration/University of the West Indies Fish Ecology Research Project 1969–1973. Part VI. 52 pp.

Munro, J.L. (1977) Actual and potential production from the coralline shelves of the Caribbean Sea. *FAO Fish. Rep.*, **200**, 301–21.

Munro, J.L. (1980). Stock assessment models: applicability and utility in tropical small-scale fisheries, in *Stock Assessment for Tropical Small-scale Fisheries* (eds S.B. Saila and P.M. Roedel), Int. Center Mar. Resour. Develop., University of Rhode Island, Kingston, RI, pp. 35–47.

Munro, J.L. (1983a) Coral reef fish and fisheries of the Caribbean Sea, in *Caribbean Coral Reef Fishery Resources*, (*ICLARM Stud. Rev.* 7) (ed. J.L. Munro), ICLARM, Manila, Philippines, pp. 1–9.

Munro, J.L. (1983b) The composition and magnitude of line catches in Jamaican waters, in *Caribbean Coral Reef Fishery Resources* (*ICLARM Stud. Rev.* 7) (ed. J.L. Munro), ICLARM, Manila, Philippines, pp. 26–32.

Munro, J.L. (1983c) The composition and magnitude of trap catches in Jamaican waters, in *Caribbean Coral Reef Fishery Resources* (*ICLARM Stud. Rev.* 7) (ed. J.L. Munro), ICLARM, Manila, Philippines, pp. 33–49.

Munro, J.L. (1983d) The biology, ecology and bionomics of the goatfishes, Mullidae, in *Caribbean Coral Reef Fishery Resources* (*ICLARM Stud. Rev.* 7) (ed. J.L. Munro), ICLARM, Manila, Philippines, pp. 142–54.

Munro, J.L. (1983e) Biological and ecological characteristics of Caribbean Reef Fishes, in *Caribbean Coral Reef Fishery Resources* (*ICLARM Stud. Rev.* 7) (ed. J.L. Munro), ICLARM, Manila, Philippines, pp. 223–31.

Munro, J.L. (1983f) Assessment of the potential productivity of Jamaican fisheries, in *Caribbean Coral Reef Fishery Resources* (*ICLARM Stud. Rev.* 7) (ed. J.L. Munro), ICLARM, Manila, Philippines, pp. 232–48.

Munro, J.L. (1983g) Epilogue: progress in coral reef fisheries research, 1973–1982, in *Caribbean Coral Reef Fishery Resources* (*ICLARM Stud. Rev.* 7) (ed. J.L. Munro), ICLARM, Manila, Philippines, pp. 249–65.

Munro, J.L. (ed.) (1983h). *Caribbean Coral Reef Fishery Resources* (*ICLARM Stud. Rev.* 7), ICLARM, Manila, Philippines, 276 pp.

Munro, J.L. (1983i) A cost-effective data acquisition system for assessment and management of tropical multispecies, multi-gear fisheries. *Fishbyte*, **1**(1), 7–12.

Munro, J.L. (1983j) Are catch and effort statistics essential? A reply. *Fishbyte*, **1**(2), 4–5.

Munro, J.L. (1984a) Yields from coral reef fisheries. *Fishbyte* **2**(3), 13–15.

Munro, J.L. (1984b) Coral reef fisheries and world fish production. *ICLARM Newsl.*, (October) 3–4.

Munro, J.L. (1986) A systems approach to stock assessment in tropical, small-scale, multispecies, multigear fisheries, in *Development and Management of Tropical Living Aquatic Resources*. Proceedings of an International Conference, 2–5 August 1983, University Pertaniau Malaysia, Serdang, Selangor, Malaysia (eds H.H. Chan, K.J. Ang, A.T. Law, M.I.b.H. Mohammed and I.b.H. Omar). Penerbit Universiti Pertanian Malaysia, Selangar, Malaysia, pp. 85–90.

Munro, J.L. and Balgos, M.C. (eds) (1995) *Artificial Reefs in the Philippines* (*ICLARM Conf. Proc.* **49**), International Center for Living Aquatic Resources Management, Manila, Philippines. 56 pp.

Munro, J.L. and Fakahau, S.T. (1993a) Management of coastal fishery resources, in *Nearshore Marine Resources of the South Pacific* (eds A. Wright and L. Hill), Forum Fisheries Agency, Honiara, Solomon Islands, pp. 55–72.

Munro, J.L. and Fakahau, S.T. (1993b) Appraisal, assessment and monitoring of small-scale coastal fisheries in the South Pacific, in *Nearshore Marine Resources of the South Pacific* (eds A. Wright and L. Hill), Forum Fisheries Agency, Honiara, Solomon Islands, pp. 15–53.

Munro, J.L. and Heslinga, G.A. (1983). Prospects for the commercial cultivation of the giant clams (Bivalvia: Tridacnidae). *Proc. Gulf Carib. Fish. Inst.*, **35**, 122–34.

Munro, J.L. and Smith, I.R (1984). Management strategies for multi-species complexes in artisanal fisheries. *Proc. Gulf Carib. Fish. Inst.*, **36**, 127–41.

Munro, J.L. and Thompson, R. (1973) The biology, ecology, exploitation and management of Caribbean reef fishes. Part II. The Jamaican fishing industry, the area investigated, the objectives and methodology of the ODA/UWI Fisheries Ecology Research Project. *Res. Rep. Zool. Dept, Univ. West Indies*, **3**, 44 pp.

Munro, J.L. and Thompson, R. (1983) The Jamaican fishing industry, in *Caribbean Coral Reef Fishery Resources* (*ICLARM Stud. Rev.* 7) (ed. J.L. Munro), ICLARM, Manila, Philippines, pp. 10–14.

Munro, J.L. and Williams, D.M. (1985) Assessment and management of coral reef fisheries: biological, environmental and socio-economic aspects. *Proc. 5th Int. Coral Reef Congr.*, **4**, 545–81.

Munro, J.L., Reeson, P.H. and Gaut, V.C. (1971) Dynamic factors affecting the performance of the Antillean fish trap. *Proc. Gulf Carib. Fish. Inst.*, **23**, 184–94.

Munro, J.L., Gaut, V.C., Thompson, R. and Reeson, P.H. (1973) The spawning seasons of Caribbean reef fishes. *J. Fish Biol.*, **5**, 69–84.

Munro, J.L., Parrish, J.D. and Talbot, F.H. (1987) The biological effects of intensive fishing upon reef fish communities, in *Human Impacts on Coral Reefs: Facts and Recommendations* (ed. B. Salvat), Antènne du Muséum Ecole Pratique des Hautes Etudes, French Polynesia, pp. 41–9.

Murai, M. (1954) Nutrition study in Micronesia. *Atoll Res. Bull.*, **27**, 1–239.

Murata, M., Miyagawa-Kohshima, K., Nakanishi, K. and Naya, Y. (1986) Characterization of compounds that induce symbiosis between sea anemone and anemonefish. *Science*, **234**, 585–7.

Murdy, E.O. and Ferraris, C.J. (1980) The contribution of coral reef fisheries to Philippine fisheries production. *ICLARM Newsl.*, **3**(1), 21–2.

Murphy, G.I. (1982) Recruitment of tropical fishes, in *Theory and Management of Tropical Fisheries* (*ICLARM Conf. Proc.* **9**) (eds D. Pauly and G.I. Murphy), ICLARM, Manila, Philippines, pp. 141–48.

Muthiga, N.A. and McClanahan, T.R. (1987) Population changes of a sea urchin (*Echinometra mathaei*) on an exploited fringing reef. *Afr. J. Ecol.*, **25**, 1–8.

Myers, R.A. and Drinkwater, K.F. (1989) Offshelf Ekman transport and larval fish survival in the Northwest Atlantic. *Biol. Oceanogr.*, **6**, 45–64.

Myers, R.A., Rosenberg, A.A., Mace, P., Barrowman, N. and Restrepo, V. (1994) In search of thresholds for recruitment overfishing. *ICES J. mar. Sci.*, **51**, 191–205.

Myrberg, A.A., jun., Montgomery, W.L. and Fishelson, L. (1988) The reproductive behavior of *Acanthurus nigrofuscus* (Forskøl) and other surgeonfishes (Fam. Acanthuridae) off Eilat, Israel (Gulf of Aqaba, Red Sea). *Ethology*, **79**, 31–61.

Nagelkerken, W.P. (1979) Biology of the graysby, *Epinephelus cruentatus*, of the coral reef of Curaçao. *Stud. Fauna Curaçao other Carib. Islands*, **186**, 118 pp.

Nakazono, A. and Kuwamura, T. (1987) *Sex Change in Fishes*, Tokai University Press, Japan.

Nason, J.D. (1971) Clan and copra: modernization on Elal Island, Eastern Caroline Islands, PhD dissertation, Dept of Anthropol, University of Washington, Seattle.

Nath, G. and Sesewa, A. (1990) Assessment of deepwater bottomfish stocks around the Fijian Republic, in *United States Agency for International Development and National Marine Fisheries Service Workshop on Tropical Fish Stock Assessment* (eds. J.J. Polovina and R.S. Shomura), *NOAA NMFS tech. Memo.* **148**, 7–28.

National Research Council (1990) *Decline of the Sea Turtles: Causes and Prevention*, National Academy Press, Washington, DC, 259 pp.

Nelson, K. and Soulé, M. (1987) Genetical conservation of exploited fishes, in *Population Genetics and Fishery Management* (eds N. Ryman, and F. Utter), University of Washington Press, Seattle and London, pp. 354–69.

Nelson, S.G. and Wilkins, S.D. (1988) Sediment processing by the surgeonfish *Ctenochaetus striatus* at Moorea, French Polynesia. *J. Fish Biol.*, **32**, 817–24.

Nichols, P.V. and Rawlinson, N.J.F. (1990) Development of the pole-and-line fishery in Solomon Islands with reference to the baitfishery and its management, in *Tuna Baitfish in the Indo–Pacific Region* (ACIAR Proc. **30**) (eds S.J.M. Blaber and J.W. Copeland), Australian Council International Agricultural Research, Canberra, pp 30–44.

Nietschmann, B. (1989) Traditional sea territories, resources and rights in Torres Strait, in *A Sea of Small Boats: Customary Law and Territoriality in the World of Inshore Fishing* (ed. J.C. Cordell), Cultural Survival, Cambridge, MA, Rep. no. 62, pp. 60–93.

Nishihira, M. and Yamazato, K. (1974) Human interference with the coral reef community and *Acanthaster* infestation in Okinawa. *Proc. 2nd Int. Coral Reef Symp.*, **1**, 577–90.

Norris, J.E. and Parrish, J.D. (1988) Predator–prey relationships among fishes in pristine coral reef communities. *Proc. 6th Int. Coral Reef Symp.*, **2**, 107–13.

Ntiba, M.J. and Jaccarini, V. (1990) Gonad maturation and spawning times of *Siganus sutor* off the Kenya coast: evidence for definite spawning seasons in a tropical fish. *J. Fish Biol.*, **37**, 315–25.

Nzioka, R.M. (1979) Observations on the spawning seasons of East African reef fishes. *J. Fish Biol.*, **14**, 329–42.

Nzioka, R.M. (1985) Aspects of the biology of the reef fish *Scolopsis bimaculatus* (Rüppell) in Kenya I. Reproduction, feeding, and length–weight relationships. *Kenya J. Sci. Technol. Ser. B*, **6**, 13–28.

Nzioka, R.M. (1990) Fish yields of Kilifi coral reef in Kenya. *Hydrobiologia*, **208**, 81–4.

Odum, E.P. (1968) The strategy of ecosystem development. *Science*, **164**, 262–70.

Odum, H.T. and Odum, E.P. (1955) Trophic structure and productivity of a windward coral reef community on Eniwetok Atoll. *Ecol. Monogr.*, **25**, 291–320.

Ofori-Danson, P.K. (1990) Reproductive ecology of the triggerfish, *Balistes capriscus* from the Ghanaian coastal waters. *Trop. Ecol.*, **31**, 1–11.

Ogden, J.C. (1977) Carbonate sediment production by parrotfish and sea-urchins on Caribbean reefs, in *Reef and Related Carbonates – Ecology and Sedimentology*, (eds S.H. Frost, M.B. Weiss and J.B. Saunders), *Am. Ass. Petrol. Geol. Stud. Geol. Ser.* **4**, pp 281–8.

Ogden, J.C. (1988) The influence of adjacent systems on the structure and function of coral reefs. *Proc. 6th Int. Coral Reef Symp.*, **1**, 123–29.

Ogden, J.C. and Gladfelter, E.H. (eds) (1983) *Coral Reefs, Seagrass Beds and Mangroves: their Interaction in the Coastal Zones of the Caribbean*. *UNESCO Rep. mar. Sci.* **23**, 133 pp.

Ogden, J.C. and Quinn, T.P. (1984) Migration in coral reef fishes: ecological significance and orientation mechanisms, in *Mechanisms of Migration in Fishes* (eds J.D. McLeave, G.P. Arnold, J.J. Dodson and W.H. Neill), Plenum, New York, pp. 293–308.

Ogden, J.C. and Zieman, J.C. (1977) Ecological aspects of coral reef–seagrass bed contacts in the Caribbean. *Proc. 3rd Int. Coral Reef Symp.*, **1**, 377–82.

Öhman, M.C., Rajasuriya, A. and Lindén, O. (1993) Human disturbances on coral reefs in Sri Lanka: a case study. *Ambio*, **22**, 474–80.

Ohtsuka, R. and Kuchikura, Y. (1984) The comparative ecology of subsistence and commercial fishing in Southwestern Japan, with special reference to maritime institutions, in *Maritime Institutions in the Western Pacific* (Senri Ethnol. Stud. **17**) (eds K. Ruddle and T. Akimichi), National Museum of Ethnology, Osaka, 121–35.

Oliver La Gorce, J. (ed.) (1939) *The Book of Fishes*, National Geographic Society, Washington, DC.

Olney, J.E. and Boehlert, G.W. (1988) Nearshore ichthyoplankton associated with seagrass beds in the lower Chesapeake Bay. *Mar. Ecol. – Progr. Ser.*, **45**, 33–43.

Olsen, D.A. and LaPlace, J.A. (1979) A study of a Virgin Islands grouper fishery based on a breeding aggregation. *Proc. Gulf Carib. Fish. Inst.*, **31**, 130–44.

Olsen, D.A., Dammann, A.E. and LaPlace, J.A. (1978) Mesh selectivity of West Indian fish traps. *Mar. Fish. Rev.*, **40**, 15–16.

Oosterhout, H. van (1987) Issues and options of child labour in the Philippines: a practical example of the "Muro-ami" fishing operation. Unpubl. rep., International Labor Organization, Manila, 53 pp.

Opitz, S. (1993) A quantitative model of the trophic interactions in a Caribbean coral reef ecosystem, in *Trophic Models of Aquatic Ecosystems* (*ICLARM Conf. Proc.* **26**) (eds V. Christensen and D. Pauly), ICLARM, Manila, Philippines, pp. 259–67.

Orbach, M.K. (1989) An overview of marine social science and fisheries management and development, in *Proc. Thirty-ninth Annual Gulf and Caribbean Fisheries Institute* (eds G.T. Waugh and M.H. Goodwin), Hamilton, Bermuda and Charleston, SC, pp. 105–12.

Ormond, R.F.G., Bradbury, R., Bainbridge, S., Fabricus, K., Keesing, J., De Vantier, L., Medley, P. and Steven, A. (1991) Test of a model of regulation of crown-of-thorns starfish by fish predators, in *Acanthaster and the Coral Reef: a Theoretical Perspective* (ed. R. Bradbury), Springer-Verlag, Berlin, pp. 189–207.

Otto, T. (n.d.) A sociological study of the baitfish areas in New Ireland and Manus Provinces. Unpubl. MS, Dept Primary Ind., Fish. Divn, Port Moresby, Papua New Guinea.

Overholtz, W.J., Marawski, S.A. and Foster, K.L. (1991) Impact of predatory fish, marine mammals and seabirds on the pelagic fish ecosystem of the northeastern USA. *ICES mar. Sci. Symp.* **193**, 198–208.

Paine, R.T. (1966) Food web complexity and species diversity. *Am. Nat.*, **100**, 65–75.

Pajot, G. and Weerasoorriya, K.T. (1980) Fishing trials with bottom-set longlines in Sri Lanka. Bay of Bengal Programme, *Working Paper* **6**. 14 pp.

Palomares, M.L. and Pauly, D. (1989) A multiple regression model for predicting the food consumption of marine fish populations. *Aust. J. mar. Freshwat. Res.*, **40**, 259–73.

Panayotou, T. and Jetanavanich, S. (1987) *The Economics and Management of Thai Marine Fisheries* (*ICLARM Stud. Rev.* **14**). ICLARM, Manila, Philippines and Winrock International Institute for Agricultural Development, Arkansas, USA, 82 pp.

Pannella, G. (1971) Fish otoliths: daily growth layers and periodical patterns. *Science*, **173**, 1124–7.

Pannella, G. (1974). Otolith growth patterns: an aid in age determination in temperate and tropical fishes, in *Proceedings of an International Symposium on the Ageing of Fish* (ed. T.B. Bagenal), Unwin, Surrey, England, pp. 28–39.

Pannella, G. (1980) Growth patterns in fish sagittae, in *Skeletal Growth of Aquatic Organisms* (eds D.C. Rhoads and R.A. Lutz), Plenum, New York, pp. 519–60.

Parrish, F.A. (1989) Identification of habitat of juvenile snappers in Hawaii. *Fishery Bull., U.S.*, **87**, 1001–5.

Parrish, J.D. (1982) Fishes at a Puerto Rico coral reef: distribution, behavior, and response to passive fishing gear. *Carib. J. Sci.*, **18**, 9–20.

Parrish, J.D. (1987) The trophic biology of snappers and groupers, in *Tropical Snappers and Groupers: Biology and Fisheries Management* (eds J.J. Polovina and S. Ralston), Westview Press, Boulder, CO, pp. 405–63.

Parrish, J.D. (1989) Fish communities of interacting shallow-water habitats in tropical oceanic regions. *Mar. Ecol. – Progr. Ser.*, **58**, 143–60.

Parrish, J.D., Callahan, M.W. and Norris, J.E. (1985) Fish trophic relationships that structure reef communities. *Proc. 5th Int. Coral Reef Cong.*, **4**, 73–8.

Parrish, J.D., Norris, J.E., Callahan, M.W., Magarifugi, E.J. and Schroeder, R.E. (1986) Piscivory in a coral reef community, in *Gutshop '81: Fish Food Habits and*

Studies (eds G.M. Caillet and C.A. Simenstad), Washington Sea Grant Publication, University of Washington, Seattle, pp. 73–8.

Patton, M.L., Grove, R.S. and Harman, R.F. (1985) What do natural reefs tell us about designing artificial reefs in southern California. *Bull. mar. Sci.*, **37**, 279–98.

Pauly, D. (1979) Theory and management of tropical multispecies stocks. A review, with emphasis on the Southeast Asian demersal fisheries. (*ICLARM Studies and Reviews* **1**) ICLARM, Manila, Philippines, 35 pp.

Pauly, D. (1980a) A new methodology for rapidly acquiring basic information on tropical fish stocks: growth, mortality and stock recruitment relationships, in *Stock Assessment for Tropical Small-scale Fisheries* (eds S.B. Saila and P.M. Roedel), Int. Center Mar. Resour. Develop., University of Rhode Island, Kingston, RI, pp. 154–72.

Pauly, D. (1980b) On the interrelationships between natural mortality, growth parameters and mean environmental temperature in 175 fish stocks. *J. Conseil, Cons. perm. int. Explor. Mer*, **39**, 175–92.

Pauly, D. (1984a) *Fish Population Dynamics in Tropical Waters: a Manual for Use with Programable Calculators* (*ICLARM Studies and Reviews* **8**), ICLARM, Manila, Philippines, 325 pp.

Pauly, D. (1984b) A mechanism for the juvenile-to-adult transition in fishes. *J. Conseil, Cons. perm. int. Explor. Mer*, **41**, 80–284.

Pauly, D. (1986) On improving operation and use of the ELEFAN programs. Part II. Improving the estimation of L_∞. *Fishbyte*, **4**(1), 18–20.

Pauly, D. (1987) A review of the ELEFAN system for analysis of length-frequency data in fish and aquatic invertebrates, in *Length-based Methods in Fisheries Research*, (*ICLARM Conf. Proc.* **13**), (eds D. Pauly and G.R. Morgan), ICLARM, Manila, Philippines, 468 pp.

Pauly, D. (1989) Food consumption by tropical and temperate fish populations: some generalizations. *J. Fish Biol.*, **35** (Suppl. A), 11–20.

Pauly, D. (1990a) Length-converted catch curves and the seasonal growth of fishes. *Fishbyte*, **8**(3), 33–8.

Pauly, D. (1990b) On Malthusian overfishing. *NAGA, the ICLARM Quarterly*, **13**(1), 3–4.

Pauly, D. (1993) Foreword, in *On the Dynamics of Exploited Fish Populations*, (facsimile reprint of 1959 book by R.J.H. Beverton and S.J. Holt). Chapman & Hall, London, pp. 1–3.

Pauly, D. (1994) *On the Sex of Fish and the Gender of Scientists: Collected Essays in Fisheries Science*, Chapman & Hall, London, 250 pp.

Pauly, D. (in press) Small-scale fisheries in the tropics: marginality, marginalization and some implications for fisheries science, in *Proceedings of the International Symposium on Global Trends in Fisheries Management*. Univ. Washington, Trends, Seattle, 14–16 June 1994.

Pauly, D. and Christensen, V. (1993) Stratified models of large marine ecosystems: a general approach and an application to the South China Sea, in *Stress, Mitigation and Sustainability of Large Marine Ecosystems* (eds K. Sherman, L.M. Alexander and B.D. Gold), American Association for the Advancement of Science (AAAS) Press, Washington DC, pp. 148–74.

Pauly, D. and Christensen, V. (1994) Modelling coral reef ecosystems, in *The Management of Coral Reef Resource Systems* (*ICLARM Conf. Proc.* **44**) (eds J.L. Munro and P.E. Munro), ICLARM, Manila, Philippines, pp. 58–60.

Pauly, D. and Chua, T.-E. (1988) The overfishing of marine resources: socio-economic background in Southeast Asia. *Ambio*, **17**, 200–206.

Pauly, D. and Froese, R. (1991) FISHBASE: assembling information on fish. *Naga, the ICLARM Quarterly*, **14**(3), 10–11.

Pauly, D. and Gaschütz, G. (1979) A simple method for fitting oscillating length growth data, with a program for pocket calculators. *ICES CM* 1979/G:24, 26 pp.

Pauly, D. and Ingles, J. (1981) Aspects of the growth and natural mortality of exploited reef fishes. *Proc. 4th Int. Coral Reef Symp.*, **1**, 89–98.

Pauly, D. and Morgan, G.R. (eds) (1987) *Length-based Methods in Fisheries Research* (*ICLARM Conf. Proc.* **13**), ICLARM, Manila, Philippines, 468 pp.

Pauly, D. and Murphy, G.I. (eds) (1982) *Theory and Management of Tropical Fisheries* (*ICLARM Conf. Proc.* **9**), ICLARM, Manila, Philippines and Division of Fisheries Research, CSIRO, Cronulla, Australia, 360 pp.

Pauly, D. and Pullin, R.S.V. (1988) Hatching time in spherical, pelagic, marine fish eggs in response to temperature and egg size. *Env. Biol. Fishes*, **22**, 261–71.

Pauly, D. and Soriano, M.L. (1986) Some practical extensions to Beverton and Holt's relative yield-per-recruit model, in *The First Asian Fisheries Forum* (eds J.L. Maclean, L.B. Dizo and L.V. Hosillos), Asian Fisheries Society, Manila, pp. 491–5.

Pauly, D., Silvestre, G. and Smith, I.R. (1989) On development, fisheries and dynamite: a brief review of tropical fisheries management. *Nat. Resour. Modeling*, **3**, 307–29.

Pearson, R.G. and Munro, J.L. (1991) Growth, mortality and recruitment rates of giant clams, *Tridacna gigas* and *T. derasa*, at Michaelmas Reef, central Great Barrier Reef, Australia. *Aust. J. mar. Freshwat. Res.*, **42**, 241–62.

Pennington, J.T. (1985) The ecology of fertilization of echinoid eggs: the consequences of sperm dilution, adult aggregations, and synchronous spawning. *Biol. Bull. (Woods Hole)*, **169**, 417–30.

Pepin, P. and Myers, R.A. (1991) Significance of egg and larval size to recruitment variability of temperate marine fish. *Can. J. Fish. aquat. Sci.*, **48**, 1820–28.

Persson, L., Diehl, S., Johansson, L., Andersson, G. and Hamrin, S.F. (1992) Trophic interactions in temperate lake ecosystems – a test of food-chain theory. *Am. Nat.*, **140**, 59–84.

Peterman, R.M., Bradford, M.J., Lo, N.C.H. and Methot, R.D. (1988) Contribution of early life stages to interannual variability in recruitment of northern anchovy (*Engraulis mordax*). *Can. J. Fish. aquat. Sci.*, **45**, 8–16.

Peters, R.H. (1983) *The Ecological Implications of Body Size*, Cambridge University Press, Cambridge.

Petersen, C.W., Warner, R.R., Cohen, S., Hess, H.C. and Sewell, A.T. (1992) Variable pelagic fertilization success – implications for mate choice and spatial patterns of mating. *Ecology*, **73**, 391–401.

Petit-Skinner, S. (1983) Traditional ownership of the sea in Oceania, in *Ocean Yearbook* 4 (eds E.M. Borgese and N. Ginsburg), University of Chicago Press, Chicago, pp. 308–18.

Phillips, B.F. (1986) Prediction of commercial catches of the western rock lobster *Panulirus cygnus*. *Can. J. Fish. aquat. Sci.*, **43**, 2126–30.

Phillips, B.F. and Macmillan, D.L. (1987) Antennal receptors in puerulus and post-puerulus stages of the rock lobster *Panulirus cygnus* (Decapoda: Palinuridae) and their potential role in puerulus navigation. *J. Crust. Biol.*, **7**, 122–35.

Piedra, G. (1969) Materials on the biology of the yellowtail snapper (*Ocyurus chrysurus* Bloch), in *Soviet–Cuban Fishery Research* (ed. A.S. Bojdanov), Israel Program for Scientific Translations, Jerusalem, pp. 251–69.

Pinkerton, E. (ed.) (1989) *Co-operative Management of Local Fisheries*, University of British Columbia Press, Vancouver, 299 pp.

Pitcher, T.J. and Hart, P.J.B. (1982) *Fisheries Ecology*, Croom Helm, London, 414 pp.

Plan Development Team (1990) The potential of marine fisheries reserves for reef management in the U.S. Southern Atlantic (ed. J.A. Bohnsack), *NOAA NMFS tech. Memo.* NMFS-SEFC-261, 40 pp.

Planes, S. (1993) Genetic differentiation in relation to restricted larval dispersal of the convict surgeonfish *Acanthurus triostegus* in French Polynesia. *Mar. Ecol. – Prog. Ser.*, **98**, 237–46.

Planes, S., Bonhomme, F. and Galzin, R. (1993a) Genetic structure of *Dascyllus aruanus* populations in French Polynesia. *Mar. Biol.*, **117**, 665–74.

Planes, S., Levefre, A., Legendre, P. and Galzin, R. (1993b) Spatio-temporal variability in fish recruitment to a coral reef (Moorea, French Polynesia). *Coral Reefs*, **12**, 105–13.

Polacheck, T. (1990) Year around closed areas as a management tool. *Natural Resource Modeling*, **4**, 327–54.

Policansky, D. (1982) Sex change in animals and plants. *A. Rev. Ecol. Syst.*, **13**, 471–95.

Pollnac, R. (1988) Evaluating the potential of fishermen's organizations in developing countries. Int. Center Mar. Resour. Develop., Univ. Rhode Island, Kingston, RI, 79 pp.

Pollnac, R.B. (1994) Research directed at developing local organizations for peoples' participation in fisheries management, in *Community Management and Common Property of Coastal Fisheries in Asia and the Pacific: Concepts, Methods and Experiences (ICLARM Conf. Proc.* **45**) (ed. R.S. Pomeroy), ICLARM, Manila, Philippines, pp. 94–106.

Pollock, B.R. (1982) Spawning period and growth of yellowfin bream, *Acanthopagrus australis* (Gunther), in Moreton Bay, Australia. *J. Fish Biol.*, **21**, 349–55.

Pollock, B.R. (1984a) Relations between migration, reproduction and nutrition in yellowfin bream *Acanthopagrus australis*. *Mar. Ecol. – Progr. Ser.*, **19**, 17–23.

Pollock, B.R. (1984b) The reproductive cycle of yellowfin bream, *Acanthopagrus australis* (Gunther), with particular reference to protandrous sex inversion. *J. Fish Biol.*, **26**, 301–11.

Pollock, N.J. (1992) Giant clams in Wallis: prospects for development, in *Giant Clams in the Sustainable Development of the South Pacific* (ed. C. Tisdell), Aust. Council Int. Agric. Res., Canberra, pp. 65–79.

Polovina, J.J. (1984) Model of a coral reef ecosystem. I. The ECOPATH model and its application to French Frigate Shoals. *Coral Reefs*, **3**, 1–11.

Polovina, J.J. (1986) A variable catchability version of the Leslie model with application to an intensive fishing experiment on a multispecies stock. *Fishery Bull., U.S.*, **84**, 423–8.

Polovina, J.J. (1989a) Should anyone build reefs? *Bull. mar. Sci.*, **44**, 1056–7.

Polovina, J.J. (1989b) A system of simultaneous dynamic production and forecast models for multispecies or multiarea applications. *Can. J. Fish. aquat. Sci.*, **46**, 961–3.

Polovina, J.J. (1991a) Evaluation of hatchery releases of juveniles to enhance rockfish stocks, with application to Pacific Ocean perch *Sebastes alutus*. *Fishery Bull., U.S.*, **89**, 129–36.

Polovina, J.J. (1991b) Fisheries applications and biological impacts of artificial reefs, in *Artificial Habitats for Marine and Freshwater Fisheries* (eds W. Seaman, jun. and L.M. Sprague), Academic Press, San Diego, pp. 153–76.

Polovina, J.J. (1994) The case of the missing lobsters. *Nat. Hist.*, **103**, 50–9.

Polovina, J.J and Ralston, S. (1986) An approach to yield assessment for unexploited resources with application to the deep slope fishes of the Marianas. *Fishery Bull., U.S.*, **84**, 759–70.

Polovina, J.J. and Ralston, S. (eds). (1987) *Tropical Snappers and Groupers: Biology, Fisheries and Management*, Westview Press, Boulder and London.

Polovina, J.J. and Sakai, I. (1989) Impacts of artificial reefs on fishery production in Shimamaki, Japan. *Bull. mar. Sci.*, **44**, 997–1003.

Polovina, J.J., Mitchum, G.T., Graham, N.E., Craig, M.P., DeMartini, E.E. and Flint, E.E. (1994) Physical and biological consequences of a climatic event in the Central Pacific. *Fish. Oceanogr.*, 3, 15–21.

Polunin, N.V.C. (1983) The marine resources of Indonesia. *Oceanogr. mar. Biol. A. Rev.*, **21**, 455–531.

Polunin, N.V.C. (1984) Do traditional marine "reserves" conserve? A view of Indonesian and Papua New Guinean evidence, in *Maritime Institutions in the Western Pacific* (Senri Ethnol. Stud. **17**) (eds K. Ruddle and T. Akimichi), National Museum of Ethnology, Osaka, pp. 267–283.

Polunin, N.V.C. (1986) Traditional marine practices in Indonesia and their bearing on conservation, in *Culture and Conservation: the Human Dimension in Environment Planning*, (eds J.A. McNeely and D. Pitts), Croom Helm, London, pp. 155–79.

Polunin, N.V.C. (1987) Primitive myth, *Nature, Lond.*, **328**, 106.

Polunin, N.V.C. (1988) Efficient uptake of algal production by a single resident herbivorous fish on the reef. *J. exp. mar. Biol. Ecol.*, **123**, 61–76.

Polunin, N.V.C. (1991) Delimiting nature: regulated area management in the coastal zone of Malesia, in *Resident Peoples and National Parks* (eds P.C. West and S.R. Brechin), University of Arizona Press, Tucson, pp. 107–113.

Polunin, N.V.C. and Brothers, E.B. (1989) Low efficiency of dietary carbon and nitrogen conversion to growth in an herbivorous coral-reef fish in the wild. *J. Fish Biol.*, **35**, 869–79.

Polunin, N.V.C. and Klumpp, D.W. (1992a) Algal food supply and grazer demand in a very productive coral-reef zone. *J. exp. mar. Biol. Ecol.*, **164**, 1–15.

Polunin, N.V.C. and Klumpp, D.W. (1992b) A trophodynamic model of fish production on a windward reef tract, in *Plant–Animal Interactions in the Marine Benthos*, (*Systematics Ass. Spec. Vol.* 46) (eds D.M. John, S.J. Hawkins and J.H. Price), Clarendon Press, Oxford, pp. 213–33.

Polunin, N.V.C. and Morton, R.D. (1992) Fecundity – predicting the population fecundity of local fish populations subject to varying fishing mortality. Unpubl. rep., Centre Trop. Coastal Manage. Stud., University of Newcastle, Newcastle, UK, 14 pp.

Polunin, N.V.C. and Roberts, C.M. (1993) Greater biomass and value of target coral-reef fishes in two small Caribbean marine reserves. *Mar. Ecol. – Progr. Ser.*, **100**, 167–76.

Pomeroy, R.S. (1989) Economic studies of tropical small scale fisheries: a discussion of methodologies, in *Proc. Thirty-ninth Annual Gulf and Caribbean Fisheries Institute* (eds G.T. Waugh and M.H. Goodwin), Hamilton, Bermuda and Charleston, SC, pp. 113–19.

Pomeroy, R.S. (1991) Small-scale fisheries management and development: towards a community-based approach. *Mar. Policy* (January) 39–48.

Pomeroy, R.S. (1992) Economic valuation: available methods, in *Integrative Framework and Methods for Coastal Area Management* (*ICLARM Conf. Proc.* **37**) (eds T.-E. Chua and L.F. Scura), ICLARM, Manila, Philippines, pp. 149–62.

Pomeroy, R.S. (ed.) (1994) *Community Management and Common Property of Coastal Fisheries in Asia and the Pacific: Concepts, Methods and Experiences* (*ICLARM Conference Proceedings* **45**). ICLARM, Manila, Philippines. 189 pp.

Pope, J.G. (1976) The effect of biological interaction on the theory of mixed fisheries. *Int. Comm. N. Atl. Fish. Sel. Pap.*, **1**, 157–62.

Pope, J.G. (1979a) Stock assessment in multispecies fisheries. South China Sea Fisheries Development and Coordinating Programme, FAO, Manila, SCS/DEV/79/19, 106 pp.

Pope, J.G. (1979b) A modified cohort analysis in which constant natural mortality is replaced by estimates of predation levels. *ICES CM* 1979/H:16, 7 pp.

Popper, D. and Gundermann, N. (1975) Some ecological and behavioural aspects of siganid populations in the Red Sea and Mediterranean coasts of Israel in relation to their suitability for aquaculture. *Aquaculture*, **6**, 127–41.

Popper, D., May, R.C. and Lichatowich, T. (1976) An experiment in rearing larval *Siganus vermiculatus* (Valenciennes) and some observations on its spawning cycle. *Aquaculture*, **7**, 281–90.

Popper, D.D., Pitt, R. and Zohar, Y. (1979) Experiments on the propagation of Red Sea siganids and some notes on their reproduction in nature. *Aquaculture*, **16**, 177–81.

Porch, C.E. III (1993) A numerical study of larval retention in the southern Straits of Florida, PhD dissertation, University of Miami, Coral Gables, FL, 245 pp.

Porter, J.W., Porter, K.G. and Batac-Catalan, Z. (1977). Quantitative sampling of Indo–Pacific demersal reef plankton. *Proc. 3rd Int. Coral Reef Symp.*, **1**, 105–112.

Posada, J.M. and Appeldoorn, R.S. (in press) The validity of length-based methods for estimating growth and mortality of groupers, as illustrated by a comparative assessment of the creole fish, *Paranthias furcifer* (Valenciennes) (Pisces, Serranidae), in *Proceedings of the International Workshop on Tropical Groupers and Snappers, Campeche, Mexico, 1993*.

Potts, D.C. (1977) Suppression of coral populations by filamentous algae within damselfish territories. *J. exp. mar. Biol. Ecol.*, **28**, 207–16.

Potvin, J. (rapp.) (1993) *Ecological Economics: Emergence of a New Development Paradigm*. Proc. Workshop, Nov. 7–10, 1992, Inst. Res. Environment and Economy, University of Ottawa, Ottawa, Canada, 167 pp.

Powell, D.G. (1979) Estimation of mortality and growth parameters from the length frequency of a catch. *Rapp. P.-v. Réun. Cons. perm. int. Explor. Mer*, **175**, 167–9.

Powers, D.A., Allendorf, F.W., and Chen, T. (1990) Application of molecular techniques to the study of marine recruitment problems, in *Large Marine Ecosystems: Patterns, Processes, and Yields* (eds K. Sherman, L.M. Alexander and B.D. Gold), American Association for the Advancement of Science Press, Washington, DC, pp. 104–21.

Prager, M.H., O'Brien, J.F. and Saila, S.B. (1987) Using lifetime fecundity to compare management strategies: a case history for striped bass. *N. Am. J. Fish. Manage.*, 7, 403–9.

Prince, J.D., Sluczanowski, P.R. and Tonkin, J. (1991) *AbaSim, a Graphic Fishery*. Fish Insight, South Australian Department of Fisheries, GPO Box 1625, Adelaide, South Australia. 34 pp.

Pulea, M. (1985) Customary law relating to the environment, South Pacific region 'An Overview'. Unpubl. rep., South Pacific Commission Environment Programme, Nouméa, New Caledonia.

Pulliam, H.R. (1988) Sources, sinks, and population regulation. *Am. Nat.*, **132**, 652–61.

Pyle, R.L. (1992) Marine aquarium fish. Forum Fisheries Agency Report 92/55, 29 pp.

Pyle, R.L. (1993) Marine aquarium fish, in *Nearshore Marine Resources of the South Pacific* (eds A. Wright and L. Hill). Forum Fisheries Agency, Honiara, Solomon Islands,, pp. 135–76.

Quinn, N. and Kojis, B.L. (1985) Does the presence of coral reefs in proximity to a tropical estuary affect the estuarine fish assemblage. *Proc. 5th Int. Coral Reef Congr.*, **5**, 445–50.

Quinn, N.J., Kojis, B. and Warhepa, P. (eds) (1985) *Subsistence Fishing Practices of Papua New Guinea*. University of Technology, Lae, Papua New Guinea: Approp. Technol. Develop. Inst., *Trad. Technol. Ser.*, **2**, 135 pp.

Radtke, R.L. (1988) Recruitment parameters resolved from structural and chemical components of juvenile *Dascyllus albisella* otoliths. *Proc. 6th Int. Coral Reef Symp.*, **2**, 821–6.

Radtke, R.L. (1990) Information storage capacity of otoliths: response to Neilsen and Campana. *Can. J. Fish. aquat. Sci.*, **47**, 2463–7.

Raj, D. (1968) *Sampling Theory*. McGraw-Hill, New York.

Rallu, J.-L. (1990) Les populations océaniennes aux XIXe et XXe siècles. Institut National d'Etudes Démographiques, Presses Universitaires de France, *Travaux et Documents Cahier* **128**, 348 pp.

Ralston, S. (1976) Age determination of a tropical butterflyfish utilizing daily growth rings of otoliths. *Fishery Bull., U.S.*, **74**, 990–94.

Ralston, S. (1982) Influence of hook size in the Hawaiian deep-sea hand-line fishery. *Can. J. Fish. aquat. Sci.*, **39**, 1297–1302.

Ralston, S. (1985) A novel approach to ageing tropical fish. *ICLARM Newsl.*, **8**(1), 14–15.

Ralston, S. (1987) Mortality rates of snappers and groupers, in *Tropical Snappers and Groupers: Biology and Fisheries Management* (eds J.J. Polovina and S. Ralston), Westview Press, Boulder, CO, pp. 375–404.

Ralston, S. (1989) Effect of seasonal recruitment on bias of the Beverton–Holt length-based mortality estimator. *Am. Fish. Soc. Symp.*, **6**, 190–97.

Ralston, S. (1990) Size selection of snappers (Lutjanidae) by hook and line gear. *Can. J. Fish. aquat. Sci.*, **47**, 696–700.

Ralston, S. and Polovina, J.J. (1982) A multispecies analysis of the commercial deep-sea handline fishery in Hawaii. *Fishery Bull., U.S.*, **80**, 435–48.

Ralston, S., Gooding, R.M. and Ludwig, G.M. (1986) An ecological survey and comparison of bottom fish resource assessments (submersible versus hand-line fishing) at Johnson Atoll. *Fishery Bull., U.S.*, **84**, 141–55.

Ramos-Espla, A.A. and Bayle-Sempere, J. (1990) Management of living resources in the marine reserve of Tabarca Island (Alicante, Spain). *Bull. Société zool. de France*, **114**, 41–48.

Randall, J.E. (1961) Observations on the spawning of surgeonfishes (Acanthuridae) in the Society Islands. *Copeia*, **1961**, 237–8.

Randall, J.E. (1962) Tagging reef fishes in the Virgin Islands. *Proc. Gulf Carib. Fish. Inst.*, **14**, 201–41.

Randall, J.E. (1963a) An analysis of the fish populations of artificial and natural reefs in the Virgin Islands. *Carib. J. Sci.*, **3**, 31–47.

Randall, J.E. (1963b) Additional recoveries of tagged reef fishes from the Virgin Islands. *Proc. Gulf Carib. Fish. Inst.*, **15**, 155–7.

Randall, J.E. (1967). Food habits of reef fishes in the West Indies. *Stud. Trop. Oceanogr., Univ. of Miami*, **5**, 655–847.

Randall, J.E. (1974) The effects of fishes on coral reefs. *Proc. 2nd Int. Coral Reef Symp.*, **1**, 159–65.

Randall, J.E. (1987) Collecting reef fishes for aquaria, in *Human Impacts on Coral Reefs: Facts and Recommendations* (ed. B. Salvat), Antènne Muséum E.P.H.E., French Polynesia, pp. 29–39.

Randall, J.E., and Brock, V.E. (1960) Observations on the ecology of epinepheline and lutjanid fishes of the Society Islands, with emphasis on food habits. *Trans. Am. Fish. Soc.*, **89**, 9–16.

Randall, J.E. and Randall, H.A. (1963) The spawning and early development of the Atlantic parrot fish, *Sparisoma rubripinne*, with notes on other scarid and labrid fishes. *Zoologica*, **48**, 49–60.

Rangarajan, K. (1971) Maturity and spawning of the snapper, *Lutjanus kasmira* (Forskål) from the Andaman Sea. *Indian J. Fish.*, **18**, 114–25.

Ray, G.C. (1976) Critical marine habitats, in *Proceedings of an International Conference on Marine Parks and Reserves held in Tokyo, Japan, 1974*, IUCN, Morges, Switzerland, *IUCN Publ. New Ser.* **37**, pp. 15–64.

Ray, G.C. and McCormick-Ray, M.G. (1992) Marine and estuarine protected areas: a strategy for a national representative system within Australian coastal and marine environments. Unpubl. rep., Aust. National Parks Wildl. Ser., Canberra, Australia, 52 pp.

Recksiek, C.W., Appeldoorn, R.S. and Turingan, R.G. (1991) Studies of fish traps as stock assessment devices on a shallow reef in south-western Puerto Rico. *Fish. Res.*, **10**, 177–97.

Reeson, P.H. (1983) The biology, ecology and bionomics of the surgeonfishes, Acanthuridae, in *Caribbean Coral Reef Fishery Resources* (*ICLARM Stud. Rev.* 7) (ed. J.L. Munro), ICLARM, Manila, Philippines, pp. 178–90.

Reinthal, P.N., Kensley, B. and Lewis, S.M. (1984) Dietary shifts in the Queen Triggerfish *Balistes vetula* in the absence of its primary food item *Diadema antillarum*. *Mar. Ecol.*, **5**, 191–5.

Renard, Y. (1991) Institutional challenges for community-based management in the Caribbean. *Nature and Resources*, **27**, 4–9.

Render, J.H. and Wilson, C.A. (1992) Reproductive biology of sheepshead in the northern Gulf of Mexico. *Trans. Am. Fish. Soc.*, **121**, 757–64.

Reshetnikov, Y.S. and Claro, R.M. (1976) Cycles of biological processes on tropical fishes with reference to *Lutjanus synagris*. *J. Ichthyol.*, **16**, 711–21.

Richards, A. (1993) Live reef fish exports to South-east Asia from the South Pacific. South Pacific Commission, Nouméa, New Caledonia, *Fish. Newsl.*, **67**, 34–6.

Richards, D.V. and Davis, G.E. (1993) Early warnings of modern population collapse in black analone *Haliotis cracherodii*, Leach, 1814 at the California Channel Islands. *J. Shellfish Res.*, **12**, 189–94.

Richards, W.J. and Edwards, R.E. (1986) Stocking marine species to restore or enhance fisheries, in *Fish Culture in Fisheries Management* (ed. R.H. Stroud), Am. Fish. Soc., Bethesda, MD, pp. 75–80.

Ricker, W.E. (1954) Stock and recruitment. *J. Fish. Res. Bd Can.*, **11**, 555–623.

Ricker, W.E. (1975) *Computation and Interpretation of Biological Statistics of Fish Populations. Bull. Fish. Res. Bd Can.* **191**, 382 pp.

Rigney, H. (1990) Marine reserves – blueprint for protection. *Aust. Fish.*, **49**, 18–22.

Risk, M.J. (1972) Fish diversity on a coral reef in the Virgin Islands. *Atoll Res. Bull.*, **193**, 1–6.

Roberts, C.M. (1986) Aspects of coral reef fish community structure in the Red Sea and on the Great Barrier Reef, PhD thesis, University of York, York, UK, 222 pp.

Roberts, C.M. (1991) Larval mortality and the composition of coral reef fish communities. *Trends Ecol. Evol.*, **6**, 83–7.

Roberts, C.M. (1995) Rapid buildup of fish biomass in a Caribbean marine reserve. *Conserv. Biol.*, **9**, 815–26.

Roberts, C.M. and Ormond, R.F.G. (1987) Habitat complexity and coral reef fish diversity and abundance on Red Sea fringing reefs. *Mar. Ecol. – Progr. Ser.*, **41**, 1–8.

Roberts, C.M. and Polunin, N.V.C. (1991) Are marine reserves effective in management of reef fisheries? *Rev. Fish Biol. Fish.*, **1**, 65–91.

Roberts, C.M. and Polunin, N.V.C. (1992) Effects of marine reserve protection on northern Red Sea fish populations. *Proc. 7th Int. Coral Reef Symp.*, **2**, 969–77.

Roberts, C.M. and Polunin, N.V.C. (1993) Marine reserves: simple solutions to managing complex fisheries? *Ambio*, **22**, 363–68.

Roberts, C.M. and Polunin, N.V.C. (1994) Hol Chan: demonstrating that marine reserves can be remarkably effective. *Coral Reefs*, **13**, 90.

Roberts, C.M., Dawson Shepherd, A.R. and Ormond, R.F.G. (1992) Large-scale variation in assemblage structure of Red Sea butterflyfishes and angelfishes. *J. Biogeogr.*, **19**, 239–50.

Roberts, C.M., Quinn, N., Tucker, Jr., J.W. and Woodward P.N. (1995) Introduction of hatchery-reared Nassau groupers to a coral reef environment. *N. Am. J. Fish. Manage.*, **15**, 159–64.

Roberts, T.W. (1986) Abundance and distribution of pink shrimp in and around the Tortugas sanctuary, 1981–1983. *N. Am. J. Fish. Manage.*, **6**, 311–27.

Robertson, A.I. and Duke, N.C. (1987) Mangroves as nursery sites: comparisons of the abundance and species composition of fish and crustaceans in mangroves and other nearshore habitats in tropical Australia. *Mar. Biol.*, **96**, 193–205.

Robertson, D.R. (1982) Fish feces as fish food on a Pacific coral reef. *Mar. Ecol. – Progr. Ser.*, **7**, 253–65.

Robertson, D.R. (1983) On the spawning behaviour and spawning cycles of eight surgeonfishes (Acanthuridae) from the Indo–Pacific. *Env. Biol. Fishes*, **9**, 193–223.

Robertson, D.R. (1984) Cohabitation of competing territorial damselfishes on a Caribbean coral reef. *Ecology*, **65**, 1121–35.

Robertson, D.R. (1985) Sexual size dimorphism in surgeon fishes. *Proc. 5th Int. Coral Reef Congr.*, **5**, 403–8.

Robertson, D.R. (1988) Extreme variation in settlement of the Caribbean triggerfish *Balistes vetula* in Panama. *Copeia*, **1988**, 698–703.

Robertson, D.R. (1990) Differences in the seasonalities of spawning and recruitment of some small neotropical reef fishes. *J. exp. mar. Biol. Ecol.*, **144**, 49–62.

Robertson, D.R. (1991a) The role of adult biology in the timing of spawning of tropical reef fishes, in *The Ecology of Fishes on Coral Reefs* (ed. P.F. Sale), Academic Press, San Diego, pp. 356–86.

Robertson, D.R. (1991b) Increases in surgeonfish populations after mass mortality of the sea urchin *Diadema antillarum* in Panama indicate food limitation. *Mar. Biol.*, **111**, 437–44.

Robertson, D.R. and Hoffman, S.G. (1977) The roles of female mate choice and predation in the mating systems of some tropical labroid fishes. *Z. Tierpsychol.*, **45**, 298–320.

Robertson, D.R. and Warner, R.R. (1978) Sexual patterns in the labroid fishes of the western Caribbean, II: The parrotfishes (Scaridae). *Smithsonian Cont. Zool.*, **255**, 1–26.

Robertson, D.R., Polunin, N.V.C. and Leighton, K. (1979) The behavioral ecology of three Indian Ocean surgeonfishes (*Acanthurus lineatus, A. leucosternon* and *Zebrasoma scopas*): their feeding strategies, and social and mating systems. *Env. Biol. Fishes*, **4**, 125–70.

Robertson, D.R., Green, D.G. and Victor, B.C. (1988) Temporal coupling of spawning and recruitment of larvae of a Caribbean reef fish. *Ecology*, **69**, 370–81.

Robertson, D.R., Schober, U.M. and Brawn, J.D. (1993) Comparative variation in spawning output and juvenile recruitment of some Caribbean reef fishes. *Mar. Ecol. – Progr. Ser.*, **94**, 105–13.

Rochers, K.D. (1992) Women's fishing on Kosrae: a description of past and present methods. *Micronesica*, **25**, 1–22.

Roman Cordero, A.M. (1991) Estudio sobre la dinamica reproductiva de la cachicata blanca *Haemulon plumieri* (Lacépède, 1802) (Pisces: Pomadasyidae). MSc thesis, University of Puerto Rico, Mayagüez, Puerto Rico, 120 pp.

Rooker, J.R. (1991) Ontogenetic patterns in the feeding ecology of the schoolmaster snapper (*Lutjanus apodus*). MSc thesis, University of Puerto Rico, Mayagüez, Puerto Rico, 63 pp. Now published as Rooker, J.R. (1994), q.v.

Rooker, J.R. (1995) Feeding ecology of the schoolmaster snapper, *Lutjanus apodus* (Walbaum), from southwestern Puerto Rico. *Bull. mar. Sci.*, **56**, 881–94.

Rooker, J.R. and Recksiek, C. (1992) The effects of training with fish models in estimating lengths of fish underwater. *Proc. Gulf Carib. Fish. Inst.*, **41**, 321–31.

Rosenberg, A.A., Fogarty, M.J., Sissenwine, M.P., Beddington, J.R. and Shepherd, J.G. (1993) Achieving sustainable use of renewable resources. *Science*, **262**, 828–9.

Ross, R.M. (1987) Sex change linked to growth acceleration in a coral-reef fish, *Thalassoma duperrey*. *J. exp. Zool.*, **244**, 455–61.

Ross, R.M. (1990) The evolution of sex change mechanisms in fishes. *Env. Biol. Fishes*, **29**, 81–93.

Ross, R.M., Losey, G.S. and Diamond, M. (1983) Sex change in a coral reef fish: dependence of stimulation and inhibition on relative size. *Science*, **221**, 574–5.

Rothschild, B.J. (1986) *Dynamics of Marine Fish Populations*, Harvard University Press, Cambridge, MA.

Rothschild, B.J. and Ault, J.S. (1991) Statistical sampling design analysis of the Puerto Rico fishery-independent survey and design recommendations for a long-term reef fish monitoring program in Puerto Rico and the U.S. Virgin Islands. SEAMAP-Caribbean Report, Caribbean Fishery Management Council, Hato Rey, Puerto Rico, 28 pp.

Rougerie, F. and Wauthy, B. (1986) Le concept d'endo-upwelling dans le fonctionnement des atolls-oasis. *Oceanol. Acta*, **9**, 133–48.

Rowley, R.J. (1992) Impacts of marine reserves on fisheries: a report and review of the literature. NZ Dept Conserv., *Sci. Res. Ser.*, **51**, 1–50.

Rubec, P.J. (1986) The effects of sodium cyanide on coral reefs and marine fish in the Philippines, in *The First Asian Fisheries Forum* (eds J.L McLean, L.B. Dizon and L.V. Hosillos), Asian Fisheries Society, Manila, Philippines, pp. 297–302.

Ruddle, K. (1985) The continuity of traditional management practices: the case of Japanese coastal fisheries, in *The Traditional Knowledge and Management of Coastal Systems in Asia and the Pacific* (eds K. Ruddle and R.E. Johannes), UNESCO–ROSTSEA, Jakarta, pp. 157–79.

Ruddle, K. (1987a) The management of coral reef fish resources in the Yaeyama Archipelago, southwestern Okinawa. *Galaxea*, **6**, 209–35.

Ruddle, K. (1987b) Administration and conflict management in Japanese coastal fisheries. *FAO Fish. tech. Pap.*, **273**.

Ruddle, K. (1988a) Social principles underlying traditional inshore fisheries management systems in the Pacific basin. *Mar. Resour. Econ.*, **5**, 351–363.

Ruddle, K. (1988b) A framework for research on the traditional knowledge and management of coastal systems, with particular reference to coral reef fisheries. *Galaxea*, **7**, 179–84.

Ruddle, K. (1993) External forces and change in traditional community-based fishery management systems in the Asia–Pacific region. *Marit. Anthropol. Stud.*, **6**(1–2), 1–37.

Ruddle, K. (1994a) Traditional community-based marine resource management systems in the Asia–Pacific region. Status and potential. Unpubl. MS [refer to author].

Ruddle, K. (1994b) A guide to the literature on traditional community-based fisheries management systems in the tropics of the Asia–Pacific region. FAO, Rome, FIPP/C869; *FAO Fish. Circ.* **869**, 114 pp.

Ruddle, K. (1994c) Local knowledge in the folk management of fisheries and coastal-marine environments, in *Folk Management in the World Fisheries* (eds C.L. Dyer and J.R. McGoodwin), University of Colorado Press, Boulder, CO, pp. 161–206.

Ruddle, K. (in press) Local knowledge in the future management of inshore tropical marine resources and environments. *Nature and Resources*, **29**(4).

Ruddle, K. and Akimichi, T. (1984) Introduction, in *Maritime Institutions in the Western Pacific* (*Senri Ethnol. Stud.* **17**) (eds K. Ruddle and T. Akimichi), National Museum of Ethnology, Osaka, 1–10.

Ruddle, K. and Akimichi, T. (1989) Sea tenure in Japan and the southwestern Ryukyus, in *A Sea of Small Boats: Customary Law and Territoriality in the World of Inshore Fishing* (ed. J.C. Cordell), Cultural Survival, Cambridge, MA, Rep. no. 62, pp. 337–70.

Ruddle, K. and Johannes, R.E. (eds) (1985) *The Traditional Knowledge and Management of Coastal Systems in Asia and the Pacific*, UNESCO–ROSTSEA, Jakarta.

Ruddle, K. and Johannes, R.E. (eds) (1990) *Traditional Marine Resource Management in the Pacific Basin: an Anthology*, UNESCO–ROSTSEA, Jakarta.

Ruddle, K., Hviding, E. and Johannes, R.E. (1992) Marine resources management in the context of customary tenure. *Mar. Resour. Econ.*, **7**, 249–73.

Russ, G.R. (1984a) Distribution and abundance of herbivorous grazing fishes in the central Great Barrier Reef. II. Patterns of zonation of mid-shelf and outershelf reefs. *Mar. Ecol. – Progr. Ser.*, **20**, 35–44.

Russ, G.R. (1984b) Distribution and abundance of herbivorous grazing fishes in the central Great Barrier Reef. I. Levels of variability across the entire continental shelf. *Mar. Ecol. – Progr. Ser.*, **20**, 23–34.

Russ, G.[R.] (1984c) A review of coral reef fisheries. *UNESCO Rep. mar. Sci.*, **27**, 74–92.

Russ, G.[R.] (1985) Effects of protective management on coral reef fishes in the central Philippines. *Proc. 5th Int. Coral Reef Congr.*, **4**, 219–24.

Russ, G.R. (1987) Is rate of removal of algae by grazers reduced inside territories of tropical damselfishes? *J. exp. mar. Biol. Ecol.*, **110**, 1–17.

Russ, G.R. (1991) Coral reef fisheries: effects and yields, in *The Ecology of Fishes on Coral Reefs* (ed. P.F. Sale), Academic Press, San Diego, pp. 601–35.

Russ, G.R. and Alcala, A.C. (1989) Effects of intense fishing pressure on an assemblage of coral reef fishes. *Mar. Ecol. – Progr. Ser.*, **56**, 13–27.

Russ, G.R. and Alcala, A.C. (1994) Marine reserves: they enhance fisheries, reduce conflicts and protect resources. *Naga, The ICLARM Quarterly*, **17**(3), 4–7.

Russo, A.R. (1980) Bioerosion by two rock boring echinoids (*Echinometra mathaei* and *Echinometra aciculatus*) on Enewetak Atoll, Marshall Islands. *J. mar. Res.*, **38**, 99–110.

Rutledge, W.P. (1989) The Texas marine hatchery program – it works! *Calif. Coop. Oceanic Fish. Invest. Rep.*, **30**, 49–52.

Ruttley, H.L. (1987) Revision of the fisheries legislation in Solomon Islands: analysis of replies to a questionnaire on customary fishing rights in the Solomon Islands. Unpubl. rep., Fisheries Law Advisory Programme, W. Pacific and S. China Region, FAO, Rome.

Ryder, R.A. (1965) A method for estimating the potential fish production of north-temperate lakes. *Trans. Am. Fish. Soc.*, **94**, 214–18.

Sadovy, Y. (1989) Caribbean fisheries: problems and prospects. *Prog. Underwater Sci.*, **13**, 169–84.

Sadovy, Y. (1993a) The Nassau grouper, endangered or just unlucky? *Reef Encounter*, **13**, 10–12.

Sadovy, Y. (1993b) Spawning stock biomass per recruit: *Epinephelus guttatus* (Puerto Rico). Unpubl. rep., Caribbean Fishery Management Council, Hato Rey, Puerto Rico, 12 pp.

Sadovy, Y. (1994) Grouper stocks of the western central Atlantic: the need for management and management needs. *Proc. Gulf Carib. Fish. Inst.*, **43**, 43–64.

Sadovy, Y. (in press) The case of the disappearing grouper: *Epinephelus striatus*, the Nassau grouper in the Caribbean and western Atlantic. *Proc. Gulf Carib. Fish. Inst.*, **45**.

Sadovy, Y. and Colin, P.L. (1995) Sexual development and sexuality in the Nassau grouper, *Epinephelus striatus* (Pisces, Serranidae). *J. Fish Biol.*, **46**, 961–76.

Sadovy, Y. and Figuerola, M. (1992) The status of the red hind fishery in Puerto Rico and St. Thomas, as determined by yield-per-recruit analysis. *Proc. Gulf Carib. Fish. Inst.*, **42**(A), 23–38.

Sadovy, Y. and Shapiro, D.Y. (1987) Criteria for the diagnosis of hermaphroditism in fishes. *Copeia*, **1987**, 136–56.

Sadovy, Y., Figuerola, M. and Román, A. (1992) Age and growth of red hind *Epinephelus guttatus* in Puerto Rico and St Thomas. *Fishery Bull., U.S.*, **90**, 516–28.

Sadovy, Y., Colin, P.L. and Domeier, M. (1994b) Aggregations and spawning in the tiger grouper, *Mycteroperca tigris* (Pisces: Serranidae). *Copeia*, **1994**, 511–16.

Sadovy, Y., Rosario, A. and Román, A. (1994a) Reproduction in an aggregating grouper, the red hind, *Epinephelus guttatus*. *Env. Biol. Fishes*, **41**, 269–86.

Saetersdal, G. (1963) Selectivity of longlines, in *The Selectivity of Fishing Gear*. Int. Committee NW Atl. Fish. *Spec. Publ.* **5**, 189–92.

Saetre, R. and Paula e Silva, R. de (1979) The marine fish resources of Mozambique. Unpubl. reports on surveys with the R/V Dr Fridjof Nansen, Serviço de Investigaçoes Pesqueiras, Maputo, Mazambique and Institute of Marine Research, Bergen, Norway.

Safert, E.G. (1919) Kosrae, in *Results of the 1908–1910 South Seas Expedition*, Vol. 2: *Ethnography* (ed. G. Thilenius), Friederichsen, Hamburg.

SAFMC (1991) Amendment Number 4, Regulatory impact review, initial regulatory flexibility analysis and environmental assessment for the fishery management plan for the snapper grouper fishery of the south Atlantic region. South Atlantic Fishery Management Council, Charleston, SC.

Sahlins, M. (1962) *Moala. Culture and Nature on a Fijian Island.* University of Michigan Press, Ann Arbor.

Saila, S.B. (1982) Markov models in fish community studies – some basic concepts and suggested applications, in *The Biological Bases for Reef Fishery Management* (eds G.R. Huntsman, W.R. Nicholson and W.W. Fox, jun.), *NOAA NMFS tech. Memo.* NMFS-SEFC-80, pp. 202–9.

Saila, S.B. (1992) Application of fuzzy graph theory to successional analysis of a multispecies trawl fishery. *Trans. Am. Fish. Soc.*, **121**, 211–33.

Saila, S.B. and Erzini, K. (1987) Empirical approach to multispecies stock assessment. *Trans. Am. Fish. Soc.*, **116**, 601–11.

Saila, S.B. and Roedel, P.M. (eds) (1980) *Stock Assessment for Tropical Small-scale Fisheries*, Int. Center Mar. Resour. Develop., University of Rhode Island, Kingston, RI.

Saila, S.B., Recksiek, C.W. and Prager, M.H. (1988) *Basic Fishery Science Programs*, Elsevier, Amsterdam.

Saila, S.B., Kocic, V.L. and McManus, J.W. (1993) Modeling the effects of destructive fishing practices on tropical coral reefs. *Mar. Ecol. – Progr. Ser.*, **94**, 51–60.

Sainsbury, K.J. (1984) Optimal mesh size for tropical multispecies fisheries. *J. Conseil, Cons. perm. int. Explor. Mer*, **41**, 129–39.

Sainsbury, K.J. (1987) Assessment and management of the demersal fishery on the continental shelf of northwestern Australia, in *Tropical Snappers and Groupers: Biology, Fisheries and Management* (eds J.J. Polovina and S. Ralston), Westview Press, Boulder and London, pp. 465–504.

Saito, K. (1984) Ocean ranching of abalones and scallops in northern Japan. *Aquaculture*, **39**, 361–73.

Salazar-Ruíz, A. and Sánchez-Chávez, J.A. (1992) Aspectos biológico pesqueros de mero (*Epinephelus morio*) de la flota artesanal de las costas de Yucatan, Mexico. *Proc. Gulf Carib. Fish. Inst.*, **41**, 422–30.

Sale, P.F. (1970) Distribution of larval Acanthuridae off Hawaii. *Copeia*, **1970**, 765–6.

Sale, P.F. (1974) Mechanisms of co-existence in a guild of territorial fishes at Heron Island. *Proc. 2nd Int. Coral Reef Symp.*, 1, 193–206.

Sale, P.F. (1977) Maintenance of high diversity in coral reef fish communities. *Am. Nat.*, **111**, 337–59.

Sale, P.F. (1978) Coexistence of coral reef fishes – a lottery for living space. *Env. Biol. Fishes*, **3**, 85–101.

Sale, P.F. (1980) The ecology of fishes on coral reefs. *Oceanogr. mar. Biol. A. Rev.*, **18**, 367–421.

Sale, P.F. (1985) Patterns of recruitment in coral reef fishes. *Proc. 5th Int. Coral Reef Congr.*, **5**, 391–96.

Sale, P.F. (ed.) (1991a) *The Ecology of Fishes on Coral Reefs*, Academic Press, San Diego, 754 pp.

Sale, P.F. and Douglas, W.A. (1981) Precision and accuracy of visual census technique for fish assemblages on coral patch reefs. *Env. Biol. Fishes*, **6**, 333–9.

Sale, P.F. and Ferrell, D.J. (1988) Early survivorship of juvenile coral reef fishes. *Coral Reefs*, **7**, 117–24.

Sale, P.F. and Sharp, B.J. (1983) Correction for bias in visual transect censuses of coral reef fishes. *Coral Reefs*, **2**, 37–42.

Sale, P.F., Douglas, W.A. and Doherty, P.J. (1984) Choice of microhabitats by coral reef fishes at settlement. *Coral Reefs*, **3**, 91–9.

Salvat, B. (ed.) (1987) *Human Impacts on Coral Reefs: Facts and Recommendations*, Antènne Muséum Ecole Pratique d'Hautes Etudes, French Polynesia, 253 pp.

Salvat, B. (1992) Coral reefs – a challenging ecosystem for human societies. *Global Env. Change*, **2**, 12–18.

Salvat, B., Richard, G., Rougerie, F. and Coeroli, M. (1985) Atoll de Takapoto, Archipel des Tuamotu. *Proc. 5th Int. Coral Reef Congr.*, **4**, 323–77.

Sammarco, P.W. (1980) *Diadema* and its relationship with coral spat mortality: grazing, competition and biological disturbance. *J. exp. mar. Biol. Ecol.*, **45**, 245–72.

Sammarco, P.W. and Andrews, J.C. (1989) The Helix experiment: differential localized dispersal and recruitment patterns in Great Barrier Reef corals. *Limnol. Oceanogr.*, **34**, 896–912.

Samoilys, M. (1988) Abundance and species richness of coral reef fish on the Kenyan coast: the effects of protective management and fishing. *Proc. 6th Int. Coral Reef Symp.*, **2**, 261–6.

Samoilys, M. (1992) Review of the underwater visual census method developed by the QDPI/ACIAR project: visual assessment of reef fish stocks. Dept Primary Ind., Brisbane, *Conf. Workshop Ser.* QC92006, 55 pp.

Samoilys, M.A. and Squire, L.C. (1994) Preliminary observations on the spawning behavior of coral trout, *Plectropomus leopardus* (Pisces: Serranidae), on the Great Barrier Reef. *Bull. mar. Sci.*, **54**, 332–42M.

Samuel, M. and Mathews, C.P. (1987) Growth and mortality of four *Acanthopagrus* species. *Kuwait Bull. mar. Sci.*, **9**, 159–72.

Sanders, M.J. and Kedidi, S.M. (1984) Catches, fishing efforts, catches per fishing effort, and fishing locations for the Gulf of Suez and Egyptian Red Sea fishery for reef associated fish during 1979 to 1982. Unpubl. rep., UNDP/FAO, Cairo, Project for Development of Fisheries in the Areas of the Red Sea and Gulf of Aden, FAO/UNDP RAB/83/023/02, 65 pp.

Sandt, V.J. and Stoner, A.W. (1993) Ontogenetic shift in habitat by early juvenile queen conch, *Strombus gigas*: patterns and potential mechanisms. *Fishery Bull., US*, **91**, 516–25.

Sano, M., Shimizu, M. and Nose Y. (1984a) Changes in structure of coral reef fish communities by destruction of hermatypic corals: observational and experimental views. *Pac. Sci.*, **38**, 51–79.

Sano, M., Shimizu, M. and Nose, Y. (1984b) *Food Habits of Teleostean Reef Fishes in Okinawa Island, Southern Japan*. University Museum, University of Tokyo, *Bulletin* **25**, 128 pp.

Sano, M., Shimizu, M. and Nose, Y. (1987) Long-term effects of destruction of hermatypic corals by *Acanthaster planci* infestation on reef fish communities at Iriomote Island, Japan. *Mar. Ecol. – Progr. Ser.*, **37**, 191–9.

Sasekumar, A., Chong, V.C., Leh, M.U. and D'Cruz, R. (1992) Mangroves as habitat for fish and prawns. *Hydrobiologia*, **247**, 195–207.

Savina, G.C. and White, A.T. (1986) Reef fish yields and non-reef catch of Pamilacan Island, Bohol, Philippines, in *The First Asian Fisheries Forum*, Vol. 1, (eds J.L. Maclean, L.B. Dizon and L.V. Hosillos), Asian Fisheries Society, Manila, pp. 497–500.

Sawyer, D.A. (1992) Taka Bone Rate: management, development and resource valuation of an Indonesian Atoll, MA thesis, Dept Economics, Dalhousie University, Halifax, Nova Scotia.

Schaefer, M.B. (1954) Some aspects of the dynamics of populations relevant to the management of marine fisheries. *Inter-Am. Trop. Tuna Comm. Bull.*, **1**, 27–56.

Schaefer, M.B. (1957) A study of the dynamics of the fisheries for yellowfin tuna in the eastern Pacific Ocean. *Inter-Am. Trop. Tuna Comm. Bull.*, **2**, 247–68.

Scheltema, R.S. (1986) Long-distance dispersal by planktonic larvae of shoal-water benthic invertebrates among central Pacific Islands. *Bull. mar. Sci.*, **39**, 241–56.

Schlesinger, D.A. and Regier, H.A. (1982) Climatic and morphoedaphic indices of fish yields from natural lakes. *Trans. Am. Fish. Soc.*, **111**, 141–50.

Schmitt, P.D. (1986) Feeding by larvae of *Hypoatherina tropicalis* (Pisces: Atherinidae) and its relation to prey availability in One Tree Lagoon, Great Barrier Reef, Australia. *Env. Biol. Fishes*, **16**, 79–94.

Schnute, J. (1977) Improved estimates from the Schaefer production model: theoretical considerations. *J. Fish. Res. Bd Can.*, **34**, 583–603.

Schoeffel, P. (1983) Women's associations in the rural economy of the South Pacific: case studies from Western Samoa and East New Britain Province, Papua New Guinea. South Pacific Commission, Nouméa, New Caledonia, *Occ. Pap.* **19**.

Schoeninger, M.J. and DeNiro, M.J. (1984) Nitrogen and carbon isotopic composition of bone collagen from marine and terrestrial animals. *Geochim. Cosmochim. Acta*, **48**, 625–39.

Scholander, P.F., Flagg, W., Walters, V. and Irving, L. (1953) Climatic adaptations in Arctic and tropical poikilotherms. *Physiol. Zool.*, **26**, 67–92.

Schot, J.G. (1883) De Battam-Archipel. *Indische Gids*, **4**(1), 205–11, 462–79; **4**(2), 25–54, 161–88, 476–79, 617–25.

Schroeder, R.E. (1989) The ecology of patch reef fishes in a subtropical Pacific atoll: recruitment variability, community structure and effects of fishing predators. PhD thesis, University of Hawaii, Honolulu, HI, 321 pp.

Schultz, E.T. and Cowen, R.K. (1994) Recruitment of coral-reef fishes to Bermuda: local retention or long-distance transport? *Mar. Ecol. – Progr. Ser.* **109**, 15–28.

Scoffin, T.P., Stearn, C.W., Boucher, D., Frydl, P., Hawkins, C.M., Hunter, I.G. and MacGeachy, J.K. (1980) Calcium carbonate budget of a fringing reef on the west coast of Barbados. Part 2. Erosion, sediments and internal structure. *Bull. mar. Sci.*, **30**, 475–508.

Scura, L.F., Chua, T.-E., Pido, M.D. and Paw, J.N. (1992) Lessons for integrated coastal zone management: the ASEAN experience, in *Integrative Framework and Methods for Coastal Area Management* (*ICLARM Conf. Proc.* **37**) (eds T.-E. Chua and L.F. Scura), ICLARM, Manila, Philippines, pp. 1–70.

Seaman, W.J., jun. and Sprague, L.M. (eds) (1991) *Artificial Habitats for Marine and Freshwater Fisheries*, Academic Press, San Diego, 285 pp.

Seber, G.A.F. (1982) *The Estimation of Animal Abundance and Related Parameters*, 2nd edn, MacMillan, New York.

Seeb, L.W., Seeb, J.E. and Polovina, J.J. (1990) Genetic variation in highly exploited spiny lobster *Panulirus marginatus* populations from the Hawaiian archipelago. *Fishery Bull., U.S.*, **88**, 713–18.

Selvaraj, G.S.D. and Rajagopalan, M. (1973) Some observations on the fecundity and spawning habits of the rock cod, *Epinephelus tauvina* (Forskøl). *Indian J. Fish.*, **20**, 668–71.

Semper, K. (1873) *Die Palau – Inseln im Stillen Ozean*, Leipzig.

Seneca, E.D. and Broome, S.W. (1992) Restoring tidal marshes in North Carolina and France, in *Restoring the Nation's Marine Environment* (ed. G. Thayer), Maryland Sea Grant, College Park, MD, pp. 53–78.

Shafer, C.L. (1990) *Nature Reserves: Island Theory and Conservation Practice*, Smithsonian Institution Press, Washington, DC, 189 pp.

Shaklee, J.B. (1984) Genetic variation and population structure in the damselfish, *Stegastes fasciolatus*, throughout the Hawaiian Archipelago. *Copeia*, **1984**, 629–40.

Shaklee, J.B. and Samollow, P.B. (1984) Genetic variation and population structure in a deepwater snapper, *Pristipomoides filamentosus*, in the Hawaiian Archipelago. *Fishery Bull., U.S.*, **82**, 703–12.

Shapiro, D.Y. (1980) Serial female sex changes after simultaneous removal of males from social groups of a coral reef fish. *Science*, **209**, 1136–7.

Shapiro, D.Y. (1987a) Reproduction in groupers, in *Tropical Snappers and Groupers: Biology and Fisheries Management* (eds J.J. Polovina and S. Ralston), Westview Press, Boulder, CO, pp. 295–328.

Shapiro, D.Y. (1987b) Inferring larval recruitment strategies from the distributional ecology of settled individuals of a coral reef fish. *Bull. mar. Sci.*, **41**, 289–95.

Shapiro, D.Y. (1989) Sex change as an alternative life-history style, in *Alternative Life-History Styles of Animals* (ed. M.N. Bruton), Kluwer Academic, Dordrecht, The Netherlands, pp. 177–95.

Shapiro, D.Y. and Lubbock, R. (1980) Group sex ratio and sex reversal. *J. theor. Biol.*, **82**, 411–26.

Shapiro, D.Y., Hensley, D.A. and Appeldoorn, R.S. (1988) Pelagic spawning and egg transport in the coral-reef fishes: a skeptical overview. *Env. Biol. Fishes*, **22**, 3–14.

Shapiro, D.Y., Sadovy, Y. and McGehee, M A. (1993a) Periodicity of sex change and reproduction in the red hind, *Epinephelus guttatus*, a protogynous grouper. *Bull. mar. Sci.*, **53**, 399–406.

Shapiro, D.Y., Sadovy, Y. and McGehee, M.A. (1993b) Size, composition, and spatial structure of the annual spawning aggregation of the red hind, *Epinephelus guttatus* (Pisces: Serranidae). *Copeia*, **1993**, 367–74.

Shapiro, D.Y., Garcia-Moliner, G. and Sadovy, Y. (1994) Social system of an inshore population of the red hind grouper, *Epinephelus guttatus* (Pisces: Serranidae). *Env. Biol. Fishes*, **41**, 415–22.

Shapiro, D.Y., Marconato, A. and Yoshikawa, T. (1994, in press) Sperm economy in a coral reef fish *Thalassoma bifasciatum*. *Ecology*, **75**, 1334–44.

Shaul, W. and Reifsteck, D. (1991) Comparison of catches by bamboo and wire mesh fish traps in Jamaica. *Proc. Gulf Carib. Fish. Inst.*, **40**, 99–107.

Shenker, J.M., Maddox, E.D., Wishinski, E., Pearl, A., Thorrold, S.R. and Smith, N. (1993) Onshore transport of settlement-stage nassau grouper *Epinephelus striatus* and other fishes in Exuma Sound, Bahamas. *Mar. Ecol. – Progr. Ser.*, **98**, 31–43.

Shepherd, J.G. (1987) A weakly parametric method for estimating growth parameters from length composition data, in *Length-based Methods in Fisheries Research* (*ICLARM Conf. Proc.* 13) (eds D. Pauly and G.R. Morgan), ICLARM, Manila, Philippines, pp. 113–19.

Shepherd, J.G. and Cushing, D.H. (1990) Regulation in fish populations: myth or mirage? *Phil. Trans. R. Soc. Lond.*, **330B**, 151–64.

Shepherd, S.A. and Brown, L.D. (1993) What is an abalone stock: Implications for the role of refugia in conservation. *Can. J. Fish. aquat. Sci.*, **50**, 2001–9.

Shepherd, S.A., McComb, A.J., Bulthuis, D.A., Neverauskas, V., Steffensen, D.A. and West, R. (1989) Decline of seagrasses, in *Biology of Seagrasses* (eds A.W.D. Larkum, J.A. McComb and S.A. Shepherd), Elsevier, Amsterdam, pp. 346–93.

Sheppard, C.R.C., Price, A.R.G. and Roberts, C.M. (1992) *Marine Ecology of the Arabian Region: Patterns and Processes in Extreme Tropical Environments*, Academic Press, London, 359 pp.

Shomura, R.S. (1977) (ed.) *Collection of Tuna Baitfish Papers*. U.S. Dept Commerce, NOAA, *Nat. Mar. Fish. Serv. Circ.* **408**, 167 pp.

Shorthouse, B. (1990) The Great Barrier Reef Marine Park: How does it work for fishermen? *Aust. Fish.*, **49**, 16–17.

Shpigel, M. and Fishelson, L. (1991) Experimental removal of piscivorous groupers of the genus *Cephalopholis* (Serranidae) from coral habitats in the Gulf of Aqaba (Red Sea). *Env. Biol. Fishes*, **31**, 131–38.

Shulman, M.J. (1984) Resource limitation and recruitment patterns in a coral reef assemblage. *J. exp. mar. Biol. Ecol.*, **74**, 85–109.

Shulman, M.J. (1985) Recruitment of coral reef fishes: effects of distribution of predators and shelter. *Ecology*, **66**, 1056–66.

Shulman, M.J. and Ogden, J.C. (1987) What controls tropical reef fish populations: recruitment or benthic mortality? An example in the Caribbean reef fish *Haemulon flavolineatum*. *Mar. Ecol. – Progr. Ser.*, **39**, 233–42.

Shulman, M.J., Ogden, J.C., Ebersole, J.P., McFarland, W.N., Miller, S.L. and Wolf, N.G. (1983) Priority effects in the recruitment of juvenile coral reef fishes. *Ecology*, **64**, 1508–13.

Siau, Y. (1994) Population structure, reproduction and sex-change in a tropical East Atlantic grouper. *J. Fish Biol.*, **44**, 205–11.

Silvestre, G.T. (1990) Over-exploitation of the demersal stocks of the Lingayen Gulf, Philippines, in *The Second Asian Fisheries Forum* (eds R. Hirano and I. Hanyu), Asian Fisheries Society, Manila, pp. 873–6.

Silvestre, G.T., Hammer, C., Sambilay, V.C. and Torres, F. (1986) Size selection and related morphometrics of trawl-caught fish species from the Samar Sea, in *Resources, Management and Socio-economics of Philippine Marine Fisheries* (eds D. Pauly, J. Saeger and G. Silvestre), University of the Philippines in the Visayas, Quezon City, *Tech. Rep. Dept Mar. Fish.* **10**, 107–38.

Silvestre, G.T., Federizon, R., Muñoz, J. and Pauly, D. (1987) Overexploitation of the demersal resources of Manila Bay and adjacent areas, in *Proc. 22nd Session, Indo–Pacific Fisheries Commission, 16–26 February 1987, Darwin, Australia*, pp. 269–87.

Silvestre, G., Soriano, M. and Pauly, D. (1991) Sigmoid selection and the Beverton and Holt equation. *Asian Fish. Sci.*, **4**, 85–98.

Sinderman, C.J. (1993) Disease risks associated with importation of nonindigenous marine animals. *Mar. Fish. Rev.*, **54**, 1–10.

Sinoda, M. and Kobayashi, T. (1969) Studies of the fishery of Zuwai crab in the Japan Sea – IV. Efficiency of the toyama kago (a kind of crab trap) in capturing benizuwai crab. *Bull. Jap. Soc. scient. Fish.*, **35**, 948–56.

Sissenwine, M.P. (1974) Variability in recruitment and equilibrium catch of the southern New England yellowtail flounder fishery. *J. Conseil, Cons. perm. int. Explor. Mer*, **36**, 15–26.

Sissenwine, M.P., Brown, B.E., Palmer, J.E., Essig, R.J. and Smith, W. (1982) An empirical examination of population interactions for the fishery resources off the northeastern USA. *Can. spec. Publ. Fish. aquat. Sci.*, **59**, 82–94.

Skud, B.E. (1978) Factors affecting longline catch and effort: III. Bait loss and competition. *Int. Pac. Halibut Comm. Sci. Rep.* **64**, 25–50.

Slobodkin, L.B. (1972) *Growth and Regulation of Animal Populations*, Holt, Reinhart and Winston, New York.

Sluczanowski, R.P. (1984) A management oriented model of an abalone fishery whose substocks are subject to pulse fishing. *Can. J. Fish. aquat. Sci.*, **41**, 1008–14.

Sluczanowski, R.P. (1992) Computer graphics for co-management. *Naga, The ICLARM Quarterly*, **15**(3), 18.

Smale, M.J. (1988) Distribution and reproduction of the reef fish, *Petrus rupestris* (Pisces: Sparidae) off the coast of South Africa. *S. Afr. J. Zool.*, **23**, 272–87.

Smith, A. (1991) Tradition and the development of the marine resources coastal management plan for Yap State, Federated States of Micronesia, in *Adaptive Marine Resource Management Systems in the Pacific* (eds M.M.R. Freeman, Y. Matsuda and K. Ruddle), Harwood Academic Publishers, Chur, Switzerland, pp. 29–39.

Smith, A. and Dalzell, P. (1993) Fisheries resources and management investigations in Woleai Atoll, Yap State, Federated States of Micronesia. South Pacific Commission, Nouméa, New Caledonia. *Inshore Fish. Res. Proj. tech. Doc.*, **4**, xiii + 64 pp.

Smith, A.H. and Berkes, F. (1991) Solutions to the 'Tragedy of the Commons': sea-urchin management in St Lucia, West Indies. *Env. Conserv.*, **18**, 131–6.

Smith, C.L. (1972) A spawning aggregation of Nassau grouper, *Epinephelus striatus* (Bloch). *Trans. Am. Fish. Soc.*, **2**, 257–61.

Smith, C.L. (1973) Small rotenone stations: a tool for studying coral reef fish communities. *Am. Mus. Novit.*, **2512**, 2–21.

Smith, C.L. (1975) The evolution of hermaphroditism in fishes, in *Intersexuality in the Animal Kingdom* (ed. R. Reinboth), Springer Verlag, Berlin, pp. 295–310.

Smith, C.L. (1978) Coral reef fish communities: a compromise view. *Env. Biol. Fishes*, **3**, 108–28.

Smith, C.L. and Tyler, J.C. (1972) Space resource sharing in a coral reef fish community. *Bull. Nat. Hist. Mus. Los Angeles County*, **14**, 125–70.

Smith, C.L., Tyler, J.C. and Stillman, L. (1987) Inshore ichthyoplankton: a distinctive assemblage? *Bull. mar. Sci.*, **41**, 432–40.

Smith, I.R. (1979) A research framework for traditional fisheries. *ICLARM Studies and Reviews*, **2**, 45 pp.

Smith, I.R., Puzon, M.Y. and Vidal-Libunao, C.N. (1980) Philippines municipal fisheries: a review of resources, technology and socioeconomics. *ICLARM Studies and Reviews*, **4**, 87 pp.

Smith, I.R., Pauly, D. and Mines, A.N. (1983) Small-scale fisheries of San Miguel Bay, Philippines: options for management and research. *ICLARM tech. Rep.* **11**, 80 pp.

Smith, M.K. (1993) An ecological perspective on inshore fisheries in the main Hawaiian Islands. *Mar. Fish. Rev.*, **55**, 34–49.

Smith, P.J., Francis, R.I.C.C. and McVeagh, M. (1991) Loss of genetic diversity due to fishing pressure. *Fish. Res.*, **10**, 309–16.

Smith, R.O. (1947a) Fishery resources of Micronesia. U.S. Dept Interior, Fish Wildl. Serv., Washington, DC, *Fishery Leaflet* **239**.

Smith, R.O. (1947b) Survey of the fisheries of the former Japanese mandated islands. U.S. Dept Interior, Fish Wildl. Serv., Washington, DC, *Fishery Leaflet* **273**, 105 pp.

Smith, S.V. (1978) Coral-reef area and the contributions of reefs to processes and resources of the world's oceans. *Nature, Lond.*, **273**, 225–6.

Snouck Hurgronje, C. (1906) *The Achehnese*, (transl. A.W.S. O'Sullivan), Vol. 1. Brill, Leiden.

Sobel, J. (1993) Conserving biological diversity through marine protected areas. *Oceanus*, **36**, 19–26.

Soegiarto, A. and Polunin, N.V.C. (1981) The marine environment of Indonesia. Unpubl. rep., IUCN and WWF, Jakarta, ix + 257 pp.

Somers, I.F. (1988) On a seasonally-oscillating growth function. *Fishbyte* **6**(1), 8–11.

Sorgeloos, P. and Sweetman, J. (1993) Aquaculture success stories. *World Aquaculture*, **24**, 4–14.

Sorokin, Y.I. (1990) Plankton in the reef ecosystems, in *Coral Reefs* (Ecosystems of the World **25**) (ed. Z. Dubinsky), Elsevier, Amsterdam, pp. 291–327.

Souder, P.B. (1987) Guam: land tenure in a fortress, in *Land Tenure in the Pacific* (ed. R. Crocombe), University of the South Pacific, Suva, Fiji, pp. 211–25.

Soutar, A. and Isaacs, J.D. (1974) Abundance of pelagic fish during the 19th and 20th centuries as recorded in anaerobic sediment off the Californias. *Fishery Bull., U.S.*, **72**, 257–73.

Sparre, P. (1987) Computer programs for fish stock assessment. *FAO Fish. tech. Pap.* **101**, Suppl. 2, 218 pp.

Sparre, P. (1990) Can we use traditional length-based fish stock assessment when growth is seasonal? *Fishbyte* **8**(3), 29–32.

Sparre, P. (1991) Estimation of yield per recruit when growth and fishing mortality oscillate seasonally. *Fishbyte* **9**(1), 40–4.

Sparre, P., Ursin, E. and Venema, S.C. (1989) Introduction to tropical fish stock assessment. Part 1. Manual. *FAO Fish. tech. Pap.*, **306**, 337 pp.

Ssentongo, G.W. and Larkin, P.A. (1973) Some simple methods of estimating mortality rates of exploited fish populations. *J. Fish. Res. Bd Can.*, **30**, 695–8.

St John, J., Russ, G.R. and Gladstone, W. (1990) Accuracy and bias of visual estimates of numbers, size structure and biomass of a coral reef fish. *Mar. Ecol. – Progr. Ser.*, **64**, 253–62.

Stair, J. (1897) *Old Samoa, or Flotsam and Jetsam from the Pacific Ocean*, Religious Tract Society, London.

Staples, D.J., Polzin, H.G. and Heales, D.S. (1985) Habitat requirements of juvenile penaeid prawns and their relationship to offshore fisheries, in *Second Australian National Prawn Seminar, NPS2* (eds P.C. Rothlisberg, B.J. Hill and D.J. Staples), CSIRO, Cleveland, Australia, pp. 47–54.

Stearn, C.W. and Scoffin, T.P. (1977) Carbonate budget of a fringing reef, Barbados. *Proc. 3rd Ind. Coral Reef Symp.*, **2**, 471–6.

Stearns, S.C. (1976) Life-history tactics: a review of the ideas. *Q. Rev. Biol.*, **51**, 3–47.

Stearns, S.C. and Crandall, R.E. (1984) Plasticity for age and size at sexual maturity: a life-history response to unavoidable stress, in *Fish Reproduction: Strategies and Tactics* (eds G.W. Potts and R.J. Wootton), Academic Press, London, pp. 13–33.

Steele, J.H. and Henderson, E.W. (1984) Modeling long-term fluctuations in fish stocks. *Science*, **224**, 985–6.

Steneck, R.S. (1988) Herbivory on coral reefs: a synthesis. *Proc. 6th Int. Coral Reef Symp.*, **1**, 37–49.

Stevenson, D.K. (1978) Management of a tropical pot fishery for maximum sustainable yield. *Proc. Gulf Carib. Fish. Inst.*, **30**, 95–115.

Stevenson, D.K. and Marshall, N. (1974) Generalizations on the fisheries potential of coral reefs and adjacent shallow-water environments. *Proc. 2nd Int. Coral Reef Symp.*, **1**, 147–56.

Stevenson, D.K. and Stuart-Sharkey, P. (1980) Performance of wire fish traps on the western coast of Puerto Rico. *Proc. Gulf Carib. Fish. Inst.*, **32**, 173–93.

Stevenson, D.K., Pollnac, R. and Logan, P. (1982) *A Guide for the Small-scale Fishery Administrator: Information from the Harvest Sector.* Int. Center Mar. Resour. Develop., University of Rhode Island, Kingston, RI, 124 pp.

Stimson, J., Blum, S. and Brock, R. (1982) An experimental study of the influence of muraenid eels on reef fish sizes and abundance. *Hawaii Sea Grant Quarterly*, **4**, 1–6.

Stimson, J.S. (1990) Density dependent recruitment in the reef fish *Chaetodon miliaris*. *Env. Biol. Fishes*, **29**, 1–13.

Stobutzki, I.C. and Bellwood, D.R. (1994) An analysis of the sustained swimming abilities of pre-settlement and post-settlement coral reef fishes. *J. exp. mar. Biol. Ecol.*, **175**, 275–86.

Stoffle, R.W. and Halmo, D.B. (1992) The transition to mariculture: a theoretical polemic and a Caribbean case, in *Coastal Aquaculture in Developing Countries: Problems and Perspectives* (eds R.B. Pollnac and P. Weeks), Int. Center Mar. Resour. Develop., University of Rhode Island, Kingston, RI, pp. 135–61.

Stoffle, R.W., Halmo, D.B. and Stoffle, B.W. (1991) Inappropriate management of an appropriate technology: a restudy of *Mithrax* crab mariculture in the Dominican Republic, in *Small-scale Fishery Development: Sociocultural Perspectives*, (eds J.J. Poggie and R.B. Pollnac), Int. Center Mar. Resour. Develop., University of Rhode Island, Kingston, RI, pp. 131–57.

Stone, R.B. (1985) National artificial reef plan. *NOAA NMFS tech. Memo. NMFS* OF-6, 75pp.

Stoner, A.W. and Davis, M. (1994) Experimental outplanting of juvenile queen conch, *Strombus gigas*: comparison of wild and hatchery-reared stocks. *Fishery Bull., U.S.*, **92**, 390–411.

Stoner, A.W. and Sandt, V.J. (1991) Experimental analysis of habitat quality for juvenile Queen Conch in seagrass meadows. *Fishery Bull., U.S.*, **89**, 693–700.

Stoner, A.W. and Sandt, V.J. (1992) Transplanting as a test procedure before large-scale outplanting of juvenile queen conch. *Proc. Gulf Carib. Fish. Inst.*, **41**, 447–58.

Stoner, A.W. and Waite, J.W. (1990) Distribution and behavior of Queen Conch *Strombus gigas* relative to seagrass standing crop. *Fishery Bull., U.S.*, **88**, 573–85.

Stroud, R.H. (ed.) (1992) *Stemming the Tide of Coastal Fish Habitat Loss*. National Coalition for Marine Conservation, Savannah, GA, 258 pp.

Sudekum, A.E., Parrish, J.D., Radtke, R.L. and Ralston, S. (1991) Life history and ecology of large jacks in undisturbed, shallow, oceanic communities. *Fishery Bull., U.S.*, **89**, 493–513.

Sudo, K.-I. (1976) A preliminary report on social organization in the Ulul Islands, Micronesia. *A. Rep. Social Anthropol.*, **2**, 202–20. Kobundo, Tokyo.

Sudo, K.-I. (1984) Social organization and types of sea tenure in Micronesia, in *Maritime Institutions in the Western Pacific* (*Senri Ethnol. Stud.* **17**) (eds K. Ruddle and T. Akimichi), National Museum of Ethnology, Osaka, pp. 203–30.

Summerfeldt, R.C. and Hall, G.E. (eds) (1987) *Age and Growth of Fish*, Iowa State University Press, Ames, IA, 544 pp.

Sutherland, D.L. and Harper, D.E. (1983) The wire fish-trap fishery of Dade and Brouard Counties, Florida, December 1979 – September 1980. Florida Dept Nat. Resour., St Petersburg, *Florida mar. Res. Publ.* **40**, 21 pp.

Sutherland, W.J. (1990) Evolution and fisheries. *Nature, Lond.*, **344**, 814–15.

Suthers, I.M. and Frank, K.T. (1991) Comparative persistence of marine fish larvae from pelagic versus demersal eggs off southwestern Nova Scotia, Canada. *Mar. Biol.*, **108**, 175–84.

Sweatman, H.P.A. (1983) Influence of conspecifics on choice of settlement sites by larvae of two pomacentrid fishes (*Dascyllus aruanus* and *D. reticulatus*) on coral reefs. *Mar. Biol.*, **75**, 225–9.

Sweatman, H.P.A. (1984) A field study of the predatory behavior and feeding rate of a piscivorous coral reef fish, the lizardfish *Synodus englemani*. *Copeia*, **1984**, 187–94.

Sweatman, H.P.A. (1985) The influence of adults of some coral reef fishes on larval recruitment. *Ecol. Monogr.*, **55**, 469–85.

Sweatman, H.P.A. (1988) Field evidence that settling coral reef fish larvae detect resident fishes using dissolved chemical cues. *J. exp. mar. Biol. Ecol.*, **124**, 163–74.

Sweatman, H.P.A. (1995) A field study of fish predation on juvenile crown-of-thorns starfish. *Coral Reefs*, **14**, 47–53.

Swingle, N.E., Dammann, A.E. and Yntena, J.A. (1970) Survey of the commercial fishery of the Virgin Islands of the U.S. *Proc. Gulf Carib. Fish. Inst.*, **22**, 110–21.

Sylvester, J.R. and Dammann, A.E. (1972) Pot fishing in the Virgin Islands. *Mar. Fish. Rev.*, **34**, 33–5.

Tahil, A.S. (1984) Catching efficiency of fish pots of different shapes and materials. *Fish. Res. J. Philippines*, **9**(1–2), 24–31.

Talbot, F.H. (1960) Notes on the biology of the Lutjanidae (Pisces) of the east African coast, with special reference to *L. bohar* (Forskål). *Annals S. Afr. Mus.*, **45**, 549–73.

Talbot, F.H. (1965) A description of the coral structure of Tutia Reef (Tanganyika Territory, East Africa) and its fish fauna. *J. Zool.*, **145**, 431–70.

Talbot, F.H., Russell, B.C. and Anderson, G.R.V. (1978) Coral reef fish communities: unstable, high diversity systems? *Ecol. Monogr.*, **48**, 425–40.

Tan, J.G. (1993) Poverty alleviation through an integrated approach to coastal resources management: CERD-firmed experience in Daram, Samar, in *Our Sea: Our Life* (ed. L.P. de la Cruz), Voluntary Services Overseas (VSO), Quezon City, Philippines, pp. 39–49.

Tan, S.M. and Tan, K.S. (1974). Biology of the tropical grouper, *Epinephelus tauvina* (Forskål) 1. A preliminary study on hermaphroditism in *E. tauvina*. *Singapore J. Primary Ind.*, **2**, 123–33.

Tandog-Edralin, D., Cortez-Zaragosa, E., Dalzell, P. and Pauly, D. 1990 Some aspects of the biology of skipjack (*Katsuwonus pelamis*) in Philippine waters. *Asian Mar. Biol.*, **7**, 15–29.

Tashiro, J.E. and Coleman, S.E. (1977) The Cuban grouper and snapper fishery in the Gulf of Mexico. *Mar. Fish. Rev.*, **39**, 1–6.

Taurakoto, P. (1984) Customary rights to reefs and landings, in *Land Tenure in Vanuatu*, (ed. P. Lamour), Inst. Pac. Stud., Univ. S. Pacific, Suva, Fiji.

Taylor, M.H. (1984) Lunar synchronization of fish reproduction. *Trans. Am. Fish. Soc.*, **113**, 484–93.

Taylor, R.G. and McMichael, R.H. (1983) The Wire Fish-trap Fishery of Monroe and Collier Counties, Florida. Florida Dept Nat. Resour., St Petersburg, *Florida mar. Res. Publ.* **39**, 19 pp.

Tegner, M.J. (1989) The California abalone fishery: production, ecological interactions, and prospects for the future, in *Marine Invertebrate Fisheries: their Assessment and Management* (ed. J.F. Caddy), John Wiley, New York, pp. 401–20.

Tegner, M.J. (1993) Southern California abalones: can stocks be rebuilt using marine harvest refugia? *Can. J. Fish. aquat. Sci.*, **59**, 2010–18.

Teiwaki, R. (1988) *Management of Marine Resources in Kiribati*, Atoll Res. Unit and Inst. Pac. Stud., Univ. S. Pacific, Suva, Fiji.

Tetiarahi, G. (1987) The Society islands: squeezing out the Polynesians, in *Land Tenure in the Pacific* (ed. R. Crocombe), Univ. S. Pacific, Suva, Fiji, pp. 45–58.

Teulières, M.H. (1990) Traditional marine resource management among the Nenema of Northwestern New Caledonia, in *Traditional Marine Resource Management in the Pacific Basin: an Anthology* (eds K. Ruddle and R.E. Johannes), UNESCO–ROSTSEA, Jakarta, pp. 103–22.

Teulières, M.H. (1991) Melanesian and European systems of marine resource management in New Caledonia, in *Traditional and Adaptive Marine Resource Management Systems in the Western Pacific* (eds M.M.R. Freeman, Y. Matsuda and K. Ruddle), Harwood Academic Publishers, New York, pp. 41–51.

Thayer, G.W. (1992) The science of restoration: status and directions, in *Restoring the Nation's Marine Environment* (ed. G. Thayer), Maryland Sea Grant, College Park, MD, pp 1–5.

Theilacker, G.H. (1986) Starvation-induced mortality of young sea-caught jack mackerel, *Trachurus symmetricus*, determined with histological and morphological methods. *Fishery Bull., U.S.*, **84**, 1–17.

Thollot, P. (1992) Importance des mangroves pour la faune ichtyologique des récifs coralliens de Nouvelle-Caledonie. *Cybium*, **16**, 331–44.

Thomas, C.J. and Cahoon, R.L. (1993) Stable isotope analyses differentiate between different trophic pathways supporting rocky-reef fishes. *Mar. Ecol. – Progr. Ser.*, **95**, 19–24.

Thompson, L. (1940) *Southern Lau, Fiji: An Ethnography*. Bernice P. Bishop Museum, Honolulu, HI, *Bulletin* **162**.

Thompson, L. (1945) *The Native Cultures of the Marianas*. Bernice P. Bishop Museum, Honolulu, HI, *Bulletin* **185**.

Thompson, L. (1949) The relations of men, animals and plants in an island community (Fiji). *Am. Anthropol.*, **51**, 253–67.

Thompson, R. and Munro, J.L. (1978) Aspects of the biology and ecology of Caribbean reef fishes: Serranidae (hinds and groupers). *J. Fish Biol.*, **12**, 115–46.

Thompson, R. and Munro, J.L. (1983a) The biology, ecology and bionomics of the hinds and groupers, Serranidae, in *Caribbean Coral Reef Fishery Resources* (*ICLARM Stud. Rev.* 7) (ed. J.L. Munro), ICLARM, Manila, Philippines, pp. 59–81.

Thompson, R. and Munro, J.L. (1983b) The biology, ecology and bionomics of the jacks, Carangidae, in *Caribbean Coral Reef Fishery Resources* (*ICLARM Stud. Rev.* 7) (ed. J.L. Munro), ICLARM, Manila, Philippines, pp. 82–93.

Thompson, R. and Munro, J.L. (1983c) The biology, ecology and bionomics of the snappers, Lutjanidae, in *Caribbean Coral Reef Fishery Resources* (*ICLARM Stud. Rev.* 7) (ed. J.L. Munro), ICLARM, Manila, Philippines, pp. 94–109.

Thorrold, S.R., Shenker, J.M., Mojica, R., Maddox, E.D. and Wishinski, E. (1994a) Temporal patterns in the larval supply of summer-recruiting fishes to Lee-Stocking Island, Bahamas. *Mar. Ecol. – Progr. Ser.*, **112**, 75–86.

Thorrold, S.R., Shenker, J.M., Maddox, E.D., Mojica, R. and Wishinski, E. (1994b) Larval supply of shorefishes to nursery habitats around Lee Stocking Island, Bahamas. 2. Lunar and oceanographic influences. *Mar. Biol.*, **118**, 567–78.

Thresher, R.E. (1983a) Habitat effects on reproductive success in the coral reef fish *Acanthochromis polyacanthus* (Pomacentridae). *Ecology*, **64**, 1184–99.

Thresher, R.E. (1983b) Environmental correlates of the distribution of planktivorous fishes in the One Tree Reef Lagoon. *Mar. Ecol. – Progr. Ser.*, **10**, 137–45.

Thresher, R.E. (1984) *Reproduction in Reef Fishes*, T.F.H. Publications, Neptune City, NJ, 399 pp.

Thresher, R.E. (1988a) Latitudinal variation in egg sizes of tropical and subtropical North Atlantic shore fishes. *Env. Biol. Fishes*, **21**, 17–25.

Thresher, R.E. (1988b) Otolith microstructure and the demography of coral reef fishes. *Trends Ecol. Evol.*, **3**, 78–80.

Thresher, R.E. (1991) Geographic variability in the ecology of coral reef fishes: evidence, evolution, and possible implications, in *The Ecology of Fishes on Coral Reefs* (ed. P.F. Sale), Academic Press, San Diego, pp. 401–36.

Thresher, R.E. and Brothers, E.B. (1985) Reproductive ecology and biogeography of Indo–West Pacific angelfishes (Pisces, Pomacanthidae). *Evolution*, **39**, 878–87.

Thresher, R.E. and Brothers, E.B. (1989) Evidence of intra-oceanic and inter-oceanic regional differences in the early life history of reef-associated fishes. *Mar. Ecol. – Progr. Ser.*, **57**, 187–205.

Thresher, R.E. and Gunn, J.S. (1986) Comparative analysis of visual census techniques for highly mobile, reef-associated piscivores (Carangidae). *Env. Biol. Fishes*, **17**, 93–116.

Tisdell, C. and Broadus, J.M. (1989) Policy issues related to the establishment and management of marine reserves. *Coastal Manage.*, **17**, 37–53.

Titcomb, M. (1952) *Native Use of Fish in Hawaii*, University of Hawaii Press, Honolulu.

Tobin, J.A. (1952) Land tenure in the Marshall Islands. *Atoll Res. Bull.*, **11**, 1–36.

Tobin, J.A. (1958) Land tenure in the Marshall Islands, in *Land Tenure Patterns in Trust Territory of the Pacific Islands* (ed. J.E. de Young), Trust Territory Government, Agana, Guam, pp. 1–75.

Tom'tavala, D.Y. (1990) National law, international law and traditional marine claims: a case study of the Trobriand Islands, Papua New Guinea. MS thesis, Dept Law, Dalhousie University, Halifax, Nova Scotia.

Tonn, W.M., Paszkowski, C.A. and Holopainen, I.J. (1992) Piscivory and recruitment: mechanisms structuring prey populations in small lakes. *Ecology*, **73**, 951–8.

Toor, H.S. (1964) Biology and fishery of the pig-face bream, *Lethrinus lentjan* Lacépède, II, maturation and spawning. *Indian J. Fish.*, **11**, 581–96.

Torres, R., Salas M.S. and Pérez, L.E. (1992) Fecundidad y rendimiento un enfoque diferente. *Proc. Gulf Carib. Fish. Inst.*, **41**, 332–48.

Towns, D.R. and Ballantine, W.J. (1993) Conservation and restoration of New Zealand island ecosystems. *Trends Ecol. Evol.*, **8**, 452–7.

Townsley, P. (1993) *Rapid Appraisal Methods for Coastal Communities*, Bay of Bengal Programme, Madras, India, 110 pp.

Travis, J. (1993) Invader threatens Black, Azov Seas. *Science*, **262**, 1366–7.

Trenkel, V. (1993) Multivariate analysis of fish population data, MSc thesis, University of Kent, Canterbury, UK.

Trinidad, A.C. (1993) Economic exploitation in the Philippines small pelagic fishery and implications for management. *Naga, the ICLARM Quarterly*, **16**(4), 13–15.

Troadec, J.-P. (1977) Méthodes semiquantitatives d'évaluation. *FAO Fish. Circ.*, **701**, 131–41.

Tseng, W.Y., and Chan, K.L. (1982) The reproductive biology of the rabbitfish in Hong Kong. *J. World Maricult. Soc.*, **13**, 313–21.

Tucker, J.W., Parsons, J.E., Ebanks, G.C. and Bush, P.G. (1991) Induced spawning of Nassau grouper *Epinephelus striatus*. *J. World Aquacult. Soc.*, **22**, 187–91.

Tucker, J.W., Bush, P.G. and Slaybaugh, S.T. (1993) Reproductive patterns of Cayman Islands Nassau grouper (*Epinephelus striatus*) populations. *Bull. mar. Sci.*, **52**, 961–9.

Tungpalan, M.T.V., Mangahas, M.F. and Palis, M.P.E. (1991) Women in fishing villages: roles and potential for coastal resources management, in *Towards an Integrated Management of Tropical Coastal Resources* (*ICLARM Conf. Proc.* **22**) (eds L.M. Chou, T.-E. Chua, H.W. Khoo, P.E. Lim, J.N. Paw, G.T. Silvestre, M.J. Valencia, A.T. White and P.K. Wong), ICLARM, Manila, Philippines, pp. 237–43.

Turner, G.F. (1993) Teleost mating behaviour, in *Behaviour of Teleost Fishes*, 2nd edn (ed. T.J. Pitcher), Chapman & Hall, London, pp. 307–31.

Ulanowicz, R.E. (1986) *Growth and Development: Ecosystem Phenomenology*, Springer Verlag, New York, 203 pp.

Umali, A.F. (1950) Guide to the classification of Fishing Gear in the Philippines. US Dept Interior, Fish Wildl. Serv., Washington, DC, *Res. Rep.* **17**, 165 pp.

UNEP (1985) Ecological interactions between tropical coastal ecosystems. *UNEP Regional Seas Rep. Stud.*, **37**, 71 pp.

UNEP/IUCN (1988a) *Coral Reefs of the World*. Vol. 2: *Indian Ocean, Red Sea and Gulf*. UNEP Regional Seas Directories and Bibliographies, IUCN, Gland, Switzerland and Cambridge, and UNEP, Nairobi, Kenya.

UNEP/IUCN (1988b) *Coral Reefs of the World*. Vol. 3: *Central and Western Pacific*. UNEP Regional Seas Directories and Bibliographies, IUCN, Gland, Switzerland and Cambridge, and UNEP, Nairobi, Kenya.

UNEP/IUCN (1988c) *Coral Reefs of the World*. Vol. 1: *Atlantic and Caribbean*. UNEP Regional Seas Directories and Bibliographies, IUCN, Gland, Switzerland and Cambridge, and UNEP, Nairobi, Kenya.

Ursin, E. (1984) The tropical, the temperate and the arctic seas as media for fish production. *Dana*, **3**, 43–60.

Ushijima, I. (1982) The control of reefs and lagoon: some aspects of the political structure of Ulithi Atoll, in *Islanders and their World: A Report of Cultural Anthropological Research in the Caroline Islands of Micronesia in 1980–81* (ed. M. Aoyagi), St Paul's (Rikkyo) University, Tokyo, pp. 35–75.

Utanga, A. (1988) Customary tenure and traditional resource management in the Cook Islands. Paper presented at the South Pacific Commission Workshop on Customary Tenure and Traditional Resource Management. South Pacific Commission, Nouméa, New Caledonia (MS in SPC Library, Nouméa).

Uzmann, J.R., Cooper, R.A., Theroux, R.B. and Wigley, R.L. (1977) Synoptic comparison of three sampling techniques for estimating abundance and distribution of selected megafauna: submersible vs camera vs otter trawl. *Mar. Fish. Rev.*, **39**, 11–19.

Vadiya, V. (1984) Reproductive systems of *Epinephelus aeneus* and *Epinephelus alexandrinus* (Serranidae) from the southeastern Mediterranean. *J. Ichthyol.*, **24**, 77–81.

Van Der Knaap, M., Waheed, Z., Shareef, H. and Rasheed, M. (1988) Reef Fish Survey in the Maldives. Unpubl. rep., Bay of Bengal Programme, Madras, BOBP/WP/64, MDV/88/007.

Van Der Knaap, M., Waheed, Z., Shareef, H. and Rasheed, M. (1991) Reef fish resources survey in the Maldives. Bay of Bengal Programme, *Working Paper* 64, 58 pp.

van der Sande, G.A.J. (1907) *Ethnography and Anthropology. Nova Guinea* 3, 390 pp.

Van der Velde, G., Gorissen, M.W., den Hartog, C., van't Hof, T. and Meijer, G.J. (1992) Importance of the Lac-lagoon (Bonaire, Netherlands Antilles) for a selected number of fish species. *Hydrobiologia*, **247**, 139–40.

van Hoëvell, G.W.W.C. (1890) Tanimbar en Timorlaoet-eilanden. *Tijdschrift voor Indische Taal-Land-en Volkenkunde*, **33**, 160–86.

Van Pel, H. (1959) Report on the fisheries of Norfolk Island (17 January–14 February 1959). Unpubl. rep., South Pacific Commission, Nouméa, New Caledonia, 41 pp.

Van Sant, S.B., Collins, M.R and Sedberry, G.R. (1994) Preliminary evidence from a tagging study for a gag (*Mycteroperca microlepis*) spawning migration with notes on the use of oxytetracycline for chemical tagging. *Proc. Gulf Carib. Fish. Inst.*, **43**, 409–20.

van Sickle, J. (1977) Mortality rates from size distributions. *Oecologia*, **27**, 311–18.

van't Hof, T. (1985) The economic benefits of marine parks and protected areas in the Caribbean region. *Proc. 5th Int. Coral Reef Congr.*, **6**, 551–56.

Veitayaki, J. (1990) Village-level fishing in Fiji: a case study of Qoma Island, MA thesis, Univ. S. Pacific, Suva, Fiji.

Veloro, C. (1992) Complementary adaptations, contrasting images: fishing and agriculture in a peasant village. *Yakara*, **19**, 76–110.

Vicente, V.P. (1994) Structural changes and vulnerability of a coral reef (Cayo Enrique) in La Parguera, Puerto Rico, in *Proceedings, Global Aspects of Coral Reefs: Health, Hazards, and History* 1993, (ed. R.N. Ginsburg), Rosenstiel School Mar. Atmos. Sci., University of Miami, Coral Gables, FL, pp. 227–32.

Victor, B.C. (1983) Recruitment and population dynamics of a coral reef fish. *Science*, **219**, 419–20.

Victor, B.C. (1984) Coral reef fish larvae: patch size estimation and mixing in the plankton. *Limnol. Oceanogr.*, **29**, 1116–19.

Victor, B.C. (1986a) Delayed metamorphosis with reduced larval growth in a coral reef fish (*Thalassoma bifasciatum*). *Can. J. Fish. aquat. Sci.*, **43**, 1208–13.

Victor, B.C. (1986b) Duration of the planktonic larval stage of one hundred species of Pacific and Atlantic wrasses (family Labridae). *Mar. Biol.*, **90**, 317–27.

Victor, B.C. (1986c) Larval settlement and juvenile mortality in a recruitment-limited coral reef fish population. *Ecol. Monogr.*, **56**, 145–60.

Victor, B.C. (1987) Growth, dispersal, and identification of planktonic labrid and pomacentrid reef-fish larvae in the eastern Pacific Ocean. *Mar. Biol.*, **95**, 145–52.

Victor, B.C. (1991) Settlement strategies and biogeography of reef fishes, in *The Ecology of Fishes on Coral Reefs* (ed. P.F. Sale), Academic Press, San Diego, pp. 231–60.

Villanoy, C.L., Juinio, A.R. and Menez, L.A. (1988) Fishing mortality rates of giant clams (Family Tridacnidae) from the Sulu Archipelago and Southern Palawan, Philippines. *Coral Reefs*, **7**, 1–5.

Villarroel, A.J.P. (1982) Desarrollo gonadal en el mero tofia, *Epinephelus guttatus* L. (Serranidae). Trabajo Especial de Grado [thesis], Universidad Central de Venezuela. 97 pp.

Vinogradov, A.P. (1953) *The Elementary Chemical Composition of Marine Organisms. Mem. Sears Found. Mar. Res.*, **2**, xiv + 647 pp.

Vivien, M.L. (1973) Contribution à la connaissance de l'éthologie alimentaire de l'ichthyofaune du platier interne des récifs coralliens de Tuléar (Madagascar). *Téthys* (Suppl.), **5**, 221–308.

Waldner, R.E. and Robertson, D.R. (1980) Patterns of habitat partitioning by eight species of territorial Caribbean damselfishes (Pisces: Pomacentridae). *Bull. mar. Sci.*, **30**, 171–86.

Walker, D.I. and McComb, A.J. (1992) Seagrass degradation in Australian coastal waters. *Mar. Poll. Bull.*, **25**, 5–8.

Walsh, W.J. (1987) Patterns of recruitment and spawning in Hawaiian reef fishes. *Env. Biol. Fishes*, **18**, 257–76.

Walters, C.J. (1986) *Adaptive Management of Renewable Resources*, Macmillan, New York.

Waltz, C.W., Roumillat, W.A. and Wenner, C.A. (1982) Biology of the whitebone porgy, *Calamus leucosteus*, in the South Atlantic Bight. *Fishery Bull., U.S.*, **80**, 863–74.

Wanink, J.H. (1991) Survival in a purturbed environment: the effects of Nile Perch introduction on the zooplankton and fish community in Lake Victoria, in *Terrestrial and Aquatic Ecosystems: Perturbation and Recovery* (ed. O. Ravera), Ellis Horwood, New York, pp. 269–75.

Wanter, P. (1990) Property, women fishers and struggles for women's rights in Mozambique. *SAGE*, **7**, 33–7.

Wantiez, L. (1992) Importance of reef fishes among the soft bottom fish assemblages of the North Lagoon of New Caledonia. *Proc. 7th Int. Coral Reef Symp.*, **2**, 942–50.

Wantiez, L. and Kulbicki, M. (1991) Les pêches exploratoires au chalut en baie de Saint Vincent (Nouvelle Calédonie). Institut Français de Recherche Scientifique pour le Développement en Coopération ORSTOM, Centre de Nouméa, New Caledonia, *Rapp. scient. tech. Sci. Mer, Biol. Mar.* **60**, 73 pp.

Waples, R.S. (1987) A multispecies approach to the analysis of gene flow in marine shorefishes. *Evolution*, **41**, 385–400.

Ward, J. (1988) Mesh size selection in arrowhead fish traps, in *Contributions to Tropical Fisheries Biology* (eds S. Venema, J. Müller-Christensen and D. Pauly), *FAO Fish. Rep.*, **389**, 455–67.

Ware, D.M. (1975) Relation between egg size, growth, and natural mortality of larval fish. *J. Fish. Res. Bd Can.*, **32**, 2503–12.

Warner, R.R. (1978) The evolution of hermaphroditism and unisexuality in aquatic and terrestrial vertebrates, in *Contrasts in Behaviour* (eds E.S. Reese and F.J. Lighter), Wiley Interscience, New York, pp. 78–95.

Warner, R.R. (1984) Mating behavior and hermaphroditism in coral reef fishes. *Am. Scient.*, **72**, 128–36.

Warner, R.R. (1988) Traditionality of mating-site preference in a coral reef fish. *Nature, Lond.*, **335**, 719–21.

Warner, R.R. (1990) Resource assessment versus traditionality in mating site determination. *Am. Nat.*, **135**, 205–17.

Warner, R.R. (1991) The use of phenotypic plasticity in coral reef fishes as tests of theory in evolutionary ecology, in *The Ecology of Fishes on Coral Reefs* (ed. P.F. Sale), Academic Press, San Diego, pp. 387–98.

Warner, R.R. and Chesson, P.L. (1985) Coexistence mediated by recruitment fluctuations: a field guide to the storage effect. *Am. Nat.*, **125**, 769–87.

Warner, R.R. and Hughes, T.P. (1988) The population dynamics of reef fishes. *Proc. 6th Int. Coral Reef Symp.*, **1**, 149–55.

Warner, R.R. and Robertson, D.R. (1978) Sexual patterns in the labroid fishes of the western Caribbean, I: the wrasses (Labridae). *Smithsonian Contr. Zool.*, **254**, 1–27.

Wass, R.C. (1982) The shoreline fishery of American Samoa: past and present, in *Marine and Coastal Processes in the Pacific: Ecological Aspects of Coastal Zone Management* (ed. J.L. Munro), UNESCO–ROSTSEA, Jakarta, pp. 51–83.

Wassef, E. and Bawazeer, F (1992) Reproduction of longnose emperor *Lethrinus elongatus* in the Red Sea. *Asian Fish. Sci.*, **5**, 219–29.

Watson, M. and Ormond, R.F.G. (1994) Effects of an artisanal fishery on the fish and urchin populations of a Kenyan coral reef. *Mar. Ecol. – Progr. Ser.*, **109**, 115–29.

Watson, R.A., Carlos, G.M. and Samoilys, M.A. (1995) Bias introduced by the non-random movement of fish in visual transect surveys. *Ecol. Modelling*, **77**, 205–14.

Watson, W. (1974) Diel changes in the vertical distribution of some common fish larvae in southern Kaneohe Bay, Oahu, Hawaii. MSc thesis, University of Hawaii, Honolulu, HI, 175 pp.

Weatherall, J.A., Polovina, J.J. and Ralston, S. (1987) Estimating growth and mortality in steady-state fish stocks from length-frequency data, in *Length-based Methods in Fisheries Research* (*ICLARM Conf. Proc.* **13**) (eds D. Pauly and G.R. Morgan), ICLARM, Manila, Philippines and Kuwait Institute of Scientific Research, Safat, Kuwait, pp. 53–74.

Webb, K.L., DuPaul, W.D., Wiebe, W. *et al.* (1975) Enewetak (Eniwetok) Atoll: aspects of the nitrogen cycle on a coral reef. *Limnol. Oceanogr.*, **20**, 198–210.

Wedgewood, C.H. (1934) Report on research in Manam Island, Mandated Territory of New Guinea. *Oceania*, **4**, 373–403.

Weerasoorriya, K.T., Pieris, S.S.C. and Fonseka, M. (1985) Promotion of bottom set longlining in Sri Lanka. Bay of Bengal Programme, *Working Paper* **40**, 38 pp.

Weiler, D. and Suarez-Caabro, J.A. (1980) Overview of Puerto Rico's small-scale fisheries statistics, 1972–1978. *Tech. Rep., CODEMAR* **1**(1), 1–27.

Wellington, G.M. (1992) Habitat selection and juvenile persistence control the distribution of two closely related Caribbean damselfishes. *Oecologia*, **90**, 500–8.

Wellington, G.M. and Victor, B.C. (1985) El Niño mass coral mortality: a test of resource limitation in a coral reef damselfish population. *Oecologia*, **68**, 15–19.

Wellington, G.M. and Victor, B.C. (1988) Variation in components of reproductive success in an undersaturated population of a coral reef damselfish: a field perspective. *Am. Nat.*, **131**, 588–601.

Wellington, G.M. and Victor, B.C. (1989) Planktonic larval duration of one hundred species of Pacific and Atlantic damselfishes (Pomacentridae). *Mar. Biol.*, **101**, 557–67.

Wells, S. (1993) Coral reef conservation and management, progress in the South and Southeast Asian Regions. *Coastal Manage. trop. Asia*, **1**, 8–13.

Wells, S. and Hanna, N. (1992) *The Greenpeace Book of Coral Reefs*, Sterling Publ. Co., New York, 160 pp.

Wells, S. and Price, A.R.G. (1992) Coral reefs – valuable but vulnerable. WWF Discussion Paper, World Wide Fund for Nature, Cambridge, UK, 40 pp.

Werner, F. E., Page, F.H., Lynch, D.R., Loder, J.W., Lough, R.G., Perry, R.I., Greenberg, D.A. and Sinclair, M.M. (1993) Influences of mean advection and simple behavior on the distribution of cod and haddock early life stages on Georges Bank. *Fish. Oceanogr.*, **2**, 43–64.

Werren, J.H. and Charnov, E.L. (1978) Facultative sex ratios and population dynamics. *Nature, Lond.*, **171**, 349–50.

West, G. (1990) Methods of assessing ovarian development in fishes: a review. *Aust. J. mar. Freshwat. Res.*, **41**, 199–222.

Wheal, C. (1993) Clams on the move. *Fisheries*, **18**, 42.

White, A.T. (1979) Marine park management in the Philippines. *Likas-Yaman*, **2**(1), 1–60.

White, A.T. (1981) Management of Philippine marine parks. *ICLARM Newsl.*, **3**(14), 17–18.

White, A.T. (1984) Marine parks and reserves: management for Philippine, Indonesian and Malaysian coastal reef environments, PhD dissertation, University of Hawaii, Honolulu, HI. 275 pp.

White, A.T. (1986) Marine reserves: how effective as management strategies for Philippine, Indonesian and Malaysian coral reef environments? *Ocean Manage.*, **10**, 137–59.

White, A.T. (1988a) The effect of community-managed marine reserves in the Philippines on their associated coral reef fish populations. *Asian Fish. Sci.*, **2**, 27–41.

White, A.T. (1988b) *Marine Parks and Reserves: Management for Coastal Environments in Southeast Asia*, ICLARM, Manila, Philippines.

White, A.T. and Savina, G.C. (1987) Community-based marine reserves, a Philippine first, in *Coastal Zone '87*, Vol. 2 (eds O.T. Magoon, H. Converse, D. Miner, L.T. Tobin, D. Clark and G. Domurant), Am. Soc. Civil Engineers, New York, NY, pp. 2022–36.

White, A.T., Delfin, E. and Tiempo, F. (1986) The marine conservation and development program of Silliman University, Philippines. *Trop. Coastal Area Manage.*, **1**(2), 1–4.

White, A.T., Hale, L.Z., Renard, Y. and Cortesi, L. (eds) (1994) *Collaborative and Community-Based Management of Coral Reefs: Lessons from Experience*, Kumarian Press, West Hartford, CT, 130 pp.

White, J.P., Flannery, T.F., O'Brien, R., Hancock, R.V. and Palish, L. (1991) The Balof shelters, New Ireland. Dept Prehistory, Australian National University, Canberra, *Occ. Pap. Prehist.*, **20**, 46–57.

Whitehead, M., Gilmore, J., Eager, E., McGinnity, P., Craik, G.J.S. and McCleod, P. (1986) Aquarium fishes and their collection in the Great Barrier Reef Region. Great Barrier Reef Marine Park Authority, Townsville, *Tech. Memo.*, GBRMPA-TM-13, 39 pp.

Whitelaw, A.W., Sainsbury, K.J., Dews, G.J. and Campbell, R.A. (1991) Catching chracteristics of four fish-trap types on the North-West Shelf of Australia. *Aust. J. mar. Freshwat. Res.*, **42**, 369–82.

Wicklund, R. (1969) Observations on spawning of lane snapper. *Underwater Nat.*, **6**, 40.

Wiebe, P.H., Morton, A.W., Bradley, A.M., Backus, R.H., Craddock, J.E., Barber, V., Cowles, T.J. and Flierl, G.R. (1985) New developments in MOCNESS, an apparatus for sampling zooplankton and micronekton. *Mar. Biol.*, **87**, 313–23.

Wiebe, W.J. (1988) Coral reef energetics, in *Concepts of Ecosystem Ecology*, (*Ecological Studies* 67) (eds L.R. Pomeroy and J.J. Alberts), Springer, Heidelberg, pp. 231–45.

Wiebe, W.J., Johannes, R.E. and Webb, K.L. (1975) Nitrogen fixation in a coral reef community. *Science*, **188**, 257–9.

Wijkstrom, V.N. (1974) Processing and marketing marine fish: possible guidelines for the 1975–1979 periods, in *Int. Conf. Marine Resources Development in Eastern Africa* (eds A.S. Msangi and J.J. Griffin), University of Rhode Island, Kingston, RI, pp. 55–67.

Wilkinson, C.R. and Cheshire, A.C. (1988) Cross-shelf variations in coral reef structure and function: influences of land and ocean. *Proc. 6th Int. Coral Reef Symp.*, **1**, 227–33.

Williams, D.M. (1980) Dynamics of the pomacentrid community on small patch reefs in One Tree Lagoon (Great Barrier Reef). *Bull. mar. Sci.*, **30**, 159–70.

Williams, D.M. (1982) Patterns in the distribution of fish communities across the central Great Barrier Reef. *Coral Reefs*, **1**, 35–43.

Williams, D.M. (1983) Daily, monthly and yearly variability in recruitment of guild of coral reef fishes. *Mar. Ecol. – Progr. Ser.*, **10**, 231–7.

Williams, D.M. (1986) Temporal variation in the structure of reef slope fish communities (central Great Barrier Reef): short term effects of *Acanthaster planci* infestation. *Mar. Ecol. – Progr. Ser.*, **28**, 157–64.

Williams, D.M. (1991) Patterns and processes in the distribution of coral reef fishes, in *The Ecology of Fishes on Coral Reefs* (ed. P.F. Sale), Academic Press, San Diego, pp. 437–74.

Williams, D.M. and English, S. (1992) Distribution of fish larvae around a coral reef – direct detection of a meso-scale, multispecific patch. *Cont. Shelf Res.*, **12**, 923–37.

Williams, D.M. and Hatcher, A.I. (1983) Structure of fish communities on outer slopes of inshore, middle-shelf and outer shelf reefs of the Great Barrier Reef. *Mar. Ecol. – Progr. Ser.*, **10**, 239–50.

Williams, D.M., English, S. and Milicich, M.J. (1994) Annual recruitment surveys of coral reef fishes are good indicators of patterns of settlement. *Bull. mar. Sci.*, **54**, 314–31.

Williamson, H.R. (1989) Conflicting claims to the gardens of the sea: the traditional ownership of resources in the Trobriand Islands of Papua New Guinea. *Melanesian Law J.*, **17**, 26–42.

Winans, G.A. (1985) Geographic variation in the milkfish *Chanos chanos*. II. Multivariate morphological evidence. *Copeia*, **1985**, 890–8.

Winemiller, K.O. and Rose, K.A. (1993) Why do most fish produce so many tiny offspring? *Am. Nat.*, **142**, 585–603.

Wolf, R.S. and Chislett, C.R. (1974) Trap fishing explorations for snapper and related species in the Caribbean and adjacent waters. *Mar. Fish. Rev.*, **36**, 49–61.

Wong, P.P. (1991) *Coastal Tourism in Southeast Asia* (*ICLARM Education Series* **13**), ICLARM, Manila, Philippines, 40 pp.

Wood, E. (1992) Trade in tropical marine fish and invertebrates for aquaria: proposed guidelines and labelling scheme. Unpubl. rep., Marine Conservation Society, Ross-on-Wye, UK, 36 pp.

Woodland, D.J. (1960) Some notes on the commercial aspects of arrow–head trap fishing in tropical Queensland. *Univ. Queensland Pap.*, **1**(10), 243–48.

Woodley, J.D. and Clark, J.R. (1989) Rehabilitation of degraded coral reefs, in *Coastal Zone '89*, (eds O.T. Magoon, H. Converse, D. Miner, L.T. Tobin and D. Clark), Am. Soc. of Coastal Engineers, Charleston, SC, pp. 3059–75.

Wootton, R.J. (1990) *Ecology of Teleost Fishes*, Chapman & Hall, London.

WPFMC (1991) *Amendment 7: Fishery Management Plan for the Crustacean Fisheries of the Western Pacific Region*. Western Pacific Fisheries Management Council, Honolulu, HI, 110 pp.

Wright, A. and Hill, L. (eds) (1993) *Nearshore Marine Resources of the South Pacific*, Forum Fisheries Agency, Honiara, Solomon Islands, xvi + 710 pp.

Wright, A. and Richards, A.H. (1985) A multispecies fishery associated with coral reefs in the Tigak Islands, Papua New Guinea. *Asian mar. Biol.*, **2**, 69–84.

Wright, A., Dalzell, P.J. and Richards, A.H. (1986) Some aspects of the biology of the red bass, *Lutjanus bohar* (Forsskål), from the Tigak Islands, Papua New Guinea. *J. Fish Biol.*, **28**, 533–44.

Wyatt, J. (1983) The biology, ecology and bionomics of the squirrelfishes, Holocentridae, in *Caribbean Coral Reef Fishery Resources (ICLARM Stud. Rev.* 7) (ed J.L. Munro), ICLARM, Manila, Philippines, pp. 50–58.

Yamomoto, K. and Yoshioka, H. (1964) Rhythm of development in the oocyte of the medaka, *Oryzias latipes. Bull. Fac. Fish., Hokkaido Univ.*, **15**, 5–19.

Yap, H.T. and Gomez, E.D. (1985) Coral reef degradation and pollution in the East Asian Seas region. *UNEP Regional Seas Rep. Stud.*, **69**, 185–207.

Yater, L.R. (1982) The fisherman's family: economic roles of women and children, in *Small Scale Fisheries of San Miguel Bay, Philippines: Social Aspects of Production and Marketing* (*ICLARM tech. Rep.* 9) (ed. C. Bailey), ICLARM, Manila, Philippines, pp. 42–50.

Yazon, L.-B. and McManus, L.T. (1987) The coastal resources profile of Lingayen Gulf, Philippines. *Trop. Coastal Area Manage. Newsl.*, 2(3), 1–4.

Young, P.C. and Martin, R.B. (1982) Evidence for protogynous hermaphroditism in some lethrinid fishes. *J. Fish Biol.*, **21**, 475–84.

Zann, L.P. (1983) Traditional management of fisheries in Fiji. Unpubl. rep., Inst. Mar. Resour., Univ. S. Pacific, Suva, Fiji (MS in USP Library, Suva).

Zann, L.P. (1985) Traditional management and conservation of fisheries in Kiribati and Tuvalu atolls, in *The Traditional Knowledge and Management of Coastal Systems in Asia and the Pacific* (eds K. Ruddle and R.E. Johannes), UNESCO–ROSTSEA, Jakarta, pp. 53–77.

Zann, L.P., Brodie, J. and Vuki, V. (1990) History and dynamics of the crown-of-thorns starfish *Acanthaster planci* (L.) in the Suva area, Fiji. *Coral Reefs*, **9**, 135–44.

Zann, L.P., Bell, L. and Sua, T. (1991) The inshore fisheries resources of Upolu, Western Samoa: coastal inventory and database. *FAO/UNDP Field Rep.* 5, 30 pp.

Zaret, T.M. and Paine, R.T. (1973) Species introduction in a tropical lake. *Science*, **182**, 449–55.

Zerner, C. (1991) Reefs in a riptide: community management of coastal resources in the Maluku Islands, Indonesia. Paper presented at Second Ann. Meeting Int. Ass. Study Common Property, Univ. Manitoba, Winnipeg, September 26–29.

Zimmer-Faust, R.K. and Tamburri, M.N. (1994) Chemical identity and ecological implications of a waterborne, larval settlement cue. *Limnol. Oceanogr.*, **39**, 1075–87.

Species index

SCIENTIFIC NAMES

COMMON NAMES

Subject index

Page numbers in **bold** refer to figures, and those in *italics* refer to tables

MIX
Papier aus verantwortungsvollen Quellen
Paper from responsible sources
FSC® C105338

If you have any concerns about our products,
you can contact us on
ProductSafety@springernature.com

In case Publisher is established outside the EU,
the EU authorized representative is:
Springer Nature Customer Service Center GmbH
Europaplatz 3, 69115 Heidelberg, Germany

Printed by Libri Plureos GmbH
in Hamburg, Germany